Environmental, Health, and Business Opportunities in the New Meat Alternatives Market

Diana Bogueva
Curtin University, Australia

Dora Marinova
Curtin University, Australia

Talia Raphaely
Curtin University, Australia

Kurt Schmidinger
University of Vienna, Austria

A volume in the Advances in Business Strategy and Competitive Advantage (ABSCA) Book Series

IGI Global
DISSEMINATOR OF KNOWLEDGE

Published in the United States of America by
IGI Global
Business Science Reference (an imprint of IGI Global)
701 E. Chocolate Avenue
Hershey PA, USA 17033
Tel: 717-533-8845
Fax: 717-533-8661
E-mail: cust@igi-global.com
Web site: http://www.igi-global.com

Library of Congress Cataloging-in-Publication Data

Names: Bogueva, Diana, 1974- editor.
Title: Environmental, health, and business opportunities in the new meat alternatives market / Diana Bogueva [and three others], editors.
Description: Hershey, PA : Business Science Reference, [2019]
Identifiers: LCCN 2018024730| ISBN 9781522573500 (hardcover) | ISBN 9781522573517 (ebook)
Subjects: LCSH: Meat industry and trade. | Meat substitutes--Economic aspects.
Classification: LCC HD9410.5 .E58 2019 | DDC 338.4/76649--dc23 LC record available at https://lccn.loc.gov/2018024730

This book is published in the IGI Global book series Advances in Business Strategy and Competitive Advantage (ABSCA) (ISSN: 2327-3429; eISSN: 2327-3437)

British Cataloguing in Publication Data
A Cataloguing in Publication record for this book is available from the British Library.

For electronic access to this publication, please contact: eresources@igi-global.com.

Advances in Business Strategy and Competitive Advantage (ABSCA) Book Series

Patricia Ordóñez de Pablos
Universidad de Oviedo, Spain

ISSN:2327-3429
EISSN:2327-3437

Mission

Business entities are constantly seeking new ways through which to gain advantage over their competitors and strengthen their position within the business environment. With competition at an all-time high due to technological advancements allowing for competition on a global scale, firms continue to seek new ways through which to improve and strengthen their business processes, procedures, and profitability.

The **Advances in Business Strategy and Competitive Advantage (ABSCA) Book Series** is a timely series responding to the high demand for state-of-the-art research on how business strategies are created, implemented and re-designed to meet the demands of globalized competitive markets. With a focus on local and global challenges, business opportunities and the needs of society, the **ABSCA** encourages scientific discourse on doing business and managing information technologies for the creation of sustainable competitive advantage.

Coverage

- Outsourcing
- Strategic alliances
- Value Chain
- Differentiation Strategy
- Business Models
- Cost Leadership Strategy
- Ethics and Business Strategy
- Customer-Orientation Strategy
- Competitive Strategy
- Economies of Scale

IGI Global is currently accepting manuscripts for publication within this series. To submit a proposal for a volume in this series, please contact our Acquisition Editors at Acquisitions@igi-global.com or visit: http://www.igi-global.com/publish/.

The Advances in Business Strategy and Competitive Advantage (ABSCA) Book Series (ISSN 2327-3429) is published by IGI Global, 701 E. Chocolate Avenue, Hershey, PA 17033-1240, USA, www.igi-global.com. This series is composed of titles available for purchase individually; each title is edited to be contextually exclusive from any other title within the series. For pricing and ordering information please visit http://www.igi-global.com/book-series/advances-business-strategy-competitive-advantage/73672. Postmaster: Send all address changes to above address.

Titles in this Series

For a list of additional titles in this series, please visit: www.igi-global.com/book-series

Creating Business Value and Competitive Advantage With Social Entrepreneurship
N. Oyku Iyigun (Istanbul Commerce University, Turkey)
Business Science Reference • copyright 2019 • 435pp • H/C (ISBN: 9781522556879) • US $205.00 (our price)

Applying Business Intelligence Initiatives in Healthcare and Organizational Settings
Shah J. Miah (Victoria University, Australia) and William Yeoh (Deakin University, Australia)
Business Science Reference • copyright 2019 • 384pp • H/C (ISBN: 9781522557180) • US $215.00 (our price)

Strategic Optimization of Medium-Sized Enterprises in the Global Market
Milan Vemić (DAI Europe, Serbia)
Business Science Reference • copyright 2019 • 386pp • H/C (ISBN: 9781522557845) • US $215.00 (our price)

Effective Open Innovation Strategies in Modern Business Emerging Research and Opportunities
N. Raghavendra Rao (FINAIT Consultancy Services, India)
Business Science Reference • copyright 2018 • 159pp • H/C (ISBN: 9781522557210) • US $145.00 (our price)

Examining the Role of Women Entrepreneurs in Emerging Economies
David Chitakunye (London School of Commerce, UK) and Amandeep Takhar (University of Northampton, UK)
Business Science Reference • copyright 2018 • 375pp • H/C (ISBN: 9781522551126) • US $235.00 (our price)

Handbook of Research on Strategic Innovation Management for Improved Competitive Advantage
George Leal Jamil (InesTec, Portugal) João José Pinto Ferreira (InesTec, Portugal) Maria Manuela Pinto (InesTec, Portugal) Cláudio Roberto Magalhães Pessoa (Universidade Fumec, Brazil) and Alexandra Xavier (InesTec, Portugal)
Business Science Reference • copyright 2018 • 772pp • H/C (ISBN: 9781522530121) • US $465.00 (our price)

Utilizing Consumer Psychology in Business Strategy
Tevfik Dalgic (University of Texas at Dallas, USA) and Sevtap Unal (Izmir Katip Celebi University, Turkey)
Business Science Reference • copyright 2018 • 325pp • H/C (ISBN: 9781522534488) • US $215.00 (our price)

Ethical Standards and Practice in International Relations
F. Sigmund Topor (Keio University, Japan)
Information Science Reference • copyright 2018 • 364pp • H/C (ISBN: 9781522526506) • US $195.00 (our price)

IGI Global
DISSEMINATOR OF KNOWLEDGE

701 East Chocolate Avenue, Hershey, PA 17033, USA
Tel: 717-533-8845 x100 • Fax: 717-533-8661
E-Mail: cust@igi-global.com • www.igi-global.com

To all sentient beings who inhabit this beautiful planet.

Table of Contents

To all sentient beings who inhabit this beautiful planet.

Table of Contents

Detailed Table of Contents

of food. An important advantage of insect production is the high environmental safety compared to conventional livestock. Conventional animal husbandry is responsible for at least 18% of total greenhouse gas emissions and large consumption of drinking water. A much smaller amount of water is used to produce insect meat and insects require far less feed. Production of insect protein requires much less land and energy than the more widely consumed forms of animal protein. The nutritional usefulness of edible insects varies depending on the species, on the stage of development of the insect and the method of breeding and feeding. Insects have a high nutritional value. They are a rich source of protein which includes all eight essential amino acids (phenylalanine, isoleucine, leucine, lysine, methionine, threonine, tryptophan, and valine). Edible insects contain on average 10-30% of fat in dry matter and they are good source of edible oil which contains more than 50% of polyunsaturated fatty acids (PUFA) desirable for nutritional and health reasons. The average energy value of edible insects is about 400-500 kcal/100g of dry matter. Insects also contain a variety of water soluble or lipophilic vitamins and minerals. Their consumption can build a well-balanced diet. Insects can be regarded as safe, if properly managed and consumed, but international food regulations are needed.

In the forthcoming decades, insects might become an important alternative protein source for human consumption. However, what do consumers think about eating insects? The answer is still not very clear, and large differences exist between Western and Eastern societies. While the former has never really experienced edible insects as food, (some) Eastern countries have already practiced entomophagy for a long time. To better understand consumers' perception in both types of societies, a literature review was carried out. The results show that in the Western countries, the consumption of edible insects will depend primarily on availability in the market (i.e., regulatory framework and industry), product category (i.e., processed or unprocessed, familiar or unfamiliar), communication, and marketing. Nonetheless, more research studies are needed to explore Eastern consumers and the development of the edible insect market and industry in Asian countries.

This chapter provides an overview of developments in the Netherlands on new meat alternatives with a focus on plant-based meat substitutes and lab-grown meat. It devotes attention to both the supply side of the market (business activity) and the demand side (consumer appetite). The first concerns developments in the meat substitutes' innovation system since the 1990s until now. It concludes that the Netherlands has become a major player. The latter concerns the supportive purchasing power of consumers regarding the building of a viable and strong market for new meat alternatives. It is concluded that available consumer studies provide evidence for being cautiously optimistic. The closing parts of this chapter, however, bring to the fore that a transition from the current high-meat diets to more sustainable and healthier diets with more non-meat sources of proteins is anything but self-evident. However encouraging and energetic modern developments in the Netherlands are, much progress is needed as it comes to consumer acceptance of new meat alternatives, producer capacity to innovate, concentrate strengths, and capture market share, as well as governmental support for reducing the adverse effects of today's meat consumption and production levels in accordance with Sustainable Development Goal 12 concerning responsible consumption and production.

The "new market" for meat alternatives promises meaty profits and attracts enormous interest by consumers and investors alike. In this chapter, the historical development of meat alternatives is reviewed in an attempt to identify what is "new" about this particular market in the United Kingdom as an example of a Western country. Beginning in Victorian England, through the Wars into the 21st century, the societal background and developments leading into various episodes of markets for meat alternatives are discussed. Together with a description of the "new" market, historical continuities and current opportunities are outlined. It is concluded that health, environment, and business opportunities have played an important role throughout the history of the market, but the significance of this market in the commercial world is new.

Shifting consumer preferences towards meat alternatives can be attributed to the factors such as health and ecological benefits, as well as meat adulteration. Increasing consumer demand for better grade of meat alternatives is also expected to boost the market growth in the near future. Protein sources from maize, peas, rice, and chickpeas are anticipated to witness significant growth and new developments. Alternatives such as bean curd or wheat gluten are expected to be the beneficial source of protein and phosphorous. This can be attributed to the fact that 50 g of vital wheat gluten in combination with water produces 2 ounces of gluten in a solid form, which further comprises of nearly 38 g of protein in each serving. However, factors such as the higher cost of meat substitutes inhibit the market growth, particularly in developing economies, wherein the dietary awareness is expected to be lower. Further factors, related to gluten intolerance and soy allergy, are also anticipated to restrain the market growth. This chapter includes a market study of meat alternatives across the world based on analyzing, estimating, and forecasting for the 2015-2025 period. Market determinants of the meat alternatives market are also explored to analyze market drivers, restraints, challenges, opportunities, trends, and developments. The competitive landscape section includes information related to key market players with an overview of product portfolio and strategic initiatives.

Preface

Global population has been on the rise since the beginning of industrialization. Humanity has become very efficient and clever in producing food and the Malthusian predictions for un unresolvable conflict between population growth and food supply have not eventuated until now (Malthus, 1798). However, the recent changes in human diets are likely to produce such a ruthless clash between human civilization and the planet it inhabits. These dietary changes can be summarized with a very simple description – a sustained raising consumption of meat and animal-based foods. A lot has been written about the negative health and environmental consequences from such dietary changes (Bogueva, Marinova, & Raphaely, 2018) and the scientific evidence is categorical that humans are on a pathway to cause irreversible changes to the climate and ecological systems of this planet and pose existential threats to their own future as well as that of all other species on the planet as they exist now (Greenpeace, 2018).

The facts of the extent of livestock's infiltration and misuse of the planet's resources are often hidden to life in the city as modern agricultural activities happen outside of urban boundaries. Urban dwellers, who according to the latest European Commission satellite-based estimates now represent 84% of global human population (Scruggs, 2018), are predominantly unaware about the consequences from their food choices. In this day and age, 27% of all land mass available on this planet is used for meat and dairy production, including grazing and feed crops (see Figure 1). By comparison, the entire human population of 7.6 billion people (Woldometers, 2018 as at 2 September 2018) with its settlements, such as cities, town and villages, and all types of infrastructure, such as roads, power lines, industrial utilities and telecommunications, occupies only 1% of the planet's land mass (see Figure 1). In other words, people's dietary choices have triggered a large-scale conversion of the planet's natural resources to feed this global population.

Moreover, the animal-based food choices are a dominant characteristic of a western type of diet causing injustice and inequalities across the globe. Malnutrition is still prevalent in many countries and there has been regress in achieving Sustainable Development Goal 2 of No Hunger with the proportion of undernourished people increasing to 11% in 2016 from 10.6% in 2015 (UN, 2018). The UN has been warning for a long time that food production needs to drastically raise to alleviate global hunger (United Nations, 2009) but what this book argues is that the foods people eat also need to change towards more plant-based options and new alternatives to meat. Without such a transition, there is high probability that humanity will be on a road which potentially leads to conflicts over resources, such as land, water and minerals.

Population projections, such as those by the United Nations, expect population numbers to stabilise by 2100 with the medium estimate being 11.2 billion people (Roser & Ortiz-Ospina, 2018). These people will require to be fed on a healthy and nutritious diet. The current western diet is not a good

Figure 1. Land use 2017
Source: Roser and Ritchie (2017)

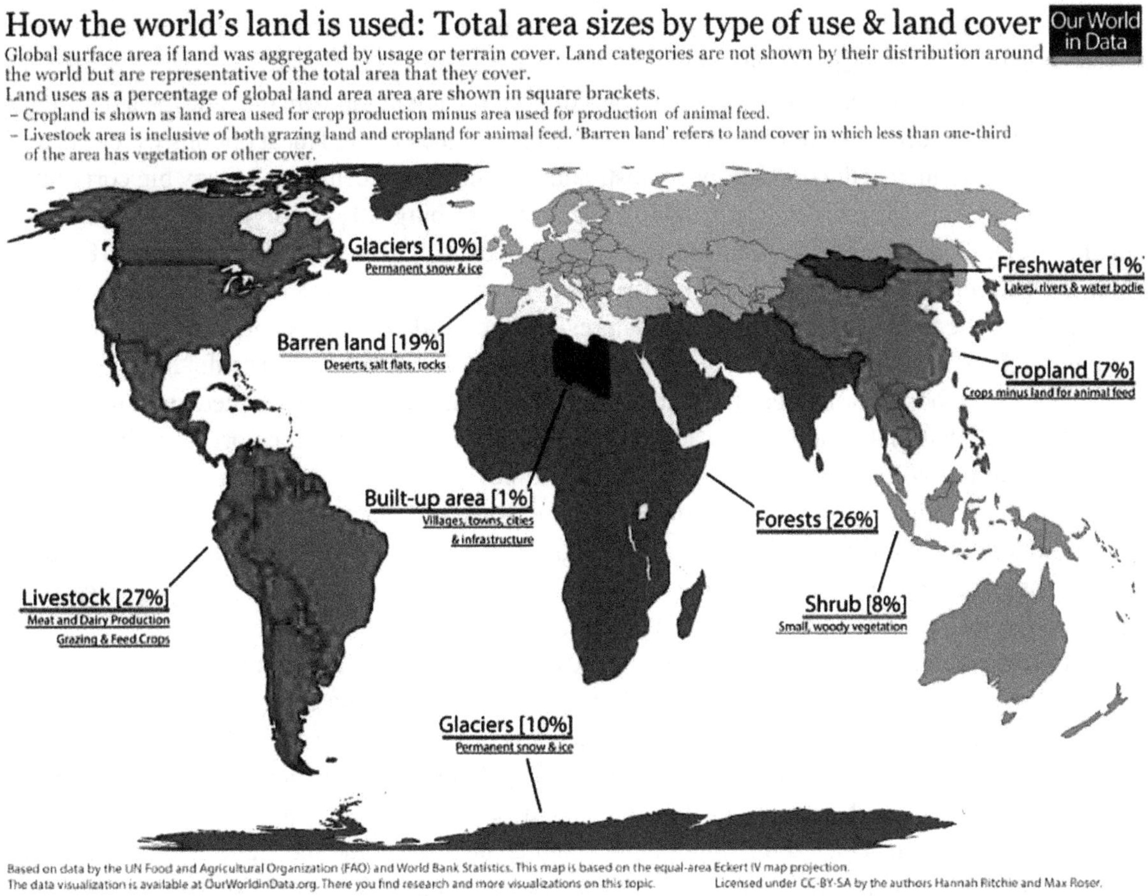

model neither for human health nor for planetary wellbeing as it relies heavily on animal products, such as meat and dairy. This diet however is being emulated across the globe, including countries such as China, where the consumption of meat and dairy used to be modest in the past. Westernisation of diets is a global phenomenon, to which restaurant chains, such as McDonald's, Kentucky Fried Chicken (KFC) and Domino's Pizza, are contributing creating a web of western dietary influence. These changing diets are seen as a symbol of wealth, fashion and prosperity. Urban dwellers across the globe are rarely aware that a western type of diet is an impossibility for the entire global population given the limitations of this planet in terms of land resources, water and clean air.

However, it will be easy to feed the human population, including the expected 3.6 billion extra people to be born between now and 2100 with a different approach to food production and consumption. Livestock options are not viable in the long run. Even now they are undesirable food choices from the point of view of health and the environment. Many believe that they are also not ethical (Singer, 1975). Humanity needs to re-think, re-action and re-evaluate how we are producing, eating and using food and urgently shift to plan B and a different food trajectory.

This book is about looking at food differently. Our aim is to offer a positive outlook for the future instead of the doom and gloom scenarios. It is also about presenting opportunities which are fast gaining momentum. There is a lot of innovation and disruptive trends to the current powerful meat and dairy industries and blind consumer obsession with livestock-based proteins in the affluent world. We have tried to capture some of these developments.

At the heart of these innovations and new ways of thinking about food are the new meat alternatives from plant-based sources to microbial proteins, edible insects and cultured meat options which currently are still in their infancy in the mind of the mainstream consumer. In recent years, many big corporations and powerful players, such as Microsoft and Bill Gates (Nickelsburg, 2017), Virgin and Richard Branson (Singh, 2017), Google and its co-founder Sergey Brin (Kowitt, 2017a. 2017b; Kotecki, 2018; Schonwald, 2014; Cheredar, 2013), are already investing billions of dollars in new meat alternatives which will transform the food industry through foods previously considered impossible. They will inevitably and forever change the livestock model of producing food. As Richard Branson said: "In 30 years or so, I believe we will look back and be shocked at what was the accepted way we killed animals en masse for food. I think that in the future clean and plant-based meat will become the norm, and in 30 years it is unlikely animals will need to be killed for food anymore" (Branson, 2018, n.p.).

Many readers might be wondering how their own tastes will react to some of the new ideas and novel products. Visual and sensory acceptance might also be problematic. Furthermore, the industry transformation and opportunities with these new food products are other aspects of innovation in this sector whose purpose is to satisfy basic existential needs.

This book which combines contributions from 30 authors from 12 countries attempts to provide some answers. We define new meat alternatives as food products which replace meat from livestock in the human diet. The book contains four parts and 16 chapters offering insights from different perspectives.

The first section, "Perceptions and Health Aspects of Alternatives to Meat," focusses on different views and health-related characteristics linked to new meat alternatives. Chapter 1 by Chrysostommos Apostolidis seeks to understand the opportunities and challenges in the meat alternatives market and discusses the main barriers and market drivers for this emerging food sector emphasizing the importance of consumer perceptions about quality, the drawbacks of unrealistic expectations as well as the increasing competition and ineffective marketing strategies. It also proposes policies to create a stable environment for investments and trade in the new meat alternatives which, combined with exemplifying, enabling, encouraging and engaging consumers, can transform the attitudes towards these foods from yuk to yam. Consumer perceptions are further explored in Chapter 2 by Diana Bogueva and Kurt Schmidinger which elaborates on the social readiness and acceptability of new meat alternatives as normal, natural, necessary and nutritious amongst Generation Y and Generation Z consumers. It shows a diversity of opposing opinions with concerns about the unnaturalness of some of the new food options. This indicates a discrepancy between the idea of natural meat production and the unnatural conditions in which livestock animals are currently being raised. The mass adoption of different, non-animal derived proteins will require a substantially distinctive transition which includes education and adequate knowledge about the alternatives. Chapter 3 by Patricia Marshall and Dora Marinova examines the health benefits of consuming more plant-based food from a dietitian's point of view. It presents an overview of the latest systematic literature reviews of the large body of studies which analyze the links between meat consumption and noncommunicable diseases, such as cardiovascular diseases, type 2 diabetes, cancers, mental health and dementia. This is supplemented with evidence about the protective role fibre and other plant materials can play for gastrointestinal health. The issue about plant-based proteins is also discussed and potential warning given that some of the problems surrounding heme iron and high presence of protein in the human diet may remain even in the case of clean lab-grown meat. A balanced plant-based diets with suggestions how this could be done is seen as an easier solution to the health and environmental problems associated with livestock production. Exploring further the benefits of plant-based protein sources, Chapter 4 by Seydi Yıkmış, Ramazan Mert Atan, Nursena Kağan, Levent Gülüm, Harun Aksu and Mehmet Alpaslan examines nine important plant-derived proteins, often seen as superfoods despite being traditional foods in parts of the planet for millennia. These nine plants are: quinoa, chia, buckwheat, teff, beans, soybeans, lentils, chickpeas and spirulina. It is interesting to note that four of these plants are legumes which in addition to providing nutritional benefits to humans can also contribute for the nitrogen fixing of the soil and improving its fertility.

Section 2, "New Meat Alternatives," presents glimpses of the vast innovative, new solutions and opportunities which exist in the food area for consumers and businesses. Clean meat is the central topic for Chapter 5 by Kurt Schmidinger. It overviews the technology behind the production of lab-grown in-vitro cultured meat and presents the major players in this business from an international perspective. The early movers in this space were from the Netherlands while now USA is increasingly taking the lead. Chapter 6 by Akash Saklani and Navneet focuses on microbial protein obtained from microorganisms, such as algae, fungi, yeast and bacteria. The authors argue due the health properties and advantages of the microbial proteins are to be seen as beneficial in replacing the conventional animal-based proteins. Since ancient times, people have been using these microorganisms but in this day and age, their low environmental impacts are becoming a significant consideration in food production. A major environmental concern or uncertainty is the use of genetically modified (GM) organisms as in the case of soybeans. Using the case of China – the world's biggest consumer and importer of soybeans, including GM varieties from Brazil and USA, Chapter 7 by Xiumei Guo, Xiaoling Shao, Shagufta Trishna, Dora Marinova and Amzad Hossain argues that if soybeans are used directly for human consumption, this would significantly reduce the stress on the natural environment to produce larger quantities, including the need to use fertilizers and GM varieties, as well as provide better prospects for the country's food self-sufficiency and reduce global greenhouse gas emissions. The high environmental safety of insect production as a food source in comparison to conventional livestock is argued in Chapter 8 by Anna Żołnierczyk. Information about the high nutritional value of insects is provided and their potential as a balanced and universal source of protein including essential amino-acids, edible oils, water soluble vitamins and minerals, is argued. Chapter 9 by Giovanni Sogari, Aijun Liu and Jie Li explores the future prospects for edible insects as food for human consumption and their acceptability by consumers in western and eastern countries in light of market availability, industry regulation and edible insects' product categories. More communication and marketing efforts are required to make insects acceptable in the West while more research studies are also needed to explore this market and its industry potential in Asian countries.

"The Business Aspects of the New Meat Alternatives" is the title of Section 3 of the book which examines some of the developments that are happening and the existing opportunities to drive change in reducing the dependence on livestock products. Chapter 10 by Hans Dagevos, Ella Tolonen and Jaco Quist analyses the developments in the new meat alternatives in the Netherlands which is a major market player in plant-based meat substitutes and lab-grown meat. Although the authors are cautiously optimistic about building a strong viable market in this country, they also warn that a transition to more sustainable and healthier diets with less animal-derived proteins is only in its infancy. Further progress is needed for the consumers to accept such new products, the producers to have the capacity to innovate and increase the market share of these better alternatives which will be in line with the United Nations Sustainable Development Goal 12 of responsible consumption and production. The next chapter in this section – Chapter 11 by Malte B. Rödl, reviews the historical development of meat alternatives in the UK from the times of Victorian England to the 21st century and outlines the health, environmental and business opportunities. It emphasises that never in the history of the UK has the significance of this market been as high as at present because of the combination between environmental and health priorities which should drive business performance. Chapter 12 by Anusha Thakur presents a 2015–2025 projection for the market of meat alternatives emphasising the expected boost in consumer demand. Given the higher cost of other meat substitutes, the importance of traditional plants, such as maize, peas, rice and chickpeas, is expected to grow as they represent solid sources of protein. Factors, such as soy allergies and gluten intolerance will also need to be anticipated with possible options included in product portfolios.

"Future Directions" in the thinking surrounding new meat alternatives is the final Section 4 of the book. Chapter 13 by Clive Philips and Matti Wilks argues that future food technology developments will end human reliance on the traditional cattle farming practice. At the moment, the cattle mass on this planet is approximately twice the size of the human population. In the face of diminishing natural resources and rising awareness about the inefficiencies and pollution created by the modern farming systems, livestock production may ultimately cease to exist in response to these major concerns. However, for the time being the livestock industry is being strongly supported by governments as well as at the grassroots with images of cultural identity and pure masculinity. The marketing discourse which represents the powerful institutions within the society has long been reinforcing the link between men and meat consumption. Chapter 14 by Diana Bogueva and Dora Marinova proposes that in this moment of time, new images of masculinity are needed which break the tradition and establish a new value system guided by influential people who have adopted vegan diets for the benefit of the planet and their own health. The new marketing discourse needs to disassociate meat consumption with masculinity and reconcile goodness with a transition to plant-based and other non-livestock derived foods. The power of the globalized networked community through the use of social media and online platforms is the topic of Chapter 15 by Weronika Kalamus. The author builds her argument on the difference between force expressed as the physical and legitimized ability to influence decisions and power as a subtler way to change the course of events through better awareness and spiritual strength. She sees the transition to a predominantly plant-based diet as a manifestation of power. Social media play an important part by exposing the cruel realities of livestock farming, influencing people to see–feel-change and ultimately modify their dietary preferences. Chapter 16 by Alexis Nagy and Dora Marinova presents an eye-opening perspective about the future perception of farm animals not as food commodities but also as sentient beings. Including animal welfare in the broader sustainability agenda based on the concept of personhood provides implications for new meat alternatives and any animals that may continue to be part of the food provision system. A historic movement is building momentum for greater animal welfare and sympathy towards other sentient beings which is similar to the origins of the global initiative to seek dignity and freedom for all human beings within the ethical framework of sustainability.

"[W]hat we eat turns out to be number one cause of global warning" – Hawken (2017, p. 37). What we eat turns out to be the number one cause for all kind of environmental deterioration of the land, water, air and soil. What we eat turns out to be number one cause of many debilitating noncommunicable diseases and health conditions.

What this book wants to show is that what we eat can also turn out to be number one opportunity for reducing and solving these problems.

Dora Marinova
Curtin University, Australia

Diana Bogueva
Curtin University, Australia

Talia Raphaely
Curtin University, Australia

Kurt Schmidinger
University of Vienna, Austria

REFERENCES

Bogueva, D., Marinova, D., & Raphaely, T. (Eds.). (2018). Handbook of research on social marketing and its influence on animal origin food product consumption. Hershey, PA: IGI Global.

Branson, R. (2018). Clean meat is the future of meat. *Virgin*. Retrieved from https://www.virgin.com/richard-branson/clean-meat-future-meat

Cheredar, T. (2013). Google's Sergey Brin buys a $325,000 synthetic hamburger. *Venture Beat*. Retrieved from https://venturebeat.com/2013/08/05/googles-sergey-brin-backs-synthetic-hamburger-meat-project/

Greenpeace. (2018). *Less is more: Reducing meat and dairy for a healthier life and planet*. Scientific background of the Greenpeace vision of the meat and dairy system towards 2050. Retrieved from https://storage.googleapis.com/p4-production-content/international/wp-content/uploads/2018/03/6942c0e6-longer-scientific-background.pdf

Hawken, P. (Ed.). (2017). *Drawdown: The most comprehensive plan ever proposed to reverse global warming*. New York, NY: Penguin Books.

Kotecki, P. (2018). The Sergey Brin-backed startup behind the first lab-grown burger just got a boost in the race to transform the meat industry. *Business Insider Australia*. Retrieved from https://www.businessinsider.com.au/sergey-brin-backed-startup-boost-race-transform-meat-industry-2018-7?r=US&IR=T

Kowitt, B. (2017a). Bill Gates and Richard Branson are investing in this clean meat startup. *Fortune*. Retrieved from: http://fortune.com/2017/08/23/bill-gates-richard-branson-invest-meat/

Kowitt, B. (2017b). Silicon Valley and the search for Meatless meat. *Fortune*. Retrieved from http://fortune.com/2017/12/19/silicon-valley-meatless-meat/

Malthus, R. (1798). *Essay on the principle of population*. Retrieved from http://www.esp.org/books/malthus/population/malthus.pdf

Nickelsburg, M. (2017). Bill Gates invests in veggie burger that 'bleeds' like beef to feed the masses and save the planet. *GeekWire*. Retrieved from https://www.geekwire.com/2017/bill-gates-invests-veggie-burger-bleeds-like-beef-feed-masses-save-planet/

Raphaely, T., & Marinova, D. (Eds.). (2016). *Impact of meat consumption on health and environmental sustainability*. Hershey, PA: IGI Global. doi:10.4018/978-1-4666-9553-5

Roser, M., & Ortiz-Ospina, E. (2018). *World population growth*. Our World in Data. Retrieved from https://ourworldindata.org/world-population-growth

Roser, M., & Ritchie, H. (2017). *Yields and land use in agriculture*. Our World in Data. Retrieved from https://ourworldindata.org/yields-and-land-use-in-agriculture

Schonwald, J. (2014). The Frankenburger is coming sooner than you think thanks to Google. *Time Magazine*. Retrieved from http://time.com/3118571/lab-grown-meat-frankenburger-google/

Scruggs, G. (2018). *'Everything we've heard about global urbanization turns out to be wrong' – researchers*. Place. Retrieved from http://www.thisisplace.org/i/?id=0150beca-e3f5-47e0-bc74-9ccc5ef1db8a&utm_source=Newsletter&utm_medium=Email&utm_campaign=Place

Singer, P. (1975). *Animal liberation*. London, UK: Harper Collins.

Singh, S. (2017). Bill Gates and Richard Branson back startup that grows 'clean meat'. *Bloomberg*. Retrieved from https://www.bloomberg.com/news/articles/2017-08-23/cargill-bill-gates-bet-on-startup-making-meat-without-slaughter

United Nations. (2009). *Food production must double by 2050 to meet demand from world's growing population, innovative strategies needed to combat hunger, experts tell second committee*. Retrieved from https://www.un.org/press/en/2009/gaef3242.doc.htm

United Nations. (2018). *The Sustainable Development Goals Report 2018*. Statistical Division. Retrieved from https://unstats.un.org/sdgs/report/2018/overview/

Woldometers. (2018). *Current world population*. Retrieved from http://www.worldometers.info/world-population/

Acknowledgment

We express our special thanks and gratitude by acknowledging all knowledgeable and committed authors and reviewers who joined us on the journey of this book. The seemingly endless process of submissions, reviews, revisions and sharing of ideas helped create this book and its vision for the future of food without livestock. We also want to thank the food animals who have been feeding us for millennia and reassure them that there are other and better alternatives.

Furthermore, we want to express our gratitude to all people who work on replacing the devastating factory farming systems – be it through calculating eco-balances, developing plant-based alternatives to animal products, working on cultured meat or by providing funding for such trendsetting innovations.

This book would not have been possible without the help of our loved ones and their unending inspiration, unconditional patience, love and support. Lily, Alexander, Iavor, Ana, Mira, Nikolay, Daniel, Alex, Ariel, Roger – you are the motivation behind our efforts and our optimism for tomorrow's future.

Section 1

Perceptions and Health Aspects of Alternatives to Meat

Chapter 1
From "Yucky" to "Yummy": Drivers and Barriers in the Meat Alternatives Market

Chrysostomos Apostolidis
Northumbria University, UK

ABSTRACT

Worldwide, a growing range of meat alternative products are being developed and introduced in the market, taking advantage of the increasing health and environmental concerns, technological advances, and the overall rising profile of meat-free diets. This chapter reviews market research and academic literature to identify the market drivers and barriers that will affect the future of new meat alternative products, from a business, consumer, and policy perspective. Key barriers discussed include lower perceived quality, increasing competition, unrealistic consumer expectations, and ineffective marketing strategies. On the other hand, raising the profile of meat alternatives, technological advances, and increasing interest from consumers, investors, and policymakers can support the market success of meat alternatives. The chapter concludes with a discussion on the way forward and the strategies and interventions that can lead to a stronger position of meat alternatives in the food market.

INTRODUCTION

During the past decade, market research reports an increasing interest in the development and marketing of new meat alternative products, due to changes in consumer lifestyles and innovative gastronomical developments (Mintel, 2017; Passport, 2016). Currently, policymakers, businesses and research institutes worldwide are working together to support the development, commercialization and wider acceptance of healthier and more sustainable meat alternatives (Passport, 2016). However, as the different actors in the food sector are trying to adapt to the spirit of the times, what are the factors that will enable (or impede) their efforts?

There are several signals that indicate that the meat alternatives market is (now more than ever) primed to come into the mainstream. From the growing health and environmental concerns (Hartmann, Ruby, Schmidt, & Siegrist, 2018; Charlebois, McCormick, & Juhasz, 2016; Van de Kamp, Seves, &

DOI: 10.4018/978-1-5225-7350-0.ch001

Temme, 2018) to the rising profile of alternative high-protein ingredients (Mintel, 2017; Heffernan, 2017), research suggests that there are several reasons for the hike in demand for meat alternatives, particularly in Western markets such as Europe and America. Hence, what started as a small group of vegetarian-focused, meat substitute products with low sensory appeal (yucky), may now have the chance to become a mainstream market of their own, creating opportunities for manufacturers and retailers to improve their bottom line, while contributing to the improvement of consumers' diets and the sustainability of the food sector (yummy).

This chapter is based on a review of the literature published in academic journals, marketing reports and public media relating to the development, marketing and consumption of meat alternatives. Although the subject is too vast to address in one book chapter, by using information from academic literature to explain findings of recent market research this chapter summarizes and explains some of the key trends, drivers and barriers in the meat alternatives market, including:

- Perceived quality and consumer expectations of meat alternatives;
- Increasing concerns over public health and environmental sustainability;
- Stronger presence of meat alternatives in the food market, and;
- Technological advances.

By presenting this information, the chapter contributes to the main theme of the book as it covers some of the key factors that are expected to influence the new meat alternatives market, and the associated opportunities and challenges that the sector may face in the future.

BACKGROUND

The world is witnessing a surge in the demand for meat alternatives. Although the meat alternatives market has seen some peaks and troughs during the past decade, there is evidence of a steadily growing interest in substitutes to livestock meat products over the years (Mintel, 2017; Passport, 2016). This interest has attracted the attention of food businesses, such as manufacturers and retailers, who are developing and introducing several new types of meat alternative products to the market (Heffernan, 2017, Wild, Czerny, Janssen, Kole, Zunabovic, & Domig, 2014). New and innovative technologies like artificial intelligence (AI), 3D food printing and cultured meat production have helped with the development (and future mass commercialization) of such novel products (Post, 2018; Bonny, Gardner, Pethick, & Hocquette, 2017).

In the literature, several definitions of the term 'meat alternatives' can be found, depending on the viewpoint of the author. For example, according to Hoek et al. (2011), the term 'meat alternatives' refers to protein-containing foods that are primarily vegetable-based and can replace the function of meat as a meal component. More recently however, in addition to the increasingly popular, 'traditional', plant-based meat alternatives based on ingredients such as tofu, seitan, tempeh and fungi (mycoprotein), food companies have started looking at the potential of insects, cultured meat and various new plant-based ingredients (such as oats) to create innovative products that can replace meat in consumers' diets (Schmidinger, Bogueva, & Marinova, 2018; Bonny, Gardner, Pethick, & Hocquette, 2017; Passport, 2016). From lab-grown burgers in Belgium to using AI to develop more realistic, plant-based alternatives in Chile, meat alternative products are being tested and hoping to reach the mass market (and the consumers' table) in the near future. Therefore, it was considered essential not to limit this chapter to a discussion on plant-based meat alternatives but cover the broader product category of alternatives to 'conventional' livestock meat products. This chapter explores some of these emerging meat alternatives and discusses the market opportunities and challenges that emerge and will affect the future success of these products.

THE MARKET FOR NEW MEAT ALTERNATIVES: BARRIERS, DRIVERS AND TRENDS

The development and marketing of meat alternatives that are popular with consumers has never been an easy task, as these products have to compete with a very popular and well-established product – meat. Recently, researchers have focused on identifying ways to support substitution of livestock meat with healthier and more sustainable products, made from alternative protein sources such as mycoprotein, plants and insects (e.g. Hartmann, Ruby, Schmidt, & Siegrist, 2018; Wild, Czerny, Janssen, Kole, Zunabovic, & Domig, 2014; Zorpette, 2013). From a marketing perspective, in order to understand better the future market and demand for these products, the barriers and drivers that may support (or impede) their success need to be determined. Therefore, the following sections will summarize some of the key factors that are expected to influence the consumption and demand of new meat alternatives. Although not an exhaustive list, this information aims to provide a good understanding of the opportunities and challenges in the meat alternatives market.

Barriers

The main barriers to the dynamic meat alternatives market are the perception of low quality, high prices, predominance of credence attributes and increasing competition. They are discussed in turn below.

Low Perceived Quality or "The Yuck Factor"

Despite the increasing interest in meat alternative products, there are still several challenges interfering with the wider acceptance of these products. During the past decade, researchers in several countries have reported that the main issue with meat alternatives is their lower perceived aesthetic and nutritional quality compared to meat. For example, based on a research in the US market, Mintel (2013) suggests that a large number of consumers avoid meat alternatives because of their taste and texture. Similar findings have been reported by studies in other countries, such as the Netherlands (Elzerman, Hoek, van Boekel, & Luning, 2015; Hoek, Elzerman, Hageman, Kok, Luning, & de Graaf, 2013) and the UK (Apostolidis & McLeay, 2016a). According to recent academic studies in the area of novel protein food, this is partly due to food neophobia, i.e. the tendency to avoid new food products, and food technology neophobia, i.e. the lower acceptance of food produced using unfamiliar technologies (Verbeke, 2015; Hoek, Elzerman, Hageman, Kok, Luning, & de Graaf, 2013). Both of these factors are expected to influence strongly the market of new meat alternatives. For example, although insects are considered a normal part of the diet in many cultures around the world, consumer acceptance of insect-based food products may be limited in many Western countries, as consumers are not familiar with the idea of insect as food or the technologies used in the production of insect-based products (Tan, van den Berg, & Stieger, 2016; Verbeke, 2015).

Several years ago, one of the (currently very popular) meat alternatives manufacturer, Quorn Foods, came across this very issue, with studies in countries such as the Netherlands and UK reporting a low acceptance of Quorn, partly influenced by consumers' evaluations of taste, texture and how well the substitute fits in with the whole meal (Apostolidis & McLeay, 2016a; Elzerman, Hoek, Van Boekel, & Luning, 2011; Elzerman, van Boekel, & Luning, 2013). During the past decade however, Quorn has invested in improving and extending its product range, promoting its products as a tasty and nutritious meal component, which led it to the top of the meat alternatives companies list in many countries (Mintel, 2017).

In addition to the issue of food (technology) neophobia, the recent technological developments in the food industry, such as in vitro cultivation of meat, led to growing consumer concerns about the naturalness of meat alternatives (Slade, 2018; Buscemi, 2014). In his research with Canadian consumers, Slade (2018) found that cultured meat is perceived as less natural compared not only to meat products but also to plant-based meat alternatives. This suggests that naturalness might create issues in the acceptance of more novel meat alternative products, particularly for consumers who hold strong views about the naturalness of their food driven by aesthetic but also by health and safety reasons (Rozin, Spranca, Krieger, Neuhaus, Surillo, Swerdlin, & Wood, 2004).

High Prices or "Are We Expecting Too Much for Too Little?"

Higher prices of meat alternatives are another commonly cited issue in the popularization of meat alternatives (e.g. Kumar, Chatli, Mehta, Singh, Malav, & Verma, 2017; Apostolidis & McLeay, 2016b; Elzerman, van Boekel, & Luning, 2013). Existing academic research in countries such as Canada and the US predicts that the probability of consumers substituting a meat product with a similar meat alternative is low, if the prices of the two products are equal (Slade, 2018; Wilks & Phillips, 2017). Although this suggests that consumers are not willing to pay more for meat alternatives than for animal meat, information from the market is inconclusive. In the UK, cases of food manufacturers losing customers to more expensive meat alternative products have been reported (Mintel, 2017), as consumers prioritise a more diverse range of products and exciting flavours. Additionally, while more than half of US shoppers surveyed by Mintel (2013) felt that meat alternative products were more expensive, there was not a big difference in consumption rates between consumers on high and low incomes. According to researchers, the influence of factors like income and price on food choices decline over time, as the market becomes more saturated, and they are replaced by other factors, such as product quality (e.g. Henchion, McCarthy, Resconi, & Troy, 2014).

This suggests that affordability may not always be the primary issue. Instead, academic literature suggests that the unwillingness of consumers to pay more for a product could be due to the expectations they have, which influence their willingness to pay, but also their satisfaction with the specific product (Yi & La, 2004). From a theoretical perspective, the expectancy/disconfirmation theory developed by Oliver (1980) suggests that consumer satisfaction depends on the comparison between the initial consumer expectations and the actual results. Satisfaction arises when expectations are met, whereas dissatisfaction arises when those expectations are not confirmed. Therefore, we might be able to conclude that successfully managing expectations can result in increased satisfaction which in turn can lead to increased loyalty and willingness to pay higher prices for a product (Homburg, Koschate, & Hoyer, 2005).

Johnston (2017) uses Canada as an example to argue that currently there might be an unrealistic win-win logic in the context of more sustainable and ethical alternatives to 'mainstream' food products, which suggests that consumers do not have to sacrifice anything (e.g. money, taste or convenience) to achieve more sustainable diets. Rödl (2018) uses several examples of meat alternative advertising, to demonstrate how meat alternatives are promoted as traditional, natural and necessary for good health food products, and explains how this approach might have negative results for their success in the market. From a consumer behaviour perspective, creating unrealistic expectations may have a detrimental effect on the long-term demand and consumption of these products. The importance of managing consumer expectations in the context of meat alternatives is not new. Several decades ago McCarney (1975, p. 194) suggested that in the case of meat alternatives:

Extravagant claims, the use of hard sell advertising copy techniques and anything smacking of gimmickry must, at all costs, be avoided. A sensitive treatment based on the communication of information presented in an attractive, compelling way should be the keynote.

One might argue that the above arguments are particularly important for new, higher-priced meat alternatives, entering an increasingly competitive market, promising "meat-like" taste and texture and healthier, more sustainable diets. Thus, managing consumer expectations of meat alternative products is important, as expectations can be a determinant of choice, consumer satisfaction and willingness to pay. Which leads us to the next issue of…

Predominance of Credence Attributes or "Marketing the Invisible"

Closely related to the aforementioned issues of food neophobia and consumer expectations, is the way that businesses market and promote meat alternatives. In the UK, although recent years have seen a stronger promotion of meat alternatives, the investment in marketing activities, such as advertising, remains limited, which may limit the potential of businesses and policymakers to build on the current momentum in the market (Mintel, 2017). In addition to the limited investment, an additional challenge for marketing meat alternative products is that their main competitive points (compared to meat) refer to credence qualities i.e. abstract product characteristics that cannot be seen or tasted by the consumer, such as healthfulness and sustainability (Fernqvist & Ekelund, 2014). To resolve this issue of unclarity and invisibility, governments, businesses and organisations have developed schemes to communicate credence attributes to the consumers in the form of food labels.

The effectiveness of this approach, however can be questioned as many consumers are unaffected by this type of information (Hoek, Pearson, James, Lawrence, & Friel, 2017; Fenger, Aschemann-Witzel, Hansen, & Grunert, 2015). In many countries (including France, Germany, Spain, Sweden, Poland and the UK), researchers have investigated the effectiveness of compulsory or optional labelling schemes, such as nutritional labels, carbon footprint, and organic logos, which have been introduced in the market aiming to inform consumers regarding the invisible, "credence" characteristics of food products (e.g. Grunert, Hieke, & Wills, 2014). In the case of meat alternatives, in a study we have conducted with UK consumers, we discovered that although providing more information through labels can influence consumer choices between products within the same category, e.g. purchasing a healthier beef mince option, their effectiveness to convince people to replace meat with meat alternatives is rather limited (Apostolidis & McLeay, 2016b).

This is because in many cases food is not just about nutrition. Food is a concept deeply embedded in people's traditions, memories and culture. Therefore, consumers are not necessarily buying products based on information provided by the businesses at the point of purchase. During food shopping, where a selection between products is usually a matter of seconds, consumers may not be motivated or able to process all the available product information (Grunert, Hieke, & Wills, 2014). In fact, many choices in settings such as grocery stores, cafeterias or canteens are relatively low involvement choices, i.e. consumers do not actively process all available information, and therefore other factors – such as in-store positioning, accessibility and visibility, play a more important role in consumer choices (Van Kleef, Otten, & van Trijp, 2012). Supporting this argument, market research indicates that the way meat alternatives are positioned in-store has an impact on consumer choices, particularly for the customers who have no or little experience with these products, as the segregation of meat alternatives from similar meat products limits their visibility (Mintel, 2017). Additionally, from a foodservice perspective, Bacon and Krpan (2018) conducted research in the use of restaurant menus in the UK and found that separating vegetarian and non-vegetarian products had a negative impact on the demand for the former, even for frequent consumers of vegetarian products, which further supports the impact of positioning and visibility on consumer choices.

Increasing Competition or "The War for Consumers' Hearts, Minds and Wallets"

The focus on the development of tastier, healthier and more sustainable meat alternative products, can hide the fact that there is an increasing number of products aiming to feed the growing population while reducing the environmental and social impact of food consumption. Although the market share of meat alternatives is still a fraction of the one of the livestock meat products, there are indications that increasing launch activity will rise the competition in the meat alternatives sector, which is expected to create difficulties for new, less well-established products (Mintel, 2017; Passport, 2016). For example, according to Hocquette (2016) products like cultured meat may be in a disadvantage once they become available to the public, as they will have to compete with products such as plant- and insect- based meat alternatives, which are already expanding in the market. To resolve this important issue, businesses should involve the consumers early in the development process. In the case of insect-based meat alternatives, Verbeke et al. (2015) suggest that involving consumers in new product development can help businesses develop more successful products and avoid competition-related issues. The authors argue that in a market where most new products fail, consumer insights are crucial for developing differentiated new products adapted to the needs of specific consumer segments.

Nevertheless, competition does not only come from other meat alternative products, since the increasing interest in healthy and sustainable protein food has not only benefitted the market for meat alternatives, but also strengthened the competitive position of healthier and more sustainable livestock meat products, such as organic meat (Hocquette, 2016; Mintel, 2017). This will become particularly relevant in the markets where meat alternatives continue to be considered as substitutes to meat products, instead of forming their own independent 'novel protein' food category.

The case of cultured meat presents an interesting example of the above. Although the use of the term 'clean meat' has been advocated for products from cultivated meat (in an attempt to emphasise on its health and sustainability benefits instead of its provenance), researchers support that cultured meat is not always more environmentally sustainable than livestock meat (Mattick, Landis, Allenby, & Genovese, 2015). Additionally, the use of the term 'clean meat' has been criticized as according to Hocquette (2016, p. 169) it:

recognises implicitly that the word «meat» represents positive values: so, for example, meat is a symbol of force (inherited from the fact that primitive hunters had to be strong to hunt wild animals) and of high nutritional value.

This association with meat and the unclarity regarding the environmental impact of the in-vitro production of cultured meat can reduce its competitiveness in the market, particularly in comparison with more environment-friendly meat products.

Drivers

The changing image of meat alternatives, consumer power, having appropriate policies in place, increased investment and technological advancement are drivers which can stimulate this market. They are discussed below.

Changing Image of Meat Alternatives or "Making It to the Mainstream"

Despite the various barriers, we need to acknowledge that with every challenge comes opportunity. Once ignored by the general public, meat alternatives are no longer a niche vegetarian product in many food cultures. In the UK for example, recent years saw a significant jump in the number of meat eaters who are also exploring alternative protein sources (also known as "flexitarians" or "meat reducers") (Mintel, 2017). Overall, the popularity of meat alternative products is expected to increase further, due to the growing number of influencers supporting or endorsing meat alternative products (such as Sir Richard Branson, Bill Gates and Biz Stone) and the increasing availability of information on meat-free cooking (e.g. food blogs and cookbooks) (Mouat & Prince, 2018; Véron, 2016). This can provide opportunities for new and existing meat alternatives to shake off the niche "substitute" product image and give food manufacturers the prospects to develop new products as part of a stand-alone product category, attract more investment and acquire more shelf space, which can lead to a stronger market presence and demand in the future.

Given their changing image, an important growth opportunity for meat alternative products is presented in the foodservice sector. As meat-free food has now started establishing itself as a cuisine in its own right in markets such as the UK (Mintel, 2010), meat alternatives manufacturers have the opportunity to develop new products targeting the foodservice sector including restaurants, canteens and cafeterias. By taking advantage of the prospects in the foodservice sector, businesses can reduce the overall impact of the foodservice sector on the environment, but also further promote healthier and more sustainable consumption patterns. For example, consumers are able to experiment with new products in canteens and restaurants, which may lead to changes in their own private consumption practices (Wahlen, Heiskanen, & Aalto, 2012).

Consumer Power Can Shape the Market or "Who's the Boss?"

In recent years, the power in many markets has shifted away from food manufacturers and retailers to shoppers. The era of "if we built it, they will come" approach to product development is long gone, and consumers now possess the information and power to shape the food sector through their purchases (Shaw, Newholm, & Dickinson, 2006). Although positive attitudes towards healthier and more sustainable products are not always translated to higher market share for these products, due to the so-called 'gap' between consumers pro-environmental/ethical attitudes and their actual purchasing behaviour (Vermeir & Verbeke, 2008), individuals are increasingly expressing their values and beliefs through their choices in the market (Shaw, Newholm, & Dickinson, 2006).

These 'empowered' consumers will most likely influence the market for new meat alternatives as well, as businesses would not generally invest in new products without some indication that consumers are going to buy them. This is evident in the case of vegan meat alternative product launches in the UK market. As the appeal of vegan products has extended beyond the limited pool of vegans, this led food manufacturers to develop and introduce in the market more vegan meat alternatives (Mintel, 2017), demonstrating that consumer preferences can be one of the main drivers of product innovation.

Further supporting the consumer-driven nature of the market, customisable meal kits, tailored according to consumers' wants and needs, are becoming increasingly popular in Europe and the US, supported by consumer interest in convenience, personalisation and home cooking. For example, French retailer Carrefour has decided to capitalise on the increased demand for personalization by using recipe boxes and meal kit deliveries, a path also taken by Walmart in the US (Mintel, 2018a; 2018b). Meal kits may offer significant opportunities for the development of new meat alternative offerings, increase product trial and long-term demand. Recently, purchases of meat alternatives in the UK has dropped with the decreasing demand for ready-made meals, due to the lower perceived healthiness and quality image of prepared food, and growing consumer interest in scratch cooking (Mintel, 2017). At the same time, in Finland, meat alternatives have been reportedly avoided by consumers, because of their lack of knowledge on how to prepare them in a satisfactory manner (Jallinoja, Niva, & Latvala, 2016). The development of meat alternative based meal kits can support time-poor consumers and consumers not familiar with new meat alternatives, by enabling them to prepare convenient meals according to their preferences without compromising on the health and sustainability impact of their food.

Policies Can Fuel Growth or "Why Should Policymakers Catch Up"

In their report for the European Commission, Probst et al. (2015) indicated that the current restrictive and unclear regulatory environment around novel protein food is holding the meat alternatives industry back. More recently however, there are signs that food policies and regulations are gradually adapted to fit the increasing interest in the replacement of livestock products with alternative sources, driven by health and sustainability concerns (Lähteenmaki-Uutela, Grmelová, Hénault-Ethier, Deschamps, Vandenberg, Zhao, ... & Nemane, 2017).

Recently, the Food and Agricultural Organisation of the United Nations (FAO) reports have been critical of the ecological impact of increased meat consumption and production (Tubiello, Salvatore, Cóndor Golec, Ferrara, Rossi, Biancalani … Flammin, 2014), while government agencies have highlighted the high levels of meat in consumer diets in several counties, such as Sweden (Brugård Konde, Bjerselius, Haglund, Jansson, Pearson, Färnstrand, & Johansson, 2015) and the USA (USDA, 2015). Food sustainability concerns led many governments around the world to develop policies and frameworks aiming to support and encourage more sustainable consumption patterns. For example, the UK's Department of Environment, Food and Rural Affairs (DEFRA) has developed a framework (named the "4Es"), categorizing potential sustainable consumption policies in four key types based on their focus (DEFRA, 2008):

- Enabling, which refers to facilitating the accessibility, affordability and availability of more sustainable products;
- Encouraging, approaches primarily focusing on price interventions (e.g. taxes or subsidies) to support consumption of more sustainable products;
- Exemplifying, which highlights the importance of governments and public bodies to act as role models in terms of choosing sustainable alternatives, and;
- Engaging, which includes more people-oriented policy approaches, focusing on encouraging consumer participation and interaction to make policies work.

As the aim of most new meat alternative products (from insect-based products to cultivated meat) is to provide consumers with healthier and more sustainable food options, policies endeavouring to encourage more sustainable food consumption will act as a catalyst in the development and commercialization of these products. Furthermore, policies and regulations can play a significant role in educating consumers and increasing awareness of meat alternative products (Vanhonacker, Van Loo, Gellynck, & Verbeke, 2013; Vinnari, 2008). Wellesley, Happer and Froggatt (2015) describe governments as the only stakeholders that possess the required resources and power to redirect public diets towards more sustainable food products. In their report for the UK's Royal Institute of International Affairs, the authors explain how governments can support meat substitution and less meat-based diets through various interventions, such as supporting research and development (R&D) in new meat alternatives and guiding consumers towards more sustainable protein sources. Probst et al. (2015) agree that a clear and coherent framework will enable meat alternative businesses to develop and bring more sustainable food products to the consumers.

Increased Investment or "Putting Money Where the Mouth ... Will Be"

In addition to the interest of consumers and policymakers, recently the news have been buzzing about the influx of investment into the meat alternative sector, with the industry focusing on the production of healthy food using fewer resources. For example, the French-based Five Seasons Ventures fund has gained access to funding of millions of Euros in 2018 (Bloomberg, 2018), with the plan to support start-up companies focusing on healthier and more sustainable food, personalised nutrition and alternative proteins. Prominent billionaires Bill Gates and Richard Branson have also reportedly invested in technologies to cultivate lab-grown meat, while large food companies are growing their meat alternative business, through research in alternative protein sources and launching new ranges of products (Mouat & Prince, 2018). Recently, Nestlé added the veggie brand Garden Gourmet to their portfolio, in an attempt to strengthen their presence in the flexitarian market (Michail, 2016). The newly-acquired company's website, explains that the company's philosophy is (Garden Gourmet, 2018):

At Garden Gourmet we want to make exploring this new Flexitarian lifestyle worthwhile. That's why we constantly develop new and surprising varieties of great tasting and nutritiously balanced meat alternatives, made from quality ingredients you know. It's made for trying.

The clear reference to flexitarianism and the development of tasty and nutritious meat alternatives using ingredients familiar to consumers, highlight that the focus of large companies is not only to invest in the development of innovative products for vegetarian or vegan consumers, but also to make meat alternatives attractive to more people, by dealing with the aforementioned issues of food neophobia and lower perceived quality. Although, time will tell how successful these investments would be, there is no question that the growing interest of major players indicates the potential that the meat alternatives sector can offer for future investment.

Technological Advances or "Ones and Zeros in the Service of Sustainability"

In 2016, Eric Schmidt, the (back then) executive chairman of Google's Alphabet, and a professional highly involved with ground-breaking technologies, mentioned that plant-based food technology will be one of the top technology trends in the future (D'Onfro, 2016). In line with his prediction, there have been extensive discussions around the use of technology to develop and market innovative meat alternative products (e.g. Post, 2018; Bonny, Gardner, Pethick, & Hocquette, 2017).

The use of technology to develop "better" meat alternatives is not new. For example, Zorpette (2013) and Wild et al. (2014) provide an overview of the technological developments that have shaped the meat alternatives market over the years. More recently, advanced technologies like 3D food printing and AI have been employed to assist businesses create cheaper, tastier, and more sustainable food for consumers (Schmidinger, Bogueva, & Marinova, 2018; Bonny, Gardner, Pethick, & Hocquette, 2017; Wild, Czerny, Janssen, Kole, Zunabovic, & Domig, 2014). An example of advanced technology in the service of meat alternatives, is the Chilean-based start-up, with the imaginative name the "Not" company. The company uses AI to produce plant-based meat alternatives that will most closely resemble the molecular structure and by that the taste and texture of meat products (Schmidinger, Bogueva, & Marinova, 2018). Giuseppe (the name the company gave to their AI software) aims to achieve wider acceptance of meat alternatives, by eliminating barriers relating to their unfamiliar taste and texture (Penarrendonda, 2017).

Technology however is not limited to the manufacturers. From a retailer perspective, the digital age has also created new capabilities and opportunities. Recently technology allowed retailers to use "big data" to target consumers with customisable messages and products. According to Carolan (2018), big data can provide food retailers with useful information on consumer preferences and the opportunity to "nudge" consumers, i.e. gently encourage them to make decisions that will improve their lives, without forcing their choices. As discussed earlier, given the nature of meat alternatives products, further attention should be paid on the way they are marketed and promoted. As the range of meat alternatives entering the market (each one with different characteristics and qualities) expands and the variety of interested consumers increases, technologies like big data will enable more effective segmentation and targeting strategies.

SOLUTIONS AND RECOMMENDATIONS

The discussion so far emphasises that, in many countries, food businesses (e.g. manufacturers, retailers and foodservice businesses) need to be prepared to meet the growing consumer and investor interest, but also the increasing competition and scepticism in meat alternatives. Further investing in new food technologies can enable businesses to develop meat alternatives that meet better consumer expectations, in terms of nutritional and aesthetic quality, and increase their competitiveness in the food market (Post, 2018; Bonny, Gardner, Pethick, & Hocquette, 2017). From a policy perspective, although some of the ingredients used in the production of new meat alternatives are regulated (either as novel or traditionally used ingredients), in several countries the current regulatory environment is rather unclear, which limits the investment and international trade in the sector (Lähteenmaki-Uutela, Grmelová, Hénault-Ethier, Deschamps, Vandenberg, Zhao... & Nemane, 2017; Probst, Frideres, Pedersen, & Amato, 2015). Future policies should aim to provide a clear and consistent legal framework that will support trade and investment in new meat alternative products internationally. This suggests that strategic planning will be required from all stakeholders to ensure the effective development and introduction of such products in the market.

Creating new meat alternatives however, is only the first challenge. Convincing consumers to eat them is another. From a consumer point of view, although food neophobia and lower perceived quality are two of the major reported obstacles of novel food acceptance (mainly in Western societies), consumer preferences can be updated as consumers are exposed to the new products and information is accumulated over time (Hoek, Elzerman, Hageman, Kok, Luning, & de Graaf, 2013; Yi & La, 2004). New meat alternatives are not an exception, as consumer exposure and learning can help address scepticism and concerns about their naturalness and overall quality, for example in the cases of insect-based (Slade, 2018) or cultured meat-based products (Bekker, Fischer, Tobi, & van Trijp, 2017). Therefore, in addition to investing and supporting product development, businesses, policymakers and organisations need to encourage information dissemination, product trial and continued use. This could be achieved by strategies and policies focusing on exemplifying, enabling, encouraging and engaging consumers (i.e. the 4Es framework) to facilitate a shift towards more sustainable food products.

As consumers in many countries are still unfamiliar with new meat alternatives, early interventions in response to consumer concerns, including efforts to educate and inform the consumer about the social and environmental benefits of these products are considered essential (Siegrist, Sütterlin, & Hartmann, 2018). As discussed earlier though, encouraging consumption of meat alternatives based on product information is challenging, as consumers are not always processing the credence attribute information available in the market place. Marketing can help overcome this consumer inertia. Better in-store placement can increase visibility of meat alternative products, while story-telling (i.e. turning information into stories) can improve the impact of credence attribute information, as it can facilitate information processing and create emotional reactions towards a product (Fenger, Aschemann-Witzel, Hansen, & Grunert, 2015; Schouteten, De Steur, De Pelsmaeker, Lagast, Juvinal, De Bourdeaudhuij,... & Gellynck, 2016). Technology can also support these efforts as businesses and policymakers today have access to information and technology that enables them to identify and target specific consumers and develop more effective interventions and strategies. Big data are increasingly used to provide better insights in consumer preferences and behaviour, in order to develop better products, inform and "nudge" consumers towards the choices that will help them reduce the impact of their diet on their health and the environment (Carolan, 2018).

Although developing more convincing substitutes to meat is the priority of the industry today, in the future, product-focused and consumer-focused policies and strategies need to be combined. As the impact of price reduces, due to increasing market saturation, managing and meeting consumer expectations will play a key role in the meat alternatives market. Meeting (realistic) consumer expectations can increase customer satisfaction, loyalty and willingness to pay for meat alternative products and therefore negate the effect of higher prices. In the long term, strategically managing consumer expectations, by informing and educating consumers, can establish meat alternative products as an independent category (possibly under a different product category name such as novel protein foods), based on their rising popularity and the increasing interest in health and sustainable food products. This can yield multiple benefits for the meat alternatives industry, as it will reduce the need to imitate specific meat products, reduce competition and offer further opportunities for marketing and developing innovative products.

FUTURE RESEARCH DIRECTIONS

The growing size of the meat alternatives market has attracted the attention of businesses, organizations and policymakers. As new products enter the market however, further research is needed to inform business strategies and policies that will enable their long-term success.

In order to inform the development of successful behavioural change strategies, researchers agree that more information is required on consumer perceptions of new meat alternative products and their potential as substitutes to meat (e.g. Slade, 2018; Siegrist, Sütterlin, & Hartmann, 2018). Additionally, more research should focus on specifying the nutritional, health and environmental benefits of new meat alternative products, to enable the promotion of these products as more sustainable food. Finally, there is currently limited research on the impact of technological advances and digitalization in the context of meat alternatives. As technologies like AI, 3D food printing and big data become more popular in the food sector, further research is required to identify the best approach to implementing these technologies, from a consumer, business and policy perspective.

CONCLUSION

Although the increasing interest and investment indicates that meat alternatives are here to stay, it is not easy to keep track of the factors that influence the success of these products. This chapter has reviewed some of the key drivers and barriers that may shape the market of new meat alternatives in the future.

The results suggest that many interesting new products are being developed currently, aiming to tap into the flexitarian, vegetarian and vegan trend. Nevertheless, following an "if we build it, they will come" approach to gain market share is not enough in an ever-changing and increasingly competitive global market. In many cases, addressing issues of lower perceived quality and unreasonable expectations through investment in new product development and marketing, can play a key role in strengthening the position of meat alternatives, supported by consistent policies and technological developments.

As the food industry is currently experiencing significant transformations – fuelled by a rise in health and sustainability concerns, changes in food culture and gastronomical advances - consumers, businesses and policymakers now face a great opportunity – and a big dilemma. Can we work together to reduce the negative impact of food consumption on our personal and global well-being and create a system that supports social, environmental and economic sustainability?

Although there is not a simple answer to this question, with the help of new meat alternatives, the future looks optimistic.

REFERENCES

Apostolidis, C., & McLeay, F. (2016a). It's not vegetarian, it's meat-free! Meat eaters, meat reducers and vegetarians and the case of Quorn in the UK. *Social Business*, *6*(3), 267–290. doi:10.1362/204440816X14811339560938

Apostolidis, C., & McLeay, F. (2016b). Should we stop meating like this? Reducing meat consumption through substitution. *Food Policy*, *65*, 74–89. doi:10.1016/j.foodpol.2016.11.002

Bacon, L., & Krpan, D. (2018). (Not) Eating for the environment: The impact of restaurant menu design on vegetarian food choice. *Appetite*, *125*, 190–200. doi:10.1016/j.appet.2018.02.006 PMID:29428546

Bekker, G. A., Fischer, A. R., Tobi, H., & van Trijp, H. C. (2017). Explicit and implicit attitude toward an emerging food technology: The case of cultured meat. *Appetite*, *108*, 245–254. doi:10.1016/j.appet.2016.10.002 PMID:27717657

Bloomberg. (2018). *Company overview of five seasons ventures sarl.* Company Report. Retrieved April 03, 2018 from https://www.bloomberg.com/profiles/companies/1578659D:FP-five-seasons-ventures-sarl

Bonny, S. P., Gardner, G. E., Pethick, D. W., & Hocquette, J. F. (2017). Artificial meat and the future of the meat industry. *Animal Production Science*, *57*(11), 2216–2223. doi:10.1071/AN17307

Brugård Konde, Å., Bjerselius, R., Haglund, L., Jansson, A., Pearson, M., Färnstrand, J. S., & Johansson, A. (2015). *Swedish dietary guidelines - risk and benefit management report.* Retrieved March 16, 2018 from http://www.fao.org/3/a-az907e.pdf

Buscemi, F. (2014). From killing cows to culturing meat. *British Food Journal*, *116*(6), 952–964. doi:10.1108/BFJ-11-2012-0288

Carolan, M. (2018). Big data and food retail: Nudging out citizens by creating dependent consumers. *Geoforum*, *90*, 142–150. doi:10.1016/j.geoforum.2018.02.006

Charlebois, S., McCormick, M., & Juhasz, M. (2016). Meat consumption and higher prices: Discrete determinants affecting meat reduction or avoidance amidst retail price volatility. *British Food Journal*, *118*(9), 2251–2270. doi:10.1108/BFJ-03-2016-0121

D'Onfro, J. (2016). *These are the future tech products that Alphabet's Eric Schmidt thinks will change the world.* Retrieved April 01, 2018 from http://uk.businessinsider.com/eric-schmidt-technologies-that-could-change-the-world-2016-6?r=US&IR=T

Department for Environment, Food & Rural Affairs (DEFRA). (2008). *A framework for proenvironmental behaviours.* Retrieved April 10, 2018 from https://assets.publishing.service.gov.uk/government/uploads/system/uploads/attachment_data/file/69277/pb13574-behaviours-report-080110.pdf

Elzerman, J. E., Hoek, A. C., Van Boekel, M. A., & Luning, P. A. (2011). Consumer acceptance and appropriateness of meat substitutes in a meal context. *Food Quality and Preference*, *22*(3), 233–240. doi:10.1016/j.foodqual.2010.10.006

Elzerman, J. E., Hoek, A. C., van Boekel, M. J., & Luning, P. A. (2015). Appropriateness, acceptance and sensory preferences based on visual information: A web-based survey on meat substitutes in a meal context. *Food Quality and Preference*, *42*, 56–65. doi:10.1016/j.foodqual.2015.01.010

Elzerman, J. E., van Boekel, M. A., & Luning, P. A. (2013). Exploring meat substitutes consumer experiences and contextual factors. *British Food Journal*, *115*(5), 700–710. doi:10.1108/00070701311331490

Fenger, M. H., Aschemann-Witzel, J., Hansen, F., & Grunert, K. G. (2015). Delicious words – Assessing the impact of short storytelling messages on consumer preferences for variations of a new processed meat product. *Food Quality and Preference*, *41*, 237–244. doi:10.1016/j.foodqual.2014.11.016

Fernqvist, F., & Ekelund, L. (2014). Credence and the effect on consumer liking of food–A review. *Food Quality and Preference*, *32*, 340–353. doi:10.1016/j.foodqual.2013.10.005

Garden Gourmet. (2018). *Philosophy*. Retrieved March 25, 2018 from https://www.gardengourmet.com/about

Grunert, K. G., Hieke, S., & Wills, J. (2014). Sustainability labels on food products: Consumer motivation, understanding and use. *Food Policy*, *44*, 177–189. doi:10.1016/j.foodpol.2013.12.001

Hartmann, C., Ruby, M. B., Schmidt, P., & Siegrist, M. (2018). Brave, health-conscious, and environmentally friendly: Positive impressions of insect food product consumers. *Food Quality and Preference*, *68*, 64–71. doi:10.1016/j.foodqual.2018.02.001

Heffernan, O. (2017). Sustainability: A meaty issue. *Nature*, *544*(7651), S18–S20. doi:10.1038/544S18a PMID:28445446

Henchion, M., McCarthy, M., Resconi, V. C., & Troy, D. (2014). Meat consumption: Trends and quality matters. *Meat Science*, *98*(3), 561–568. doi:10.1016/j.meatsci.2014.06.007 PMID:25060586

Hocquette, J. F. (2016). Is in vitro meat the solution for the future? *Meat Science*, *120*, 167–176. doi:10.1016/j.meatsci.2016.04.036 PMID:27211873

Hoek, A. C., Elzerman, J. E., Hageman, R., Kok, F. J., Luning, P. A., & de Graaf, C. (2013). Are meat substitutes liked better over time? A repeated in-home use test with meat substitutes or meat in meals. *Food Quality and Preference*, *28*(1), 253–263. doi:10.1016/j.foodqual.2012.07.002

Hoek, A. C., Pearson, D., James, S. W., Lawrence, M. A., & Friel, S. (2017). Healthy and environmentally sustainable food choices: Consumer responses to point-of-purchase actions. *Food Quality and Preference*, *58*, 94–106. doi:10.1016/j.foodqual.2016.12.008

Hoek, A. C., van Boekel, M. A., Voordouw, J., & Luning, P. A. (2011). Identification of new food alternatives: How do consumers categorize meat and meat substitutes? *Food Quality and Preference*, *22*(4), 371–383. doi:10.1016/j.foodqual.2011.01.008

Homburg, C., Koschate, N., & Hoyer, W. D. (2005). Do satisfied customers really pay more? A study of the relationship between customer satisfaction and willingness to pay. *Journal of Marketing*, *69*(2), 84–96. doi:10.1509/jmkg.69.2.84.60760

Jallinoja, P., Niva, M., & Latvala, T. (2016). Future of sustainable eating? Examining the potential for expanding bean eating in a meat-eating culture. *Futures*, *83*, 4–14. doi:10.1016/j.futures.2016.03.006

Johnston, J. (2017). Can consumers buy alternative foods at a big box supermarket? *Journal of Marketing Management*, *33*(7–8), 662–671. doi:10.1080/0267257X.2017.1297033

Kumar, P., Chatli, M. K., Mehta, N., Singh, P., Malav, O. P., & Verma, A. K. (2017). Meat analogues: Health promising sustainable meat substitutes. *Critical Reviews in Food Science and Nutrition*, *57*(5), 923–932. doi:10.1080/10408398.2014.939739 PMID:25898027

Lähteenmaki-Uutela, A., Grmelová, N., Hénault-Ethier, L., Deschamps, M. H., Vandenberg, G. W., Zhao, A., ... Nemane, V. (2017). Insects as food and feed: Laws of the European Union, United States, Canada, Mexico, Australia, and China. *European Food and Feed Law Review*, *12*(1), 22–37. Retrieved from https://www.researchgate.net/publication/316280459/download

Mattick, C. S., Landis, A. E., Allenby, B. R., & Genovese, N. J. (2015). Anticipatory life cycle analysis of in vitro biomass cultivation for cultured meat production in the United States. *Environmental Science & Technology*, *49*(19), 11941–11949. doi:10.1021/acs.est.5b01614 PMID:26383898

McCarney, L. J. (1975). Communication problems in the marketing of synthetic meats. *European Journal of Marketing*, *9*(3), 188–197. doi:10.1108/EUM0000000005067

Michail, N. (2016). *Investors worth $1.25trn urge industry to switch to sustainable plant proteins.* Foodnavigator. Retrieved March 01, 2018 from https://www.foodnavigator.com/Article/2016/09/27/Investors-worth-1.25trn-urge-industry-to-switch-to-sustainable-plant-proteins

Mintel. (2010). *Meat-free foods – UK – December 2010.* Market Research Report. Mintel. Retrieved April 11, 2018, from http://academic.mintel.com/display/480957/#

Mintel. (2013). *Meat alternatives – US – June 2013.* Market Research Report. Mintel. Retrieved April 03, 2018, from https://store.mintel.com/meat-alternatives-us-june-2013

Mintel. (2017). *Meat-free foods – UK – May 2017.* Market Research Report. Mintel. Retrieved April 04, 2018, from http://academic.mintel.com/display/796253/#

Mintel. (2018a). *France: Carrefour acquires majority stake in meal kit company.* Market Research Report. Mintel. Retrieved April 01, 2018 from http://academic.mintel.com/sinatra/oxygen_academic/display/id=887703?highlight#hit1

Mintel. (2018b). *Walmart pushes its meal kit business with plans to roll out to 2,000 stores this year.* Market Research Report. Mintel. Retrieved April 01, 2018 from http://academic.mintel.com/sinatra/oxygen_academic/display/id=885731?highlight#hit1

Mouat, M. J., & Prince, R. (2018). Cultured meat and cowless milk: On making markets for animal-free food. *Journal of Cultural Economics*, *11*(4), 315–329. doi:10.1080/17530350.2018.1452277

Oliver, R. L. (1980). A cognitive model of the antecedents and consequences of satisfaction decisions. *JMR, Journal of Marketing Research*, *17*(4), 460–469. doi:10.2307/3150499

Passport. (2016). *Future meat alternatives beyond plant-based: In-vitro or 3D printed meat.* Market Research Report. Passport. Retrieved March 28, 2018, from https://www.portal.euromonitor.com/portal/analysis/tab

Penarrendonda, J. L. (2017). *Could AI help to create a meat-free world?* Retrieved March 15, 2018 from http://www.bbc.com/future/story/20171214-could-ai-help-create-a-meat-free-world

Post, M. J. (2018). Proteins in cultured beef. In R. Y. Yada (Ed.), *Proteins in food processing* (2nd ed.; pp. 289–298). Cambridge, UK: Woodhead Publishing. doi:10.1016/B978-0-08-100722-8.00012-7

Probst, L., Frideres, L., Pedersen, P., & Amato, F. (2015). Sustainable, safe and nutritious food. New nutrient sources. *Business Innovation Observatory*. Retrieved April 01, 2018, from https://ec.europa.eu/docsroom/documents/13426/attachments/1/translations/en/renditions/native

Rödl, M. B. (2018). Marketing meat alternatives: Meat myths and their replication in advertising for plant-based meat. In D. Bogueva, D. Marinova, & T. Raphaely (Eds.), *Handbook of research on social marketing and its influence on animal origin food product consumption* (pp. 327–343). Hershey, PA: IGI Global. doi:10.4018/978-1-5225-4757-0.ch022

Rozin, P., Spranca, M., Krieger, Z., Neuhaus, R., Surillo, D., Swerdlin, A., & Wood, K. (2004). Preference for natural: Instrumental and ideational/moral motivations, and the contrast between foods and medicines. *Appetite*, *43*(2), 147–154. doi:10.1016/j.appet.2004.03.005 PMID:15458801

Schmidinger, K., Bogueva, D., & Marinova, D. (2018). New meat without livestock. In D. Bogueva, D. Marinova, & T. Raphaely (Eds.), *Handbook of research on social marketing and its influence on animal origin food product consumption* (pp. 344–361). Hershey, PA: IGI Global. doi:10.4018/978-1-5225-4757-0.ch023

Schouteten, J. J., De Steur, H., De Pelsmaeker, S., Lagast, S., Juvinal, J. G., De Bourdeaudhuij, I., ... Gellynck, X. (2016). Emotional and sensory profiling of insect-, plant-and meat-based burgers under blind, expected and informed conditions. *Food Quality and Preference*, *52*, 27–31. doi:10.1016/j.foodqual.2016.03.011

Shaw, D., Newholm, T., & Dickinson, R. (2006). Consumption as voting: An exploration of consumer empowerment. *European Journal of Marketing*, *40*(9/10), 1049–1067. doi:10.1108/03090560610681005

Siegrist, M., Sütterlin, B., & Hartmann, C. (2018). Perceived naturalness and evoked disgust influence acceptance of cultured meat. *Meat Science*, *139*, 213–219. doi:10.1016/j.meatsci.2018.02.007 PMID:29459297

Slade, P. (2018). If you build it, will they eat it? Consumer preferences for plant-based and cultured meat burgers. *Appetite*, *125*, 428–437. doi:10.1016/j.appet.2018.02.030 PMID:29501683

Tan, H. S. G., van den Berg, E., & Stieger, M. (2016). The influence of product preparation, familiarity and individual traits on the consumer acceptance of insects as food. *Food Quality and Preference*, *52*, 222–231. doi:10.1016/j.foodqual.2016.05.003

Tubiello, F. N., Salvatore, M., Cóndor Golec, R. D., Ferrara, A., Rossi, S., Biancalani, R., … Flammin, A. (2014). *Agriculture, forestry and other land use emissions by sources and removals by sinks 1990–2011 analysis*. FAO Statistics Division. Working Paper Series ESS/14-02. Retrieved February 23, 2018 from http://www.fao.org/docrep/019/i3671e/i3671e.pdf

United States Department of Agriculture (USDA). (2015). *Dietary guidelines for Americans 2015–2020* (8th ed.). Retrieved April 01, 2018 from https://health.gov/dietaryguidelines/2015/resources/2015-2020_Dietary_Guidelines.pdf

Van de Kamp, M. E., Seves, S. M., & Temme, E. H. (2018). Reducing GHG emissions while improving diet quality: Exploring the potential of reduced meat, cheese and alcoholic and soft drinks consumption at specific moments during the day. *BMC Public Health*, *18*(1), 264. doi:10.118612889-018-5132-3 PMID:29458352

Van Kleef, E., Otten, K., & van Trijp, H. C. (2012). Healthy snacks at the checkout counter: A lab and field study on the impact of shelf arrangement and assortment structure on consumer choices. *BMC Public Health*, *12*(1), 1072. doi:10.1186/1471-2458-12-1072 PMID:23231863

Vanhonacker, F., Van Loo, E. J., Gellynck, X., & Verbeke, W. (2013). Flemish consumer attitudes towards more sustainable food choices. *Appetite*, *62*, 7–16. doi:10.1016/j.appet.2012.11.003 PMID:23195711

Verbeke, W. (2015). Profiling consumers who are ready to adopt insects as a meat substitute in a Western society. *Food Quality and Preference*, *39*, 147–155. doi:10.1016/j.foodqual.2014.07.008

Verbeke, W., Marcu, A., Rutsaert, P., Gaspar, R., Seibt, B., Fletcher, D., & Barnett, J. (2015). 'Would you eat cultured meat?': Consumers' reactions and attitude formation in Belgium, Portugal and the United Kingdom. *Meat Science*, *102*, 49–58. doi:10.1016/j.meatsci.2014.11.013 PMID:25541372

Vermeir, I., & Verbeke, W. (2008). Sustainable food consumption among young adults in Belgium: Theory of planned behaviour and the role of confidence and values. *Ecological Economics*, *64*(3), 542–553. doi:10.1016/j.ecolecon.2007.03.007

Véron, O. (2016). From Seitan Bourguignon to Tofu Blanquette: Popularizing veganism in France with food blogs. In J. Catricano & R. R. Simonsen (Eds.), *Critical perspectives on veganism* (pp. 287–305). London, UK: Palgrave Macmillan. doi:10.1007/978-3-319-33419-6_13

Vinnari, M. (2008). The future of meat consumption—Expert views from Finland. *Technological Forecasting and Social Change*, *75*(6), 893–904. doi:10.1016/j.techfore.2007.02.001

Wahlen, S., Heiskanen, E., & Aalto, K. (2012). Endorsing sustainable food consumption: Prospects from public catering. *Journal of Consumer Policy*, *35*(1), 7–21. doi:10.100710603-011-9183-4

Wellesley, L., Happer, C., & Froggatt, A. (2015). *Changing climate, changing diets pathways to lower meat consumption.* Chatham House Report. Retrieved March 13, 2018 from: https://www.chathamhouse.org/sites/files/chathamhouse/publications/research/CHHJ3820%20Diet%20and%20climate%20change%2018.11.15_WEB_NEW.pdf

Wild, F., Czerny, M., Janssen, A. M., Kole, A. P., Zunabovic, M., & Domig, K. J. (2014). The evolution of a plant-based alternative to meat. From niche markets to widely accepted meat alternatives. *Agro Food Industry Hi-Tech*, *25*(1), 45–49.

Wilks, M., & Phillips, C. J. (2017). Attitudes to in vitro meat: A survey of potential consumers in the United States. *PLoS One*, *12*(2), e0171904. doi:10.1371/journal.pone.0171904 PMID:28207878

Yi, Y., & La, S. (2004). What influences the relationship between customer satisfaction and repurchase intention? Investigating the effects of adjusted expectations and customer loyalty. *Psychology and Marketing*, *21*(5), 351–373. doi:10.1002/mar.20009

Zorpette, G. (2013). Muscling out meat. *IEEE Spectrum*, *50*(6), 64–70. doi:10.1109/MSPEC.2013.6521036

ADDITIONAL READING

Apostolidis, C., & McLeay, F. (2016). It's not vegetarian, it's meat-free! Meat eaters, meat reducers and vegetarians and the case of Quorn in the UK. *Social Business*, *6*(3), 267–290. doi:10.1362/20444081 6X14811339560938

Apostolidis, C., & McLeay, F. (2016). Should we stop meating like this? Reducing meat consumption through substitution. *Food Policy*, *65*, 74–89. doi:10.1016/j.foodpol.2016.11.002

Bonny, S. P., Gardner, G. E., Pethick, D. W., & Hocquette, J. F. (2017). Artificial meat and the future of the meat industry. *Animal Production Science*, *57*(11), 2216–2223. doi:10.1071/AN17307

Hartmann, C., Ruby, M. B., Schmidt, P., & Siegrist, M. (2018). Brave, health-conscious, and environmentally friendly: Positive impressions of insect food product consumers. *Food Quality and Preference*, *68*, 64–71. doi:10.1016/j.foodqual.2018.02.001

Heffernan, O. (2017). Sustainability: A meaty issue. *Nature*, *544*(7651), S18–S20. doi:10.1038/544S18a PMID:28445446

Post, M. J. (2018). Proteins in cultured beef. In R. Y. Yada (Ed.), *Proteins in food processing* (2nd ed., pp. 289–298). Cambridge, UK: Woodhead Publishing. doi:10.1016/B978-0-08-100722-8.00012-7

Rödl, M. B. (2018). Marketing meat alternatives: Meat myths and their replication in advertising for plant-based meat. In D. Bogueva, D. Marinova, & T. Raphaely (Eds.), *Handbook of research on social marketing and its influence on animal origin food product consumption* (pp. 327–343). Hershey, PA: IGI Global. doi:10.4018/978-1-5225-4757-0.ch022

Schmidinger, K., Bogueva, D., & Marinova, D. (2018). New meat without livestock. In D. Bogueva, D. Marinova, & T. Raphaely (Eds.), *Handbook of research on social marketing and its influence on animal origin food product consumption* (pp. 344–361). Hershey, PA: IGI Global. doi:10.4018/978-1-5225-4757-0.ch023

Slade, P. (2018). If you build it, will they eat it? Consumer preferences for plant-based and cultured meat burgers. *Appetite*, *125*, 428–437. doi:10.1016/j.appet.2018.02.030 PMID:29501683

van de Kamp, M. E., Seves, S. M., & Temme, E. H. (2018). Reducing GHG emissions while improving diet quality: Exploring the potential of reduced meat, cheese and alcoholic and soft drinks consumption at specific moments during the day. *BMC Public Health*, *18*(1), 264. doi:10.118612889-018-5132-3 PMID:29458352

KEY TERMS AND DEFINITIONS

3D Food Printing: Printing three-dimensional food products layer-by-layer using edible ingredients.

Artificial Intelligence (AI): The term *artificial intelligence* refers to all kind of intelligence demonstrated by machines, mimicking human cognitive functions of learning and problem solving.

Big Data: Large and complex datasets that require advanced data processing technology in order to be analyzed.

Consumer Expectations: Consumers' preconceived ideas or perceptions regarding a product.

Cultured Meat: (also known as "clean meat") Meat grown in the laboratory from in vitro animal cells, instead of slaughtered animals.

Flexitarian: The term *flexitarian* is used to describe consumers who are intentionally reducing their meat consumption for health, environmental, or ethical reasons, without completely rejecting meat from their diets.

New Meat Alternatives: Although *newness* is a subjective term, the new meat alternatives refer to the products that are currently under development or have entered the market during the past decade.

Perceived Quality: Consumers' subjective estimations of how well a product can fulfil their requirements.

Chapter 2
Normality, Naturalness, Necessity, and Nutritiousness of the New Meat Alternatives

Diana Bogueva
Curtin University, Australia

Kurt Schmidinger
Vienna University, Austria

ABSTRACT

In the West, meat is acceptable, tasty, delicious, palatable, and enjoyable. It has a well-established position in the consumers' food habits shaping the taste of the affluent eating culture and accepted as normal, natural, necessary, and nutritious. Although recent scientific evidence recognizes that meat has a high negative environmental impact, there is still lack of attention on the fact that we live on a planet with limited resources which need to be preserved. Part of this is a transition to more sustainable consumption habits and diets. This chapter examines the social readiness and acceptability of new meat alternatives as normal, natural, necessary, and nutritious amongst Gen Y and Gen Z consumers. It concludes that a reduction in meat consumption should be an essential part of creating a more sustainable diet in light of the projected increase of the world population, expected human health benefits, and improved environmental wellbeing of the planet.

INTRODUCTION

In this day and age, consumption, and especially meat consumption, has moved beyond its primary utilitarian function of serving basic human needs. The culture of the wealthiest societies is imbued with the idea of excessive meat consumption as absolutely normal part for everyone's equal opportunities to have abundant access to meat protein, often taken for granted and constantly fulfilling consumers' voracious appetites. Over the past fifty years, global meat production and consumption have increased five to ten-fold and the trends are expected to rise by 2050 (Ritchie & Roser, 2018). In a Western diet type, the prevalent excessive, unsustainable meat eating is based on consumption levels from daily to at

DOI: 10.4018/978-1-5225-7350-0.ch002

least 4–5 days a week (Bogueva, Marinova, & Raphaely, 2017). In a wealthy country like Australia, meat consumption has reached 116 kg per person a year (Ritchie & Roser, 2018). Such consumption levels are environmentally harmful and have major repercussions on several related global crises linked with water, climate, and energy (Steinfeld, Gerber, Wassenaar, Castel, Rosales & de Haan, 2006).

It is indeed indisputable that all serious environmental problems the world is facing today, including climate change, resource depletion, degradation of the planet's ecosystems, biodiversity depletion, and pollution of air, water and soil, are human-made (Cook, Oreskes, Doran, Anderegg, Verheggen, Maibach, … Green, 2016; Raphaely & Marinova, 2016; Springmann, Mason-D'Croz, Robinson, Garnett, Godfray, Gollin, … Scarborough, 2016; Myers, Gaffikin, Golden, Ostfeld, Redford, Ricketts … Osofsky, 2013; Steinfeld, Gerber, Wassenaar, Castel, Rosales & de Haan, 2006) and connected with our consumption and production patterns. The Earth's ecosystems cannot survive without urgent changes in human behaviour.

The environmental problems are further compounded with health issues caused by people's voluntarily dietary choices of high animal protein intake. This leads to early mortality risk (Sarich, 2013), higher incidence of heart disease (Quintana Pacheco, Sookthai, Wittenbecher, Graf, Stübel, Johnson … Kühn, 2018), diabetes (Mari-Sanchis, Gea, Basterra-Gortari, Martinez-Gonzalez, Beunza, & Bes-Rastrollo, 2016; Bernard, Levin, & Trapp, 2014), cancer (Lippi, Mattiuzzi, & Cervellin, 2016), including colon cancer (Singh & Fraser, 1998; Giovannucci, Rimm, Stampfer, Colditz, Ascherio, & Willett, 1994), prostate cancer (Dagnelie, Schuurman, Goldbohm, & Van den Brandt, 2004; Giovannucci, Rimm, Colditz, Stampfer, Ascherio, Chut, & Willett, 1993; Kolonel, 1996), breast cancer (Carroll & Braden, 1985), and obesity (You & Henneberg, 2016). Future dietary change directions need to be identified to reduce the burden of diseases and predisposing factors.

There is a pressing need for re-evaluation of consumer dietary choices. As the effects from meat consumption and production are detrimental, humanity's long-term survival prospects are dependent on shifting to alternative proteins, including new plant-based meat alternatives, emerging insect or algae-based foods and lab-grown meat products (Schmidinger, Bogueva, & Marinova, 2018). Meat alternatives will have to play an essential role in replacing meat products or supplementing them, so that people consume less animal-based meat. The market for new meat alternatives is still developing. Those producing and promoting these new products are trying to influence consumers by portraying the new meat alternatives as good, sustainable options, with the hope to establish them as regular food choices.

This chapter aims to fill in the gap in understanding consumer attitudes toward what is normal, natural, necessary and nutritious in relation to meat alternatives. It also aims to explore the future prospects for their acceptability.

BACKGROUND

From soy-based tofu to chopped nuts, almond and peanut meatless meat, plant-based blood and lab-grown steaks, science continues to work to take the animal out of the flesh. It is not about culinary delights, experiments, or something special, but about looking for global solutions to global problems. According to data gathered by the Food and Agriculture Organization of the United Nations, by 2050 the world's population will grow by 34 per cent reaching 9.1 billion and will require a 70 per cent increase in food production (FAO, 2009). Scientists are trying to find food alternatives that will save humankind from starvation and feed the next generation of humans because most of the world's population is estimated to suffer from food shortages (Hincks, 2018; Breene, 2016). Besides world nutrition, these food alternatives should also solve other global problems associated with industrial mass meat production, including: alarming animal welfare issues, high water and land usage, massive contribution to global greenhouse gas emissions, rainforest destruction and conversion into feed and grazing land, water contamination, loss of biodiversity and soil erosion, risks of new global pandemics, rise of antibiotic-resistant bacteria, lifestyle diseases and many more (Schmidinger, 2012, Steinfeld, Gerber, Wassenaar, Castel, Rosales & de Haan, 2006) .

Creating animal-free meat is actually not a new invention for the consumer market. Created during the Han dynasty in China (206BC–220AD), the first meat alternative – tofu, was known as "small mutton" (Du Bois, Tan, & Mintz, 2008). In Medieval Europe, during Lent people were replacing mincemeat with chopped almonds and grapes and diced bread was made into imitation cracklings and greaves (Adamson, 2004). Around 1877, the American medical physician and inventor John Kellogg developed meat replacements from nuts, grains, and soy as an alternative to feed patients with his vegetarian Sanitarium foods (Shurtleff & Aoyagi, 2004; 2014).

CAN GEN Y AND GEN Z HELP SOLVING THE MEAT PROBLEM?

Gen Y or Millennials (born between 1977 to 1995) and Gen Z or Centennials, or iGen (born between 1996 and 2009) can definitely be instrumental in resolving the humanity's insatiable appetite for meat. Currently the Millennials and Centennials are not only the two largest generations in the world considered as the present and future buying and decision-making power, but they are also the imminent leaders (Bresman & Rao, 2017) and the world's most environmentally, health and well-being conscientious generations ever born (Nielsen, 2015; Chang, 2017; Bogueva & Marinova, 2018). Anxiously they attribute global warming to human activity, want to be drivers in health and environmental discoveries and support environmentally friendly policies (Pew Research Centre, 2011). They are armed with their duty and a desire to protect the environment. In exploring the best possible innovative ways to do so, as adopting meat alternatives in their diets, these two generations are the hope to solve the carnivore's dilemma of humanity.

METHODOLOGY

Due to the massive expansion of the internet and social media into all aspects of the digital life of Gen Y and Gen Z target market population, the research study was conducted through an online-based survey. The online transfer of traditional research methods and techniques was used to adapt the whole study to the new technological environment in which the target groups are naturally habituating. Using the online research method, we were able to obtain information in a cost-effective and quick way, and were able to easily target and reach the specific respondent groups. The used qualitative online survey consisted of predominantly open-ended questions as we were interested to explore the opinions, knowledge, perceptions, and concerns of individuals in regard to meat alternatives. These open-ended questions allowed the participants to provide unstructured responses which became an important part of the research. They allowed the participants to share their own explanations, unique opinions and attitudes, feelings, provide additional comments that they considered relevant to the research in relation to its specific topic.

Indicators for the quality of the online survey are the value of the data obtained and the participants' satisfaction with the included questions. The survey had a high response rate of 75.6% – it was successfully completed by 227 respondents out of 300 invited people. We targeted adult people born between 1980 and 2000 and employed in both part or full-time work. The number of participants in this study was purposely limited for data processing purposes as the survey included a lot of open-ended questions.

The research explored in this chapter is based on four assumptions in regard to new meat alternatives. We assumed that for most participants, meat alternatives are abnormal, unnatural, unnecessary and not nutritious and we were seeking approval or disapproval of these assumptions. The central aim of the research was to discover what is considered and accepted as normal, natural, necessary and nutritious by the participants who are representatives of Millennial and Centennial consumers. The research is inspired by similar 3N or 4Ns classifications used for common rationalisation in defence of meat eating with the justification that it is natural, necessary, normal and nice (Joy, 2010, p. 97; Piazza, Ruby, Loughnan, Luong, Kulik, Watkins, & Seigerman, 2015). The three persistent reasonings individuals employ to diffuse their guilt when consuming animal products, described by Joy as part of a carnism ideology are: Normal, Natural, and Necessary (Joy, 2010). The additional N added in a study lead by Piazza is "Nice" and the defending justifications behind each of the 4Ns of meat-eating are: Natural – "humans are natural carnivores", Necessary – "meat provides essential nutrients", Normal – "I was raised eating meat" and Nice – "It's delicious" (Piazza, Ruby, Loughnan, Luong, Kulik, Watkins, & Seigerman, 2015).

The current study looks at psychological perception and barriers among consumers in regard only to meat alternatives and their food content, taste and quality as well as previous consumer experience. Together they form a concept of the ways the acceptability of new meat alternatives could be enhanced.

FINDINGS

When it comes to food, all people believe they are experts in what is normal, natural, necessary and nutritious. In total, 227 respondents – 111 females and 116 males, born between 1980 and 2000 and representatives of the two generations (Table 1), shared their expertise about meat alternatives and gave very thoughtful and informative responses in regard to the researched topic.

The majority of participants were consuming meat at a different frequency ranging from daily consumption (44.5%), few times per week (34.8%), sometimes or occasionally (12.8%) and 7.9% were abstaining from meat (Table 2).

Meat alternatives are not a novelty, especially plant-based ones, and this was clearly stated by the majority of the survey participants who were asked to explain what they thought new meat alternatives

Table 1. Survey demographics

Generation	Number Male	Number Female	Total Number	Total Percentage
Gen Y Millennials	61	58	119	52.4%
Gen Z Centennials	55	53	108	47.6%
Total	116	111	227	100%

Table 2. Meat consumption frequency

Frequency	Number	Percentage
Never, I don't eat meat	18	7.9%
Yes, sometimes on occasion	29	12.8%
Yes, few times per week	79	34.8%
Yes, daily	101	44.5%
Total:	227	100%

were. A large number, namely 199 (87.7%) of the participants, demonstrated and shared a clear idea about the essence of new meat alternatives. The presented answers not only named some of the meat alternatives available worldwide and on the Australian market but also stated their ingredients, such as tofu, tempeh, quorn, lentils, mushrooms, beans, chickpeas etc. They also offered some focused and meaningful explanations as shown in Table 3.

In addition to the informed majority, there was a small number of 28 (12.3%) participants who were lacking knowledge about the nature of the new meat alternatives as shown in Table 4. Interestingly, many of them linked the new meat alternatives to other, not so common in mass consumption internal animal parts or meat from wild animals.

It was clear that the attitudes and understanding demonstrated by the survey participants about new meat alternatives was positioned at the intersection of the concepts of normality, naturalness, necessity and nutrition. These four concepts are discussed in turn below.

ARE MEAT ALTERNATIVES NORMAL?

If something is claimed to be normal, it means it is perceived as typical, expected, established, standard by which behaviours are measured and being consistent with the traditional norm within a culture and society. The survey participants are equally divided around the idea whether consumption of meat alternatives is normal or abnormal. Just below half of the sample – 112 (49.3%) of the survey participants, consider meat alternative as a normal part of their diet compared to 115 (50.7%) who believe the consumption of meat alternatives is not normal. These opinions are also relatively evenly distributed between the two generations as shown in Figure 1. However, a slightly higher proportion of Gen Z considered eating meat alternatives as not normal.

A substantial number of the participants claim that they are open-minded about new meat alternatives and they are willing to try them if they are not already eating them as part of their diet. Many participants expressed united view about the idea that meat alternatives are essential for humankind because of their environmental benefits. One of the participants stated: "Meat alternatives are a good alternative, especially looking at the environmental benefits we often ignore (when choosing food) and prefer to use meat instead of creating a difference". Selected other opinions and reasons representing the normality of consuming meat alternatives are presented in Table 5.

Table 3. What do you think new meat alternatives are?

1	Meat alternatives should be healthy (which meat-free does not necessarily equate to), sustainable and environmentally friendly and also taste good
2	Food that does not contain animal products but that tastes or is similar to a meat product, e.g. vegetarian sausages and fake bacon
3	An alternative to meat that offers similar qualities to real meat, i.e. tofu
4	Products that can either be used to substitute meat or provide an alternative with similar nutritional properties
5	Vegetable-based ingredients made to taste like meat or something exotic like insects
6	Vegan options that fit meat profile, offering similar/close to nutritional level
7	Meat produced by science that does not need animals
8	Vegetarian meat, manufactured/artificial meat
9	I would see it 2 ways – as other food groups that provide the same nutritional value as meat or it could also be an actual product that vegans/vegos could use to replace meat, i.e. Quorn
10	Protein based alternatives to meat (vegan, vegetarian options) which can include anything from nuts to vegetables etc.
11	Substitutes for meat – either imitation or just an alternative that has the same nutrients
12	Something that can provide the same nutritional content as meat. Needs to be high in protein, it needs to mix well with other food groups and needs to taste good
13	Lab-grown meats
14	Plant-based products made to look, smell and taste like meat
15	Both – alternative ingredients that are made to mimic meat, e.g. soy-based protein sausages, and substitutes, e.g. mushroom instead of meat in a burger
16	Products that mimic the taste, texture and protein content of meat but are in fact plant-based
17	Food items with similar nutritional value to meat, similar flavour profiles and textures that are made from non-animal sources of protein, such as soy or gluten. These products may also look like meat
18	New meat alternatives are the food we can use and can taste as if there were meat in your dish but in fact there is not meat
19	Plant-based foods appropriate for those looking to adopt a meat-free diet
20	Chemically produced food that mimics the taste, smell and texture of meat, also aimed to reduce overconsumption of animals
21	Any meat substitutes like plant-based food, insects, anything that is not from animal origin
22	Innovative solutions to replace meat as the main source of protein and other nutrients
23	Meat alternatives replicate the feel, texture and taste of meat, like Quorn, fake bacon, fake chicken, etc.
24	Vegan products that are made to taste similar to traditional counterparts, the same taste without the impact
25	Possible alternatives to meat, such as tofu, and vegetable products used to imitate the nutrients, flavours and textures that are present in some meat products
26	Meat obtained from other sources, such as plants or even insects.

As one of the participants in the study stated: "What is normal today may be not so normal tomorrow as we are evolving in our food preferences". If positioned well, meat alternatives could be accepted as normal as they could serve various valuable purposes, including helping a person in a transition from a meat diet to a more plant-based diet, because of health reasons, animal welfare and environmental concerns. Meat alternatives can help with the reduction or elimination of meat consumption, can be comfort food, for vegetarian and vegan socialising purposes around the BBQ gatherings or to blend in at Australia Day or Thanksgiving.

Table 4. What do you think new meat alternatives are?

1	Introduction of some less common animal parts
2	No idea. Some type of meat perhaps
3	Different animals / different part of the animal
4	Maybe internal organs like liver, kidney as alternative to meat
5	New combinations
6	Other internal parts of meat that are given to the animals
7	Kangaroo
8	Feral animals and other uncommon animal parts
9	Different animals
10	Not sure what you mean by meat alternatives
11	Some rubbish food with no real meat in it like kidney and liver
12	Dairy products, flour made products
13	Perhaps some wild meat
14	No idea at all, I think wild meat
15	Internal parts of meat, dog's food
16	New way of cooking
17	No idea what meat alternatives are
18	Internal organs that are not for human consumption yet
19	Something that real human can't consume such as animal intestines or it will be artificial
20	Perhaps new options from different types of meat
21	Different animal parts not so popular among normal people
22	No idea at all, perhaps different type of meat cuts
23	I think it is some different part of the animal
24	Internal parts of meat used as type of food some nations around the world traditionally eat
25	Kangaroo, other not popular meat parts
26	Feral animals, kangaroo
27	Different animal parts not eaten by human before
28	Not sure what meat alternatives are. I hope some other meaty parts like kidneys, tripe, liver.

ARE MEAT ALTERNATIVES NATURAL?

Despite the main meaning of natural as something derived from nature and created without human intervention, in relation to food and especially meat, the word natural is more complex in nature and conveys much broader meanings. It is used as a metaphor for good and wholesome, and is also culturally defined. Although half of the survey participants are accepting meat alternatives as normal, only 78 (34.4%) of them think meat alternatives are natural to consume. The remaining participants – 149 (65.6%), consider meat alternatives as unnatural. Some of the popular arguments in both directions are presented in Table 6.

Figure 1. Is consumption of meat alternatives normal?

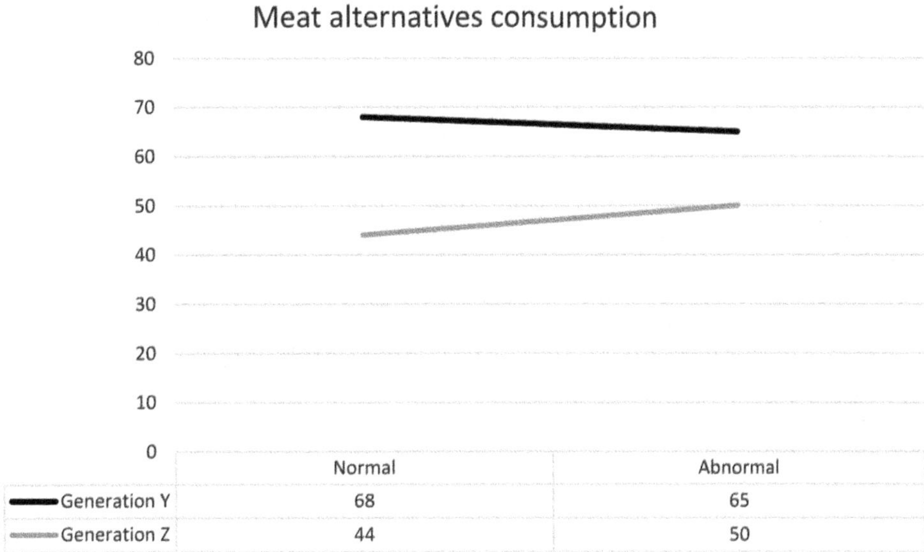

Table 5. Reasons for meat alternatives being normal to consume

1	It must be normal as we are all overeating not only meat, but vegetables as well and soon we will be left without any food
2	Normal as I make an effort to not eat too much meat, so I like alternatives. My flatmates are vegetarian, so we all get veggie burgers, tofu bacon and even quorn 'chicken' nuggets. I have tried crickets once but that's not something I can imagine eating all the time. But who knows…
3	Fairly normal, I am reducing my meat intake and branching out into vegetarian foods, so like to try new alternatives
4	It's becoming normal and quite fashionable, especially amongst young people
5	It's normal and about adjusting one's mentality to the endless possibilities that nature offers
6	Now meat alternatives are absolutely normal. I have only recently stopped eating meat and the reason for this was primarily because of the impact meat industry has on the environment – having always previously eaten meat, I like to have options to have meals I enjoy with meat alternatives
7	It is normal to eat them as we need to change the way we are eating, it is necessary if we want to save our environment, change human impact on our planet
8	For me is normal as one of my favourite restaurants has soybean meat alternatives and it mimics actual meat products so well I can't really tell the difference. I'm always interested to see how food can be manipulated into different forms
9	Eating plant-based products is normal for me, I buy some occasionally to avoid eating meat
10	Normal for me, but not normal for the greater part of humanity as people have lots of prejudice about what they are putting into their mouth
11	They are normal, high in protein and good sources of iron and zinc. I also find them tasty and recently living with a vegetarian, I have enjoyed eating these meat alternatives and adding them to my everyday salads and diet
12	More normal as we need a variety of food and we all need to be willing to try new products
13	It is absolutely normal to consume them if this is to change the way we are eating to save our environment and our planet
14	Consumption of substitutes of meat is becoming more and more normal for many. Maybe there is a need of a push or advertising of it. Even when I eat meat only few times a week I still have the taste and the respect for it.
15	It is normal to eat alternatives to meat and to change the way we are eating, and it is not normal to hurt animals for our food
16	I eat tofu and tempeh sometimes and I found them normal, although they have not much taste and need some supplements to be edible
17	Sometimes is normal to eat in some circumstances, but I don't think it is normal for everyone in our society as people love eating meat

Table 6. Natural/unnatural to consume meat alternatives

Natural	Unnatural
I think it is natural and it's a really sustainable and smart solution – I would rather eat crickets than cows	Not natural. Our ancestors survived eating natural stuff and our generations are spoilt by scientific advancement that may or may not actually be beneficial for the human race and human health. People in the old days didn't have all these allergies and health problems. We don't need another chemically produced food, but to return back to nature and eat as our ancestors.
It would be as natural as eating meat itself; if it's insects then it's a living being same as a cow, chicken etc. I don't see any of those options being unnatural to consume.	Maybe some of the meat alternatives are unnatural. I like to eat healthier if possible and I have problem with the chemically processed food.
Natural and important for the sustainability and future of our planet.	Not normal. The normality of eating meat alternatives I believe comes with the societal acceptance and at the moment this is not the norm. I prefer to be part of the society and to follow its norms.
Natural food made of plants or insects. People may not like them, but they are good healthy protein alternative.	Abnormal right now as meat alternatives are pretty new and undiscovered yet, but with time perhaps they become more natural and common to eat.
It is very natural as we are an adapting world and not everyone wants to eat the same products anymore. As long as we are getting our nutrients from somewhere that's most important	It is not natural for us, because we are used to eat meat and we are taught to eat meat as something very natural for humans. My family was always fed on meat and we never questioned it.
I feel it's a natural progression as livestock is not sustainable in the foreseeable future	Not sure what do you mean by meat substitutes to be natural. They never will be as people will associate them with engineering and modifications.
If it tasted like the most popular meats like beef, lamb, and it became the social norm then it would become more "natural" for humans to adopt it. As people are becoming more socially aware of the consequences of eating meat, it is becoming more and more popular.	Not necessarily natural, but certainly more beneficial in times of meat scarcity.
I think we are seeing a big cultural shift towards people being curious about meat alternatives and I think it is quite natural for many of us to want to gravitate towards a more plant-based diet	Nothing natural about these foods. We are living in a civilised society not in the jungle searching for some insects and larvae.
Probably a good idea as meat could become expensive in the future as well as plant-based food can help reduce cholesterol	Natural is to eat meat, I grew up with this idea, we have to eat meat and to have two veggies with it, now everybody is pro meat alternatives.
It is natural. I feel like it makes sense for us to move away from consuming meat now that we know the negative effects it has on the environment	Better for the environment maybe, but not natural for your tummy. All meat alternatives are heavily processed to mimic meat.
All young people like me, my friends lately are very much interested in meat alternatives and this is a natural process of acceptance of a plant-based diet. The rest of the alternatives, cultured meat especially and the insects to some extent as we are all eating seafood which is similar stuff to crickets, but in the sea.	Not considering them as natural as they are chemically produced to imitate meat
Meat alternatives are important for us to know and enjoy as the humanity is concentrating too much on meat	It is not natural for humans as we consume meat as a natural thing. My mum was telling me: Eat your meat darling:)
Completely natural, there is no reason to consume anything that sustains as long as it doesn't harm anyone else. If anything, eating alternatives is more natural than our current system of mass farming. Technology in food is a natural human advancement, but the cost of it is not worth it	Pretty natural for some people, but for me not natural because I can't abandon my meat for other food. I am a man and I like the taste, the smell, the blood if you want of the real juicy meat, not the imitation.
I don't see why it wouldn't be natural. Cultures all around the world eat various things, something what is 'natural' is just what we have become accustomed to. If people only ate the things suggested, these would be deemed 'natural'.	People think they are natural, I am still not sure about it. Maybe I need more time, but also, I feel I am forced to accept meat alternatives as natural, even I am uncertain about it. I feel there is a push from my peers, sometimes I think it is fashionable.
We really need more of those smart solutions to become natural for all the people especially in Australia, even embracing the crickets.	Natural ingredients, I believe, they use to make it, but the final result is not so natural to eat. Probably need some improvements in the presentation also as it is totally sick looking, shrank and ugly when not freshly made.
Fairly natural as people these days are developing new allergies and the growing research that shows diets high in meat aren't always good.	Meat alternatives can be natural for some nations and people with specific dietary requirements, but not natural for the majority of the Australians and the normal food consuming society worldwide.

The participants in the study are clearly divided into two opposite poles in terms of their understanding whether it is natural or not natural to consume meat alternatives. While the first group, the proponents of the naturalness of consuming meat alternatives opinionated that embracing such diet is good for human health, animal welfare and the sustainability of our planet, the second group was not so categorical in the unnaturalness of eating meat alternatives. Some key concerns highlighted were about the unnaturalness of meat alternatives because they are a product of scientific advancement, chemicals, engineering and modifications, which was not perceived positively by the participants.

ARE MEAT ALTERNATIVES NECESSARY?

The understanding of necessity as essential, fundamental, indispensable for maintaining a minimum standard and an important element for human survival is the meaning shared by the research partakers. The majority of the participants – 198 (87.2%), are united and believe in the idea that meat alternatives are necessary to consume for the sake of our planet's future, for sustainable animal raising, to deal with resource scarcity, for feeding the growing population and for maintaining humankind's physical health (Table 7).

The rest of the participants – 29 (12.8%), who consider the consumption of meat alternatives unnecessary, base their arguments on the abundance of meat in Australia and the country being one of the largest meat exporters in the world. Some also believed that the meat scarcity scenario is still far away.

Table 7. Necessity of meat alternative consumption

1	I definitely think there needs to be some sort of change within human feeding habits. The production and distribution of the meat industry are extremely detrimental to the environment, let alone the lack of rights given to the animals themselves. Thus, it is quite necessary for humanity to embrace new-meat alternatives that are more sustainable for our future.
2	I believe given the current population growth and food production issues it will become an increasing necessity for humanity to explore and consume these new types of meat alternatives.
3	Consuming new meat alternatives is definitely a vital part of the future. Humans cannot continue to consume and deliver meat as there is an exponential growth in human population. There isn't enough land to be able to cater for all meat types ethically. Alternatives are critical for us to save Earth in the future and humanity.
4	I think it's important for lots of different reasons; we evolve and, in some ways, go back to basics in finding meat substitutes. We live in a world where we are aware more than ever of our wellbeing, heath and the part our diet plays in this. Also, meat alternatives bring up lifestyle, health but also sustainability issues and simple choice
5	Eating less meat and choosing meat alternatives would be a benefit to lessen inappropriate farming methods
6	Necessary is to make cultured meat as this will be in huge need in near future when we will create food wars because of not enough meat and other food resources.
7	Humans need to look at new protein sources in order to increase biodiversity and be more environmentally sustainable
8	Essential. We know that meat as it is processed today is unsustainable for the planet. We need alternatives if we want to keep some meat in our diet without destroying everything.
9	Depends on availability, to be honest. If nothing else is available, there is no doubt you will eat it.
10	With the increasing amount of added hormones in meat, it seems somewhat natural to try other products.

ARE MEAT ALTERNATIVES NUTRITIOUS?

Nourishing, healthy and efficient as food is the meaning behind nutritious. The majority of the survey participants – 188 (82.8%), shared immense uncertainty about the meat alternatives' nutritional values. Their arguments gravitated around two major concerns. One of them is the vitamins and mineral content of the meat alternatives, especially the sufficiency of iron, zinc and vitamin B12. The other concern the participants gravitated around is the protein content in comparison to real meat. This was complimented by the fear for their own health based on the view that mimicked meat is too processed and chemically produced to be nutritious. Table 8 shows some of the shared opinions.

Table 8. Nutritiousness of meat alternative consumption

Yes	No	Unsure
Meat alternatives have essential nutrients we all can benefit from.	Chemically produced foods can mimic being nutritious, but actually they are not.	I have no idea, but I imagine that they are nutritious otherwise why they will produce them as alternatives to meat.
The ingredients are nutritious, so I suppose they are nutritious.	Meat alternatives can't beat the nutritional benefits of meat like iron, zinc, B12.	I have no idea about the nutritious component of the meat alternatives. Maybe they are good, but I will doubt it because of the way they are produced.
As long as they share the characteristics similar to meat protein then all meat alternatives are acceptable and nutritious.	If meat alternatives are nutritious for someone it's okay as this is their choice, but personally for me they are not nutritious, more likely a combination of who knows what type of chemicals.	I'm not knowledgeable about the nutritional value. I sometimes read the ingredients on the packages and I can't recall anything bad there, especially in the meat alternatives I tried. I can't be certain I read it 100% properly, but I assume.
The nutrition depends on the nutrition of the alternative protein made. Often they are all good.	Nutritious is maybe too much to be used to describe meat alternatives.	I have absolutely no idea if these alternatives are having similar iron, zinc and magnesium content to say if they are nutritious.
Depending on the ingredients included, it can be more nutritious than meat.	It's not nutritious. What nutrients you can find in crickets and bugs? The whole thing is too much under question.	They must have the proper nutrition and I am not sure about that and not sure how they will make me feel after I try them.
I don't think vegan meat is very nutritious, but the insects were used by the Aborigines and they are certainly nutritious.	Can be nutritious, but I have a big doubt about it.	It could be nutritious, I don't really know, as they are so much meaty looking for me. I don't like it when people think I am vegetarian and I'll enjoy them, and they cook me a veggie sausage at the BBQ. It is not the problem they are plant-based made, but the look they have imitating real meat stuff.
Meat alternatives are nutritious with similar flavour, taste, textures and look like meat.	I think real veggies are more nutritious to consume. Meat alternatives look too processed and not nutritious at all.	Nutritious aspect is difficult to describe as I am not familiar with all the meat alternatives on the market.
I believe they are nutritious enough, but people don't know about meat alternatives. They are a less popular choice and not enough advertised as a correct choice.	Can't beat the nutritional benefits of meat like iron, zinc, B12.	I have no idea about it as I am not eating them. I prefer to cook veggie food instead from raw ingredients.
Highly nutritious and tasty for us as consumers to eat.	Not nutritious I think. They definitely can't have the properties of meat with enough iron and B12 like meat.	More unsure of the nutritional value of alternative meat then to be certain about it. Maybe if I have more info… I'll have some opinion about it.

The existing knowledge gap about meat alternatives is worth filling in as most of the participants are convinced about the necessity of meat alternatives because of their environmental benefits for the planet and human health.

MANLINESS AND MEATINESS OF MEAT ALTERNATIVES

The research revealed the specific need of reconceptualization the appearance of meat alternatives. When issues related to manliness and meatiness were mentioned, the look of the meat alternatives was in the centre of the discourse. Quite a few of the male survey participants, predominantly representatives of generation Y, found meat alternatives not so masculine to consume. The meat issue as a genuine consumer concern was clearly outlined in the respondents' answers (Table 9).

Although the less meaty appearance of meat alternatives bothered some male participants, the resemblance to meat in the appearance of meat alternatives emerged as an issue for some female participants (Table 10).

THE INSECTS DILEMMA

Most of the participants objected to the eating of insects. Being an unpopular meal dictated by our affluent culture, the insect dilemma was pretty much discussed as a choice or necessity in this research. A solid number – 112 (75.2%) out of 149 (65.6%) of the survey participants sharing the opinion that meat alternatives are not natural to consume, feel some sort of disgust when hearing the word insects (Table 11). Contrary to the opinion shared by the survey participants, many scientists believe that entomophagy (known as insect-eating) will not only benefit human health, but also the planetary health

Table 9. Manliness and meat alternatives

1	Not natural and not masculine at all for me as real men eat real meat.
2	I reckon these (meat alternatives) are food for pussies. They are like veggies.
3	These are not meat at all as they taste, look and smell like no meat and there is no blood in it when you want to cook them.
4	For humans is natural to eat meat not some alternatives to it. We are hunter-gatherers, not pussy.
5	I am a man and I like the taste, the smell, the blood, if you want, of the real juicy meat, not the imitation.

Table 10. Meatiness of meat alternatives

1	I have a real problem with their look mimicking meat and when they are cooked they look not so appealing.
2	If they didn't duplicate the look of normal meat I believe they would be more acceptable by people with my believes against animal killing.
3	I never will eat them, they are made to imitate meat and I don't like it as I care about the animal welfare.
4	I feel kind of uncomfortable with the similar to meat appearance of meat alternatives.
5	It's fine for meat eater, I reckon, not for vegetarians like me as I prefer not to eat something that seems like meat

Table 11. Attitude to insect consumption

Against Insect Consumption	Pro Insect Consumption
I lived most my childhood in China. I ate a lot of weird things when I was little – water beetles, snakes, rabbits, racoons. I might not feel comfortable eating something that looks like insect anymore, but I'm willing to try things that does not look like insects.	Regarding insects, I have tried on holidays in Asia, but never considered to make them part of my diet. I'm not reluctant and don't think this is weirder than eating other animals but I wouldn't know where to buy them.
… the crickets and larvae are totally sick	I think it's natural enough to eat insects and bugs, as people have done all over the world for hundreds of thousands of years. However, I think the more chemically-based manufactured 'meats' are, perhaps less natural for humans to eat.
I am not in favour of the idea to eat insects, they are kind of gross. I think people are not grass feeders, so insects and cultured meat are something we all have to be reluctant about to consume, as we have better options with plant-based alternatives.	People just need to get over the stigma of eating insects.
If it is insects I don't feel that it's very human to eat at all.	Sounds lovely as insects are a good source of protein.
Not normal for us to eat gross insects and larvae.	Insects have a branding problem.
The insect options are under big question. We have to starve ourselves before trying it.	Crickets and larvae, algae can be highly nutritional and beneficial for us to eat.

as insects are an environmentally friendly source of human protein (Sogari, 2015; van Huis, Itterbeeck, Klunder, Mertens, Haloran, Muir, & Vantomme, 2013; Bennington-Castro, 2017), with low production cost, short life cycle, low space requirement (van Huis, 2013) and good nutritional quality (Rumpold & Schlüter, 2013).

Making people look at insects simply as a source of food will be an incredible challenge. While 80% of the world's population traditionally, freely and regularly eats insects as normal, natural, necessary and nutritious (Carrington, 2010; Guynup & Ruggia, 2004), in the Western world, this is perceived more as a strange delicacy, which more often provokes negative feelings (Verbeke, 2015; Hartmann & Siegrist, 2017) and disgust rather than any positive reactions (Ruby, Rozin, & Chan, 2015), and draws food neophobia (Gere, Székely, Kovács, Kokai, & Sipos, 2017). Maybe we should think of a new name for edible insects to eliminate the disgusting factor or presenting them in more familiar forms to enhance the willingness of people to try (Megido, Gierts, Blecker, Brostaux, Haubruge, Alabi, & Fransis, 2016; Tan, Verbaan, & Stieger, 2017) as a solution to the insect dilemma.

FUTURE RESEARCH DIRECTIONS

What used to be normal, natural, necessary and nutritious, should no longer be perceived in the same way, because in our ever-changing world we must be able to assess the pressing problems, including our over-consumption, overuse of resources, the human and the planetary health, and to learn how to resolve them quickly and in a timely manner. Humankind needs to be open to new ideas and to break through its traditional beliefs in order to learn from the mistakes of the past and remove the barriers to offering new opportunities. Attention to what the generations in power want, their awareness of the necessity of new meat alternatives and the actions they are willing and ready to take should be at the centre of any future discussions and research.

CONCLUSION

New meat alternatives are regarded as normal and necessary by most of the survey participants, but are unnatural, and not nutritious for the majority of them. Especially the perception about meat alternatives being "unnatural" obviously in contrast to "natural" meat, shows a notable discrepancy between consumers' ideas of a natural meat production versus the actual unnatural reality in industrial intensive livestock facilities as described in many publications (such as Ewbank, Ray, Kim-Madslien, & Hart, 1999 or Webster, 2010). Replacement with new meat alternatives does not have the same respect in people's minds and palates. The idea of mass use as different protein is still at the beginning of a long journey before being accepted as normal, natural, necessary and nutritious.

Over the next few decades, people will have to change their eating habits, food-related gender norms and stereotypes, and give up their prejudice about food because the health of the planet and their own health are likely to remain a big issue.

REFERENCES

Adamson, M. W. (2004). *Food in medieval times*. Westport, CT: Greenwood Press.

Bennington-Castro, J. (2017). How crickets could help save the planet. *NBC News*. Retrieved from https://www.nbcnews.com/mach/environment/how-eating-crickets-could-help-save-planet-n721416

Bernard, N., Levin, S., & Trapp, C. (2014). Meat consumption as a risk factor for type 2 diabetes. *Nutrients*, *6*(2), 897–891. doi:10.3390/nu6020897 PMID:24566443

Bogueva, D., & Marinova, D. (2018). What is more important perception of masculinity or personal health and the environment? In D. Bogueva, D. Marinova, & T. Raphaely (Eds.), *Handbook of research on social marketing and its influence on animal origin food product consumption* (pp. 148–162). Hershey, PA: IGI Global. doi:10.4018/978-1-5225-4757-0.ch010

Bogueva, D., Marinova, D., & Raphaely, T. (2017). Reducing meat consumption: The case for social marketing. *Asia Pacific Journal of Marketing and Logistics*, *29*(3), 477–500. doi:10.1108/APJML-08-2016-0139

Breene, K. (2016). *Food security and why it matters*. World Economic Forum. Retrieved from https://www.weforum.org/agenda/2016/01/food-security-and-why-it-matters/

Bresman, H., & Rao, V. D. (2017). A survey of 19 countries shows how generations X, Y, and Z are — and aren't — different. *Harvard Business Review*. Retrieved from https://hbr.org/2017/08/a-survey-of-19-countries-shows-how-generations-x-y-and-z-are-and-arent-different

Carrington, D. (2010). Insects could be the key to meeting food needs of growing global population. *The Guardian*. Retrieved from https://www.theguardian.com/environment/2010/aug/01/insects-food-emissions

Carroll, K. K., & Braden, L. M. (1985). Dietary fat and mammary carcinogenesis. *Nutrition and Cancer*, *6*(4), 254–259. doi:10.1080/01635588509513831 PMID:6443636

Chang, R. (2017). Survey: Generation Z's best students seek careers in STEM and healthcare. *The Journal*. Retrieved from https://thejournal.com/articles/2017/07/12/survey-generation-zs-best-students-seek-careers-in-stem-and-healthcare.aspx

Cook, J., Oreskes, N., Doran, P. T., Anderegg, W. R. L., Verheggen, B., Maibach, E. W., ... Green, S. A. (2016). Consensus on consensus: A synthesis of consensus estimates on human-caused global warming. *Environmental Research Letters*, *11*(4), 048002. doi:10.1088/1748-9326/11/4/048002

Dagnelie, P. C., Schuurman, A. G., Goldbohm, R. A., & Van den Brandt, P. A. (2004). Diet, anthropometric measures and prostate cancer risk: A review of prospective cohort and intervention studies. *BJU International*, *93*(8), 1139–1150. doi:10.1111/j.1464-410X.2004.04795.x PMID:15142129

Du Bois, C. M., Tan, C.-B., & Mintz, S. W. (2008). *The world of soy*. National University of Singapore: National University of Singapore Press.

Ewbank, R., Ray, P., Kim-Madslien, F., & Hart, B. (Eds.). (1999). *Management and welfare of farm animals*. Wheathampstead, UK: Universities Federation for Animal Welfare.

Food and Agriculture Organization of the United Nations (FAO). (2009). *How to feed the world in 2015*. Retrieved from http://www.fao.org/fileadmin/templates/wsfs/docs/expert_paper/How_to_Feed_the_World_in_2050.pdf

Gere, A., Székely, G., Kovács, S., Kokai, Z., & Sipos, L. (2017). Readiness to adopt insects in Hungary: A case study. *Food Quality and Preference*, *59*, 81–86. doi:10.1016/j.foodqual.2017.02.005

Giovannucci, E., Rimm, E. B., Colditz, G. A., Stampfer, M. J., Ascherio, A., Chut, C. C., & Willett, W. C. (1993). A prospective study of dietary fat and risk of prostate cancer. *Journal of the National Cancer Institute*, *85*(19), 1571–1579. doi:10.1093/jnci/85.19.1571 PMID:8105097

Giovannucci, E., Rimm, E. B., Stampfer, M. J., Colditz, G. A., Ascherio, A., & Willett, W. C. (1994). Intake of fat, meat, and fiber in relation to risk of colon cancer in men. *Cancer Research Journal*, *54*(9), 2390–2397. PMID:8162586

Guynup, S., & Ruggia, N. (2004). *For most people eating bugs is only natural*. National Geographic Channel, National Geographic. Retrieved from https://news.nationalgeographic.com/news/2004/07/0715_040715_tvinsectfood.html

Hartmann, C., & Siegrist, M. (2017). Insects as food: Perception and acceptance. Findings from current research. *Ernährungs-Umschau*, *64*(3), 44–50. doi:10.4455/eu.2017.010

Hincks, J. (2018). The world is headed for a food security crisis. Here's how we can avert it. *Time Magazine*. Retrieved from http://time.com/5216532/global-food-security-richard-deverell/

Joy, M. (2010). *Why we love dogs, eat pigs and wear cows: An introduction to carnism*. San Francisco, CA: Red Wheel/Weiser.

Kolonel, L. N. (1996). Nutrition and prostate cancer. *Cancer Causes & Control*, *7*(1), 83–44. doi:10.1007/BF00115640 PMID:8850437

Lippi, G., Mattiuzzi, C., & Cervellin, G. (2016). Meat consumption and cancer risk: A critical review of published meta-analysis. *Critical Reviews in Oncology/Hematology, 97*, 1–14. doi:10.1016/j.critrevonc.2015.11.008 PMID:26633248

Mari-Sanchis, A., Gea, A., Basterra-Gortari, F. J., Martinez-Gonzalez, M. A., Beunza, J. J., & Bes-Rastrollo, M. (2016). Meat consumption and risk of developing type 2 diabetes in the SUN Project: A highly educated middle-class population. *PLoS One, 11*(7), e0157990. doi:10.1371/journal.pone.0157990 PMID:27437689

Megido, R. C., Gierts, C., Blecker, C., Brostaux, Y., Haubruge, E., Alabi, T., & Fransis, F. (2016). Consumer acceptance of insectbased alternative meat products in Western countries. *Food Quality and Preference, 52*, 237–243. doi:10.1016/j.foodqual.2016.05.004

Myers, S. S., Gaffikin, L., Golden, C. D., Ostfeld, R. S., Redford, K. H., & Ricketts, T. H. … Osofsky, S. A. (2013). Human health impact of ecosystem alteration. *Proceedings of the National Academy of Sciences of the United States of America, 110*(47), 18753–18760. 10.1073/pnas.1218656110

Nielsen. (2015). *The sustainability imperative*. Nielsen Research. Retrieved from http://www.nielsen.com/au/en/insights/reports/2015/the-sustainability-imperative.html

Pew Research Centre. (2011). *The generation gap and the 2012 election*. Retrieved from http://www.people-press.org/2011/11/03/section-8-domestic-and-foreign-policy-views/

Piazza, J., Ruby, M. B., Loughnan, S., Luong, M., Kulik, J., Watkins, H. M., & Seigerman, M. (2015). Rationalizing meat consumption. The 4Ns. *Appetite, 91*, 114–128. doi:10.1016/j.appet.2015.04.011 PMID:25865663

Quintana Pacheco, D. A., Sookthai, D., Wittenbecher, C., Graf, M. E., Stübel, R., Johnson, T., … Kühn, T. (2018). Red meat consumption and risk of cardiovascular diseases—is increased iron load a possible link? *The American Journal of Clinical Nutrition, 107*(1), 113–119. doi:10.1093/ajcn/nqx014 PMID:29381787

Raphaely, T., & Marinova, D. (Eds.). (2016). *Impact of meat consumption on health and environmental sustainability*. Hershey, PA: IGI Global. doi:10.4018/978-1-4666-9553-5

Ritchie, H., & Roser, M. (2018). *Meat and seafood production & consumption*. Retrieved from https://ourworldindata.org/meat-and-seafood-production-consumption

Ruby, M. B., Rozin, P., & Chan, C. (2015). Determinants of willingness to eat insects in the USA and India. *Journal of Insects as Food and Feed, 1*(3), 215–225. doi:10.3920/JIFF2015.0029

Rumpold, B. A., & Schlüter, O. K. (2013). Nutritional composition and safety aspects of edible insects. *Molecular Nutrition & Food Research, 57*(5), 802–823. doi:10.1002/mnfr.201200735 PMID:23471778

Sarich, C. (2013). Harvard says reducing red meat consumption can extend life by 20%. *Natural Society*. Retrieved from http://naturalsociety.com/reducing-red-meat-consumption-extend-life-20-percent/#ixzz2zdnAhdzY

Schmidinger, K. (2012). *Worldwide alternatives to animal derived foods – Overview and evaluation models – Solutions to global problems caused by livestock* (Doctoral thesis). University of Natural Resources and Life Sciences, Vienna, Austria.

Schmidinger, K., Bogueva, D., & Marinova, D. (2018). New meat without livestock. In D. Bogueva, D. Marinova, & T. Raphaely (Eds.), *Handbook of research on social marketing and its influence on animal origin food product consumption* (pp. 344–361). Hershey, PA: IGI Global. doi:10.4018/978-1-5225-4757-0.ch023

Shurtleff, W., & Aoyagi, A. (2004). *Dr. John Harvey Kellogg and Battle Creek Foods: Work with soy.* Lafayette, CA: Soyinfo Center. Retrieved from http://www.soyinfocenter.com/HSS/john_kellogg_and_battle_creek_foods.php

Shurtleff, W., & Aoyagi, A. (2014). *History of meat alternatives (965 CE to 2014): Extensively annotated bibliography and sourcebook.* Lafayette, CA: Soyinfo Center. Retrieved from http://www.soyinfocenter.com/pdf/179/MAL.pdf

Singh, P. N., & Fraser, G. E. (1998). Dietary risk factors for colon cancer in a low-risk population. *American Journal of Epidemiology, 148*(8), 761–774. doi:10.1093/oxfordjournals.aje.a009697 PMID:9786231

Sogari, G. (2015). Entomophagy and Italian consumers: An exploratory analysis. *Progress in Nutrition, 17*(4), 311–316.

Springmann, M., Mason-D'Croz, D., Robinson, S., Garnett, T., Godfray, H. C. J., Gollin, D., ... Scarborough, P. (2016). Global and regional health effects of future food production under climate change: A modelling study. *Lancet, 387*(10031), 1937–1346. doi:10.1016/S0140-6736(15)01156-3 PMID:26947322

Steinfeld, H., Gerber, P., Wassenaar, T., Castel, V., Rosales, M., & de Haan, C. (2006). *Livestock's long shadow: Environmental issues and options.* Rome, Italy: Food and Agricultural Organization of the United Nations. Retrieved from www.virtualcentre.org/en/library/key_pub/longshad/A0701E00.pdf

Tan, H., Verbaan, Y., & Stieger, M. (2017). How will better products improve the sensory-liking and willingness to buy insect-based foods? *Food Research International, 92*, 95–105. doi:10.1016/j.foodres.2016.12.021 PMID:28290303

van Huis, A. (2013). Potential of insects as food and feed in assuring food security. *Annual Review of Entomology, 58*(1), 563–583. doi:10.1146/annurev-ento-120811-153704 PMID:23020616

van Huis, A., Itterbeeck, J. V., Klunder, H., Mertens, E., Haloran, A., Muir, G., & Vantomme, P. (2013). *Edible insects: Future prospects for food and feed security.* Rome, Italy: Food and Agriculture Organization of the United Nations (FAO). Retrieved from https://www.researchgate.net/publication/311424459_Edible_insects_Future_prospects_for_food_and_feed_security_Food_and_Agriculture_Organiation_of_the_United_Nations_FAO_Rome_Italy

Verbeke, W. (2015). Profiling consumers who are ready to adopt insects as a meat substitute in a Western society. *Food Quality and Preference, 39*, 147–155. doi:10.1016/j.foodqual.2014.07.008

Webster, J. (2010). *Management and welfare of farm animals.* London, UK: John Wiley and Sons.

You, W., & Henneberg, M. (2016). Meat consumption providing a surplus energy in modern diet contributes to obesity prevalence: An ecological analysis. *BMC. Nutrition (Burbank, Los Angeles County, Calif.), 2*, 22. doi:10.118640795-016-0063-9

ADDITIONAL READING

Bogueva, D., Marinova, D., & Raphaely, T. (2017). Reducing meat consumption: The case for social marketing. *Asia Pacific Journal of Marketing and Logistics*, *29*(3), 477–500. doi:10.1108/APJML-08-2016-0139

Breene, K. (2016). *Food security and why it matters*. World Economic Forum. Retrieved from https://www.weforum.org/agenda/2016/01/food-security-and-why-it-matters/

Du Bois, C. M., Tan, C.-B., & Mintz, S. W. (2008). *The world of soy*. National University of Singapore: National University of Singapore Press.

Joy, M. (2010). *Why we love dogs, eat pigs and wear cows: An introduction to carnism*. San Francisco, CA: Red Wheel/Weiser.

Piazza, J., Ruby, M. B., Loughnan, S., Luong, M., Kulik, J., Watkins, H. M., & Seigerman, M. (2015). Rationalizing meat consumption. The 4Ns. *Appetite*, *91*, 114–128. doi:10.1016/j.appet.2015.04.011 PMID:25865663

Schmidinger, K., Bogueva, D., & Marinova, D. (2018). New meat without livestock. In D. Bogueva, D. Marinova, & T. Raphaely (Eds.), *Handbook of research on social marketing and its influence on animal origin food product consumption* (pp. 344–361). Hershey, PA: IGI Global. doi:10.4018/978-1-5225-4757-0.ch023

Springmann, M., Mason-D'Croz, D., Robinson, S., Garnett, T., Godfray, H. C. J., Gollin, D., ... Scarborough, P. (2016). Global and regional health effects of future food production under climate change: A modelling study. *Lancet*, *387*(10031), 1937–1346. doi:10.1016/S0140-6736(15)01156-3 PMID:26947322

KEY TERMS AND DEFINITIONS

Generation Y (Gen Y): (referred also as the Millennials) People born between 1977 and 1995; they have grown with technologies such as the internet, computers, and video games and are considered to be technologically savvy.

Generation Z (Gen Z): (referred also as the Centennials or iGen) People born between 1996 and 2009 (although the end of this generation is not clearly defined); they have grown with social media and are considered independent and entrepreneurial.

Natural: Derived from nature, not made or caused by humankind.

Necessary: Required, compulsory, mandatory, inevitable.

New Meat Alternatives: A meat analogue, substitute, vegetarian meat, or vegan meat, a food product which replaces nutritionally animal meat and may or may not imitate meat qualities, such as taste, texture, flavor, and appearance.

Normal: Conforming to a standard; usual, typical, or expected.

Nutritious: Efficient as food to provide essential nutrient; nourishing.

Chapter 3
Health Benefits of Eating More Plant Foods and Less Meat

Patricia Marshall
Curtin University, Australia & Australian Diabetes Educators Association, Australia

Dora Marinova
Curtin University, Australia

ABSTRACT

The health benefits of eating more plant-based foods and less meat are scientifically proven. This chapter examines the evidence in relation to common health and medical conditions, such as cardiovascular diseases, type 2 diabetes, cancers, mental health, and dementia. It also analyzes the issues related to gastrointestinal health and diet in light of the presence of fiber and other plant materials. Although the environmental benefits of a plant-based diet are well-established, there are some concerns about the ability of such food choices to supply essential nutrients to the human body, such as protein, iron, vitamin B12, and Omega 3 fatty acids. They are discussed within the framework of a healthy diet. Some of the disadvantages of diets rich on animal proteins, such as heme iron, are highlighted with a warning that the consumption of lab-grown meat may carry similar risks. A balanced plant-rich diet seems a better and easier choice.

INTRODUCTION

Global meat supply has been on the rise since the 1960s as so has been global human population. The rate of increase of meat consumption however has significantly outpaced that of population (Raphaely & Marinova, 2016b). In 1961, the average annual per capita meat consumption was 23 kg per person and by 2013, this figure reached 43 kg per person (Ritchie & Roser, 2017). Hence, population growth cannot explain the significant increase in supply of meat while changing human diets can. For many people "eating meat is an expression of strength, masculinity, social status, prosperity and prestige" (Bogueva, Marinova, & Raphaely, 2017), and as incomes rise, demand for meat increases (WHO, 2003). Although there have been some shifts in dietary preferences between 1961 and 2013 with per capita beef

DOI: 10.4018/978-1-5225-7350-0.ch003

consumption decreasing by 1%, pork increasing by 100% and poultry by 520% during the same period (FAO, 2018), humans are simply fixated and hooked on meat consumption.

This comes at the expense of environmental harm and misappropriation of common resources, such as water, clean air and land use, but also with negative health consequences. What this chapter does is examine the health benefits from eating more plant foods and less meat-based products. Plant-based choices can be in the form of direct intake, through consuming cooked and processed options or as ingredients of the newly emerging alternatives to meat, such as vegan sausages, steaks, patties or mince. Irrespective of the form the individual consumer would opt for, the health benefits from eating more plants and less meat are distinct and scientifically proven. The objectives of this chapter are to provide an overview from a dietitian's point of view of the benefits from a change to such dietary choices as they relate to common medical and health conditions, such as obesity, cardiovascular disease, cancer, type 2 diabetes and dementia as well as pre-condition for good health, such as improved gut health and supply of essential nutrients. Finally, the chapter provides some information about essential nutrients in plant-based foods which can help people transition from a diet rich in animal products to being a partial vegetarian – the overall recommendation the chapter makes for the many people who are not yet prepared to give up completely meat and dairy.

REDUCED RISK OF OBESITY, CARDIOVASCULAR DISEASE, TYPE 2 DIABETES, WEIGHT GAIN, CANCER, AUTOIMMUNE DISEASES AND DEMENTIA

There is a large body of scientific evidence which shows that the risk for many of the most common threats to human health in wealthy societies can be substantially reduced by improving people's diets. In this day and age, these threats to human health are mainly associated with non-communicable diseases (NCDs). In fact, NCDs are now responsible for 71% of all deaths globally killing 41 million people each year (WHO, 2018). The majority of these deaths (44%) are due to cardiovascular diseases, followed by different types of cancers (22%) and diabetes (4%) (WHO, 2018). Most of these health conditions, in particular obesity, heart disease, stroke, type 2 diabetes, cancer and dementia, are partially attributed to diet.

The World Health Organization (WHO, 2014) cites evidence that reduced fruit and vegetable consumption is linked to increased risk of noncommunicable diseases. Although the exact mechanism of how this occurs is unclear, "[f]ruit and vegetable consumption may reduce the risk of NCDs through the increased availability of various nutrients and their ability to modulate associated risk factors" (WHO, 2014, n.p.).

Many reviews of the health implications of plant-based diets present significant evidence that they are associated with lower rates of obesity, cardiovascular diseases, type 2 diabetes, some cancers, gallstones, kidney stones, constipation, and diverticular disease, and overall increase longevity (e.g. Marsh, Zeuschner, & Saunders, 2012; 2016). There is also some evidence that vegetarian eating can reduce gout, rheumatoid arthritis and kidney disease (Marsh, Zeuschner, & Saunders, 2012). A review of the health advantages of a vegan food style, which is entirely plant-based, presents evidence that in some cases it is even more effective than a vegetarian diet in reducing the risk of obesity, hypertension, type 2 diabetes and cardiovascular disease (Glick-Bauer & Yeh, 2014).

Notwithstanding this scientific evidence, many people and health practitioners continue to consider meat as a healthy choice. The current research and industry innovations related to lab-grown or cultured meat is likely to result in a significant decrease in the environmental impacts of livestock-based food products. Some of the health problems associated with the intake of meat however are likely to remain. This is particularly likely to be the case when people's diet is rich in meat and poor in plant-based foods. Below we examine evidence about a range of health problems and their alleviation through the lenses of more plants and less meat.

More Plants

There is clear evidence that diets high in fruits, vegetables, nuts, seeds and legumes reduce the risk of non-communicable diseases. A meta-analysis of prospective cohort studies provided evidence that a higher consumption of fruit and vegetables is associated with a lower risk of all-cause mortality, particularly cardiovascular mortality (Wang, Ouyang, Liu, Zhu, Zhao, Bao, & Hu, 2014). However, it is interesting to unpack this overall message and examine the biggest human killer diseases one by one.

Cardiovascular Diseases

Cardiovascular diseases (CVDs) are the leading cause of death globally affecting people in high-, middle- and low-income countries (WHO, 2017a). Most CVDs can be prevented by addressing behavioral risk factors and improving people's lifestyle, including their diets (WHO, 2017a).

The review of the health implications of a vegetarian diet presents significant evidence that vegetarians have reduced risk factors for CVDs in lower total cholesterol, and in particular lower "bad" low-density lipoprotein (LDL) cholesterol levels, lower bodyweight and lower hypertension, possibly due to less intake of saturated fat and higher consumption of soluble fibre (Marsh, Zeuschner, & Saunders, 2012). This review also provides evidence that a high intake of wholegrains, legumes and nuts and low in meat, such as in the Mediterranean diet, is associated with lower risk of CVD (Marsh, Zeuschner, & Saunders, 2012). As Haddad (2018, p.45) succinctly explains, the "benefits do not simply stem from foods avoided by vegetarians but more importantly from foods consumed by vegetarians".

A critical review of red meat, diseases and healthier alternatives showed nut consumption to be associated with a lower risk of coronary heart disease (CHD) through lower total cholesterol, lower LDL and lower ratio between the "bad" and "good" (high-density lipoprotein or HDL) cholesterol, and lower blood pressure (Ekmekcioglu, Wallner, Kundi, Weisz, Haas, & Hutter, 2018). This is possibly because nuts are high in potassium, low in saturated fatty acids and high in monounsaturated fatty acids. The same review also found that legumes reduced blood pressure, total cholesterol, LDL cholesterol and triglycerides. Furthermore, nuts and legumes both lowered the risk of ischemic heart disease (Ekmekcioglu, Wallner, Kundi, Weisz, Haas, & Hutter, 2018). A meta-analysis of fruits and vegetables consumption and stroke provided evidence of an inverse relationship between intake and risk of stroke (Hu, Huang, Wang, Zhang, & Qu, 2014).

Sodium occurs naturally in many foods especially meat, seafood and milk, it is added to processed food and is consumed as table salt (sodium chloride). A systematic review and meta-analysis of the effect of sodium on health provided evidence that reduced intake of this chemical element, which humans mainly consumed as sodium chlorate or table salt, decreased blood pressure and lowered the risk of fatal coronary heart disease and stroke (Aburto, Ziolkovska, Hooper, Elliott, Cappuccio, & Meerpohl, 2013b). Compared to fruit and vegetables, meat and other animal products usually are associated with high intake of sodium – including sodium chlorate (table salt), but also from other sodium compounds, such as monosodium glutamate, sodium nitrate and sodium bisulphate (American Heart Association, 2017). Conversely, a similar systematic review and meta-analysis of the effect of increased potassium intake on CVD risk "provided evidence that increased potassium intake reduces blood pressure in people with hypertension" (Aburto, Hanson, Gutierrez, Hooper, Elliott, & Cappuccio, 2013a, n.p.). It also provided some evidence that a higher potassium intake was associated with a lower risk of stroke (Aburto, Hanson, Gutierrez, Hooper, Elliott, & Cappuccio, 2013a, n.p.) with "no adverse effects on blood lipid concentrations, catecholamine concentrations, or renal function in either case". The top 10 foods that are high in this essential element are plant-based (Jones, 2017). A survey in the US found that only 3% of the Americans have the necessary intake of potassium due to low consumption of fruit and vegetables (Weaver, 2013).

The World Health Organization (WHO, 2016) recommends increasing potassium intake to reduce blood pressure in adults through consuming a variety of unrefined foods, including beans, chickpeas, peas, lentils, nuts as well as leafy green vegetables such as spinach, cabbage, kale and parsley and fruits such as berries, bananas, paw-paw, pomegranates and dates. A diet high in processed foods and low in fresh fruits and vegetables is often lacking in potassium as food processing reduces the amount of potassium in many food products (WHO, 2016).

Although the environmental benefits from all plant-based options are indisputable compared to animal-based foods (Aleksandrowicz, Green, Joy, Smith, & Haines, 2016; Springmann, Godfray, Rayner, & Scarborough, 2016; Clark & Tilman, 2017; Shepon, Eshel, Noor, & Milo, 2018), from a health point of view the quality of plant foods is critical. Only healthy choices, such as whole grains, fruits, vegetables, nuts, legumes, oils, tea and coffee, are associated with a lower cardiovascular disease risk (Satija, Bhupathiraju, Donna Spiegelman, Chiuve, Manson, Willett, Rexrode, Rimm, & Hu, 2017). For example, an intake of whole grains of at least three servings per day is associated with a reduction in risk of death from cardiovascular disease (Wei, Gao, Liang, Li, Hao, & Liu, 2016).

Type 2 Diabetes

The reviews of the health implications of a plant-based diet also show that such a food style is beneficial in managing type 2 diabetes through weight reduction, lower glycosylated haemoglobin, lower blood fats, better blood glucose control and reduced medication use (Marsh, Zeuschner, & Saunders, 2012; 2016). When animal protein was replaced by plant protein, the risk and progression of renal disease in type 2 diabetes was reduced as well (Marsh, Zeuschner, & Saunders, 2012).

Two meta-analyses of prospective cohort studies found that higher fruit or green leafy vegetables intake is associated with a reduced risk of type 2 diabetes compared to refined grains (Cooper, Forouhi, Ye, Buijsse, Arriola, Balkau. … & Wareham, 2012; Li, Fan, Zhang, Hou, & Tang, 2014), while a critical review of red meat, diseases and healthier alternatives showed nut consumption to also be associated with a lower risk of diabetes (Ekmekcioglu, Wallner, Kundi, Weisz, Haas, & Hutter, 2018). Intake of whole grains reduces the risk of type 2 diabetes (Ekmekcioglu, Wallner, Kundi, Weisz, Haas, & Hutter, 2018) as they usually have a lower glycaemic index, meaning that they have less impact on blood glucose. The lower glycaemic index contributes to a reduced risk of diabetes and cardiovascular disease (Barclay, Petocz, McMillan-Price, Flood, Prvan, Mitchell, & Brand-Miller, 2008; Greenwood, Threapleton, Evans, Cleghorn, Nykjaer, Woodhead, & Burley, 2013; Bhupathiraju, Tobias, Malik, Pan, Hruby, Manson, … & Hu, 2014). A critical review of red meat, diseases and healthier alternatives also found that legumes lower blood glucose and glycosylated haemoglobin and increase insulin sensitivity (Ekmekcioglu, Wallner, Kundi, Weisz, Haas, & Hutter, 2018).

Weight Gain

Weight gain is a contributing factor for type 2 diabetes as well as CVDs. Increased intake of fruit and vegetables, even in the absence of decreased consumption of other foods, appears unlikely to lead to weight gain in the short-term and may have a role in weight maintenance or loss (Mytton, Nnoaham, Eyles, Scarborough, & Ni Mhurchu, 2014). Numerous studies based on different populations confirm that plant-based diets are better for weight management and prevention of morbidity and mortality. Examples include the studies of the Seventh-Day Adventists, cohorts in the UK and Europe, including the European Prospective Investigation into Cancer and Morbidity (EPIC), Australian groups, cohorts in India, Taiwan, South Korea and Pakistan as well as white and black populations in the USA (Heskey, 2018). A possible mechanism explaining this is the influence different nutrients have on the gut microbiota.

Cancer

A review of research on wholegrains, fruit and vegetables and the risk of cancer as part of a global perspective on the role of diets (WCRF/AICR, 2018) concluded that there is strong evidence that wholegrains and other foods containing dietary fibre decrease the risk of colorectal cancer. There is also some evidence of a protective effect of fruits and non-starchy vegetables on other cancers (WCRF/AICR, 2018). These findings are based on the latest scientific evidence from the world's largest and most authoritative source on cancer prevention – the World Cancer Research Fund (WCRF) and American Institute for Cancer Research's (AICR) Continuous Update Project. Their recommendation is to make wholegrains, vegetables, fruit and legumes (or pulses), such as beans, lentils, chickpeas, soybeans and peanuts, a major part of the daily diet because they protect against cancer.

The critical review of red meat, diseases and healthier alternatives also concluded that whole grain and legume intake was associated with reduced risk of colorectal (or colon) cancer (Ekmekcioglu, Wallner, Kundi, Weisz, Haas, & Hutter, 2018). As the digestive tract is constantly exposed to the ingredients provided by the food we consume, the link between what we eat and gastrointestinal cancer is wide-ranging. Colorectal cancer in particular is one of the most common cancers worldwide (WCRF, 2012) and overall cancer is more frequent in countries with high red meat intake (Wilson, 2012).

Autoimmune Diseases

As our understanding of the importance of gut microbiota improves, evidence emerges that plant-based foods also reduce the risk of autoimmune diseases. A review of the health advantages of a vegan diet provides evidence that such food choices, but not vegetarian, protect against autoimmune diseases, including hypothyroid disease and rheumatoid arthritis (Glick-Bauer & Yeh, 2014).

Mental Health and Dementia

While cognitive impairment and dementia often occur with age-related chronic diseases, some conditions such as hypertension and diabetes increase the risk of cognitive impairment and dementia (Ross,

2018). Reducing the occurrence of these conditions through dietary modification will clearly reduce the risk of cognitive impairment and dementia. Systematic reviews of the role of a Mediterranean-style diet, which is high in fruit, vegetables, legumes and wholegrain cereals, conclude that this eating pattern can slow the rate of cognitive decline and decrease the risk of dementia (Sara, Ralston, & Walker, 2013; Petersson & Philippou, 2016). A review of diet, cognition and Alzheimer's disease examined whether there are any potentially modifiable behavioural factors that could protect from this chronic neurodegenerative ailment (Otaegui-Arrazola, Amiano, Elbusto, Urdaneta, & Martínez-Lage, 2014). It provided evidence that a variety of diets which are based on consuming significant amounts of fruits, vegetables, nuts and other foods high in antioxidants and vitamins may prevent cognitive aging and Alzheimer's disease (Otaegui-Arrazola, Amiano, Elbusto, Urdaneta, & Martínez-Lage, 2014). This is also supported by further evidence from a critical review of vegetables and fruit in the prevention of chronic disease which shows that cognitive impairment and risk of dementia are increased with low fruit and vegetable intake (Boeing, Bechthold, Bub, Ellinger, Haller, Kroke, … & Watzl, 2012).

A systematic review of cohort studies also concluded that a high intake of vegetables is associated with slower cognitive decline and a lower risk of dementia in older age (Loef & Walach, 2012). As recommended for other chronic diseases, a diet high in plant-based foods, especially green leafy vegetables, and low in animal and high saturated fats is recommended (Ross, 2018). The Mediterranean-DASH (Dietary Approaches to Stop Hypertension) Intervention for Neurodegenerative Delay or MIND Diet, which was developed specifically for brain health with a high content of plant-based foods, especially berries and green leafy vegetables, and low amounts of animal foods and saturated fat, was shown to be associated with a lower risk of developing Alzheimer's disease (Morris, Tangney, Wang, Sacks, Bennett, & Aggarwal, 2015; Hosking, Eramudugolla, & Anstey, 2017).

Another aspect of the link between diet and mental health is simply how people feel, what their mood and outlook on life are. Many cross-sectional studies confirm that there is correlation between the intake of fruit and vegetables and emotional wellbeing, including feelings of happiness and optimism. For example, a study of young British adults showed that a daily consumption of 7–8 servings of fruit and vegetables results in positive affective experiences (White, Horwath, & Conner, 2013). Similar results are reported from studies in other parts of the world, such as Chile (Piqueras, Kuhne, Vera–Villarroel, van Straten, & Cuijpers, 2011), South Korea (Kye & Park, 2014) and USA (Boehm, Williams, Rimm, Ryff, & Kubzansky, 2013; Blanchflower, Oswald, & Stewart–Brown, 2013). According to Blanchflower et al. (2012, p. 1), "happiness and mental health rise in an approximately dose-response way with the number of daily portions of fruit and vegetables. The pattern is remarkably robust to adjustment for a large number of other demographic, social and economic variables."

Less Meat

Eating less meat is also very important for those who continue to include this food in their diets. Multiple evidence exists (NutritionFacts, 2018) that heme iron – the type of iron found in animal blood and muscle tissue, is linked to many serious medical conditions, such as coronary heart disease, stroke, atherosclerosis, type 2 diabetes, cancer, Alzheimer's and Parkinson's diseases as well as metabolic syndrome – a condition which combines a collection of components, such as obesity, high blood pressure, high blood triglycerides, low levels of the "good" HDL cholesterol and insulin resistance (Victoria State Government, 2018). Although heme iron is more easily absorbed by the human body, there are no mechanisms for any excess quantities to be expelled or removed (NutritionFacts, 2018). By comparison, the human body naturally regulates the absorption of iron from plants, such as whole grains, dark green leafy vegetables, legumes, dried fruits, nuts and seeds.

Too much iron in the human body can trigger oxidative stress (inflammation) and can cause DNA damage (NutritionFacts, 2018). Studies (in rats) show that the major promoter of colon carcinogenesis is heme iron via fat peroxidation and ask whether we should give up red meat or make it safer for humans to consume by adding substances which can suppress the toxic effects of heme iron (Corpet, 2011). The toxicity and associated heme iron effects will be similar with clean or lab-grown meat. It may have a better environmental footprint, but the challenges for human health will remain.

It is not surprising then that the review of the health implications of diets rich in plants and low in meat intake are linked to greater longevity. Red meat and processed meat are linked to increased risk of obesity, type 2 diabetes, gestational diabetes, CVDs and some types of cancer (Marsh, Zeuschner, & Saunders, 2012; 2016; 2018; Micha, Michas, & Mozaffarian, 2012). A meta-analysis of prospective cohort studies of red and processed meat consumption and mortality also concluded that higher consumption of red meat and processed meat is associated with an increased risk of total, cardiovascular and cancer mortality (Wang, Lin, Ouyang, Liu, Zhao, Pan, & Hu, 2016). Again, let's examine the specific diseases.

Cardiovascular Diseases

While the high heme iron content of meat, especially red meat, can be of benefit in maintaining haemoglobin levels, a critical review of red meat, diseases and healthier alternatives found high iron stores to be associated with a higher risk of CVDs (Ekmekcioglu, Wallner, Kundi, Weisz, Haas, & Hutter, 2018). It also found that the high sodium content of processed meat contributes to higher blood pressure (hypertension) (Ekmekcioglu, Wallner, Kundi, Weisz, Haas, & Hutter, 2018) – a common precursor of cardiovascular diseases, especially heart attack and stroke. In addition to table salt, higher amounts of sodium are contained in processed meats like bacon, corned beef and sausages (WHO, 2017b). The systematic review and meta-analysis of the effect of decreased consumption of sodium provides evidence that lower intake reduces blood pressure and the risk of fatal coronary heart disease and stroke (Aburto, Ziolkovska, Hooper, Elliott, Cappuccio, & Meerpohl, 2013b).

Type 2 Diabetes

The critical review of red meat, diseases and healthier alternatives found red meat to be linked to obesity and increased waist circumference, both risk factors for type 2 diabetes (Ekmekcioglu, Wallner, Kundi, Weisz, Haas, & Hutter, 2018). A study on the impact of meat on the risk of diabetes on a multiethnic cohort found that intake of red meat, especially processed red meat, was positively associated with diabetes (Steinbrecher, Erber, Grandinetti, Kolonel, & Maskarinec, 2010). Saturated fatty acids correlate to decreased insulin sensitivity, a factor in type 2 diabetes, and in rats, nitrosamines, which are used in processed meats, also cause insulin resistance (Ekmekcioglu, Wallner, Kundi, Weisz, Haas, & Hutter, 2018).

Furthermore, there is evidence that animal proteins themselves are associated with an increased risk of type 2 diabetes (Malik, Li, Tobias, Pan, & Hu, 2016; Shang, Scott, Hodge, English, Giles, Ebeling, & Sanders, 2016; Marsh, Zeuschner, & Saunders, 2018). The reviews of the health implications of a vegetarian diet present significant evidence that a high meat and animal product intake is related to development of type 2 diabetes (Marsh, Zeuschner, & Saunders, 2012; 2016; 2018). Replacing red meat with vegetable protein improved renal function, for both – people with type 1 or type 2 diabetes, and lowered blood fats (Marsh, Zeuschner, & Saunders, 2012). In people without diabetes, lower animal foods intake reduced markers of renal risk (Marsh, Zeuschner, & Saunders, 2012).

Cancer

The review of research into diet, nutrition, physical activity and cancer as part of the WCRF/AICR's Continuous Update Project (WCRF/AICR, 2018) showed convincing evidence that red meat, and especially processed meat, increases the risk of colorectal (or colon) cancer. Furthermore, the review of the health implications of a vegetarian diet suggested that red meat cooked at high temperatures and high fat processed meats may contain cancerous compounds (Marsh, Zeuschner, & Saunders, 2012).

Other cancers, such as stomach, liver, lung, bladder, pancreas and esophagus are similarly related to the consumption of red meat (Boada, Henríquez-Hernández, & Luzardo, 2016; Lippi, Mattiuzzi, & Cervellin, 2016; Wang, Lin, Ouyang, Liu, Zhao, Pan, & Hu, 2016). High iron intake may increase the formation of free radicals which stimulate cancer formation (Marsh, Zeuschner, & Saunders, 2012). Processed meat is particularly toxic to the human body as in addition to heme iron, it contains harmful compounds and chemicals which have been used in the process of preservation (for example, this includes curing, salting, smoking, drying or canning of meat). Therefore, the World Health Organization (2015) categorized processed meats as Group 1 carcinogenic to humans. Red meat is classified in Group 2A as probably carcinogenic to humans based on the evidence provided by the International Agency on Research on Cancer (IARC, 2015; Bouvard, Loomis, Guyton, Grosse, El Ghissassi, Benbrahim-Tallaa, … Straif, 2015).

Mental Health and Dementia

A review of diet, cognition and Alzheimer's disease provided evidence that a variety of diets high in meat, butter, high-fat dairy products as well as refined sugars has been shown to increase the risk of Alzheimer's disease (Otaegui-Arrazola, Amiano, Elbusto, Urdaneta, & Martínez-Lage, 2014). On the other hand, systematic reviews of the impact of a Mediterranean-style diet, which is low in meat and saturated fat, concluded that this eating pattern can slow the rate of cognitive decline and decrease the risk of dementia (Sara, Ralston, & Walker, 2013; Petersson & Philippou, 2016).

The above section examined the scientific evidence about the link between diets represented by more plants and less meat and health conditions. According to Hippocrates – the father of medicine, human health depends on the bacteria in the human gastrointestinal tract putting it at the centre of human wellbeing. The section to follow examines the latest evidence about the role of plants in maintaining gut health.

IMPROVED GUT HEALTH

A microbiota is a term which describes the population of microorganisms, such as bacteria, fungi, viruses and mites, which live on the human body (Acapsil, n.d.). Inside the human gastrointestinal tract, the microbiota is represented by millions of different bacteria estimated to represent about 10^{13}–10^{14} microorganisms (Byrne, Chambers, Morrison, & Frost, 2015) and between 300 and 1000 species (Guarner & Malagelada, 2003). The activity of these bacteria in the human gut influences the digestive processes, energy metabolism, immune modulation and inflammation responses of the host, with consequences on the risk of allergies, cancer, diabetes, blood pressure, obesity, and even the mental health of the host (Nie, Luo, & Lin, 2018).

Dahl et al. (2018) cited evidence of the association between diets high in fibre and lower risk of chronic diseases such as cardiovascular disease, type 2 diabetes and some cancers in their review of the health benefits of fibre and its effect in the biome. They concluded that fibre, through its effect on the microbiota, is one of the most important factors in this relationship. Hence, they recommended that an adequate intake of dietary fibre should be encouraged.

The bacteria in the biome can both produce and reduce harmful metabolites. When fibre and other plant materials which are not digested in the upper digestive tract of the human gut reach the colon they are fermented to produce metabolites such as short-chain fatty acids that have beneficial effects on the host's health, including the immune system (Birt, Boylston, Hendrich, Jane, Hollis, Li, … & Whitley, 2013). Short-chain fatty acids help with digestive disorders, inflammatory bowel disease, prevent cancer development, improve blood sugar control, especially for people with diabetes, assist energy regulation and the hormones in the gut which control appetite (Brown, 2016). The diet of the host determines both the composition of the biome and the metabolic outputs of its bacterial members (Flint, Scott, Louis, & Duncan, 2012). Eating a lot of fibre-rich foods helps the friendly bacteria ferment the fibre and affects the production of short-chain fatty acids.

In their earlier review into the health advantages of a vegan diet, Glick- Bauer and Yeh (2014) proposed that a high fibre diet such as vegan appeared to increase beneficial bacteria which confer protection against metabolic disease, such as diabetes, intestinal disorders, inflammation and obesity. This high fibre diet also appears to decrease pathobionts which have been implicated in triggering low-grade inflammation and exerting pathogenic effects on the host (Zechner, 2017). Inflammation may be the critical factor that influences glucose tolerance and lipid metabolism. During a vegan diet rich in fibre, faecal enzymes associated with toxic and inflammatory products diminished and rheumatoid arthritis improved (Glick- Bauer & Yeh, 2014).

A review of the health effects of whole cereal grains quoted evidence that their consumption is linked to reduced metabolic syndromes and several chronic diseases such as diabetes. The authors identified fibre and other components of whole cereal grains as beneficially influencing the microbiota and thereby providing pathways for these health benefits (Gong, Cao, Chi, Wang, Zhang, Liu, & Sun, 2018)

Studies indicate that diets high in protein intake, particularly from animal sources, are detrimental to the gut microbiota (Zhang, Ju, & Zuo, 2018). Dietary L-carnitine, a trimethylamine found in red meat, produces trimethylamine-N-oxide (TMAO) in the gut, which has been shown to promote atherosclerosis (Glick-Bauer & Yeh, 2014). Vegans and vegetarians have low levels of TMAO (Glick-Bauer & Yeh, 2014).

Overall, there is a lot of conclusive evidence that plant-rich diets are better alternatives to meat and animal products. Many people however are concerned that by avoiding meat they will not get enough protein, vitamins and minerals. This is an unjustified concern as a healthy diet can provide all these nutrients, even with little or no meat, as well as deliver all the other benefits already discussed.

PLANT-BASED NUTRIENTS

Protein, vitamins and minerals are essential nutrients for the functioning of the human body. Any deficiency can increase disease risk; therefore people are concerned how to achieve a balanced diet.

Protein

The second component of human cells after water is protein. It is used by the body to make muscle, blood, skin, hair, fingernails, hormones, haemoglobin, enzymes and antibodies. Protein is made of building blocks called amino acids which are linked together. These structures do not last forever – they are broken down and new ones are created from the amino acid pool. As the body cannot store the amino acids, they need to be generated each day through the food we consume. The amino acids from the broken-down proteins are also added to the pool and recycled.

Out of the 20 amino acids in the human body, 11 are non-essential and 9 are essential. The amino acids which can be synthesized within the body or made in the body by modifying existing amino acids are called non-essential. They do not need to be provided through food. The amino acids which cannot be made by the body are called essential. Animal-based foods contain amino acids in similar proportions to what the human body requires and are referred to as complete proteins. However, there are plant-based options, such as buckwheat, quinoa and soy which also contain all nine essential amino acids. If there is some concern about incomplete or lower quality vegetable proteins, this problem can be rectified by combining a plant protein low in one or more essential amino acids with another that is high in those amino acids. For example, when legumes like beans, lentils and peanuts are combined with grains like wheat, rice and corn, the outcome is a complete protein. This allows for a balanced diet to be achieved.

Since the body is very efficient at recycling and modifying amino acids, adults only need relatively small amounts of protein. The adult requirement is about 0.6 g/kg body weight/day for women and 0.7 g/kg for men. Therefore, a woman of 60 kg needs about 36 g protein/day, while a man of 80 kg needs about 56 g protein/day. To ensure that this requirement is met, the Australian recommended intake is 0.75 g/kg/day for women and 0.84 g/kg/day for men (NHMRC, 2006). People over 70 years of age are advised to consume slightly more. The USDA Guidelines recommend an intake of 0.8 g/kg/day, or 46 g/day for women and 56 g/day for men (Busch, 2018). The list in Table 1 shows how easy it is to achieve these protein intakes on a fully vegetarian diet.

Although legumes, including beans, lentils and peanuts, are not complete proteins, they can easily be combined with other grains, such as rice or wheat, to achieve the full requirement of essential amino acids (see some suggestions in Table 1). The nutritional profile of legumes is extremely rich and they reduce the risk for many chronic diseases and metabolic syndrome. According to Pribis (2018, p.191), beans in particular "have cholesterol-lowering, blood-glucose-lowering and anti-inflammatory effects". From an environmental point of view, legumes are one of the foods with the lowest carbon footprint (Green Eatz, n.d.; Environmental Working Group, 2011; Clune, Crossin, & Verghese, 2017). They are a perfect example of co-benefits between human and environmental health.

Iron

Iron is needed to make the haemoglobin that carries oxygen in the red blood cells, as well as a number of enzymes. When the red blood cells come to the end of their life, the haemoglobin is broken down and the iron cycled to make new haemoglobin. However, blood loss can deplete the body's iron stores. For this reason, women who menstruate regularly need to replenish their iron stores. This is a modern problem, as in earlier times women spent much of their reproductive years pregnant or breastfeeding, so were able to retain their iron stores.

Table 1. Protein content in selected plant-based foods

Food	Amount	Weight (g)	Protein (g)	Complete Protein	New in Western Diet
Almonds	½ cup	75	16.5	No	No
Banana, sliced	1 small	100	1	No	No
Broccoli	1 medium stalk	300	4	No	No
Buckwheat	1 cup	150	6	Yes	No
Chia seeds	1 tablespoon	12	2	Yes	Yes
Chickpeas	½ cup	75	7	No	No
Dried fruit	2 tablespoons	30	2	No	No
Ezekiel bread (made of sprouted wheat, barley, beans, lentils, millet, and spelt)	2 slices	70	8	Yes	No
Hemp seeds	1 tablespoon	12	5	Yes	Yes
Kale	1 cup	150	2	No	No
Lentils	½ cup	75	9	No	No
Mixed beans, canned	1 cup	100g	6	No	No
Mushrooms	5 medium	150	3	No	No
Mycoprotein (fungal protein, Quorn)	½ cup	75	13	Yes	Yes
Nuts	2 tablespoons	15	3	No	No
Peanuts	½ cup	75	20.5	No	No
Peanut butter sandwich	2 slices with 2 tablespoons	102 (70 plus 32)	15	Yes	No
Peas, boiled	½ cup	75	4	No	No
Potato, baked	3 small (egg size)	150	5	No	No
Pumpkin, baked	2 small pieces	75	2	No	No
Quinoa	1 cup	150	8	Yes	Yes
Rice and beans	1 cup	150	7	Yes	No
Rolled oats, cooked in water	½ cup	180	4	No	No
Salad	1 cup	75	1	No	No
Seitan (wheat gluten)	1/3 cup	50	21	Yes	Yes
Soy (edamame, immature soybeans)	½ cup	75	8.5	Yes	Yes
Soy (firm tofu, unfermented soybean)	½ cup	75	10	Yes	Yes
Soy (nattõ, fermented soybean)	½ cup	75	15	Yes	Yes
Soy (tempeh, fermented soybean)	½ cup	75	15	Yes	Yes
Spirulina (algae)	2 tablespoons	15	8	No	Yes
Spirulina with grains or nuts	1 tablespoon	7	4	Yes	Yes
TVP (textured vegetable protein from soybeans)	¼ cup	24	12	Yes	Yes
Wholegrain bread	2 slices	74	7	No	No

Source of data: English (2017) and Villines (2018).

Absorption can also be enhanced by consuming high iron foods with a good source of vitamin C such as fruit or raw vegetables. While eating less meat means consuming less iron, especially the more absorbable heme iron, the human body is well-positioned to absorb and regulate this element directly from plants (Marsh, Zeuschner, & Saunders, 2012). Also, absorption is increased when iron status is lower. The same mechanism applies for zinc and calcium.

According to Craig and Mangels (2009), a vegetarian diet can provide the recommended levels of iron. Usually the bodies of vegetarians have lower stores of iron which may not be a disadvantage as higher ferritin levels (representing stored iron) are associated with increased risk of chronic disease and lower ferritin levels – with reduced risk of type 2 diabetes (Craig & Saunders, 2018).

Vitamin B12

Vitamin B12 is only found in animal products, but as it is stored in the body for up to four years and is required in small amounts, deficiency takes many years to develop (Craig & Saunders, 2018). Those who consume eggs and dairy products will consume sufficient vitamin B12, but vegans, especially pregnant and breast-feeding women, need to take B12 supplements to avoid deficiency (Marsh, Zeuschner, & Saunders, 2012; Tucker, 2014) which can irreversibly affect the nervous system as well as the bone density and the risk of osteoporosis. The vegan versions of B12 supplements are produced commercially through bacterial fermentation synthesis (Nordqvist, 2017).

Calcium

Calcium intake is generally lower on a plant-based diet. For example, Craig (2010) expressed concern that vegans consume about 20% less calcium than lacto-ovo vegetarians. When the intake of calcium is lower than 800 mg per day, there is a significantly higher risk of hip fracture (Warensjö et al., 2011) with others arguing that this threshold is even lower (Kurpad & Vaz, 2000; Darling, Millward, Torgerson, Hewitt, & Lanham-New, 2009; Shams-White, Chung, Du, Fu, Insogna, Karlsen, … Weaver, 2017). The intake of calcium however can be increased to adequate levels by consumption of plant-based milk alternatives, such as soy milk as well as soy-based products such as tofu and tempeh. Fortified foods and supplements can also be used to maintain the levels of nutrients required for good bone health (Giudici & Weaver, 2018).

Low-oxalate leafy green vegetables, such as bok choy, Chinese cabbage, broccoli, cabbage, collard, kale, dandelion and watercress, as well as almonds, sesame seeds, tahini, figs and blackstrap molasses are rich in calcium (Hever & Cronise, 2017). Calcium intake is particularly important for growing adolescents and athletes as it helps build and repair of bone tissue as well as regulates muscle functions, normal blood clotting and the condition of the nerve system (Larson-Meyer, 2018).

Omega 3

Omega-3 fatty acids (alpha-linolenic or ALA, which is converted to docosahexaenoic acid or DHA, and eicosapentaenoic acid or EPA) contribute to cardiovascular health, neurological development and eye development, and reduce inflammation. While fish and seafood are considered main sources (Willis, 2018), chia and hemp seeds, linseeds, walnuts, soy products, green leafy vegetables, such as Brussels sprouts, and algal oil are similarly significant sources (Link, 2017).

Usually vegetarians have a higher intake of Omega-6 fatty acids as contained in linoleic acid (LA) found in sunflower, corn and safflower oils. Craig and Saunders (2018) recommend for a portion of the LA-rich vegetable oils to be replaced in the diet by monounsaturated oils, such as olive oil or macadamia oil.

How to Eat a Balanced Diet

A vegetarian or vegan meal is not one that simply leaves out the meat. From a health perspective, it needs to include whole foods rather than processed foods (see Table 2). Highly processed foods have lost important nutrients, especially fibre, iron and zinc. Some processed foods are also much higher in saturated fat, sugar and salt than in their original form.

Protein sources such as lentils, chickpeas, kidney beans, nuts, seeds and soy products like tofu or tempeh are important for a plant-based diet. Nuts and seeds also provide valuable unsaturated fatty acids. A good way to include plenty of vegetables in a meal is to start with a hearty vegetable soup in winter or a crisp salad in summer. Fruit is also very refreshing any time of the year. Another advice is to cook at home from the raw ingredients rather than use ready and pre-prepared meals or restaurant food. In this way, you have full control over the food you eat and preparing a nice balanced meal can actually be a very relaxing and rewarding exercise. Remember that the vegetables and fruits together with any legumes and nuts should be the foundation of any nutritious meal.

CONCLUSION

According to Marsh et al. (2018, p.18):

Despite significant evidence showing health benefits of plant-based diets and negative health effects of diets high in animal foods, there continues to be a focus on what is 'missing' when animal foods are removed from the diet. Yet, particularly in the Western world, chronic disease is a much bigger contributor to morbidity and mortality than nutrient deficiencies. Furthermore, plant-based diets more closely match dietary recommendations for good health and prevention of chronic disease, and there is good evidence to show that plant-based diets containing little or no animal foods are associated with a reduced risk of chronic disease.

Table 2. Examples of whole and processed foods

Whole Foods	Processed Foods
Avocado	Avocado dip
Baked potato	Chips
Brown rice	White rice
Fresh fruit	Fruit juices
Tomato	Tomato sauce or ketchup
Vegetables	Vegetable juices
Wholegrain bread	White bread
Wholemeal pasta	White pasta

There have been numerous systematic and meta-reviews which confirm these statements. The medical community is now fully aware of the benefits from a diet high in plants and low in meat. Dietary guidelines have also accepted and recommend balanced plant-based diets (vegetarian and vegan) as healthy options, including in relation to providing proteins, iron and other essential nutrients and with suggestions how to deal with B12 supply which appear a minor problem compared to the detrimental effects of overconsumption of meat. In fact, by setting limits for the intake of meat – around 70 g per day for red meat and 0 g for processed meat, dietary guidelines also warn against the potential toxicity of these foods and encourage all plant-based food options.

Personal health has always been a matter of individual selection of the preferred foods. The role of dietitians, health professionals and the medical community is to inform and properly direct this choice. Things have now drastically changed with the expansion and power of the livestock industry. In light of the overwhelming evidence about the negative environmental impacts of animal-sources foods, such as meat and dairy (e.g. Raphaely & Marinova, 2016a; Bogueva, Marinova, & Raphaely, 2018; Greenpeace, 2018), our diets no longer represent an individual choice. They affect the air, land, water and use of minerals on this planet as well as all species. The food choices of the wealthy misuse the common resources and deprive others of decent opportunities and livelihoods.

The scientific evidence is clear that we need to consume more plant-based foods and reduce the presence of meat and other animal products in the human diet. Some of the properties associated with animal flesh, such as heme iron and high presence of proteins, are likely to remain a health concern in the case of lab-grown meat. Whilst the majority of people are not yet willing to give up their meat, they and the public health system need to be prepared to also deal with the risks of chronic diseases as explained in this chapter. A balanced plant-rich diet seems like a better and easier solution.

REFERENCES

Aburto, N. J., Hanson, S., Gutierrez, H., Hooper, L., Elliott, P., & Cappuccio, F. P. (2013a). Effect of increased potassium intake on cardiovascular risk factors and disease: Systematic review and meta-analyses. *British Medical Journal, 346*, f1326. doi:10.1136/bmj.f1326 PMID:23558164

Aburto, N. J., Ziolkovska, A., Hooper, L., Elliott, P., Cappuccio, F. P., & Meerpohl, J. J. (2013b). Effect of lower sodium intake on health: Systematic review and meta-analyses. *British Medical Journal, 346*(apr03 3), f1326. doi:10.1136/bmj.f1326 PMID:23558163

Acapsil. (n.d.). *What is the difference between microbiome, microbiota and microflora in relation to wounds?* Retrieved from http://acapsil.com/en-gb/2016/12/28/what-is-the-difference-between-microbiome-microbiota-and-microflora-in-relation-to-wounds/

Aleksandrowicz, L., Green, R., Joy, E. J. M., Smith, P., & Haines, A. (2016). The impacts of dietary change on greenhouse gas emissions, land use, water use, and health: A systematic review. *PLoS One, 11*(11), e0165797. doi:10.1371/journal.pone.0165797 PMID:27812156

American Heart Association. (2017). *Top 25 foods that add the most sodium to your diet.* Retrieved from https://sodiumbreakup.heart.org/top_25_foods_that_add_the_most_sodium_to_your_diet

Barclay, A. W., Petocz, P., McMillan-Price, J., Flood, V. M., Prvan, T., Mitchell, P., & Brand-Miller, J. C. (2008). Glycemic index, glycemic load, and chronic disease risk: A meta-analysis of observational studies. *The American Journal of Clinical Nutrition*, *87*(3), 627–637. doi:10.1093/ajcn/87.3.627 PMID:18326601

Bhupathiraju, S. N., Tobias, D. K., Malik, V. S., Pan, A., Hruby, A., Manson, J. E., ... Hu, F. B. (2014). Glycemic index, glycemic load, and risk of type 2 diabetes: Results from 3 large US cohorts and an updated meta-analysis. *The American Journal of Clinical Nutrition*, *100*(1), 218–232. doi:10.3945/ajcn.113.079533 PMID:24787496

Birt, D. F., Boylston, T., Hendrich, S., Jane, J. L., Hollis, J., Li, L., ... Whitley, E. M. (2013). Resistant starch: Promise for improving human health. *Advances in Nutrition*, *4*(6), 587–601. doi:10.3945/an.113.004325 PMID:24228189

Blanchflower, D. G., Oswald, A. J., & Stewart–Brown, S. (2012). *Is psychological well-being linked to the consumption of fruit and vegetables? Working Paper 18469*. Cambridge, MA: National Bureau of Economic Research Working Paper Series. Retrieved from http://www.nber.org/papers/w18469.pdf doi:10.3386/w18469

Blanchflower, D. G., Oswald, A. J., & Stewart–Brown, S. (2012). Is psychological well-being linked to the consumption of fruit and vegetables? *Social Indicators Research*, *114*(3), 785–801. doi:10.100711205-012-0173-y

Boada, L. D., Henríquez-Hernández, L. A., & Luzardo, O. P. (2016). The impact of red and processed meat consumption on cancer and other health outcomes: Epidemiological evidences. *Food and Chemical Toxicology*, *92*, 236–244. doi:10.1016/j.fct.2016.04.008 PMID:27106137

Boehm, J. K., Williams, D. R., Rimm, E. B., Ryff, C., & Kubzansky, L. D. (2013). Association between optimism and serum antioxidants in the midlife in the United States study. *Psychosomatic Medicine*, *75*(1), 2–10. doi:10.1097/PSY.0b013e31827c08a9 PMID:23257932

Boeing, H., Bechthold, A., Bub, A., Ellinger, S., Haller, D., Kroke, A., ... Watzl, B. (2012). Critical review: Vegetables and fruit in the prevention of chronic diseases. *European Journal of Nutrition*, *51*(6), 637–663. doi:10.100700394-012-0380-y PMID:22684631

Bogueva, D., Marinova, D., & Raphaely, T. (2017). Reducing meat consumption: The case for social marketing. *Asia Pacific Journal of Marketing and Logistics*, *29*(3), 477–500. doi:10.1108/APJML-08-2016-0139

Bogueva, D., Marinova, D., & Raphaely, T. (Eds.). (2018). *Handbook of research on social marketing and its influence on animal origin food product consumption*. Hershey, PA: IGI Global. doi:10.4018/978-1-5225-4757-0

Bouvard, V., Loomis, D., Guyton, K. Z., Grosse, Y., El Ghissassi, F., & Benbrahim-Tallaa, L. (1599–1600). … Straif, K. on behalf of the International Agency for Research on Cancer Monograph Working Group (2015) Carcinogenicity of consumption of red and processed meat. *Lancet*, *16*(16). doi:10.1016/S1470-2045(15)00444-1

Brown, M. J. (2016). How short-chain fatty acids affect health and weight. *Healthline*. Retrieved from https://www.healthline.com/nutrition/short-chain-fatty-acids-101

Busch, S. (2018). USDA protein requirements in grams. *Healthy Eating, SF Gate*. Retrieved from http://healthyeating.sfgate.com/usda-protein-requirements-grams-8619.html

Byrne, C. S., Chambers, E. S., Morrison, D. J., & Frost, G. (2015). The role of short chain fatty acids in appetite regulation and energy homeostasis. *International Journal of Obesity*, *39*(9), 1331–1338. doi:10.1038/ijo.2015.84 PMID:25971927

Clark, M., & Tilman, D. (2017). Comparative analysis of environmental impacts of agricultural production systems, agricultural input efficiency, and food choice. *Environmental Research Letters*, *12*(6), 064016. doi:10.1088/1748-9326/aa6cd5

Clune, S., Crossin, E., & Verghese, K. (2017). Systematic review of greenhouse gas emissions for different fresh food categories. *Journal of Cleaner Production*, *140*(Part 2), 766–783. doi:10.1016/j.jclepro.2016.04.082

Cooper, A. J., Forouhi, N. G., Ye, Z., Buijsse, B., Arriola, L., Balkau, B., ... Wareham, N. J. (2012). Fruit and vegetable intake and type 2 diabetes: EPIC-InterAct prospective study and meta-analysis. *European Journal of Clinical Nutrition*, *66*(10), 1082–1092. doi:10.1038/ejcn.2012.85 PMID:22854878

Corpet, D. E. (2011). Red meat and colon cancer: Should we become vegetarians, or can we make meat safer? *Meat Science*, *89*(3), 310–316. doi:10.1016/j.meatsci.2011.04.009 PMID:21558046

Craig, W. J. (2010). Nutrition concerns and health effects of vegetarian diets. *Nutrition in Clinical Practice*, *25*(6), 613–620. doi:10.1177/0884533610385707 PMID:21139125

Craig, W. J., & Mangels, A. R. (2009). Position of the American Dietetic Association: Vegetarian diets. *Journal of the American Dietetic Association*, *109*(7), 1266–1282. doi:10.1016/j.jada.2009.05.027 PMID:19562864

Craig, W. J., & Saunders, A. V. (2018). Critical nutrients in plant-based diet. In W. J. Craig (Ed.), *Vegetarian nutrition and wellness* (pp. 213–230). Boca Raton, FL: CRC Press. doi:10.1201/b22003

Dahl, W. J., Agro, N. C., Eliasson, A. M., Mialki, K. L., Olivera, J. D., Rusch, C. T., & Young, C. N. (2017). Health benefits of fiber fermentation. *Journal of the American College of Nutrition*, *36*(2), 127–136. doi:10.1080/07315724.2016.1188737 PMID:28067588

Darling, A. L., Millward, D. J., Torgerson, D. J., Hewitt, C. E., & Lanham-New, S. A. (2009). Dietary protein and bone health: A systematic review and meta-analysis. *The American Journal of Clinical Nutrition*, *90*(6), 1674–1692. doi:10.3945/ajcn.2009.27799 PMID:19889822

Ekmekcioglu, C., Wallner, P., Kundi, M., Weisz, U., Haas, W., & Hutter, H.-P. (2018). Red meat, diseases, and healthy alternatives: A critical review. *Critical Reviews in Food Science and Nutrition*, *58*(2), 247–261. doi:10.1080/10408398.2016.1158148 PMID:27128451

English, N. (2017). 10 complete proteins vegans need to know about. *Greatist*. Retrieved from https://greatist.com/health/complete-vegetarian-proteins

Environmental Working Group. (2011). *Meat-eater's guide to climate change + health. Lifecycle assessments: Methodology and results*. Retrieved from http://static.ewg.org/reports/2011/meateaters/pdf/methodology_ewg_meat_eaters_guide_to_health_and_climate_2011.pdf

Flint, H. J., Scott, K. P., Louis, P., & Duncan, S. H. (2012). The role of the gut microbiota in nutrition and health. *Nature Reviews. Gastroenterology & Hepatology*, *9*(10), 577–589. doi:10.1038/nrgastro.2012.156 PMID:22945443

Food and Agriculture Organization of the United Nations (FAO). (2018). *FAOSTATS: Food Supply - Livestock and Fish Primary Equivalent*. Retrieved from http://www.fao.org/faostat/en/#data/CL

Giudici, K. V., & Weaver, C. M. (2018). Plant-based diets and risk of osteoporosis. In W. J. Craig (Ed.), *Vegetarian nutrition and wellness* (pp. 93–111). Boca Raton, FL: CRC Press.

Glick-Bauer, M., & Yeh, M. (2014). The health advantage of a vegan diet: Exploring the gut microbiota connection. *Nutrients*, *6*(11), 4822–4838. doi:10.3390/nu6114822 PMID:25365383

Gong, L., Cao, W., Chi, H., Wang, J., Zhang, H., Liu, J., & Sun, B. (2018). Whole cereal grains and potential health effects: Involvement of the gut microbiota. *Food Research International*, *103*, 84–102. doi:10.1016/j.foodres.2017.10.025 PMID:29389647

Green Eatz. (n.d.). *Food's carbon footprint*. Retrieved from http://www.greeneatz.com/foods-carbon-footprint.html

Greenpeace. (2018). *Less is more: Reducing meat and dairy for a healthier life and planet.* Scientific background of the Greenpeace vision of the meat and dairy system towards 2050. Retrieved from https://storage.googleapis.com/p4-production-content/international/wp-content/uploads/2018/03/6942c0e6-longer-scientific-background.pdf

Greenwood, D. C., Threapleton, D. E., Evans, C. E., Cleghorn, C. L., Nykjaer, C., Woodhead, C., & Burley, V. J. (2013). Glycemic index, glycemic load, carbohydrates, and type 2 diabetes: Systematic review and dose-response meta-analysis of prospective studies. *Diabetes Care*, *36*(12), 166–171. doi:10.2337/dc13-0325 PMID:24265366

Guarner, F., & Malagelada, J.-R. (2003). Gur flora in health and disease. *Lancet*, *361*(9356), 512–519. doi:10.1016/S0140-6736(03)12489-0 PMID:12583961

Haddad, E. H. (2018). Vegetarian diet and risk of cardiovascular disease. In W. J. Craig (Ed.), *Vegetarian nutrition and wellness* (pp. 45–70). Boca Raton, FL: CRC Press.

Heskey, C. E. (2018). The role of vegetarian diets in weight management. In W. J. Craig (Ed.), *Vegetarian nutrition and wellness* (pp. 113–169). Boca Raton, FL: CRC Press.

Hever, J., & Cronise, R. J. (2017). Plant-based nutrition for healthcare professionals: Implementing diet as a primary modality in the prevention and treatment of chronic disease. *Journal of Geriatric Cardiology: JGC*, *14*(5), 355–368. doi:10.11909/j.issn.1671-5411.2017.05.012 PMID:28630615

Hosking, D. E., Eramudugolla, R., & Anstey, K. J. (2017). The MIND diet is associated with reduced incidence of 12-year cognitive impairment in an Australian setting. *Alzheimer's & Dementia*, *13*(7), 1429–1430. doi:10.1016/j.jalz.2017.06.2234

Hu, D., Huang, J., Wang, Y., Zhang, D., & Qu, Y. (2014). Fruits and vegetables consumption and risk of stroke: A meta-analysis of prospective cohort studies. *Stroke*, *45*(6), 1613–1619. doi:10.1161/STROKEAHA.114.004836 PMID:24811336

International Agency for Research on Cancer (IARC). (2015). *IARC Monographs evaluate consumption of red meat and processed meat*. International Agency for Research on Cancer. Press release No. 240. World Health Organization. Retrieved from https://www.iarc.fr/en/media-centre/pr/2015/pdfs/pr240_E.pdf

Jones, T. (2017). *14 healthy foods that are high in potassium*. Healthline. Retrieved from https://www.healthline.com/nutrition/high-potassium-foods#section3

Kurpad, A. V., & Vaz, M. (2000). Protein and animo acid requirements in the elderly. *European Journal of Clinical Nutrition, 54*(S3), S131–S142. doi:10.1038j.ejcn.1601035 PMID:11041085

Kye, S. Y., & Park, K. (2014). Health-related determinants of happiness in Korean adults. *International Journal of Public Health, 59*(5), 731–738. doi:10.100700038-014-0588-0 PMID:25033934

Larson-Meyer, D. E. (2018). Nutritionally adequate vegetarian diets and athletic performance. In W. J. Craig (Ed.), *Vegetarian nutrition and wellness* (pp. 279–302). Boca Raton, FL: CRC Press.

Li, M., Fan, Y., Zhang, X., Hou, W., & Tang, Z. (2014). Fruit and vegetable intake and risk of type 2 diabetes mellitus: Meta-analysis of prospective cohort studies. *BMJ Open, 4*(11), e005497. doi:10.1136/bmjopen-2014-005497 PMID:25377009

Link, R. (2017). The 7 best plant sources of omega-3 fatty acids. *Healthline*. Retrieved from https://www.healthline.com/nutrition/7-plant-sources-of-omega-3s

Lippi, G., Mattiuzzi, C., & Cervellin, G. (2016). Meat consumption and cancer risk: A critical review of published meta-analyses. *Critical Reviews in Oncology/Hematology, 97*, 1–14. doi:10.1016/j.critrevonc.2015.11.008 PMID:26633248

Loef, M., & Walach, H. (2012). Fruit, vegetables and prevention of cognitive decline or dementia: A systematic review of cohort studies. *The Journal of Nutrition, Health & Aging, 16*(7), 626–630. doi:10.100712603-012-0097-x PMID:22836704

Malik, V. S., Li, Y., Tobias, D. K., Pan, A., & Hu, F. B. (2016). Dietary protein intake and risk of type 2 diabetes in US men and women. *American Journal of Epidemiology, 183*(8), 715–728. doi:10.1093/aje/kwv268 PMID:27022032

Marsh, K., Zeuschner, C., & Saunders, A. (2012). Health implications of a vegetarian diet: A review. *American Journal of Lifestyle Medicine, 6*(3), 250–267. doi:10.1177/1559827611425762

Marsh, K., Zeuschner, C., & Saunders, A. (2016). Red meat and health: Evidence regarding red meat, health, and chronic disease risk. In T. Raphaely & D. Marinova (Eds.), *Impact of meat consumption on health and environmental sustainability* (pp. 131–177). Hershey, PA: IGI Global. doi:10.4018/978-1-4666-9553-5.ch008

Marsh, K., Zeuschner, C., & Saunders, A. (2018). The health impact of eating foods of animal origin: Evidence regarding animal foods, health, and disease risk. In D. Bogueva, D. Marinova, & T. Raphaely (Eds.), *Handbook of research on social marketing and its influence on animal origin food product consumption* (pp. 283–297). Hershey, PA: IGI Global. doi:10.4018/978-1-5225-4757-0.ch002

Micha, R., Michas, G., & Mozaffarian, D. (2012). Unprocessed red and processed meats and risk of coronary artery disease and type 2 diabetes–an updated review of the evidence. *Current Atherosclerosis Reports*, *14*(6), 515–524. doi:10.100711883-012-0282-8 PMID:23001745

Morris, M. C., Tangney, C. C., Wang, Y., Sacks, F. M., Bennett, D. A., & Aggarwal, N. T. (2015). MIND diet associated with reduced incidence of Alzheimer's disease. *Alzheimer's & Dementia*, *11*(9), 1007–1014. doi:10.1016/j.jalz.2014.11.009 PMID:25681666

Mytton, O., Nnoaham, K., Eyles, H., Scarborough, P., & Ni Mhurchu, C. (2014). Systematic review and meta-analysis of the effect of increased vegetable and fruit consumption on body weight and energy intake. *BMC Public Health*, *14*(1), 886–897. doi:10.1186/1471-2458-14-886 PMID:25168465

National Health and Medical Research Council (NHMRC). (2006). *Nutrient reference values for Australia and New Zealand (2006)*. Canberra, Australia: National Health and Medical Research Council, Australian Government Department of Health and Ageing, New Zealand Ministry of Health.

Nie, Y., Luo, F., & Lin, Q. (2018). Dietary nutrition and gut microflora: A promising target for treating disease. *Trends in Food Science & Technology*, *75*, 72–80. doi:10.1016/j.tifs.2018.03.002

Nordqvist, C. (2017). Everything you need to know about vitamin B-12. *Medical News Today*. Retrieved from https://www.medicalnewstoday.com/articles/219822.php?sr

NutritionFacts. (2018). *Heme iron*. Retrieved from https://nutritionfacts.org/topics/heme-iron/

Otaegui-Arrazola, A., Amiano, P., Elbusto, A., Urdaneta, E., & Martínez-Lage, P. (2014). Diet, cognition, and Alzheimer's disease: Food for thought. *European Journal of Nutrition*, *53*(1), 1–23. doi:10.100700394-013-0561-3 PMID:23892520

Petersson, S., & Philippou, E. (2016). Mediterranean diet, cognitive function, and dementia: A systematic review of the evidence. *Advances in Nutrition*, *7*(5), 889–904. doi:10.3945/an.116.012138 PMID:27633105

Piqueras, J. A., Kuhne, W., Vera–Villarroel, P., van Straten, A., & Cuijpers, P. (2011). Happiness and health behaviours in Chilean college students: A cross-sectional survey. *BMC Public Health*, *11*(1), 443. doi:10.1186/1471-2458-11-443 PMID:21649907

Pribis, P. (2018). The role of legumes in maintaining health. In W. J. Craig (Ed.), *Vegetarian nutrition and wellness* (pp. 191–209). Boca Raton, FL: CRC Press.

Raphaely, T., & Marinova, D. (2016b). Preface. In T. Raphaely & D. Marinova (Eds.), *Impact of meat consumption on health and environmental sustainability* (pp. xvii–xxiii). Hershey, PA: IGI Global. doi:10.4018/978-1-4666-9553-5

Raphaely & Marinova. (Eds.). (2016a). *Impact of meat consumption on health and environmental sustainability*. Hershey, PA: IGI Global.

Ritchie, H., & Roser, M. (2017). *Meat and seafood production & consumption*. Our World in Data. Retrieved from https://ourworldindata.org/meat-and-seafood-production-consumption

Ross, S. M. (2018). The nutrition-brain connection: Nutritional status and cognitive decline. *Holistic Nursing Practice*, *32*(3), 169–171. doi:10.1097/HNP.0000000000000270 PMID:29642132

Sara, R., Ralston, R. A., & Walker, K. Z. (2013). Adherence to a Mediterranean-style diet can slow the rate of cognitive decline and decrease the risk of dementia: A systematic review. *Nutrition & Dietetics: the Journal of the Dietitians Association of Australia, 70*, 206–217. doi:10.1111/1747-0080.12016

Satija, A., Bhupathiraju, S. N., Donna Spiegelman, D., Chiuve, S. E., Manson, J. E., Willett, W., ... Hu, F. B. (2017). Healthful and unhealthful plant-based diets and the risk of coronary heart disease in U.S. adults. *Journal of the American College of Cardiology*, *70*(4), 411–422. doi:10.1016/j.jacc.2017.05.047 PMID:28728684

Shams-White, M. M., Chung, M., Du, M., Fu, Z., Insogna, K. L., Karlsen, M. C., ... Weaver, C. M. (2017). Dietary protein and bone health: A systematic review and meta-analysis from the National Osteoporosis Foundation. *The American Journal of Clinical Nutrition, 105*(6), 1528–1543. doi:10.3945/ajcn.116.145110 PMID:28404575

Shang, X., Scott, D., Hodge, A. M., English, D. R., Giles, G. G., Ebeling, P. R., & Sanders, K. M. (2016). Dietary protein intake and risk of type 2 diabetes: Results from the Melbourne Collaborative Cohort Study and a meta-analysis of prospective studies. *The American Journal of Clinical Nutrition, 104*(5), 1352–1365. doi:10.3945/ajcn.116.140954 PMID:27629053

Shepon, A., Eshel, G., Noor, E., & Milo, R. (2018). The opportunity cost of animal based diets exceeds all food losses. *Proceedings of the National Academy of Sciences of the United States of America, 15*(15), 3804–3809. doi:10.1073/pnas.1713820115 PMID:29581251

Springmann, M., Godfray, H. C. J., Rayner, M., & Scarborough, P. (2016). Analysis and valuation of the health and climate change cobenefits of dietary change. *Proceedings of the National Academy of Sciences of the United States of America, 113*(15), 4146–4151. doi:10.1073/pnas.1523119113 PMID:27001851

Steinbrecher, A., Erber, E., Grandinetti, A., Kolonel, L. N., & Maskarinec, G. (2010). Meat consumption and risk of type 2 diabetes: The Multiethnic Cohort. *Public Health Nutrition, 14*(4), 568–574. doi:10.1017/S1368980010002004 PMID:20624337

Tucker, K. L. (2014). Vegetarian diets and bone status. *The American Journal of Clinical Nutrition, 100*(Supplement 1), 329S–335S. doi:10.3945/ajcn.113.071621 PMID:24898237

Victoria State Government. (2018). Metabolic syndrome. *Better Health Channel*. Retrieved from https://www.betterhealth.vic.gov.au/health/conditionsandtreatments/metabolic-syndrome

Villines, Z. (2018). Top 15 sources of plant-based protein. *Medical News Today*. Retrieved from https://www.medicalnewstoday.com/articles/321474.php

Wang, X., Lin, X., Ouyang, Y. Y., Liu, J., Zhao, G., Pan, A., & Hu, F. B. (2016). Red and processed meat consumption and mortality: Dose–response meta-analysis of prospective cohort studies. *Public Health Nutrition, 19*(5), 893–905. doi:10.1017/S1368980015002062 PMID:26143683

Wang, X., Ouyang, Y., Liu, J., Zhu, M., Zhao, G., Bao, W., & Hu, F. B. (2014). Fruit and vegetable consumption and mortality from all causes, cardiovascular disease, and cancer: Systematic review and dose-response meta-analysis of prospective cohort studies. *British Medical Journal, 349*(7969). doi:10.1136/bmj.g4490 PMID:25073782

Warensjö, E., Byberg, L., Melhus, H., Gedeborg, R., Mallmin, H., Wolk, A., & Michaëlsson, K. (2011). Dietary calcium intake and risk of fracture and osteoporosis: Prospective longitudinal cohort study. *BMJ (Clinical Research Ed.), 342*(1), d1473. doi:10.1136/bmj.d1473 PMID:21610048

Weaver, C. M. (2013). Potassium and health. *Advances in Nutrition, 4*(3), 368S–377S. doi:10.3945/an.112.003533 PMID:23674806

Wei, H., Gao, Z., Liang, R., Li, Z., Hao, H., & Liu, X. (2016). Whole-grain consumption and the risk of all-cause, CVD and cancer mortality: A meta-analysis of prospective cohort studies. *British Journal of Nutrition, 116*(03), 514–525. doi:10.1017/S0007114516001975 PMID:27215285

White, B. A., Horwath, C. C., & Conner, T. S. (2013). Many apples a day keep the blues away--daily experiences of negative and positive affect and food consumption in young adults. *British Journal of Health Psychology, 18*(4), 782–798. doi:10.1111/bjhp.12021 PMID:23347122

Willis, O. (2018). Omega-3: What are the health benefits and how should you eat it? *ABC News Health.* Retrieved from http://www.abc.net.au/news/health/2018-02-15/what-are-the-health-benefits-of-omega-3/9446528

Wilson, A. (2012). Meat and cancer – country comparisons. *Cancer News, Articles, & Information.* Retrieved from http://healthhubs.net/cancer/meat-and-cancer-country-comparisons/

World Cancer Research Fund. (2012). *Worldwide data.* Retrieved from https://www.wcrf.org/int/cancer-facts-figures/worldwide-data

World Cancer Research Fund/American Institute for Cancer Research (WCRF/AICR). (2018). *Diet, nutrition, physical activity and cancer: A global perspective.* Continuous Update Project Expert Report 2018. Retrieved from https://www.dietandcancerreport.org

World Health Organization (WHO). (2003). *Diet, nutrition and the prevention of chronic diseases: Report of a joint WHO/FAO Expert Consultation.* WHO Technical Report Series 916. Retrieved from http://www.who.int/dietphysicalactivity/publications/trs916/en/

World Health Organization (WHO). (2014). *Increasing fruit and vegetable consumption to reduce the risk of noncommunicable diseases: Biological, behavioural and contextual rationale.* Retrieved from http://www.who.int/elena/titles/bbc/fruit_vegetables_ncds/en/

World Health Organization (WHO). (2015). *Q&A on the carcinogenicity of the consumption of red meat and processed meat.* Retrieved from http://www.who.int/features/qa/cancer-red-meat/en/

World Health Organization (WHO). (2016). *Increasing potassium intake to reduce blood pressure and risk of cardiovascular diseases in adults.* Retrieved from http://www.who.int/elena/titles/potassium_cvd_adults/en/

World Health Organization (WHO). (2017a). *Cardiovascular diseases (CDVs): Key facts.* Retrieved from http://www.who.int/en/news-room/fact-sheets/detail/cardiovascular-diseases-(cvds)

World Health Organization (WHO). (2017b). *Reducing sodium intake to reduce blood pressure and risk of cardiovascular diseases in adults.* Retrieved from http://www.who.int/elena/titles/sodium_cvd_adults/en/

World Health Organization (WHO). (2018). *Noncommunicable diseases: Fact sheet*. Retrieved from http://www.who.int/news-room/fact-sheets/detail/noncommunicable-diseases

Zechner, E. (2017). Inflammatory disease caused by intestinal pathobionts. *Current Opinion in Microbiology*, *35*, 64–69. doi:10.1016/j.mib.2017.01.011 PMID:28189956

Zhang, N., Ju, Z., & Zuo, T. (2018). Time for food: The impact of diet on gut microbiota and human health. *Nutrition (Burbank, Los Angeles County, Calif.)*, *51-52*, 80–85. doi:10.1016/j.nut.2017.12.005 PMID:29621737

ADDITIONAL READING

Blanchflower, D. G., Oswald, A. J., & Stewart–Brown, S. (2012). Is psychological well-being linked to the consumption of fruit and vegetables? *Social Indicators Research*, *114*(3), 785–801. doi:10.100711205-012-0173-y

Marsh, K., Zeuschner, C., & Saunders, A. (2016). Red meat and health: Evidence regarding red meat, health, and chronic disease risk. In T. Raphaely & D. Marinova (Eds.), *Impact of meat consumption on health and environmental sustainability* (pp. 131–177). Hershey, PA: IGI Global. doi:10.4018/978-1-4666-9553-5.ch008

Marsh, K., Zeuschner, C., & Saunders, A. (2018). The health impact of eating foods of animal origin: Evidence regarding animal foods, health, and disease risk. In D. Bogueva, D. Marinova, & T. Raphaely (Eds.), *Handbook of research on social marketing and its influence on animal origin food product consumption* (pp. 283–297). Hershey, PA: IGI Global. doi:10.4018/978-1-5225-4757-0.ch002

Wang, X., Ouyang, Y., Liu, J., Zhu, M., Zhao, G., Bao, W., & Hu, F. B. (2014). Fruit and vegetable consumption and mortality from all causes, cardiovascular disease, and cancer: Systematic review and dose-response meta-analysis of prospective cohort studies. *British Medical Journal*, *349*(7969). doi:10.1136/bmj.g4490 PMID:25073782

Wei, H., Gao, Z., Liang, R., Li, Z., Hao, H., & Liu, X. (2016). Whole-grain consumption and the risk of all-cause, CVD and cancer mortality: A meta-analysis of prospective cohort studies. *British Journal of Nutrition*, *116*(03), 514–525. doi:10.1017/S0007114516001975 PMID:27215285

World Cancer Research Fund/American Institute for Cancer Research (WCRF/AICR). (2018). *Diet, nutrition, physical activity and cancer: A global perspective*. Continuous Update Project Expert Report 2018. Retrieved from https://www.dietandcancerreport.org

World Health Organization (WHO). (2014). *Increasing fruit and vegetable consumption to reduce the risk of noncommunicable diseases: Biological, behavioural and contextual rationale*. Retrieved from http://www.who.int/elena/titles/bbc/fruit_vegetables_ncds/en/

World Health Organization (WHO). (2015). *Q&A on the carcinogenicity of the consumption of red meat and processed meat*. Retrieved from http://www.who.int/features/qa/cancer-red-meat/en/

Zhang, N., Ju, Z., & Zuo, T. (2018). Time for food: The impact of diet on gut microbiota and human health. *Nutrition (Burbank, Los Angeles County, Calif.)*, *51-52*, 80–85. doi:10.1016/j.nut.2017.12.005 PMID:29621737

KEY TERMS AND DEFINITIONS

Alzheimer's Disease: The most common form of dementia or impairment of a person's thinking, memory, and behavior as a result of changes in the brain; it is named after Dr Alzheimer who first recognized and described this incurable debilitating condition which initially leads to loss of enthusiasm for normal activities and then to complete dependence and eventually death.

Cardiovascular Diseases (CVD): A group of diseases affecting the heart and the blood vessels; the most common diseases within this group are stroke, infarct, coronary heart disease, and heart failure.

Coronary Heart Disease (CHD): A disease resulting in the narrowing of coronary arteries – the blood vessels supplying blood to the heart, due to the gradual building up of fatty deposits; it is a precursor to angina or heart attack.

Diet: The food consumed by a person or an organism; the word also has an implied meaning that the food consumption is with a health-related aims although ethical considerations can also affect the choice of foods to be consumed.

Glycemic Index (GI): This index ranks carbohydrates depending on their effect on the glucose levels in the blood; the lower the GI, the slower the rise in blood glucose levels when this particular food is eaten; the recommendation for people with diabetes is to eat food with a lower GI.

Heme Iron: A type of iron which is supplied to the human body by animal-based foods, such as red meat, poultry, fish and seafood; it is absorbed easier than the non-heme iron found in plants, dairy, eggs, and also red meat, poultry, fish, and seafood; the body has no mechanism to release the excess quantities of heme iron and its storage represents a risk factor for many chronic diseases, such as cardiovascular, type 2 diabetes, atherosclerosis, cancer, Alzheimer's, as well as metabolic syndrome.

High-Density Lipoprotein (HDL): Good cholesterol which carries cholesterol from other parts of the body to the liver for it to be removed.

Legumes: A class of plants grown as a food sources for their seeds (called pulses); legumes have nitrogen-fixing bacteria in their root noodles and play a key role in crop rotation; they also have one of lowest carbon footprint among foods.

Low-Density Lipoprotein (LDL) Cholesterol: The bad cholesterol which is deposited inside the arteries causing them to narrow and increasing the risk of coronary artery disease.

Metabolic Syndrome: A human health condition which combines a collection of symptoms, such as obesity, high blood pressure, high blood triglycerides, low levels of the "good" HDL cholesterol and insulin resistance, which lead to increased risk of diabetes, stroke, and heart disease.

Metabolite: A substance formed during metabolism or necessary for metabolism.

Microbiome: The microorganisms in a particular environment, including the human body.

Pathobionts: Disease causing organisms or microbes which live symbiotically in the human body and represent a risk factor for inflammation.

Protein: An organic substance – polymer chains of amino acids, considered an essential nutrient for the human body; there are 20 types of amino acids representing the building blocks for the human proteins – 11 are non-essential which can be synthesized by the human organism and 9 are essential which need to be provided by food.

Type 2 Diabetes: A medical condition associated with the body becoming resistant to the normal effects of insulin and eventually losing the capacity to produce insulin in the pancreas; it is associated with lifestyle factors, such as obesity and lack of physical activity; it requires proper management to prevent complications and life-threatening situations.

Chapter 4
Nutritional Benefits of Selected Plant-Based Proteins as Meat Alternatives

Seydi Yıkmış
Tekirdağ Namık Kemal University, Turkey

Ramazan Mert Atan
Bandırma Onyedi Eylül University, Turkey

Nursena Kağan
Tekirdağ Namık Kemal University, Turkey

Levent Gülüm
Abant İzzet Baysal University, Turkey

Harun Aksu
Istanbul University – Cerrahpaşa, Turkey

Mehmet Alpaslan
Tekirdağ Namık Kemal University, Turkey

ABSTRACT

Humans meet their nutritional requirements by consuming food, and our body uses naturally sufficient amounts of all necessary nutrients to maintain its functioning. Proteins form the basis of the human diet because they are necessary for immune responses, cell signals, muscle masses, and the repair of damaged cells. Animal and plant food products are the main protein sources in the human diet. Based on scientific evidence, proteins derived from animals recently started to be replaced by plant-based options as prefered proteins for a range of reasons. Consumption of non-meat protein sources being shown to be healthy and environmentally friendly is a major consideration. Plant-based protein is helping minimize high cholesterol, type 2 diabetes, high blood pressure, obesity, certain types of cancer, including colorectal, ovarian, and breast cancers, and a diet based on non-animal proteins could increase life expectancy and decrease greenhouse gases emissions from livestock as less resources are used for plant production. The chapter describes the nutritional benefits and current uses of nine non-animal protein sources and the health benefits arising from replacing animal protein.

DOI: 10.4018/978-1-5225-7350-0.ch004

INTRODUCTION

Nutrients play a crucial role in maintaining overall human health. Protein is the most important nutrient required for growth and development (Besler, Rakıcıoglu, Ayaz, Demirel Büyüktuncer, Özel, Samur Eroğlu, … & Yürük, 2015). The structure and function of our bodies, the regulations of cells, tissues and organs depend on proteins. Approximately 16% of the adult human body is composed of protein. As proteins in the body do not form storage depots, we need to get enough of them daily through a balanced diet. Proteins are long chains of amino acids, which are their building blocks created, formed and synthesised for our human body to function correctly. Although the properties of the amino acids vary between animal and plant sourced proteins, they both can supply the needed and recommended daily requirement for protein of 0.8 g/day, considered sufficient for almost all healthy adult individuals (Pasiakos, Agarwal, Lieberman, & Fulgoni, 2015).

Protein sources in a diet come from animal and vegetable sources (Lin, Lu, Kelly, Zhang, Zheng, & Miao, 2017). Common examples of animal protein sources are meat, poultry, fish and eggs, and common examples of plant–derived proteins are beans, lentil and soybean (Nehete, Bhambar, Narkhede, & Gawali, 2013). In the majority of industrialized countries, the main protein source of dietary protein is animal foods. However, increased consumption of red meat and processed meat along with other animal products has been shown to be associated with obesity, coronary heart disease, high blood pressure, cancer, elevated serum and urinary uric acid levels (Møller, Sluik, Ritz, Mikkilä, Raitakari, Hutri–Kähönen, … Raben, 2017). Because of the adverse effects of animal protein on health, vegetable protein sources seem to be a wise alternative in meeting proteins needs. Vegetable protein sources are increasingly being recommended because of their positive effects on health (Chen, Song, Chen, Ding, Peng, & Mao, 2016; Comerford & Pasin, 2016; Wu, Zeng, Huang, Li, Zhang, Ho, & Zheng, 2016). In order to provide human protein requirements, it is necessary to support the production of plant proteins which can replace the sources of animal protein (Comerford & Pasin, 2016). Vegetable proteins can meet the essential amino acids that people need (López, Galante, Robson, Boeris, & Spelzini, 2018). Such sources are known to provide greater saturation than animal protein sources because of their low energy content and high fibre content (Nielsen, Kristensen, Klingenberg, Ritz, Belza, Astrup, & Raben, 2018). It is known that consuming foods with high fibre content enhances insulin sensitivity and provides glycemic control (Moorthi, Vorland, & Hill Gallant, 2017). There is also evidence that the risk of cardiovascular diseases can be reduced by a flexitarian dietary model which involves the consumption of more vegetative proteins rather than a meat–rich diet (Richter, Skulas–Ray, Champagne, & Kris–Etherton, 2015).

This chapter presents an evaluation of the nutritional properties and the many human health advantages of nine important plant–derived proteins obtained from beans, soybean, chickpea, lentil, quinoa, buckwheat, chia, teff and spirulina in term of their physiological benefits. Amino acid composition, nutritional aspects, functional properties and their role in promoting good human health are examined. Although these plants are well-known, only recently nutritionists started drawing attention to them as superfoods and alternatives in meat replacement.

HEALTH BENEFITS AND DRAWBACKS OF PLANT AND MEAT CONSUMPTION

The health benefits and drawbacks of both plant and meat consumption are multifaceted and complicated. Evidentially meat production and overconsumption attract more negative health and environmental effects in contrast to the production and consumption of plants (Garnett, 2014; Raphaely & Marinova, 2016; Bogueva, Marinova, & Raphaely, 2017). Being central to our traditional dietary source of protein, meat tends to deliver all necessary amino–acids we need for our body to function properly in comparison with plant–based protein sources which may lack one or more essential amino acids. Meat is also a rich source of beneficial minerals and essential nutrients, B vitamins, iron and zinc; however, it can contain high amounts of saturated fat, claimed to link cholesterol to cardiovascular disease (Newby, 2009) and processed meats can be high in sodium and cancerogenic (WHO, 2015). Meat consumption and particularly red meat are associated with an increased risk of non–communicable diseases including cardiovascular, cancer, obesity and type 2 diabetes (Newby, 2009; Marsh, Zeuschner, & Saunders, 2016; 2018).

Compared to the meat–rich Western diet, from a nutritional perspective, plant–based diets are associated with health promotion and disease prevention. They contain nutrients and vitamins, unsaturated fatty acids and beneficial fibre. Eating a plant–based diet has been linked to lower risk of obesity and many chronic diseases, such as heart disease, type 2 diabetes (Satija, Bhupathiraju, Rimm, Spiegelman, Chiuve, Borgi,… Hu, 2016), inflammation and cancer (Mattisson, 2004). Plant–based diets are also associated with higher metabolic rates (Montalcini, De Bonis, Ferro, Carè, Mazza, Accattato, … Pujia, 2015). Additionally, eating little to no meat may increase your life expectancy, to have lower body weights compared to their meat–eating counterparts, and lower risk of obesity (Newby, 2009).

NINE PLANT–DERIVED PROTEINS

Protein obtained from plant–based sources plays a special role in vegetarian and plant–based diets. The nine plant–derived protein sources presented in this chapter are chosen particularly because of their highly valued nutritional properties and health advantages. They are more desirable than meat consumption because of their nutritional physiology discussed in this section. Quinoa is an important source of plant protein because it contains all essential amino acids. In quinoa, leucine and isoleucine branched-chain amino acids, also known as limiting amino acids, are present in significant amounts. The chia seed is known as a rich omega–3 source. It regulates blood sugar because of the high content of soluble fibre. Buckwheat has high biologically valuable protein with all essential amino acids. Quinoa and chia are alternatives for those with gluten sensitivity because they do not contain gluten. Dried beans, lentils, chickpeas, beans, kidney beans etc. are important protein and fibre sources, along with economic benefits. They are rich in calcium, iron, zinc, magnesium minerals, all B vitamins except B12 and vitamin E. Spirulina, an algae species of the blue–green algae, contains high amounts of β–carotene, B 12 vitamins and iron. According to the studies done, spirulina is an important antioxidant which reduces oxidative stress. Fatty seeds rich in unsaturated fatty acids also give diets diversity thanks to the protein and fibre they contain. Teff has an attractive nutritional profile because of the fact that most of the carbohydrates it contains are complex and gluten–free. It is also noted for its high iron, calcium and rich polyphenol content.

The overview in Table 1 presents the nine plants and their nutrient contents. All plants are discussed individually in the remainder of this section.

Table 1. Nutritional content of plant–based protein sources

Plant	Protein Content	Carbohydrate Content	Fat Content	Fibre Content
Quinoa (*Chenopodium quinoa Willd.*)	13.1–16.7%	58–68% starch and 5% sugar	2–9.5%	10%
Chia (*Salvia hispanica L.*)	15–25%	26–41%	30–33%	18–30%
Buckwheat (*Fagopyrum esculentum*)	12%	59–70% starch	1.7–4.0%	12.7–17.8%
Teff (*Eragrostis tef*)	8–11%	73%	23–32%	3–4.5%
Beans (*Phaseolus Vulgaris L.*)	20–30%	50–60%	2.5%	15–19%
Soybean	36.5%	30%	15.6%	9.3%
Lentils (*Lens Culinaris*)	23–27%	64–74%	2%	4–9%
Chickpea (*Ciger Arietinum*)	18–28%	50–70%	6–9%	10–20%
Spirulina	60–70%	8%	12%	14%

Quinoa (*Chenopodium Quinoa Willd.*)

The quinoa plant (*Chenopodium quinoa Willd.*) belongs to the family of Chenopodiaceae, which also includes spinach and beet. It is native to South America and has around 250 species all over the world. People who live in the Andes, especially in Peru and Bolivia, began to produce quinoa thousands of years ago by domesticating the wild species. In the local languages, the plant is called quinua and quinoa, especially in Bolivia, Peru, Ecuador, Argentina and Chile, while different names such as suphan, suba, jupha and dahue are also used. Although quinoa does not exhibit grain characteristics, it is considered to be a pseudo–grain and even a pseudo–seed because it does not belong to the Gramineae family (Abugoch James, 2009; Li & Zhu, 2018; Navruz–Varli & Sanlier, 2016).

Quinoa has a very good adaptation to different ecological agricultural conditions. It is a fertile plant that is resistant to moist soil and harvests at acceptable levels with 100–200 mm rainfall. It can grow at a relative humidity range of 40% –88% and withstands temperatures between –4 ° C and 38 ° C (Bojanic, 2011; Li & Zhu, 2018).

Similar to rice, the seeds are used for making soup where they absorb water and inflate; in grinding cereals for cookies, bread, biscuits, pasta, chips; bakery products are also produced, such as tortilla and flatbread (Navruz–Varli & Sanlier, 2016). In addition, quinoa is traditionally used in South America for making sweets, pastries, drinks and dry snacks (Bojanic, 2011). Quinoa seeds are fermented to make beer, used in the making of an alcoholic beverage called 'chicha' which is consumed in traditional ceremonies in South America (Vilcacundo & Hernández–Ledesma, 2017).

Nutritionally, quinoa is included in the whole grain category (Graf, Rojas–Silva, Rojo, Delatorre–Herrera, Baldeón, & Raskin, 2015). The quinoa seed is a source of starch, protein, dietary fibre, fat, minerals, polyphenols and vitamins (Li & Zhu, 2018). The superiority of quinoa compared to other cereals (such as rye, barley and oats) is due to its rich protein, lipid and ash content (Vilcacundo & Hernández–Ledesma, 2017). It contains 368 kcal energy per 100 g (Navruz–Varli & Sanlier, 2016).

The protein content of quinoa seeds (expressed as g/100 g of edible material) ranges from 13.1% to 16.7%. Quinoa's protein content is higher than the protein content of rice, barley, maize and rye, and is close to the protein content of wheat (Vilcacundo & Hernández–Ledesma, 2017). Similarly to animal-based products, it contains all essential amino acids. Quinoa is rich in histidine and lysine, essential amino acids found in many different cereals. The lysine amino acid aids in the formation of antibodies, enhances immunity, increases gastric function, assists in cell repair, participates in the metabolism of fatty acids, helps calcium absorption and transport, and slows or even prevents cancer metastasis with vitamin C (Bojanic, 2011). Preliminary studies on protein fractions show that the main proteins in quinoa seeds are albumin and globulin (about 77% of total proteins). The rest consists essentially of prolamins (Fischer, Wilckens, Jara, Aranda, Valdivia, Bustamante, … Obal, 2017). Quinoa protein is particularly well balanced in amino acid composition and does not contain gluten (Zhang, Li, Ma, Gao, Du, Han, … Qiao, 2017) which makes it a suitable food for celiac diet (Repo–Carrasco–Valencia & Serna, 2011).

Quinoa is an ideal energy source. Its seed's carbohydrates contain 58–68% starch and 5% sugar, which is released into the body slowly due to the high fibre content of 10% (Bojanic, 2011). Dietary fibre is essential for optimal digestive health and at the same time provides various functional benefits, such as facilitating satiety, reducing cholesterol and lipid absorption, regulating postprandial insulin response, converting endogenous cholesterol to bile acids, regulating intestinal microbiotics, gastrointestinal infection and reducing inflammation and severity.

Due to the quality and quantity of the lipid fraction, quinoa is considered an alternative oil seed. The fat content is between 2.0% and 9.5% and is rich in essential fatty acids such as linoleic and alpha–linolenic acids (Navruz–Varli & Sanlier, 2016). The essential fatty acids play an important role in brain development, insulin sensitivity, cardiovascular health, prostaglandin metabolism, immune, inflammation and membrane function (Graf, Rojas–Silva, Rojo, Delatorre–Herrera, Baldeón, & Raskin, 2015).

Quinoa helps to raise the good (high density lipoproteins or HDL) cholesterol in the body due to its omega 3 and omega 6 content and to lower the bad (high density lipoproteins – LDL) cholesterol (Bojanic, 2011). It is also rich in vitamins and minerals, especially calcium, phosphorus and iron (Bojanic, 2011).

In addition to its high nutritional value and its gluten–free properties, quinoa is reported to be beneficial to children, elderly, lactose intolerant persons and to consumers in high–risk groups such as anemia, diabetes, obesity, dyslipidemia and celiac disease (Vilcacundo & Hernández–Ledesma, 2017).

Chia (*Salvia Hispanica L.*)

Chia (*Salvia hispanica L.*) is a herbaceous plant that has been raised for centuries by ancient Mexican Aztecs, whose roots come from the Lamiaceae family, based on southern Mexico (Alican, 2017; de Campo, dos Santos, Costa, Paese, Guterres, Rios, de & Flôres, 2017; López, Galante, Robson, Boeris & Spelzini, 2018). It is grown in Argentina, Australia, Bolivia, Colombia, Guatemala, Mexico, Peru and Southeast Asia (Karim, Ashrafuzzaman, & Hossain, 2016).

The chia seed contains 15–25% protein, 30–33% fat, 26–41% carbohydrate, 18–30% high dietary fibre, 4–5% ash and 90–93% minerals, vitamins and dry substance. Because it contains high amounts of polyunsaturated fatty acids (w–3 and w–6) (60% alpha linolenic acid and 20% linoleic acid), it is an important source of essential fatty acids for the body (Alican, 2017). These two essential fatty acids make chia seed oil one of the healthiest oils with a more than 80% of the fatty acid composition (Timilsena, Vongsvivut, Adhikari, & Adhikari, 2017). Omega 3 fatty acids have strong anti–inflammatory properties, helping to reduce cardiovascular disease and blood pressure. It also reduces sleep deprivation and decreases the risk of depression (Giaretta, Lima, & Carpes, 2017). Chia seed has the highest content of omega–3 α–linolenic acid (C18: 3, ALA, up to 68%) when compared with flaxseed (50.6%), rapeseed

(8.1%), soybean (7.6%) and sunflower (estimated at 0.2%) and it is seen as a plant with a high antioxidant potential (Marineli, da S., Lenquiste, Moraes, & Maróstica, 2015).

Unlike many vegetable protein sources, chia seeds contain all essential amino acids, so they have a better protein quality than cereal and other fatty seeds (Caruso, Favati, Di Cairano, Galgano, Labella, Scarpa, & Condelli, 2018; López, Galante, Robson, Boeris, & Spelzini, 2018). Essential amino acids such as leucine, isoleucine and valine account for 42.2–42.9% of the total amount of amino acids in the chia seed. The chia seed is also rich in non–essential amino acids such as glutamic acid, arginine and aspartic acids. It is known that glutamic acid regulates the immunoregulatory response and enhances athletic performance; therefore, it is considered an important amino acid in the diet. Arginine is known to play a role in preventing heart disease (Timilsena, Adhikari, Barrow, & Adhikari, 2016).

Chia seeds are rich in dietary fibre, with a total content ranging from 34% to 50%, and the fibre content is higher than other grains such as flaxseed. Dietary fibre is known to have various biological effects such as delaying the release of glucose from foods and thus reducing postprandial glycosemia. These factors are certainly effective in reducing the incidence of diabetes and cardiovascular diseases (Menga, Amato, Phillips, Angelino, Morreale, & Fares, 2017). It has been reported that the intake of chia seeds significantly reduces serum triglycerides (TG) and low density lipoprotein (LDL) and increases high density lipoprotein (HDL) (Timilsena, Vongsvivut, Adhikari, & Adhikari, 2017).

Due to the diversity of phenolic compounds, chia seeds have excellent antioxidant capacity as well as antimicrobial activity and are also used against various pathological disorders such as atherosclerosis, brain dysfunction and cancer (Rahman, de Camargo, & Shahidi, 2017). Caffeic and rosmarinic acids are among the phenolic compounds currently identified in chia products and these acids prevent and play a role in different neurological disorders such as epilepsy (Oliveira–Alves, Vendramini–Costa, Betim Cazarin, Maróstica Júnior, Borges Ferreira, Silva, … Bronze, 2017).

Due to its nutritional value and chemical composition, the chia seed has been attributed different medical features and is considered as a new functional ingredient (Mesías, Holgado, Márquez–Ruiz, & Morales, 2016). For people on a plant-based diet, it can play an important role as an alternative to meat.

Buckwheat (*Fagopyrum Esculentum*)

Buckwheat is found in the pseudo–grain group of the genus Fagopyrum of the Polygonaceae family (Giménez–Bastida, Piskuła, & Zieliński, 2015; Zhu, 2016b). The most grown species are Fagopyrum esculentum and tartary karabuğday, commonly known as buckwheat (*Fagopyrum esculentum*). Tartary buckwheat is also known as bitter buckwheat because of the bitter taste found in the seeds and the high content of flavonoids (Zhu, 2016a). It is grown mostly in Asia (China, Bhutan, Nepal and India). Common buckwheat grows widely in Asia, Europe and America, while in low quantities it also grows in Europe (Luxembourg, Germany and Belgium) (Zhu, 2016b). Buckwheat has excellent ecological compatibility and can grow in harsh climatic conditions and unusual soil (Zhu, 2016b). Products made from buckwheat such as noodles, pancakes and cabbage are consumed in many countries, especially in China, Japan, Korea, Nepal, and also in European countries, such as Ukraine and Russia (Giménez–Bastida, Piskuła, & Zieliński, 2015; Sytar, Brestic, & Rai, 2013).

Buckwheat is a rich source of starch, protein, dietary fibre, vitamins (thiamine, riboflavin, pyridoxine), antioxidants and minerals as well as antioxidative substances such as rutin, quercetin, hyperin and catechin (Sun, Li, Hu, Zhou, Ji, Yu, … Luan, 2018). It is known for its wide flavonoid content characterized by health benefits such as lowering cholesterol, inhibiting tumors, regulating hypertension, controlling inflammation, carcinogenesis and regulating diabetes (Sytar, Brestic, & Rai, 2013).

The amount of protein in buckwheat is about 12% and is therefore similar to the protein content in wheat (Zhang, Zhou, Tang, Li, Tang, Shao, … Wu, 2012). Buckwheat protein has balanced amino acids and a high level of lysine, arginine in its aminoacid content delivering quality nutritional value (Sun et al., 2018). It has been reported that buckwheat protein has many unique physiological functions such as curing chronic diseases, lowering blood cholesterol, inhibiting breast cancer caused by 7,12–dimethyl-benzene and gallstone inhibitor (Zhang, Zhou, Tang, Li, Tang, Shao, … Wu, 2012).

As in all cereals, the starch found in buckwheat most commonly contains 59–70% of the dry matter of buckwheat. The amount of resistant starch in buckwheat corn is between 7 and 37%, and the decrease in glycemic index is due to the increase in the amount of this resistant starch (Elif, 2017). The raw fibre concentration of buckwheat is very high, 12.7–17.8% (Zhang, Zhou, Tang, Li, Tang, Shao, … Wu, 2012). Dietary fibre contributes to physiological functions such as cholesterol and fat–stripping, reduction of blood glucose levels, prevention of constipation and regulation of colonic health (Zhu, Du, Li, & Li, 2014).

Unsaturated fatty acids constitute 80% of the total, with a total fat content of 40% of the buckwheat and 1.7–4.0%, polyunsaturated fatty acid (Elif, 2017). Compared to common buckwheat, the nutraceutical effect is higher in tartary buckwheat due to higher vitamin B content and antioxidants (Zhang, Li, Ma, Gao, Du, Han, … Qiao, 2017).

The buckwheat dietary fibre performs functions of cholesterol–lowering, antihypertensive effect, constipation and obesity–reducing effect (Sun, Li, Hu, Zhou, Ji, Yu, … Luan, 2018). It regulates bowel movements in the body. Buckwheat does not contain gluten and can be consumed on a celiac diet (Molinari, Costantini, Timperio, Lelli, Bonafaccia, Bonafaccia, & Merendino, 2017). It is also considered a prebiotic nutrient source because it contains lactic acid bacteria such as Bifidobacter and Lactobacillus. It is effective in protecting against radiation by increasing body resistance (Hande, 2015).

Teff (*Eragrostis Tef*)

Teff (*Eragrostis tef*) is a small tropical grain originating from Ethiopia, typically used for the production of traditionally fermented wheat flour, injera (Marti, Marengo, Bonomi, Casiraghi, Franzetti, Pagani, & Iametti, 2017). In Ethiopia and Eritrea, it is an important food product used for the production of traditional foods and drinks such as injera (food basement), kitta (unleavened bread) and tella (opaque beer). Teff granules do not contain gluten and have good potential to be formulated in many food and beverage products that can be used by celiac patients (Zhu, 2018).

Being a very adaptable plant, teff can be grown in changing environmental conditions such as drought and humidity (Zhu, 2018). Its seeds can survive for several years provided direct contact with moisture and sun is avoided. Compared to other common cereals, the teff seed is more resistant to attacks by harmful insects and other storage pests. Thus, it can be safely stored under conventional storage conditions without chemical protection (Gebremariam, Zarnkow, & Becker, 2012).

Teff varieties are defined according to the colors of the grains, flowers, flowering form and size of the plants. Different varieties are known as netch (white), qey (red/brown) and sergegna (mixed) according to the color of the beans and available in the market (Gebremariam, Zarnkow, & Becker, 2012). The grain mass (0.2–0.4 mg) is the smallest among carbohydrate–rich seeds (Gebremariam, Zarnkow, & Becker, 2012).

In recent years, teff has gained popularity in the world because of its quite attractive nutritional properties. Growing has been successfully adapted to other parts of the world, such as the US, India and Australia (Zhu, 2018).

As a result of its unique chemical composition and form of the whole grains, teff has been associated with a number of health benefits. For example, studies have shown that teff helps to prevent the incidence of malaria, anemia and diabetes by showing in vitro antioxidative activities, increasing the level of hemoglobin in the human body (Zhu, 2018).

It is also a very nutritious plant with 100 grams of raw teff containing 367 kcal of energy. Starch is the main component of the teff cereal and accounts for more than 70% of its dry weight (Zhu, 2018). Whole–grain teffin constitutes 9.8% of the dry weight of the dietary fibre (Zhu, 2018).

Teff is rich in carbohydrates, fibre and essential amino acids. It has a starch content of approximately 73%, making teff a starchy cereal, a protein content of 8–11% and a 3–4.5% fibre content (Baye, 2014). Teff contains a high amount of iron and has also higher contents of calcium, copper and zinc than other grains (Campo, del Arco, Urtasun, Oria, & Ferrer–Mairal, 2016). In the teff protein, the basic storage components are glutelin and albumin. The amino acid composition of teff flour is convenient and the teff protein is readily digestible compared to cereals, such as corn and sorghum, because it contains the most digestible types of major protein fractions such as albumin, glutelin and globulin. Teff grains are rich in unsaturated fatty acids – oleic acid 32.41%, and linoleic acid 23.83% (Gebremariam, Zarnkow, & Becker, 2012).

Recently, the teff Trolor [Eragrostis tef (zuccini.)] is used as a raw material for gluten–free alternatives. In addition, the presence of a nutritional property linked to essential amino acids, high mineral, polyphenol and dietary fibre content is the cause of this dissemination of the idea (Di Ghionno, Marconi, Sileoni, De Francesco, Perretti, 2017).

Beans (*Phaseolus Vulgaris L.*)

Beans are among the most consumed pulses in the world. Some of the most important consumers are: South America (9.3 kg/person/year), the Caribbean (9.1 kg/person/year), Central America (8.8 kg/person/year) and Central Africa (8.0 kg/person/year) (Luna–Vital, Mojica, González de Mejía, Mendoza, & Loarca–Piña, 2015). They are a source of energy for millions of people, especially in developing countries. Beans contain dietary fibre and are important dietary protein sources. They are a cheaper source of protein compared to foods of animal origin and can be added to different food formulations (Santiago–Ramos, Figueroa–Cárdenas, de, Véles–Medina, & Salazar, 2018).

Historically, beans have been an important component of the tropical and subtropical cuisine. It is the most important plant protein in the American and African continent, where animal proteins are limited due to economic, religious and cultural reasons. Besides, beans are closely related to the improvement of health. They have functional properties due to their chemical composition and are recommended for dietary treatment of diseases such as cardiovascular disease, diabetes mellitus, obesity and cancer (Oliveira, Mateó, dos Fioroto, Oliveira, de, & Naozuka, 2018). Beans contain 20–30% proteins, 2.5% fats and 50–60% carbohydrates (Hayat, Ahmad, Masud, Ahmed, & Bashir, 2014).

For the last decade, beans have been defined as nutraceutical foods due to bioactive compound contents, such as polyphenols, resistant starch, oligosaccharides, digestible fractions and bioactive peptides. Beans consist mainly of carbohydrates. The amount of protein is approximately 16–33% and is considered to be a good source of protein. In Central and South America, the amount of protein from bean consumption is about 5–6 g/person/day (Luna–Vital, Mojica, González de Mejía, Mendoza, & Loarca–Piña, 2015). In addition to the important nutritive value of beans, low levels of methionine and cysteine content and high resistance to proteolysis should not be overlooked (Carrasco–Castilla, Hernández–Álvarez, Jiménez–Martínez, Jacinto–Hernández, Alaiz, Girón–Calle, … Dávila–Ortiz, 2012).

Beans are the main sources of dietary fibre and contain 15–19% dietary fibre as raw (Ganesan & Xu, 2017). Per 100 grams of edible parts, beans have two to three times more fibre than other dietary sources. In recent years, the consumption of beans in developed countries has been replaced by other foods, and the proportion of fibre in people's diets has decreased. However, the resistant starch and the dietary fibre found in the bean paste have been associated with the protection of the digestive system of humans, particularly colon health. In addition, the beans are lower in the glycemic index due to the higher proportion of slower digestible starch in beans compared with carbohydrate–rich foods containing other dietary fibre. Thus, adding beans to foods and using them in the formulation of processed foods can reduce glycemic load and can bring significant advantages to human health (Los, Zielinski, Wojeicchowski, Nogueira, & Demiate, 2018).

After the beans are cooked, more than 70% of the copper and iron minerals are shown to be insoluble due to protein denaturation, polyphenols and phytate associations (Naozuka & Oliveira, 2012). Cooking, on the other hand, softens the food matrix and releases the substances bound to the protein, thus facilitating protein absorption. In addition, the heating of foods alters the natural factors which prevent mineral absorption, such as phytate and dietary fibre. For this reason, the positive effect of cooking on the chemical composition of the beans is evident (Oliveira, de, Mateó, dos, Fioroto, Oliveira, de, & Naozuka, 2018).

In most studies, beans are associated with improved health. It has been shown that diseases are positively affected by the reduction of risk of metabolic and cardiovascular disease, the decrease of serum cholesterol level and hyperglycemia, the prevention of colon, breast and prostate cancer by the non–nutritional substances in beans (Hall, Hillen, & Garden Robinson, 2017).

Soybean

Soybean (*Glycinemax*) is one of the most widely consumed legume products in the world (Vagadia, Vanga, & Raghavan, 2017). For centuries, soya has been grown in the eastern Asian countries as a crop. The ability to grow in a wide range of soil and climatic conditions makes it a versatile crop and one of the most commonly grown greasy seed products. In addition to the supply of vegetable oil for human consumption, soybean is one of the best sources of protein (Al Loman & Ju, 2017). Soy proteins are widely used to form foodstuffs with the goal of improving nutritional and functional qualities through a high protein level and a well–balanced amino acid composition (Vagadia, Vanga, & Raghavan, 2017). All essential amino acids found in animal proteins are also found in soy proteins. In addition, the nutritional value of soy protein equals animal protein with high biological value (Singh, Vij, & Hati, 2014). According to the United States Department of Agriculture's (USDA) nutrition database, soybean seeds contain approximately 36.5% protein, 19.9% lipid, 30% carbohydrate and 9.3% dietary fibre, 15.6% total saturated fatty acid, 57.7% total polyunsaturated fatty acid and 22.8% monounsaturated fatty acid (Vagadia, Vanga, & Raghavan, 2017).

Known locally as Bhatmala in Nepalese, soybean paste is traditionally used to prepare a variety of fermented and non–fermented recipes in Nepal, India and the Eastern Himalayan regions of Bhutan

(Tamang, 2015). Especially fermented soybean products have become an important part of the Korean diet, used daily in spices and consumed in side dishes and soups (Shin & Jeong, 2015).

Soybean consumption has increased over the last few years due to its positive effects on human health. It also serves as the main protein source for people who follow a vegan diet around the world. Commercial products derived from soya beans are very diverse: soybean sprouts and nuts, soybean flour, soy protein isolates and protein concentrates, soybean oil, soy milk, tofu, okara, tempeh, soy sauce, non–dairy desserts, soy sauce and textured meat products.

Soybean meal is an important source of phytochemicals such as isoflavones, phytosterols and lecithins. In addition, soluble fibres, saponins and polysaccharides can act collectively or through independent mechanisms to provide unique health benefits. For example, soy lecithins and saponins play a role in lipid metabolism; phytosterols and linoleic acid produce hypocholesterolemic effects and soy fibres have been shown to promote weight loss (Ramdath, Padhi, Sarfaraz, Renwick, & Duncan, 2017).

Epidemiological studies have shown that soybean meal consumption plays an important role in the prevention and treatment of a variety of chronic diseases including cardiovascular diseases, reduction of plasma cholesterol, protection against intestinal and kidney diseases and osteoporosis (Vagadia, Vanga, & Raghavan, 2017). Because it does not contain cholesterol, gluten, and lactose, it is a convenient food for vegetarians, people with lactose intolerance and milk allergies (Singh, Vij, & Hati, 2014).

Lentils (*Lens Culinaris*)

Lentils are the oldest grown crops among legumes. According to evidence from archaeological finds, their use dates back to 7500–6500 BC (Cokkizgin & Shtaya, 2013). Currently the annual lentil production is about 5 million tons, the largest production being in Western Canada (38%), followed by India (23%), Turkey (8%), Australia (7%) and USA (5%). More than 90% of the lentils produced in Canada, USA and Australia are exported to South East Asia, the Middle East and Africa (FAO, 2018).

Lentil are a rich source of protein containing a balanced amino acid profile, abundant low digestible carbohydrates and a variety of essential micronutrients. The nutritional content of 100 g of lentils is 2 g fat, 4–9 g dietary fibre, 23–27 g protein and 64–74 g carbohydrate (Chung, Liu, Hoover, Warkentin, & Vandenberg, 2008). There are 39.3 g of essential amino acids per 100 grams of protein in the lentil. The limiting amino acids in the lentil protein are sulfurized amino acids, tryptophan and threonine. For this reason, the consumption of lentil, rice, corn, potatoes and other root and tuber plants ensures that all necessary amino acids are met (Joshi, Timilsena, & Adhikari, 2017).

Similar to other legumes, lentils contain some anti–nutritional factors. Of these anti–nutritional factors, trypsin inhibitors inactivate key digestive enzymes. Tannins, another anti–nutritional factor, can alter protein bioavailability by complexing diet proteins to reduce digestibility of cholimetry (Nosworthy, Medina, Franczyk, Neufeld, Appah, Utioh, … House, 2018). However, studies have shown that culinary firing reduces the activity and concentration of anti–nutritional factors such as trypsin inhibitors, tannins and phytic acid (Hefnawy, 2011; Wang, Hatcher, Toews, & Gawalko, 2009). The reduction of the anti–nutritional factors increases the digestibility of the dietary protein and thus increases the bioavailability of the lentils (Nosworthy, Medina, Franczyk, Neufeld, Appah, Utioh, … House, 2018).

The lentil is a nutritious high medium energy pulp and contains various micronutrient ingredients, including 3.7–4.5 mg iron, 2.2–2.7 mg zinc, 22–34 μg selenium, 50–250 μg β–carotene and 216–290 μg folate (Siva, Thavarajah, Johnson, Duckett, Jesch, & Thavarajah, 2017). Unlike other grains, the lentil is very low in terms of phytic acid content (2.5–4.4 mg / g) linking iron and zinc, thus making these nutrients available (Thavarajah, Thavarajah, & Vandenberg, 2009).

Most carbohydrates found in lentils are starch and this starch refers to non–structural carbohydrates containing 47–52 g total starch in 100 g lentils. Lentil starch is composed of amylose (a few branched linear glucans) and amylopectin (a larger, highly branched molecule). The amylose ratio is higher than the amylopectin ratio, and the digestibility of the liquor is slower because of the character of the outermost layers of the crystallization grade or starch granule of amylose starch (Siva, Thavarajah, Johnson, Duckett, Jesch, & Thavarajah, 2017).

According to recent studies, lentils may be a good source of prebiotic carbohydrates. Total prebiotic carbohydrate concentrations indicate that more than 13 g of prebiotic can be provided by 100 g culinary portions (Johnson, Thavarajah, Combs, & Thavarajah, 2013). Prebiotics in the diet are defined as selectively fermented substances that cause specific changes in the gastrointestinal microbiotinin composition and/or activity, thus benefiting the host's health (Valcheva & Dieleman, 2016). These prebiotic carbohydrates in the lentils are associated with the hunger–toughness mechanism. They may decrease the rate and degree of starch digestibility and may lead to better management of body weight, decreased glycemic response and insulin resistance (Siva, Thavarajah, Johnson, Duckett, Jesch, & Thavarajah, 2017).

Chickpea (*Ciger Arietinum*)

Chickpea (*Cicer arietinum*) is the third most important species among legumes after soybeans and peas and the second most important legume grown in Asia, the Mediterranean regions, Australia, Canada, USA and Africa (Acharjee & Sarmah, 2013). It is one of the earliest grown vegetables and is thought to have originated in the Middle East about 7450 years ago (Roy, Boye, & Simpson, 2010). Worldwide, chickpeas are grown in 12 million hectares with 11 million tons produced. The South Asian region, including Iran, is the largest chickpea producer in the world and covers 76% of the total production (FAO, 2010).

There are two main species of chickpeas grown in the world – desi with smaller seeds and kabuli with larger seeds. The desi chickpea seed is dark, irregularly shaped and grown in semi–arid areas. Kabuli chickpeas (Garbanzo bean) have a light–colored seed coat, and are normally grown in temperate regions of the world (Roy, Boye, & Simpson, 2010).

Factors, such as climate, soil, nutrition, biotic and abiotic stress affect the nutritional composition of chickpeas. There are about 367 kcal per 100 g of chickpea seed. The composition of carbohydrate ranges from 50 to 70%. Generally, chickpeas contain more lipid (2–8%) and fibre (10–20%) compared to other legumes (Acharjee & Sarmah, 2013).

Chickpeas contain high quality protein. In addition, carbohydrates are a good source of vitamins (thiamine and niacin) and minerals (calcium, phosphorus, iron, magnesium and potassium). Their fat content is rich in linoleic base oil acidity. The quality of chickpea's protein is similar to soybean protein, but it contains eight essential amino acid residues.

Lysine, the limiting amino acid in grains, is found in the chickpea. Hence, when chickpeas are consumed together with cereals, they complement each other and provide balanced nutrition. Also, chickpeas contain vitamins such as vitamin B–complex, vitamin C, vitamin A and vitamin K, which are necessary for various metabolic pathways. Chickpea seeds are also a mineral-rich source for calcium, phosphorus, zinc and iron (Bar–El Dadon, Abbo, & Reifen, 2017).

Studies have reported that the amount of protein in chickpea varies between 18% and 28% and of fat between 6% and 9% (Alajaji & El–Adawy, 2006; Ghavidel & Prakash, 2006). Chickpeas are an important source of amino acids. However, they contain limited amounts of sulfur–containing amino acids. Accordingly, the content of methionine is 1.3–1.6% and the content of cysteine is 2.5–3.0% (Acharjee & Sarmah, 2013).

Chickpeas have some unwanted features. These are phenolic compounds that have long cooking times, contain enzyme inhibitors and phytates, form gas problems and must be removed for effective use (Milan–Carrillo, Valdez–Alarcon, Gutierrez–Dorado, Cárdenas–Valenzuela, Mora–Escobedo, Garzón–Tiznado, & Reyes–Moreno, 2007).

The consumption of chickpeas has been associated with the prevention of cardiovascular disease, the management of type 2 diabetes and lowering of LDL–cholesterol levels. While insoluble dietary fibre in chickpea is associated with a reduction in the incidence of colon cancer, it has been shown that soluble fibre has a beneficial effect on weight loss and weight management. They are used in stews, soups, salads and dips and can be processed into flour (Roy, Boye, & Simpson, 2010). Because chickpeas contain no gluten, they can provide excellent cooking characteristics in gluten–free cereal products (Shaabani, Yarmand, Kiani, & Emam–Djomeh, 2018).

Spirulina

Algae are photosynthetic organisms that convert light energy from the sun into chemical energy by photosynthesis and have a simple reproduction structure. Their biomass contains various compounds with diversified structures and functions. Algal biotechnology is divided into microalgae, macroalgae and cyanobacteria (Soni, Sudhakar, & Rana, 2017).

Spirulina was known in the past as filamentous spiral–shaped blue–green algae. Nowadays it is better known as photosynthetic bacteria (Arthrospira). This microorganism is regarded as an important food source for humans and the most popular microalgae, described by the World Health Organization as one of the world's most important superb foods (Deamici, Santos, & Costa, 2018).

It was first discovered in 1519 by the Spanish scientist Hernando Cortez who observed that spirulina was consumed by the Aztecs during a visit to Lake Texcoco in the Mexico Valley. Pierre Dangeard discovered spirulina's health benefits by observing that flamingos survived by consuming blue–green algae. The botanist Jean Leonard supported the findings of Dangeard and people soon began commercializing spirulina. It is a microalgae species that naturally grows in subtropical climates and saline lakes (Belay, Ota, Miyakawa, & Shimamatsu, 1993; Soni, Sudhakar, & Rana, 2017).

Spirulina is a concentrated food with antioxidants, phytotoxins, probiotics and nutraceuticals, which has a health–enhancing effect on people. It can quickly meet the various needs of people due to its nutritional composition. According to NASA and the European Space Agency, spirulina is one of the main things that can be consumed in long–term space missions. It is seen as "one of the best protein sources" (Soni, Sudhakar, & Rana, 2017).

Currently, spirulina is used in malnutrition treatment in many countries. According to the study on malnutrition treatment, the administering of spirulina at a dose of 10g/day improved the nutritional status of malnourished children in the intervention group compared to the control group – the malnutrition rate was 30% before spirulina and 20% after the spirulina intervention (Matondo, Takaisi, Nkuadiolandu, Kazadi Lukusa, & Aloni, 2016).

The two most important species of spirulina are Spirulina maxima and Spirulina platensis. Protein constitutes 60–70% of their content, and 47% of this protein contains all of the essential amino acids (Hannon, Gimpel, Tran, Rasala, & Mayfield, 2010). It also contains vitamins and minerals such as vitamin A, vitamin C, vitamin E, iron, calcium, chromium, copper, magnesium, manganese, phosphorus, potassium, sodium and zinc. Vitamin B12, which is not found in sufficient amounts in plant sources, is more than 4 times the amount found in raw liver. In addition, the content of iron is high and 20 times more than wheat. In addition, the content of β–carotene is unusually high and about 30 times higher than that of carrots (Soni, Sudhakar, & Rana, 2017).

Spirulina's main fatty acid is one of the best anti–inflammatories in the world because it is gamma–linolenic acid (GLA) with fat content of 12% and 14% fibre (Sathasivam Radhakrishnan, Hashem, & Abd–Allah, 2017). Anticancer, antioxidant and hepatoprotective agents have been shown in studies. Also, studies have shown that spirulina has a positive effect on cardiovascular diseases, hyperglycaemia, hyperlipidemia, immunodeficiency, inflammatory processes and improves the immune system's resistance to various types of cancer as well as on the treatment of HIV and other viral diseases (Ovando, Carvalho, de, Vinícius de Melo Pereira, Jacques, Soccol, & Soccol, 2018). It is a functional food in the immune system that feeds the intestinal flora, including Lactobacillus and Bifidus (Sathasivam Radhakrishnan, Hashem, & Abd–Allah, 2017).

CONCLUSION

Proteins are the basis of the human diet because of their role for the repair of immune responses, cell signals, muscle masses and damaged cells. Animal and plant proteins are the main protein sources in human diet. Plant-derived proteins can meet all our protein needs with a balanced diet without the need to consume meat or other animal-based products. The benefits from proteins derived from plant–based sources compared with animal–based food sources are far–reaching and can protect against chronic disease, promote overall bodily health and can have positive health effects on preventing cardiovascular disease, diabetes, cancer, weight maintenance and other health risk factors.

REFERENCES

Abugoch James, L. E. (2009). Quinoa (Chenopodium quinoa Willd.): Composition, chemistry, nutritional, and functional properties. *Advances in Food and Nutrition Research*, *58*, 1–31. doi:10.1016/S1043-4526(09)58001-1 PMID:19878856

Acharjee, S., & Sarmah, B. K. (2013). Biotechnologically generating 'super chickpea' for food and nutritional security. *Plant Science*, *207*, 108–116. doi:10.1016/j.plantsci.2013.02.003 PMID:23602105

Al Loman, A., & Ju, L. K. (2017). Enzyme–based processing of soybean carbohydrate: Recent developments and future prospects. *Enzyme and Microbial Technology*, *106*, 35–47. doi:10.1016/j.enzmictec.2017.06.013 PMID:28859808

Alajaji, S. A., & El–Adawy, T. A. (2006). Nutritional composition of chickpea (Cicer arietinum L.) as affected by microwave cooking and other traditional cooking methods. *Journal of Food Composition and Analysis*, *19*(8), 806–812. doi:10.1016/j.jfca.2006.03.015

Alican, A. (2017). *Soğuk pres tekniği ile elde edilen chia tohumu atıklarının salata sosu üretiminde kullanılması*. Yıldız Teknik Üniversitesi.

Bar–El Dadon, S., Abbo, S., & Reifen, R. (2017). Leveraging traditional crops for better nutrition and health – The case of chickpea. *Trends in Food Science & Technology*, *64*, 39–47. doi:10.1016/j.tifs.2017.04.002

Baye, K. (2014). *Teff: Nutrient composition and health benefits*. Ethiopia Strategy Support Program. Workig Paper 67. Retrieved from https://www.researchgate.net/profile/Kaleab_Baye2/publication/266316373_Teff_Nutrient_Composition_and_Health_Benefits/links/542c24320cf277d58e8b0b42.pdf

Belay, A., Ota, Y., Miyakawa, K., & Shimamatsu, H. (1993). Current knowledge on potential health benefits of Spirulina. *Journal of Applied Phycology*, *5*(2), 235–241. doi:10.1007/BF00004024

Besler, T., Rakıcıoglu, N., Ayaz, A., Demirel Büyüktuncer, Z., Özel, G. H., Samur Eroğlu, G., . . . Yürük, A. (2015). *Türkiye'ye özgü besin ve beslenme rehberi*. Ankara. Retrieved from http://www.bdb.hacettepe.edu.tr/TOBR_kitap.pdf

Bogueva, D., Marinova, D., & Raphaely, T. (Eds.). (2018). Handbook of research on social marketing and its influence on animal origin food product consumption. Hershey, PA: IGI Global.

Bojanic, A. (2011). *La quinua : Cultivo milenario para contribuir a la seguridad alimentaria mundial*. Proinpa, 58. Food and Agriculture Organization of the United Nations.

Campo, E., del Arco, L., Urtasun, L., Oria, R., & Ferrer–Mairal, A. (2016). Impact of sourdough on sensory properties and consumers' preference of gluten–free breads enriched with teff flour. *Journal of Cereal Science*, *67*, 75–82. doi:10.1016/j.jcs.2015.09.010

Carrasco–Castilla, J., Hernández–Álvarez, A. J., Jiménez–Martínez, C., Jacinto–Hernández, C., Alaiz, M., Girón–Calle, J., ... Dávila–Ortiz, G. (2012). Antioxidant and metal chelating activities of peptide fractions from phaseolin and bean protein hydrolysates. *Food Chemistry*, *135*(3), 1789–1795. doi:10.1016/j.foodchem.2012.06.016 PMID:22953924

Caruso, M. C., Favati, F., Di Cairano, M., Galgano, F., Labella, R., Scarpa, T., & Condelli, N. (2018). Shelf–life evaluation and nutraceutical properties of chia seeds from a recent long–day flowering genotype cultivated in Mediterranean area. *Lebensmittel-Wissenschaft + Technologie*, *87*, 400–405. doi:10.1016/j.lwt.2017.09.015

Chen, J.-H., Song, J., Chen, Y., Ding, Q., Peng, A., & Mao, L. (2016). The effect of vegan protein–based diets on metabolic parameters, expressions of adiponectin and its receptors in wistar rats. *Nutrients*, *8*(10), 643. doi:10.3390/nu8100643 PMID:27763537

Chung, H.-J., Liu, Q., Hoover, R., Warkentin, T. D., & Vandenberg, B. (2008). In vitro starch digestibility, expected glycemic index, and thermal and pasting properties of flours from pea, lentil and chickpea cultivars. *Food Chemistry*, *111*(2), 316–321. doi:10.1016/j.foodchem.2008.03.062 PMID:26047429

Cokkizgin, A., & Shtaya, M. J. Y. (2013). Lentil: Origin, cultivation techniques, utilization and advances in transformation. *Agricultural Science*, *1*(1), 55–62. doi:10.12735/as.v1i1p55

Comerford, K. B., & Pasin, G. (2016). Emerging evidence for the importance of dietary protein source on glucoregulatory markers and type 2 diabetes: Different effects of dairy, meat, fish, egg, and plant protein foods. *Nutrients*, *8*(8), 446. doi:10.3390/nu8080446 PMID:27455320 de Campo, C., dos Santos, P. P., Costa, T. M. H., Paese, K., Guterres, S. S., Rios, A. de O., & Flôres, S. H. (2017). Nanoencapsulation of chia seed oil with chia mucilage (Salvia hispanica L.) as wall material: Characterization and stability evaluation. *Food Chemistry, 234*, 1–9. doi:10.1016/j.foodchem.2017.04.153 PMID:28551210

Deamici, K. M., Santos, L. O., & Costa, J. A. V. (2018). Magnetic field action on outdoor and indoor cultures of Spirulina: Evaluation of growth, medium consumption and protein profile. *Bioresource Technology, 249*, 168–174. doi:10.1016/j.biortech.2017.09.185 PMID:29040851

Di Ghionno, L., Marconi, O., Sileoni, V., De Francesco, G., & Perretti, G. (2017). Brewing with prolyl endopeptidase from Aspergillus niger: The impact of enzymatic treatment on gluten levels, quality attributes and sensory profile. *International Journal of Food Science & Technology, 52*(6), 1367–1374. doi:10.1111/ijfs.13375

Elif, Ö. (2017). *Erişte üretiminde farklı oran ve kombinasyonlarda karabuğday, amarant ve kinoa unlarının kullanım imkanları*. Necmettin Erbakan Üniversitesi.

Fischer, S., Wilckens, R., Jara, J., Aranda, M., Valdivia, W., Bustamante, L., ... Obal, I. (2017). Protein and antioxidant composition of quinoa (Chenopodium quinoa Willd.) sprout from seeds submitted to water stress, salinity and light conditions. *Industrial Crops and Products, 107*, 558–564. doi:10.1016/j.indcrop.2017.04.035

Food and Agriculture Organization of the United Nations (FAO). (2010). *The state of addressing food insecurity in protracted crises*. Retrieved from http://www.fao.org/catalog/inter–e.htm

Food and Agriculture Organization of the United Nations (FAO). (2018). *Food and agriculture data*. Retrieved from http://www.fao.org/faostat/en/#home

Ganesan, K., & Xu, B. (2017). Polyphenol–rich dry common beans (Phaseolus vulgaris L.) and their health benefits. *International Journal of Molecular Sciences, 18*(11), 2331. doi:10.3390/ijms18112331 PMID:29113066

Garnett, T. (2014). *What is a sustainable healthy diet?* Food Climate Research Network. Retrieved from https://www.fcrn.org.uk/sites/default/files/fcrn_what_is_a_sustainable_healthy_diet_final.pdf

Gebremariam, M. M., Zarnkow, M., & Becker, T. (2012). Teff (Eragrostis tef) as a raw material for malting, brewing and manufacturing of gluten–free foods and beverages: A review. *Journal of Food Science and Technology, 51*(11), 2881–2895. doi:10.100713197-012-0745-5 PMID:26396284

Ghavidel, R. A., & Prakash, J. (2006). Effect of germination and dehulling on functional properties of legume flours. *Journal of the Science of Food and Agriculture, 86*(8), 1189–1195. doi:10.1002/jsfa.2460

Giaretta, D., Lima, V. A., & Carpes, S. T. (2017). (in press). Improvement of fatty acid profile in breads supplemented with Kinako flour and chia seed. *Innovative Food Science & Emerging Technologies*. doi:10.1016/j.ifset.2017.11.010

Giménez–Bastida, J. A., Piskuła, M., & Zieliński, H. (2015). Recent advances in development of gluten–free buckwheat products. *Trends in Food Science & Technology, 44*(1), 58–65. doi:10.1016/j.tifs.2015.02.013

Graf, B. L., Rojas–Silva, P., Rojo, L. E., Delatorre–Herrera, J., Baldeón, M. E., & Raskin, I. (2015). Innovations in health value and functional food development of quinoa (Chenopodium quinoa Willd.). *Comprehensive Reviews in Food Science and Food Safety*, *14*(4), 431–445. doi:10.1111/1541-4337.12135 PMID:27453695

Hall, C., Hillen, C., & Garden Robinson, J. (2017). Composition, nutritional value, and health benefits of pulses. *Cereal Chemistry Journal*, *94*(1), 11–31. doi:10.1094/CCHEM-03-16-0069-FI

Hande, A. (2015). *Hidrotermal işlemlerin karabuğday nişastasının fizikokimyasal özellikleri ve dirençli nişasta miktarı üzerindeki etkisi*. İstanbul Teknik Üniversitesi.

Hannon, M., Gimpel, J., Tran, M., Rasala, B., & Mayfield, S. (2010). Biofuels from algae: Challenges and potential. *Biofuels*, *1*(5), 763–784. doi:10.4155/bfs.10.44 PMID:21833344

Hayat, I., Ahmad, A., Masud, T., Ahmed, A., & Bashir, S. (2014). Nutritional and health perspectives of beans (Phaseolus vulgaris L.): An overview. *Critical Reviews in Food Science and Nutrition*, *54*(5), 580–592. doi:10.1080/10408398.2011.596639 PMID:24261533

Hefnawy, T. H. (2011). Effect of processing methods on nutritional composition and anti–nutritional factors in lentils (Lens culinaris). *Annals of Agricultural Science*, *56*(2), 57–61. doi:10.1016/j.aoas.2011.07.001

Johnson, C. R., Thavarajah, D., Combs, G. F. Jr, & Thavarajah, P. (2013). Lentil (Lens culinaris L.): A prebiotic–rich whole food legume. *Food Research International*, *51*(1), 107–113. doi:10.1016/j.foodres.2012.11.025

Joshi, M., Timilsena, Y., & Adhikari, B. (2017). Global production, processing and utilization of lentil: A review. *Journal of Integrative Agriculture*, *16*(12), 2898–2913. doi:10.1016/S2095-3119(17)61793-3

Karim, M. M., Ashrafuzzaman, M., & Hossain, M. A. (2016). Effect of planting time on the growth and yield of chia (Salvia hispanica L.). *Asian Journal of Medical and Biological Research*, *1*(3), 502. doi:10.3329/ajmbr.v1i3.26469

Li, G., & Zhu, F. (2018). Quinoa starch: Structure, properties, and applications. *Carbohydrate Polymers*, *181*, 851–861. doi:10.1016/j.carbpol.2017.11.067 PMID:29254045

Lin, D., Lu, W., Kelly, A. L., Zhang, L., Zheng, B., & Miao, S. (2017). Interactions of vegetable proteins with other polymers: Structure–function relationships and applications in the food industry. *Trends in Food Science & Technology*, *68*, 130–144. doi:10.1016/j.tifs.2017.08.006

López, D. N., Galante, M., Robson, M., Boeris, V., & Spelzini, D. (2018). Amaranth, quinoa and chia protein isolates: Physicochemical and structural properties. *International Journal of Biological Macromolecules*, *109*, 152–159. doi:10.1016/j.ijbiomac.2017.12.080 PMID:29247732

Los, F. G. B., Zielinski, A. A. F., Wojeicchowski, J. P., Nogueira, A., & Demiate, I. M. (2018). Beans (Phaseolus vulgaris L.): Whole seeds with complex chemical composition. *Current Opinion in Food Science*, *19*, 63–71. doi:10.1016/j.cofs.2018.01.010

Luna–Vital, D. A., Mojica, L., González de Mejía, E., Mendoza, S., & Loarca–Piña, G. (2015). Biological potential of protein hydrolysates and peptides from common bean (Phaseolus vulgaris L.): A review. *Food Research International*, *76*, 39–50. doi:10.1016/j.foodres.2014.11.024

Marineli, R. da S., Lenquiste, S. A., Moraes, É. A., & Maróstica, M. R. Jr. (2015). Antioxidant potential of dietary chia seed and oil (Salvia hispanica L.) in diet–induced obese rats. *Food Research International, 76*, 666–674. doi:10.1016/j.foodres.2015.07.039 PMID:28455051

Marsh, K., Zeuschner, C., & Saunders, A. (2016). Red meat and health: Evidence regarding red meat, health, and chronic disease risk. In T. Raphaely & D. Marinova (Eds.), *Impact of meat consumption on health and environmental sustainability* (pp. 131–177). Hershey, PA: IGI Global. doi:10.4018/978-1-4666-9553-5.ch008

Marsh, K., Zeuschner, C., & Saunders, A. (2018). The health impact of eating foods of animal origin: Evidence regarding animal foods, health, and disease risk. In D. Bogueva, D. Marinova, & T. Raphaely (Eds.), *Handbook of research on social marketing and its influence on animal origin food product consumption* (pp. 283–297). Hershey, PA: IGI Global. doi:10.4018/978-1-5225-4757-0.ch002

Marti, A., Marengo, M., Bonomi, F., Casiraghi, M. C., Franzetti, L., Pagani, M. A., & Iametti, S. (2017). Molecular features of fermented teff flour relate to its suitability for the production of enriched gluten–free bread. *Lebensmittel-Wissenschaft + Technologie, 78*, 296–302. doi:10.1016/j.lwt.2016.12.042

Matondo, F. K., Takaisi, K., Nkuadiolandu, A. B., Kazadi Lukusa, A., & Aloni, M. N. (2016). Spirulina supplements improved the nutritional status of undernourished children quickly and significantly: Experience from Kisantu, the Democratic Republic of the Congo. *International Journal of Pediatrics, 1296414*, 1–5. doi:10.1155/2016/1296414 PMID:27777589

Mattisson, I., Wirfält, E., Johansson, U., Gullberg, B., Olsson, H., & Berglund, G. (2004). Intakes of plant foods, fibre and fat and risk of breast cancer – a prospective study in the Malmö Diet and Cancer cohort. *British Journal of Cancer, 90*(1), 122–127. doi:10.1038j.bjc.6601516 PMID:14710218

Menga, V., Amato, M., Phillips, T. D., Angelino, D., Morreale, F., & Fares, C. (2017). Gluten–free pasta incorporating chia (Salvia hispanica L.) as thickening agent: An approach to naturally improve the nutritional profile and the in vitro carbohydrate digestibility. *Food Chemistry, 221*, 1954–1961. doi:10.1016/j.foodchem.2016.11.151 PMID:27979185

Mesías, M., Holgado, F., Márquez–Ruiz, G., & Morales, F. J. (2016). Risk/benefit considerations of a new formulation of wheat–based biscuit supplemented with different amounts of chia flour. *Lebensmittel-Wissenschaft + Technologie, 73*, 528–535. doi:10.1016/j.lwt.2016.06.056

Milan–Carrillo, J., Valdez–Alarcon, C., Gutierrez–Dorado, R., Cárdenas–Valenzuela, O. G., Mora–Escobedo, R., Garzón–Tiznado, J. A., & Reyes–Moreno, C. (2007). Nutritional properties of quality protein maize and chickpea extruded based weaning food. *Plant Foods for Human Nutrition (Dordrecht, Netherlands), 62*(1), 31–37. doi:10.100711130-006-0039-z PMID:17243010

Molinari, R., Costantini, L., Timperio, A. M., Lelli, V., Bonafaccia, F., Bonafaccia, G., & Merendino, N. (2017). Tartary buckwheat malt as ingredient of gluten–free cookies. *Journal of Cereal Science, 80*, 37–43. doi:10.1016/j.jcs.2017.11.011

Møller, G., Sluik, D., Ritz, C., Mikkilä, V., Raitakari, O. T., Hutri–Kähönen, N., ... Raben, A. (2017). A protein diet score, including plant and animal protein, investigating the association with HbA1c and eGFR–The PREVIEW Project. *Nutrients, 9*(7), 763. doi:10.3390/nu9070763 PMID:28714926

Montalcini, T., De Bonis, D., Ferro, Y., Carè, I., Mazza, E., Accattato, F., ... Pujia, A. (2015). High vegetable fats intake is associated with high resting energy expenditure in vegetarians. *Nutrients*, *7*(7), 5933–5947. doi:10.3390/nu7075259 PMID:26193314

Moorthi, R. N., Vorland, C. J., & Hill Gallant, K. M. (2017). Diet and diabetic kidney disease: Plant versus animal protein. *Current Diabetes Reports*, *17*(3), 15. doi:10.100711892-017-0843-x PMID:28271467

Naozuka, J., & Oliveira, P. V. (2012). Cooking effects on iron and proteins content of beans (Phaseolus Vulgaris L.) by GF AAS and MALDI–TOF MS. *Journal of the Brazilian Chemical Society*, *23*(1), 156–162. doi:10.1590/S0103-50532012000100022

Navruz–Varli, S., & Sanlier, N. (2016). Nutritional and health benefits of quinoa (Chenopodium quinoa Willd.). *Journal of Cereal Science*, *69*, 371–376. doi:10.1016/j.jcs.2016.05.004

Nehete, J. Y., Bhambar, R. S., Narkhede, M. R., & Gawali, S. R. (2013). Natural proteins: Sources, isolation, characterization and applications. *Pharmacognosy Reviews*, *7*(14), 107–116. doi:10.4103/0973-7847.120508 PMID:24347918

Newby, P. K. (2009). Plant foods and plant-based diets: Protective against childhood obesity? *The American Journal of Clinical Nutrition*, *89*(5), 1572S–1587S. doi:10.3945/ajcn.2009.26736G

Nielsen, L. V., Kristensen, M. D., Klingenberg, L., Ritz, C., Belza, A., Astrup, A., & Raben, A. (2018). Protein from meat or vegetable sources in meals matched for fibre content has similar effects on subjective appetite sensations and energy intake – a randomized acute cross–over meal test study. *Nutrients*, *10*(1). doi:10.3390/nu10010096

Nosworthy, M. G., Medina, G., Franczyk, A. J., Neufeld, J., Appah, P., Utioh, A., ... House, J. D. (2018). Effect of processing on the in vitro and in vivo protein quality of red and green lentils (Lens culinaris). *Food Chemistry*, *240*, 588–593. doi:10.1016/j.foodchem.2017.07.129 PMID:28946315

Oliveira, A. P., de, & Mateó, B. (2018). Effect of cooking on the bioaccessibility of essential elements in different varieties of beans (Phaseolus vulgaris L.). *Journal of Food Composition and Analysis*, *67*, 135–140. doi:10.1016/j.jfca.2018.01.012

Oliveira–Alves, S. C., Vendramini–Costa, D. B., Betim Cazarin, C. B., Maróstica Júnior, M. R., Borges Ferreira, J. P., Silva, A. B., ... Bronze, M. R. (2017). Characterization of phenolic compounds in chia (Salvia hispanica L.) seeds, fibre flour and oil. *Food Chemistry*, *232*, 295–305. doi:10.1016/j.foodchem.2017.04.002 PMID:28490078

Ovando, C. A., de Carvalho, J. C., Vinícius de Melo Pereira, G., Jacques, P., Soccol, V. T., & Soccol, C. R. (2018). Functional properties and health benefits of bioactive peptides derived from Spirulina: A review. *Food Reviews International*, *34*(1), 34–51. doi:10.1080/87559129.2016.1210632

Pasiakos, S. M., Agarwal, S., Lieberman, H. R., & Fulgoni, V. L. III. (2015). Sources and amounts of animal, dairy, and plant protein intake of US adults in 2007–2010. *Nutrients*, *7*(8), 7058–7069. doi:10.3390/nu7085322 PMID:26308049

Rahman, M. J., de Camargo, A. C., & Shahidi, F. (2017). Phenolic and polyphenolic profiles of chia seeds and their in vitro biological activities. *Journal of Functional Foods*, *35*, 622–634. doi:10.1016/j.jff.2017.06.044

Ramdath, D., Padhi, E., Sarfaraz, S., Renwick, S., & Duncan, A. (2017). Beyond the cholesterol–lowering effect of soy protein: A review of the effects of dietary soy and its constituents on risk factors for cardiovascular disease. *Nutrients*, *9*(4), 324. doi:10.3390/nu9040324 PMID:28338639

Raphaely, T., & Marinova, D. (Eds.). (2016). *Impact of meat consumption on health and environmental sustainability*. Hershey, PA: IGI Global. doi:10.4018/978-1-4666-9553-5

Repo–Carrasco–Valencia, R. A.-M., & Serna, L. A. (2011). Quinoa (Chenopodium quinoa, Willd.) as a source of dietary fibre and other functional components. *Food Science and Technology (Campinas)*, *31*(1), 225–230. doi:10.1590/S0101-20612011000100035

Richter, C. K., Skulas–Ray, A. C., Champagne, C. M., & Kris–Etherton, P. M. (2015). Plant protein and animal proteins: Do they differentially affect cardiovascular disease risk? *Advances in Nutrition*, *6*(6), 712–728. doi:10.3945/an.115.009654 PMID:26567196

Roy, F., Boye, J. I., & Simpson, B. K. (2010). Bioactive proteins and peptides in pulse crops: Pea, chickpea and lentil. *Food Research International*, *43*(2), 432–442. doi:10.1016/j.foodres.2009.09.002

Santiago–Ramos, D., Figueroa–Cárdenas, J. de D., Véles–Medina, J. J., & Salazar, R. (2018). Physicochemical properties of nixtamalized black bean (Phaseolus vulgaris L.) flours. *Food Chemistry*, *240*, 456–462. doi:10.1016/j.foodchem.2017.07.156 PMID:28946298

Sathasivam, R., Radhakrishnan, R., Hashem, A., & Abd–Allah, E. F. (2017). (in press). Microalgae metabolites: A rich source for food and medicine. *Saudi Journal of Biological Sciences*. doi:10.1016/j.sjbs.2017.11.003

Satija, A., Bhupathiraju, A. N., Rimm, E. B., Spiegelman, D., Chiuve, S. E., Borgi, L., ... Hu, F. B. (2016). Plant–based dietary patterns and incidence of type 2 diabetes in us men and women: Results from three prospective cohort studies. *PLoS Medicine*, *13*(6), e1002039. doi:10.1371/journal.pmed.1002039 PMID:27299701

Shaabani, S., Yarmand, M. S., Kiani, H., & Emam–Djomeh, Z. (2018). The effect of chickpea protein isolate in combination with transglutaminase and xanthan on the physical and rheological characteristics of gluten free muffins and batter based on millet flour. *Lebensmittel-Wissenschaft + Technologie*, *90*, 362–372. doi:10.1016/j.lwt.2017.12.023

Shin, D., & Jeong, D. (2015). Korean traditional fermented soybean products: Jang. *Journal of Ethnic Foods*, *2*(1), 2–7. doi:10.1016/j.jef.2015.02.002

Singh, B. P., Vij, S., & Hati, S. (2014). Functional significance of bioactive peptides derived from soybean. *Peptides*, *54*, 171–179. doi:10.1016/j.peptides.2014.01.022 PMID:24508378

Siva, N., Thavarajah, D., Johnson, C. R., Duckett, S., Jesch, E. D., & Thavarajah, P. (2017). Can lentil (Lens culinaris Medikus) reduce the risk of obesity? *Journal of Functional Foods*, *38*, 706–715. doi:10.1016/j.jff.2017.02.017

Soni, R. A., Sudhakar, K., & Rana, R. S. (2017). Spirulina – from growth to nutritional product: A review. *Trends in Food Science & Technology*, *69*, 157–171. doi:10.1016/j.tifs.2017.09.010

Sun, X., Li, W., Hu, Y., Zhou, X., Ji, M., Yu, D., ... Luan, G. (2018). Comparison of pregelatinization methods on physicochemical, functional and structural properties of tartary buckwheat flour and noodle quality. *Journal of Cereal Science*, *80*, 63–71. doi:10.1016/j.jcs.2018.01.016

Sytar, O., Brestic, M., & Rai, M. (2013). Possible ways of fagopyrin biosynthesis and production in buckwheat plants. *Fitoterapia*, *84*(1), 72–79. doi:10.1016/j.fitote.2012.10.008 PMID:23103298

Tamang, J. P. (2015). Naturally fermented ethnic soybean foods of India. *Journal of Ethnic Foods*, *2*(1), 8–17. doi:10.1016/j.jef.2015.02.003

Thavarajah, P., Thavarajah, D., & Vandenberg, A. (2009). Low phytic acid lentils (Lens culinaris L.): A potential solution for increased micronutrient bioavailability. *Journal of Agricultural and Food Chemistry*, *57*(19), 9044–9049. doi:10.1021/jf901636p PMID:19725537

Timilsena, Y. P., Adhikari, R., Barrow, C. J., & Adhikari, B. (2016). Physicochemical and functional properties of protein isolate produced from Australian chia seeds. *Food Chemistry*, *212*, 648–656. doi:10.1016/j.foodchem.2016.06.017 PMID:27374580

Timilsena, Y. P., Vongsvivut, J., Adhikari, R., & Adhikari, B. (2017). Physicochemical and thermal characteristics of Australian chia seed oil. *Food Chemistry*, *228*, 394–402. doi:10.1016/j.foodchem.2017.02.021 PMID:28317740

Vagadia, B. H., Vanga, S. K., & Raghavan, V. (2017). Inactivation methods of soybean trypsin inhibitor – A review. *Trends in Food Science & Technology*, *64*, 115–125. doi:10.1016/j.tifs.2017.02.003

Valcheva, R., & Dieleman, L. A. (2016). Prebiotics: Definition and protective mechanisms. *Best Practice & Research. Clinical Gastroenterology*, *30*(1), 27–37. doi:10.1016/j.bpg.2016.02.008 PMID:27048894

Vilcacundo, R., & Hernández–Ledesma, B. (2017). Nutritional and biological value of quinoa (Chenopodium quinoa Willd.). *Current Opinion in Food Science*, *14*, 1–6. doi:10.1016/j.cofs.2016.11.007

Wang, N., Hatcher, D. W., Toews, R., & Gawalko, E. J. (2009). Influence of cooking and dehulling on nutritional composition of several varieties of lentils (Lens culinaris). *Lebensmittel-Wissenschaft + Technologie*, *42*(4), 842–848. doi:10.1016/j.lwt.2008.10.007

World Health Organization (WHO). (2015). *Q&A on the carcinogenicity of the consumption of red meat and processed meat*. Retrieved from http://www.who.int/features/qa/cancer–red–meat/en/

Wu, J., Zeng, R., Huang, J., Li, X., Zhang, J., Ho, J. C.-M., & Zheng, Y. (2016). Dietary protein sources and incidence of breast cancer: A dose–response meta–analysis of prospective studies. *Nutrients*, *8*(11), 730. doi:10.3390/nu8110730 PMID:27869663

Zhang, L., Li, X., Ma, B., Gao, Q., Du, H., Han, Y., ... Qiao, Z. (2017). The tartary buckwheat genome provides insights into rutin biosynthesis and abiotic stress tolerance. *Molecular Plant*, *10*(9), 1224–1237. doi:10.1016/j.molp.2017.08.013 PMID:28866080

Zhang, Z. L., Zhou, M. L., Tang, Y., Li, F. L., Tang, Y. X., Shao, J. R., ... Wu, Y. M. (2012). Bioactive compounds in functional buckwheat food. *Food Research International*, *49*(1), 389–395. doi:10.1016/j.foodres.2012.07.035

Zhu, F. (2016a). Buckwheat starch: Structures, properties, and applications. *Trends in Food Science & Technology*, *49*, 121–135. doi:10.1016/j.tifs.2015.12.002

Zhu, F. (2016b). Chemical composition and health effects of Tartary buckwheat. *Food Chemistry*, *203*, 231–245. doi:10.1016/j.foodchem.2016.02.050 PMID:26948610

Zhu, F. (2018). Chemical composition and food uses of teff (Eragrostis tef). *Food Chemistry*, *239*, 402–415. doi:10.1016/j.foodchem.2017.06.101 PMID:28873585

Zhu, F., Du, B., Li, R., & Li, J. (2014). Effect of micronization technology on physicochemical and antioxidant properties of dietary fibre from buckwheat hulls. *Biocatalysis and Agricultural Biotechnology*, *3*(3), 30–34. doi:10.1016/j.bcab.2013.12.009

ADDITIONAL READING

Abugoch James, L. E. (2009). Quinoa (Chenopodium quinoa Willd.): Composition, chemistry, nutritional, and functional properties. *Advances in Food and Nutrition Research*, *58*, 1–31. doi:10.1016/S1043-4526(09)58001-1 PMID:19878856

Bar–El Dadon, S., Abbo, S., & Reifen, R. (2017). Leveraging traditional crops for better nutrition and health – The case of chickpea. *Trends in Food Science & Technology*, *64*, 39–47. doi:10.1016/j.tifs.2017.04.002

Baye, K. (2014). *Teff: Nutrient composition and health benefits*. Ethiopia Strategy Support Program. Workig Paper 67. Retrieved from https://www.researchgate.net/profile/Kaleab_Baye2/publication/266316373_Teff_Nutrient_Composition_and_Health_Benefits/links/542c24320cf277d58e8b0b42.pdf

Caruso, M. C., Favati, F., Di Cairano, M., Galgano, F., Labella, R., Scarpa, T., & Condelli, N. (2018). Shelf–life evaluation and nutraceutical properties of chia seeds from a recent long–day flowering genotype cultivated in Mediterranean area. *Lebensmittel-Wissenschaft + Technologie*, *87*, 400–405. doi:10.1016/j.lwt.2017.09.015

Cokkizgin, A., & Shtaya, M. J. Y. (2013). Lentil: Origin, cultivation techniques, utilization and advances in transformation. *Agricultural Science*, *1*(1), 55–62. Retrieved from http://www.todayscience.org/AS/article/as.v1i1p55.pdf. doi:10.12735/as.v1i1p55

Hayat, I., Ahmad, A., Masud, T., Ahmed, A., & Bashir, S. (2014). Nutritional and health perspectives of beans (Phaseolus vulgaris L.): An overview. *Critical Reviews in Food Science and Nutrition*, *54*(5), 580–592. doi:10.1080/10408398.2011.596639 PMID:24261533

Marsh, K., Zeuschner, C., & Saunders, A. (2018). The health impact of eating foods of animal origin: Evidence regarding animal foods, health, and disease risk. In D. Bogueva, D. Marinova, & T. Raphaely (Eds.), *Handbook of research on social marketing and its influence on animal origin food product consumption* (pp. 283–297). Hershey, PA: IGI Global. doi:10.4018/978-1-5225-4757-0.ch002

Ovando, C. A., de Carvalho, J. C., Vinícius de Melo Pereira, G., Jacques, P., Soccol, V. T., & Soccol, C. R. (2018). Functional properties and health benefits of bioactive peptides derived from Spirulina: A review. *Food Reviews International*, *34*(1), 34–51. doi:10.1080/87559129.2016.1210632

Ramdath, D., Padhi, E., Sarfaraz, S., Renwick, S., & Duncan, A. (2017). Beyond the cholesterol–lowering effect of soy protein: A review of the effects of dietary soy and its constituents on risk factors for cardiovascular disease. *Nutrients*, *9*(4), 324. doi:10.3390/nu9040324 PMID:28338639

Zhu, F. (2016a). Buckwheat starch: Structures, properties, and applications. *Trends in Food Science & Technology*, *49*, 121–135. doi:10.1016/j.tifs.2015.12.002

KEY TERMS AND DEFINITIONS

Beans: Edible, nutritious, widely grown seeds of various plants of the legume family, especially of the genus *Phaseolus*, usually oval or kidney-shaped.

Buckwheat: An edible plant (*Fagopyrum esculentum*) with high nutritional properties cultivated for its triangular grain-like seeds used as a food source.

Chia: An annual edible plant (*Salvia hispanica*) of the mint family, native to South America, and used for its nutritional values; its seeds are predominantly used as a food source.

Chickpea: A round yellowish edible seed (*Cicer arietinum*), a legume of the family *Fabaceae*, widely used as a pulse.

Lentils: A high-protein pulse, widely cultivated in Eurasian countries, an annual leguminous plant (*Lens culinaris*) with flattened edible seeds.

Nutritional Physiology: Deals with the study of nutrients, their role in the growth and health of different types of food and their effect on methabolism.

Plant-Derived Proteins: Foods obtained from plant sources, including vegetables, whole grains, nuts, seeds, legumes, and fruits, with no animal products.

Quinoa: A plant of the goosefoot family originally found in the Andes, where it is widely cultivated for its small, edible, starchy, ivory-colored seed, which is used as a food staple.

Soybean: Annual Asian legume (*Glycine max*) widely grown for its oil-rich proteinaceous seeds, for forage and soil improvement.

Spirulina: A biomass of cyanobacteria (blue-green algae) that can be consumed as food or nutritional supplement.

Teff: An African cereal native to Ethiopia, cultivated as a staple food crop, for flour and used for making traditional fermented breads.

Section 2
New Meat Alternatives

Chapter 5
Clean Meat:
Will We Brew Our Steaks in the Near Future Without Killing Animals?

Kurt Schmidinger
University Vienna, Austria & Institute for Philosophy, Austria

ABSTRACT

This chapter presents the production of real animal meat, which is grown outside of an animal. Starting cells are grown to meat products with the aid of tissue engineering techniques, a process with many names: "Lab meat," "in vitro meat," "cultured meat," or "clean meat." The chapter gives an overview of the technology and—maybe even more interesting for many readers—shows who were and who are the major players behind clean meat, with many well-known persons among them. Finally, the chapter shows in which ways clean meat could outperform conventional animal-derived meat and so overcome the obstacles of little consumer acceptance, which can be expected initially.

INTRODUCTION

The current mass production of livestock causes serious problems to the world – to the environment and climate, to global health and individual health, to world nutrition and last but not least, it causes suffering of billions of sentient beings that are ruthlessly reduced to meat production units. Environmentally, livestock production is a, or the, leading factor in land use, water consumption, water pollution, rainforest destruction, climate change, loss of biodiversity and soil erosion (Steinfeld, Gerber, Wassenaar, Castel, Rosales, & de Haan, 2006). Most of this can be easily explained by looking at the food chain of industrial livestock production systems. On average, approximately 7 calories that go into this system are converted into only 1 calorie of meat, dairy or eggs in these systems. Also, as the vast majority of livestock is nowadays exclusively fed with food we could as well use for direct human consumption, this is the largest loss of food calories in our food system after the harvest of the crops (see the impressive Figure 11.9 on page 836 in IPCC, 2014).

DOI: 10.4018/978-1-5225-7350-0.ch005

Such overwhelming loads of problems associated with the current production of animal products have motivated masterminds and financially potent patrons to search for revolutionary alternative nutritional concepts. Among these are many technologies aimed to simulate and replace meat, dairy and eggs with plant based products. Start-ups like the Chilean "Not Company" make use of artificial intelligence to design new plant based foods to replace animal products. There are also ideas and concepts to convert straw and harvest wastes, in other words to convert mainly inedible cellulose, into something edible for humans by using algae or other microorganisms in so called "biofermenters" (Schmidinger, 2012).

However, another approach is presented in this chapter – the production of real animal meat, which is grown outside of an animal. Starting cells form meat products with the aid of tissue engineering techniques, a process with many names: "lab meat", "in vitro meat", "cultured meat" or "clean meat" (see Figure 1). The term "clean meat" will be used in this chapter, which gives an overview of the technology and – maybe even more interesting for many readers – shows who are the major players behind clean meat.

CLEAN MEAT: WILL WE BREW OR PRINT OUR STEAKS IN THE NEAR FUTURE?

This chapter gives an overview of the state of the art of technical approaches to produce clean meat out of cells from animal origin using tissue engineering technologies. It then shows who works in this futuristic field and who supports financially the research. Finally, a discussion of the obstacles gives a picture of how and under which circumstances clean meat might revolutionize the nutrition of the upcoming human generations.

Figure 1. An image picture of lab grown meat (this is not really what it looks like) used by the Austrian web initiative futurefood.org

Source: (http://www.futurefood.org/in-vitro-meat/index_en.php; design C. Braun & A. Schmidt, 2006)

Cells, Media, Growth Factors, Bioreactor, Scaffolds, Bioprinter and More: The Technology Behind Clean Meat

The first very detailed description of the technology of clean meat production was done in 2009 in the Netherlands, the leading country of research in the first decade of this century (Haagsman, Hellingwerf, & Roelen, 2009). At the time, this was "a completely new idea to produce edible skeletal muscle (i.e. meat) by culturing and differentiating stem cells of farm animal species to skeletal muscle cells" (Haagsman, Hellingwerf, & Roelen, 2009, p. 4).

The making of processed meats, such as burgers, nuggets or sausages is easier to do in vitro (Datar & Betti, 2010), whereas a steak with its complicated fibrous structures is supposed to be a much more complex application. With the status quo as in spring 2018, only processed in vitro meats such as burgers and nuggets have already been produced in laboratories, no steaks or schnitzels yet. According to Datar and Betti (2010, p. 13), the search for alternative ways to produce meat "is driven by the growing demand for meat and the shrinking resources available to produce it by current methods. A recent summary of the technological basics for in vitro meat production is provided by the Good Food Institute (see Specht, 2017).

The schema of the process is shown in Figure 2 (see also Woll & Böhm, 2018). It starts with cells from animal species from which clean meat should be produced. Ideally, the starter cells are taken from living animals by biopsy without killing or even severely harming the individual animal at all. In the process of creating clean meat, different forms of cells have to be distinguished. Stem cells have the ability of proliferating indefinitely and differentiating into multiple cell types. Hence, they are a good option as starter cells. Within the clean meat process, cells are then differentiated and matured into the cell types required for meat, that is mainly muscle cells. Another option for cells used within the process of the clean meat production would be cells between the extremes of stem cells and fully defined muscle cells. Such cells, for example, myoblast cells, proliferate at an acceptable pace (but not as much as stem cells) and are at the same time sufficiently differentiated (but not as much as muscle cells).

Genetic modification is one possibility to direct cells to divide more quickly and to stimulate them to differentiate according to the clean meat requirements, such as the need to produce certain cell types or the requirement for cells to prosper in certain environmental conditions. However, for consumer acceptance reasons, alternative ways of directing cells without the use of genetic modification need to be explored in more detail.

Once obtained, the cells have to be fed with something in which to grow and proliferate. This brings us to the next issue – the culture media. These media must be cheap and efficient enough to let the cells grow effectively. Because clean meat grows without the animal, its digestive organs and the enzymes there, the medium must supply the cells directly with the required nutrients. Also, it's best for these culture media not to include animal based ingredients. Hence, the culture media contain proteins or amino acids, fats, salts and pH buffers and should also guarantee a healthier meat compared to conventional meat from farmed animals. Higher vitamin contents, less cholesterol or saturated fatty acids, less arachidonic acid, but more omega 3 fatty acids – these are just a few examples of the way clean meat could and should be grown. Part of the culture medium are the so-called growth factors. These growth factors are signalling the molecules that direct the cells in certain ways of development. For example, they guide the cells to become muscle cells or fat cells. Bovine serum has been mainly used so far for this purpose, but increasingly it is being phased out in tissue engineering. For example, it is being replaced in regenerative medicine – a progress that also helps in the clean meat production, where animal serums

Figure 2. Schematic process of clean meat production
Source: (Visionen von In-vi- tro-Fleisch (VIF). Project at the Institute for Technology Assessment and Systems Analysis, Karlsruhe Institute of Technology)

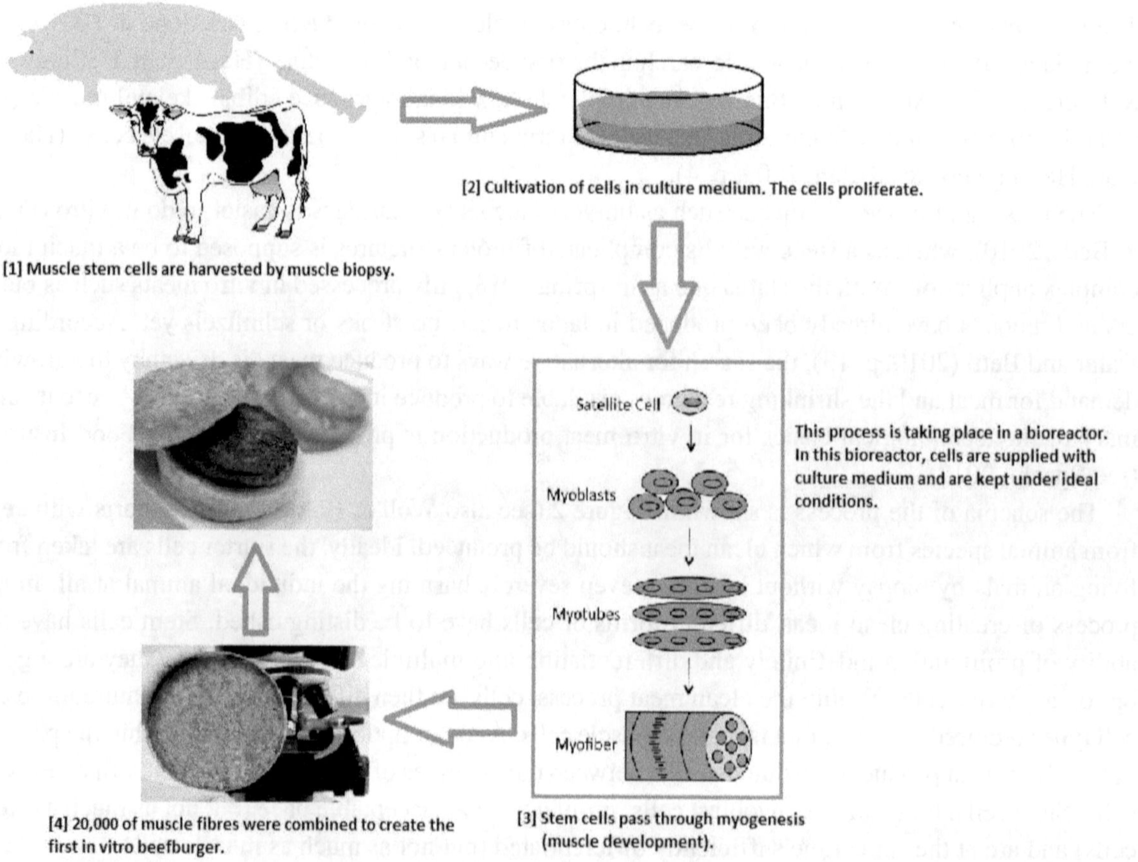

are a no-go. The growth factors are a relevant economic parameter for clean meat, and unlike in medical application, costs are an overwhelmingly important consideration in clean meat production. Clean meat will have to compete with factory farmed meat, which is produced extremely cheaply. Hence, alternatives to growth factors, such as peptides or synthetic biology approaches, are being discussed (Specht, 2017).

The next issue are scaffolds. How do the nutrients reach the cells? Also, what helps the cells to form three-dimensional structures? In nature, a cm^3 of meat is permeated by hundreds of metres up to several kilometres of small blood vessels or capillaries. To simulate something that is so complicated in vitro would be extremely challenging and would make clean meat unaffordable. So, for clean meat production, other concepts have to be found to bring the nutrients directly to the cells where they are required. One option is an edible scaffold with a huge surface that would not need to be extracted from the end product, that is, an edible material that is cheap and abundantly available. To simulate the muscle exercise, the scaffold might shift its form. Stimuli-sensitive scaffolds made of non-animal-source alginate, chitosan or collagen are possible options. Changes in temperature or the pH level can induce such periodical stretches of the clean meat muscle fibres. Engineered hydrogels might also assist in the cellular differentiation process, and therefore variations in such hydrogels might assist to achieve marbled structures similar to that in a steak.

Bioprinting cells layer for layer could be an alternative to the use of the scaffolds that also ensures that the media can reach the cells. It can also be used to produce scaffolds with variable pore sizes and microstructures.

Finally, a place is required, where the cells, the media and growth factors together with the scaffolds can be brewed into the final clean meat product. This place is the *bioreactor*, where small and large fibres of muscle cells are produced plus connective tissue which generats collagen and elastin as well as taste affecting fat cells form clean meat. There might be more than one type of bioreactor involved in the clean meat production. The first bioreactor is for the proliferation of cells, probably a big stirred tank reactor. The cells are then harvested from this bioreactor and transferred to another one, used to perfuse media through cell-seeded scaffolds and produce clean meat (see Figure 3).

One requirement for such bioreactors is media recycling to reduce inputs, waste and costs. Sensors and software are also essential to monitor the whole bioreactor content, the media components, temperature and pH and to filter out waste as well as to react to maintain the optimum process conditions.

HISTORICAL PROTAGONISTS OF THE IDEA OF CLEAN MEAT

In this historical review, the term "cultured meat" is mainly used instead of "clean meat", to make the point that by that time, the latter term was not yet coined. Early patents on cultured meat have been secured already in the last century and the beginning of this century by Dutch visionaries (e.g. Vein, 2004; Van Eelen, 2007), with Van Eelen's patent dating back to 1997 (Böhm, Ferrari, & Woll, 2017).

However, the early moments of the vision of clean meat go far back to the 20th century. Winston Churchill is often quoted of saying: "Fifty years hence, we shall escape the absurdity of growing a whole chicken in order to eat the breast or wing by growing these parts separately under a suitable medium" (Churchill, 1931).

Figure 3. Schematic use of two kinds of bioreactors – a cell proliferation bioreactor to harvest cells for the clean meat production, and a tissue perfusion bioreactor; in the latter, the cells are perfused to grow the final product (that is, clean meat) on scaffolds
Source: (Good Food Institute, Specht, 2017)

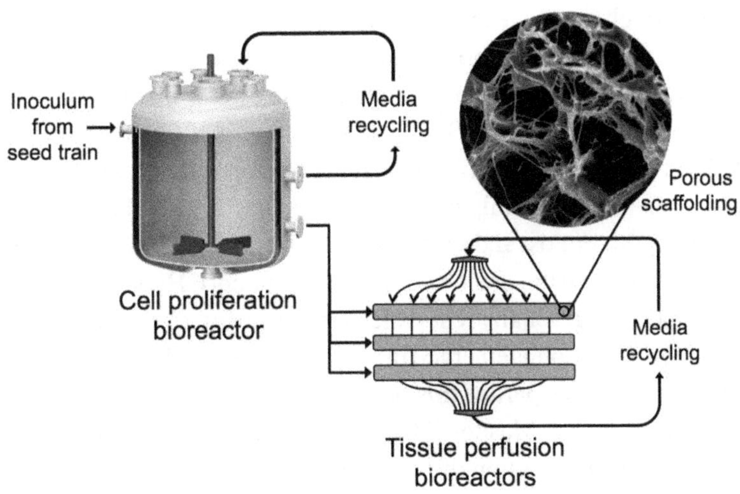

Around the millennium, different researchers appeared in articles on the scene, the majority of them being from the Netherlands. In the years until 2014, the Netherlands remained the leading country of in vitro meat research, with public grants for studies awarded to researchers like Henk Haagsman and Bernard Roelen. In parallel, Vladimir Mironov, Nick Genovese and others presented theoretical concepts for cultured meat in the first years of the 21st century, also introducing the idea of bioprinting meat and many other concepts to a mainly scientific community. Nick Genovese later on cofounded the actual leading company in clean meat – "Memphis Meats" (refer to next section of this chapter).

In the years after the millennium, there was no prototype yet to present to the public, which made it hard for symposiums to gain much public attention, and also difficult for researchers to find funding. Consequently, other concepts were chosen, such as showing micro pieces of lab grown meat as an artist group, like the Australians Oron Catts and Ionat Zurr did with their "Tissue Culture and Art" project (The Tissue Culture and Art Project, n.d.).

Around 2005, the first non-profit projects and websites on cultured meat went on air, which were "New Harvest" in the US by Jason Matheny and "Future Food" in Austria, led by Kurt Schmidinger (the authors of this chapter) – both were run by volunteers. "New Harvest" later hired employees and has grown to be a major non-profit organisation in the field (further details about their research will be presented in the next section of this chapter).

In April 2008, the first in vitro meat symposium was held in Aas, Norway, which was organised by a university team led by Stig Omholt. "In vitro meat" was the most used expression, but from that time on it was decided to use "cultured meat" as the preferred term. It has since become the most popular term. The Norwegian attempts to become pioneers in the field however did not last long, the same happened to the Swedish attempts a few years later.

The milestone into a new era of the clean meat research was the presentation by Mark Post of the first burger produced in vitro in London in August 2013 (The Guardian, 2013). As Mark Post is still among the most active players in this field on the planet, this is a good crossover from recent history to the present times of the clean meat research.

ACTUAL MAJOR PLAYERS IN THE CLEAN MEAT DEVELOPMENT

As mentioned above, a milestone in the clean meat development was the presentation of the first real burger produced in vitro, which was done by Mark Post and his team together with some guests at a large forum in London in August 2013. Mark Post was supported by the Google co-founder Sergey Brin, but that had been kept a secret until the public presentation. The cost of this first in-vitro-meat burger was published in the worldwide media at over US$ 300 000 (reported as €250,000 in The Guardian, 2013).

The media hype about this worldwide first real meat burger that had not been made from a killed animal was enormous. It is likely that this media coverage has heavily helped the funding of further research, and not just that of Mark Post and his team in the Netherlands. From that time on, a new era of research groups emerged, with new dimensions of financial support. However, aside from Mark Post, in recent years the Netherlands has gradually lost its leadership position on clean meat, mainly to the US.

Many companies and research groups emerged in recent years, especially in the US. "Modern Meadow" with Andras and Gabor Forgacs was part of this new wave of researchers, backed by PayPal cofounder the Thiel-foundation. First, "Modern Meadow" focused on the idea to bioprint meat, more recently, the company has moved over to the production of leather in vitro.

Most of the US-research now happens in California. IndieBio in San Francisco has become a real hotspot in the development of food in vitro, with "Clara Foods" – a company that works on real eggs without hens, or "Finless Foods" – a company that works on fish meat without killing fish, located there. IndieBio also helped in the starting phase of "Perfect Day Foods", a company which tries to produce real milk without cows. It was also the birthplace of "Memphis Meats", the company which by the spring of 2018 is seen as the global leader in the field of meat produced in vitro. Since 2015, "Memphis Meats" has come up with several cell grown meatballs, chicken parts and more, and has presented more clean meat samples than any other company on the globe up to now. The costs for the product has gone down significantly to below one percent compared to the initial products from Mark Post. Yet, a few hundred US$ per kg of meat is still far too expensive, but the tendency is promising. The CEO of "Memphis Meats" Uma Valeti holds speeches all over the world. In August 2017, the start-up received a US$ 17 million grant from the Microsoft founder Bill Gates, the Virgin Group founder Richard Branson, the agricultural giant Cargill Inc and others.

The gold rush atmosphere in clean meat research in California is also heated up by the announcement of the CEO of "Hampton Creek" Josh Tetrick in the summer of 2017 to bring clean meat to the market in 2018. "Hampton Creek" is famous for its brand "Beyond Eggs" and its remarkable plant based egg replacement products, and is also backed by Bill Gates and others. As they have not been visible in the clean meat field before, this announcement was quite surprising for the public.

Finally, in the most recent years Israel has become another major player in the clean meat field. The progress that the "Modern Agriculture Foundation" made on clean meat development has led to the creation of a company named "SuperMeat", led by the bioengineering professor Yaakow Nahmias. A second research group in Israel is "TheKitchenHub" and their clean meat project is "Meat the Future".

The clean meat-development is accompanied by several non-profit organisations. One outstanding example is the already mentioned group "New Harvest", which has helped much in the start-up process of most of the US-based companies. "New Harvest" only focuses on cultured meat, egg and dairy technologies. Another more recent organisation is the also already cited "Good Food Institute" (GFI). The GFI focuses on both, plant based alternatives to animal products as well as clean meat. Among many other activities, the GFI has established open source materials that should encourage new scientists to join these seminal scientific fields.

WILL CLEAN MEAT COME OUT ON THE TOP AND IF SO, WHEN?

There are many reasons for clean meat to become the dominant form of meat alternatives. They relate to environmental and animal welfare concerns, health considerations and the cost of the products.

Effects on Ecology and Animal Welfare

Clean meat is supposed to solve many problems as mentioned in the introduction of this chapter. In terms of ecology, useful ecobalances are not yet at hand, simply as the production process of clean meat is not well established. Ecobalances like life cycle assessments are only applicable to existing processes for which they are calculated. Thus, existing ecobalances on clean meat can be seen more like rough estimates. Such estimates, as done by Tuomisto et al. (2011), Toumisto et al. (2014) and Mattick et al.

(2015), indicate huge reductions of greenhouse gas (GHG) emissions, land use or water use for clean meat production compared to conventional beef, pork or poultry.

In terms of animal welfare, clean meat would dramatically reduce the numbers of required animals. There are no certain figures about that, and it is not yet clear, if still animals will need to be killed or if biopsy of starting cells from living animals will be sufficient. However, it is clear that the living conditions of the used animals can improve considerably without a major effect on the costs of the clean meat end products.

If we assume very conservatively that the donor cells taken from one pig can be used to replace the amount of meat nowadays derived from one million of pigs, then this one donor-pig would help producing approximately 60 million kilograms of meat. Even, if one invested US$ 60000 in excellent housing and feed for this donor pig, this perfect treatment of the donor pig would only add US$ 0.001 to the price of one kilogram of clean pig meat, which is an irrelevant sum. As the costs of quality housing and feed for donor animals play a negligible role in the price of the final clean meat product, it only makes sense to keep them under best conditions, and secure a major plus factor for clean meat in the public perception without relevant costs in the end product.

Health Issues

Another advantage of clean meat over conventional meat from livestock should be health. First of all, clean meat developers aim at making the process independent of antibiotics, avoiding the rise of antibiotic resistant germs. Industrial livestock farming heavily relies on antibiotic usage to keep the animals relatively healthy despite the bad conditions in which they have to live (The Pew Charitable Trusts, 2013; EFSA, 2011). Also, livestock farming is assumed a central source of new pandemics (Capua & Alexander, 2006), which also can be avoided with clean meat.

On a less global and more individual health level, the ingredients of clean meat can be influenced much easier compared to livestock meat. It is obvious that it is easier to change clean meat by altering the culture media than to change livestock meat by altering the feeding and by using slow breeding selections. So, health goals of clean meat could and should be: to avoid or reduce the unhealthy components of conventional meat, such as cholesterol, saturated fatty acids, arachidonic acid, Neu5GC, purines, free radicals (causing oxidative stress), pathogenic germs, etc.; as well as to improve the contents of certain vitamins, minerals, or essential fatty acids like omega 3s, maybe even optimize the amino acid-composition and thus the protein-quality.

Considering all this might offer tremendous advantages and justification for the consumption of clean meat. However, is the public ready to accept it?

How Much Will Clean Meat Cost? What Will It Taste Like? How Will the Public Respond?

The price of clean meat has gone down in recent years from US$ 325000 for a hamburger to US$ 11.36 at Mark Post's team (Böhm, Ferrari, & Woll, 2017). Still, as the prices come from the high end, initial applications such as pet food will be no option for clean meat, as most owners prefer rather cheap options for pet food. It is more useful to think about applications that would justify a high price audience – something fancy, something which cannot be offered in that form by conventional livestock meat production. An example could be a crocodile-kangaroo-burger, or – with the advancement of the available technology– a crocodile-kangaroo-steak. Of course, this will be achieved without harming any crocodile or kangaroo or cattle, just taking cells from these species via biopsy and letting them grow to fancy combinations that justify a high price and attract celebrities who then make clean meat presentable and also ensure large and positive media coverage.

From a technological point of view, the first applications will probably be clean meat hybrid products or burgers and nuggets without complex structures. Scaffolds can be simple here and only proliferation bioreactors are required. By contrast, a steak will be much more of an effort, so it will be a later application. A steak requires different cell types and structures, different starter cells or a variable differentiation of cells to form complex structures on specialized scaffolds in a more advanced bioreactor that can achieve this.

Considering the mentioned environmental, ethical or health advantages of clean meat might help to overcome a technophobe public attitude towards new forms of meat production. However, like anywhere in the marketing industry, the production process will not be in the focus of advertising anyway. Nowadays industrial meat production would be considered totally unnatural, if the real production process in industrial livestock facilities or slaughterhouses were realistically portrayed in commercials. Instead, meat is marketed with images that are far from these production processes. So, clean meat marketing is likely to do the same, transporting images and feelings, instead of portraying the production process in the bioreactors. Still, criticism of "techno food" will arise, and the clean meat community has taken the line to deal with that fact. Calling the product "clean meat" and not "in vitro meat" is such a strategy. Getting rid of the "laboratory feeling" is another. New blends or sorts of beer are prototyped in a lab, but later produced in a brewery. No one would therefore call it "lab beer". Clean meat will have a similar development, it will be prototyped in laboratories, and later on brewed in bioreactors that can be compared with brewery-like tanks. So, if beer is not called "lab beer", clean meat should not be called "lab meat".

These are just a few measures that show a way how to overcome the expected huge scepticism about clean meat (the Eurobarometer-survey in the EU exemplarily shows such public scepticism, see European Commission, 2005). Finally, taste, texture, aroma, price, image, advertising and health issues will highly influence the success of clean meat!

DISCUSSION AND CONCLUSION

Taking into concern all the technical and economic challenges and the expectable initial public distrust as well as the status quo of research described in this chapter, some scenarios for an introduction of clean meat into the global markets are quite realistic and others are not. They are briefly outlined below:

- Starting with pet food, for example, is unrealistic as the clean meat technology will be quite expensive at first. So, cheap applications like pet food are not an option for a first introduction of clean meat.
- Starting with complex meat applications like steaks is also unrealistic due to the technical challenges associated with the complex structure and texture of steaks.
- Starting with normal processed meats like common sausages or burgers will be too unspectacular to justifiy initial high prices and to attract celebrities to act as pioneer consumers.

Processed, fancy meats from species that are normally not available for consumption like kangaroo or crocodile or maybe even cruelty free foie gras could be applications that are technically feasible and not too complex. On the other hand, they can justify a high price and could be attractive for opinion setters and multipliers like celebrieties. Concerns could arise from the animal rights or species conservation movement, if foods that are normally produced in a very cruel way or that normally threaten biodiversity, all of a sudden become accepted and widespread. The qualms could be that such clean meat products could as well place a public scope on the animal derived "originals" of foie gras or crocodile meat and push their consumption, especially if the clean meat versions turn out to fail the expectations on taste, aroma and texture.

Hence, the ideal clean meat product for the first market introduction will need to be selected carefully.

What has already been agreed is that it will be important to communicate the massive advantages of clean meat over livestock meat in terms of ecological balances (water consumption, land area demand, greenhouse gases, etc.), world nutrition (more efficient), health (reduced global risk of new pandemics or antibtiotic resistant germs, individual issues such as reduction of lifestyle diseases) and of course animal welfare. Very important in overcoming public scepticism is also the wording and attempts to get away from the "laboratory-feeling" of clean meat and make it a normal food option.

REFERENCES

Böhm, I., Ferrari, A., & Woll, S. (2017). *In-vitro-Fleisch: Eine technische Vision zur Lösung der Probleme der heutigen Fleischproduktion und des Fleischkonsums? Institut für Technikfolgenabschätzung und Systemanalyse*. Karlruhe, Germany: ITAS.

Capua, I., & Alexander, D. J. (2006). The challenge of avian influenza to the veterinary community. *Avian Pathology, 35*(3), 189–205. doi:10.1080/03079450600717174 PMID:16753610

Churchill, W. (1931). Fifty years hence. *Strand Magazine, UK*. Retrieved Sepember 10, 2018 from http://rolandanderson.se/Winston_Churchill/Fifty_Years_Hence.php

Datar, I., & Betti, M. (2010). Possibilities for an in vitro meat production system. *Innovative Food Science & Emerging Technologies, 11*(1), 13–22. doi:10.1016/j.ifset.2009.10.007

European Commission. (2005). *Social values, science and technology*. Brussels: Directorate General Research, Directorate General Press and Communication. Retrieved September 10, 2018 from http://ec.europa.eu/commfrontoffice/publicopinion/archives/ebs/ebs_225_report_en.pdf

European Food Safety Authority (EFSA). (2011). Scientific opinion on the public health risks of bacterial strains producing extended-spectrum β-lactamases and/or AmpC β-lactamases in food and food-producing animals. *EFSA Journal, 9*(8), 95. doi:10.2903/j.efsa.2011.2322

Haagsman, H. P., Hellingwerf, K. J., & Roelen, B. A. J. (2009). *Production of animal proteins by cell systems. Desk study on cultured meat ("Kweekvlees")*. Utrecht: University Utrecht, Faculteit DiergeneeskFaculty of Veterarinary Medicine. Retrieved September 10, 2018 from http://citeseerx.ist.psu.edu/viewdoc/download?doi=10.1.1.732.9877&rep=rep1&type=pdf

Intergovernmental Panel on Climate Change (IPCC). (2014). *Climate Change 2014: Mitigation of Climate Change. Contribution of Working Group III to the Fifth Assessment Report of the Intergovernmental Panel on Climate Change*. Cambridge, UK: Cambridge University Press. Retrieved from http://www.ipcc.ch/pdf/assessment-report/ar5/wg3/ipcc_wg3_ar5_full.pdf

Mattick, C., Landis, A., Allenby, B., & Genovese, N. (2015). Anticipatory life cycle analysis of in vitro biomass cultivation for cultured meat production in the United States. *Environmental Science & Technology*, *49*(19), 11941–11949. doi:10.1021/acs.est.5b01614 PMID:26383898

Schmidinger, K. (2012). *Worldwide alternatives to animal derived foods – overview and evaluation models: Solutions to global problems caused by livestock* (Doctoral dissertation). University of Natural Resources and Life Sciences, Vienna, Austria.

Specht, L. (2017). *Mapping emerging industries. Opportunities in clean meat.* The Good Food Institute. Retrived December 13, 2017, from www.gfi.org/images/uploads/2017/06/Mapping-Emerging-Industries.pdf

Steinfeld, H., Gerber, P., Wassenaar, T., Castel, V., Rosales, M., & de Haan, C. (2006). *Livestock's long shadow. Environmental issues and options.* Rome, Italy: Food and Agriculture Organization of the United Nations (FAO). Retrieved September 10, 2018 from http://www.fao.org/docrep/010/a0701e/a0701e00.HTM

The Guardian. (2013). First lab-grown hamburger gets full marks for 'mouth feel'. *The Guardian Australian Edition.* Retrieved September 10, 2018 from https://www.theguardian.com/science/2013/aug/05/world-first-synthetic-hamburger-mouth-feel

The Pew Charitable Trusts. (2013). *Industrial food animal production in America.* Retrived December 27, 2017 from https://www.jhsph.edu/research/centers-and-institutes/johns-hopkins-center-for-a-livable-future/_pdf/research/clf_reports/CLF-PEW-for%20Web.pdf

The Tissue Culture and Art Project. (n.d.). Retrieved Septemver 10, 2018 from http://lab.anhb.uwa.edu.au/tca/

Tuomisto, H., Ellis, M., & Haastrup, P. (2014). Environmental impacts of cultured meat: Alternative production scenarios. *Proceedings of the 9th International Conference on Life Cycle Assessment in the Agri-Food Sector*, 1360–1366. Retrieved September 10, 2018 from http://lcafood2014.org/papers/132.pdf

Tuomisto, H., & Teixeira de Mattos, M. J. (2011). Environmental impacts of cultured meat production. *Environmental Science & Technology*, *45*(14), 6117–6123. doi:10.1021/es200130u PMID:21682287

Van Eelen, W. F. (2007). *Industrial scale production of meat using cell culture methods*. US Patent Number US7270829B2. Retrieved September 10, 2018 from https://patents.google.com/patent/US7270829B2/en

Vein, J. (2004). *Method for producing tissue engineered meat for consumption.* US Patent Number US6835390B1. Retrieved September 10, 2018 from https://patents.google.com/patent/US6835390B1/en

Woll, S., & Böhm, I. (2018). In-vitro meat: A solution for problems of meat production and meat consumption? *Ernaehrungs Umschau International, 65*(1), 12–21. Retrieved September 10, 2018 from https://www.ernaehrungs-umschau.de/fileadmin/Ernaehrungs-Umschau/pdfs/pdf_2018/01_18/EU01_2018_Special_invitro_englisch.pdf

ADDITIONAL READING

Datar, I., & Betti, M. (2010). Possibilities for an in vitro meat production system. *Innovative Food Science & Emerging Technologies, 11*(1), 13–22. doi:10.1016/j.ifset.2009.10.007

Mattick, C., Landis, A., Allenby, B., & Genovese, N. (2015). Anticipatory life cycle analysis of in vitro biomass cultivation for cultured meat production in the United States. *Environmental Science & Technology, 49*(19), 11941–11949. doi:10.1021/acs.est.5b01614 PMID:26383898

Schmidinger, K. (2012). *Worldwide alternatives to animal derived foods – overview and evaluation models: Solutions to global problems caused by livestock* (Doctoral dissertation). University of Natural Resources and Life Sciences, Vienna, Austria.

Schmidinger, K., Bogueva, D., & Marinova, D. (2018). New meat without livestock. In D. Bogueva, D. Marinova, & T. Raphaely (Eds.), *Handbook of research on social marketing and its influence on animal origin food product consumption* (pp. 344–361). Hershey, PA: IGI Global. doi:10.4018/978-1-5225-4757-0.ch023

Specht, L. (2017). *Mapping emerging industries. Opportunities in clean meat.* The Good Food Institute, Washington DC, USA. Retrived December 13, 2017, from www.gfi.org/images/uploads/2017/06/Mapping-Emerging-Industries.pdf

The Guardian. (2013). *First lab-grown hamburger gets full marks for 'mouth feel'*. The Guardian Australian Edition. Retrieved September 10, 2018 from https://www.theguardian.com/science/2013/aug/05/world-first-synthetic-hamburger-mouth-feel

The Pew Charitable Trusts. (2013). *Industrial food animal production in America.* Retrived December 27, 2017 from https://www.jhsph.edu/research/centers-and-institutes/johns-hopkins-center-for-a-livable-future/_pdf/research/clf_reports/CLF-PEW-for%20Web.pdf

Tuomisto, H., Ellis, M., & Haastrup, P. (2014). Environmental impacts of cultured meat: Alternative production scenarios. *Proceedings of the 9th International Conference on Life Cycle Assessment in the Agri-Food Sector* (pp. 1360–1366). Retrieved September 10, 2018 from http://lcafood2014.org/papers/132.pdf

Woll, S., & Böhm, I. (2018). In-vitro meat: A solution for problems of meat production and meat consumption? Ernaehrungs Umschau International, 1, 12–21. Retrieved from https://www.ernaehrungs-umschau.de/fileadmin/Ernaehrungs-Umschau/pdfs/pdf_2018/01_18/EU01_2018_Special_invitro_englisch.pdf

KEY TERMS AND DEFINITIONS

Clara Foods: US-based startup that works on the productions of in vitro egg proteins.

Clean Meat: Meat grown from animal cells in vitro and not derived from slaughtered animals.

Cultured Meat: Refer to clean meat.

Finless Foods: US-based startup that works on clean fish meat.

Future Food: Austrian initiative that supports alternatives to animal products (also including clean meat).

Good Food Institute: US-based organisation that supports alternatives to animal products (also including clean meat).

Hampton Creek: US-based company that produces alternatives to egg products and also works on clean meat.

In Vitro Meat: Refer to clean meat.

Mark Post: Scientist who was the first to produce and present a burger based on clean meat.

Memphis Meats: US-based startup that works on clean meat.

Modern Meadow: US-based startup that works on lab grown leather as well as on clean meat.

New Harvest: US-based organisation that supports the development of clean meat.

Perfect Day: US-based startup that works on the production of in vitro milk proteins.

Supermeat: Project and startup in Israel that works on clean meat.

Vegan: A human diet which excludes all animal-based products.

Vegetarian: A human diet which excludes all kinds of meat and fish.

Chapter 6
Microbial Protein:
An Essential Component for Future Food Security

Akash Saklani
Gurukula Kangri University, India

Navneet
Gurukula Kangri University, India

ABSTRACT

In the next 30–40 years, the food production system will face the challenge of increasing the production and availability of food products. Compared to the other foods, meat products are the least environmentally friendly. This chapter explores the opportunities of including microbial protein—the dried cells of microorganisms, such as algae, fungi, actinomycetes, and bacteria—in the food system to improve food security. Since ancient times, different microbes have been used as part of the diet all over the world. Recently, the term single cell protein gained popularity to describe the dibble single-cell microorganisms. The health benefits of such products are well-known, and the environmental impacts of their production are low. Emerging meat substitutes based on microbial proteins combined with the right technologies is one of the promising trends in food production that is analyzed in comparison with conventional proteins.

INTRODUCTION

Microbial protein is a form of dehydrated microbial culture of cells or a purified form of protein which is obtained from microorganisms, such as algae, fungi, yeast and bacteria. This source has the potential to provide protein to humans[1] and is already marketed as dehydrated, low fat content but rich of vitamins food (Garcia-Garibay, Gómez-Ruiz, Cruz-Guerrero, & Bárzana, 2014). Microbes always play a significant role in the processing of food by converting various fibres into edible form, for example, in fermenting dough, making sauerkraut, miso, tempeh as well as beer and many other fermented products (Caplice & Fitzgerald, 1999). Microbial proteins, as in the case of fungi and algae, have been used as a direct food source for millennia (Anupama & Ravindra, 2000). Microorganisms have been applied

DOI: 10.4018/978-1-5225-7350-0.ch006

in the processing of foods and to produce different types of compounds, such as organic acids, as well as enzymes from fermented food products. At present, microbes are applied in the manufacturing of biotechnological compounds ranging from antibiotics and industrial alcohol to cellular proteins. The ability of microbes to change less useful substrates into something that has higher value is an important approach in enhancing the quantity and quality of food.

With growing global population, projected to reach 10 billion by 2050 (Ezeh, Bongaarts, & Mberu, 2012), it is estimated that the world will need to produce 70% more food compared to 2006 (Ranganathan, 2013). This is not an easy task under the current food practices.

Food security is achieved “when all people, at all times, have physical, social and economic access to sufficient, safe and nutritious food that meets their dietary needs and food preferences for an active and healthy life” (Committee on World Food Security, 2012, p. 4). Many social and political issues can escalate when there is food insecurity and malnutrition, particularly in a developing country. Developing countries are also faced with continuously increasing population numbers, which makes it difficult to secure sufficient levels of high quality food. In some instances, this may result in a high percentage of malnourished people and food insecurity which ranges from food shortages to full-scale famines and when people live in fear of starvation and hunger.

Food security can be guaranteed only with adequate food that contains all nutrients and energy required by the human body. Food also should be safe and free from any contaminant or toxic materials. Considering the world population growth and available resources on this planet, the increase in food production must be achieved by using less labour, less water and less cultivated land. Microbial resources, used to make or modify food, have potential in improving the global food security scenario. They enhance properties, such nutritional value, taste, texture and shelf life and microorganisms are a natural source available everywhere in the world.

Microbial resources, however, are still considered a non-conventional source of food and proteins, especially in developing countries, but the demand for them is now on the increase. In comparison with other food sources, microbial protein is a great source of additional proteins in the human diet without representing a food material on their own. The microorganisms primarily act as agents of production which transform the principal raw material and improve its protein content generating high biological value (Kirsop,1985).

The main property of the microbial cell is that it can grow rapidly and accumulate a high amount of protein. This creates a stable source of protein and can be used for the cultivation of microorganisms under *in vitro* conditions yielding high amounts of biomass. Another advantage of microbial protein is that the process is less dependent on variations in weather, soil and climate conditions (Kuhad, 1997). The protein obtained from microbes has different nutritional value depending on the production processes which also need to be monitored for food safety. Although animal protein might be considered of high quality (Saima, Khan, Anjum, Ahmed, Rizwan, & Ijaz, 2008), there are numerous health and environmental problems associated with the production of this protein. By comparison, microbial protein has a very short growth cycle and higher protein content leading to rapid biomass production (Bekatorou, Psarianos, & Koutinas, 2006).

Microbial protein is also referred to as single-cell protein (SCP) because it is obtained from single-cell organisms which feed (or grow) on low-cost substrates (raw material). For example to grow protein, yeasts feed on carbohydrates (e.g. sugar) (Bekatorou, Psarianos, & Koutinas, 2006) while in the case of algae, this happens through the use of carbon dioxide (CO_2) and light (Garcia-Garibay, Gómez-Ruiz, Cruz-Guerrero, & Bárzana, 2014). Microbial protein appears to be a promising and reliable alternative source to strengthen future food security while minimizing the impact on the environment. It has two main advantages compared to animal-based proteins, namely it grows faster and the microbes utilise simpler feed (Goldberg, 1985).

Since the 1980s, attempts have been made to develop the production of high-quality protein from microbes. The Imperial Chemical Industries were the first to obtain a full-scale industrial production of Pruteen which was microbial protein produced from methanol oxidation as a new feed source for pigs (Braude, Hosking, Mitchell, Plonka, & Sambrook, 1977). However, the high content of nucleic acid made such products not suitable for human consumption. At present, research and development of microbial food is regaining momentum, both in the scientific and industrial domains, with the main focus being on direct human consumption. The environmental pressure associated with meat and dairy production coupled with the high demand for soybeans across the globe justifies the re-examination of the microbial alternative (Greenpeace, 2018).

Considering the importance of microbial proteins, this study aims to shed light on the development activities that have taken place by considering biotechnological and scientific approaches. The chapter describes the microbial protein as a treasure in relation to its potential for future food security by examining the importance of fungi, algae and yeast. Some plausible guidelines are put forward for the development of this non-conventional source of protein by considering its promise as a catalyst in developing food security.

METHODOLOGY

This chapter is based on a comparative review of various data. The reviewed literature was selected from sources, such as Web of Science, Google Scholar, and Scopus using searches with keywords related to microbial or single-cell protein. Particular attention is given to microbial protein developed from fungi, algae and yeast and its relevance for future food security. Their roles are evaluated in the terms of biotechnological approaches and effectiveness with an emphasis placed on improving the effectiveness of use of microbial protein.

WHY MICROBIAL PROTEIN FOR FUTURE FOOD SECURITY?

The growing world population can no longer depend on the current food production methods which prominently feature animal husbandry and fisheries as sources of protein. While developed countries are increasingly experiencing obesity problems with related to these non-communicable diseases, the developing world is still facing the problems of malnutrition, starvation, food insecurity and hunger-related diseases (Gabriel, Victor, & du Preez James, 2014). According to the World Health Organization up to 12 million people die from hunger and starvation each year across the globe, with the majority of deaths occurring in the developing world (Miller, 1985). The global number of undernourished people was 821 million in 2017 with the depth of the food deficit, that is, the difference between the required and available calories, being felt the most again in the developing world (Roser & Ritchie, 2017). Moreover, the number of undernourished people worldwide has increased in the last years due to war conflicts compounded with the effects of climate change (FAO, IFAD, UNICEF, WFP & WHO, 2017).

Against this bleak picture, the use of microbial protein requires to be explored as alternative sources of nutritious food. Tables 1 and 2 present an overview of the nutritional values of fungi, algae and bacteria as well as their present production volumes, costs and global markets. The total value of the microbial protein market is estimated to be around EUR 12 billion which is quite significant but much less than the global market values for animal proteins. Beef alone is projected to reach a global market of EUR 1850 billion by 2020 (Grand View Research, 2015) despite being the worst environmental choice for food production (Eshel, Shepon, Makov, & Milo, 2014).

MICROBIAL PROTEIN AS A SOURCE OF FOOD

In recent years, microbes are serving as an alternative source of high-quality protein and are able to replace animal protein, including red meat. However, people have been using microbes, including algae, fungi and yeast, for millennia as part of their food and as animal feed. Algae are reported to have supported the life of ancient populations living very close to the sea providing them with the best source of vitamins and protein. Since ancient times, different types of microorganisms have been used as a part of human diet. Fermented milk and cheese produced by *Lactobacillus* and *Streptococcus* were used by Bulgarians, Greeks and Egyptian Greeks around 50–100 B.C. (Frey, 1930). In 1900, the concept of single-cell protein (SCP) – the protein obtained from microbial source (Vincent, 1969; Becker & Venktaraman, 1982), came into the light. It is referring to mixed protein extracted from pure or mixed culture of single-cell

Table 1. Overview of microorganisms involved in microbial protein and added-value product formation

Microorganism	Average Crude Protein Content (% CDW)	Important Nutritional Value	Added Value by-Products (% CDW)	Reference
Fungi	30–70	Amino acids, polysaccharides; the digestibility of fungal protein is similar to egg and milk	Pullulan, carbohydrates, xylitol, astaxanthin	Thrane et al. (2007)
Algae	40–60	Can be favourably compared to soy, egg and animal protein	Microbial oil (about 50–70%), carbohydrates (about 70%) and vitamins	Draaisma et al. (2013); Harun et al. (2010)

Note: CDW – cellular dry weight

Table 2. Analysis of current production volumes and market prices of microbial proteins

Organisms	Production Volume (tonne DM/y)	Production Costs (Euro/kg DM)	Global Market Value (Billion Euro)	Important Points	References
Microalgae	9000	4–25	2.4	Used as food; derivatives also used in different medicines	Enzing et al. (2014)
Yeast	3 000 000	-	9.2	Commercialized mainly in the bakery industry and for ethanol fermentation	Kellershohn & Russell (2015)
Fungi	25 000	-	0.214	Investment for a 22000 tonnes/year plant made in 2015	Beer (2015)

Note: DM – dry matter

organisms, such as fungi, algae, bacteria and yeast. The term "single-cell protein" however was coined by Carol L. Wilson in 1966 and later Nevin S. Scrimshaw started referring to the microbial cells grown and harvested to accomplish the food requirements of humans and animals as novel proteins due to their high protein content and to avoid negative associations with bacteria and microbes (Adedayo, Ajiboye, Akintunde, & Odaibo, 2011). This protein can also be branded as mini food.

The microorganisms have special ability to upgrade low quality organic material into highly nutritive proteinaceous food and this is used by the food industry. Large-scale production commenced in the 20th century and operated in Germany during the First World War (1914-1915). Baker's yeast – a special strain of yeast, started to be produced in an aerated molasses medium supplemented with ammonium salts (Litchfield, 1983; Boze, Moulin, & Galzy, 1992; Gabriel, Gómez-Ruiz, Cruz-Guerrero, & Bárzana, 2014). In the Second World War (1939–1945), *Candida utilis*, which is generally an aerobic yeast, was used as food and feed supplement in Germany (Gabriel, Gómez-Ruiz, Cruz-Guerrero, & Bárzana, 2014). Yeast became very popular through massive production and food supply first for the army and then to the whole population (Khachatourians & Arora, 2002). Different types of methods have been developed across Europe since the Second World War for the mass cultivation of single-cell protein by formulating diverse types of growth media and improved culture of microorganisms (Gabriel, Gómez-Ruiz, Cruz-Guerrero, & Bárzana, 2014).

In the 1950s, some of the oil industry took interest in growing microorganisms on alkanes and thereafter in the 1960s, natural substrates and industrial waste products were used for cultivating microorganisms (Grewal, Kalra, & Kahlom, 1990; Osho, 1995). Different natural products were also used as a carbon source (Kuzmanova, Vandeska, Dimitrovski, & Doneva, 1989), such as cashew, apple juice, grape juice and cellulose (Osho, 1995; Haider & El-Hassy, 2000; Azzam, 1992; Pessoa, Mancilha, & Sato, 1997; Bozakouk, 2002; Zubi, 2005). Many food processing companies around the world are producing single cell protein including Kanegafuichi (Japan), BP (UK) and Liquichimica (Italy). The other substrates for single cell protein include sulphite waste liquor, citrus wastes, molasses, animal manure, and sewage (Gour, Mathur, Singh, & Pradeep, 2015).

Algae, and microalgae in particular (which are single-cell or groups of joint single-cell organisms), recently started to be used as food and food supplements in the food industry (Anupama & Ravindra, 2000; Becker, 2007), with the global production achieving 9000 tonnes dry matter per year and a market value estimated at about EUR 2.4 billion with projected yearly growth of 10% (Enzing, Ploeg, Barbosa, & Sijtsma, 2014). The role of yeast in the food industry is also of great significance. Baker's yeast and fermentation of alcohols are the two main processes in which the yeast is employed with a projected global market value for 2019 of up to EUR 9.2 billion and an annual growth forecast of 7.9% (Matassa, Boon, Pikaar, & Verstraete, 2016).

Table 3 presents the list of microbes which are utilized in the production of single-cell protein. The following sections look specifically at fungi, algae and yeast in terms of their composition and role in food security.

FUNGI

Fungi are heterotrophic thallophytes which have good protein content – approximately 20–30% of the dry matter. Their chitinous cell walls are a source of dietary fibre and their fat content is low. These are important features of a good diet. Mushrooms and bracket fungi have been used since before recorded

Table 3. Microorganisms involved in the production of single-cell protein (SCP)

	Category of Microorganisms	Carbon Source	Microorganisms Involved
1	Fungi	Brewery waste	*Calvatia gigantean*
		Cellulose	*Trichoderma viride, Chaetomium cellulolyticum*
		Glucose	*Agaricus campestris, A. blazei*
		Starch	*Fusarium graminearum, Aspergillus niger*
2	Algae	CO_2	*Dunaliella bardawil, Chlorella pyrenoidosa, C. sorokiniana, C. regularis, Spirullina maxima, Oocystis polymorpha, Scenedesmus quadricauda*
3	Yeasts	Methanol	*Kloeckera* sp., *Candida lipolytica, C. guilliermondii, C. tropicalis.*
		Sulfite waste liquor	*Candida.tropicalis, C. utilis,*
		Cellulose	*Candida utilis,*
		Ethanol	*Candida enthanothermophilum, C. Kruzei, C. utilis*
		Cane molasses	*Saccharomyces cerevisiae*
		Whey	*Candida intermedia, Kluyveromyces fragilis*
		Lipids	*Candida rugosa, C. Parapsilosis, C. deformans, C. curvata, C. lipolytica, C. blankii, C. utilis, Lipomyces kononenkoe*
4	Actinomycetes and Bacteria	Sulfite waste liquor	*Pseudomonas denitrificans*
		Ethanol	*Acinetobacter calcoaceticus,*
		Cellulosic wastes	*Thermomonospora fusca*
		Methane	*Nocardia paraffinica, Corynebacterium hydrocarbonoclastus, Hyphomicrobium* sp., *Acinetobacter* sp.

Source: Modified from (Boze, Moulin, & Galzy, 1992; Gabriel, Gómez-Ruiz, Cruz-Guerrero, & Bárzana, 2014)

history for both medicinal purposes and for food. There are more than 80,000 species in the fungus kingdom and yeast is one of them (Prabhat, 2017). Archaeological records show the use of fungi more than 4,000 years ago with Egyptian murals and tomb ornaments depicting bread and wine making.

Use of Fungi

Europe began making efforts to produce protein supplements from *Fusarium* and *Rhizopus* cultures around the Second World War (Gour, Mathur, Singh, & Pradeep, 2015). The inoculum of *Rhizopus arrhizus* and *Aspergillus oryzae* was chosen due to its harmless nature (Riviere, 1977). A property of many saprophytic fungi is to grow up on complex food organic substrates and convert them into simpler forms resulting into the production of high amounts of biomass.

Another important feature of fungus cells is their low-cost production as they can be cultivated on industrial or agricultural waste products. Mushrooms are essential food from fungal mycelia which need less care and grow on agricultural waste. The agricultural industry generates a lot of waste products because not all crop is actually used (the used crop is only 7% in sugarcane plantations, 5% in palm and coconut plantations, 2% in sisal plantations, etc.). Cultivating mushrooms is a way to tackle waste. Different species of mushrooms similar to the Oyster mushroom species (*Pleurotus sajor-caju,Pleurotus ostreatus* or *Pleurotus cystidiosus*) grow easily on cotton wastes; the straw mushroom (*Volvariella volvacea*) is traditionally grown on rice straw in South East Asia.

The process of bringing fungal protein into the market place began long before 1985 when the first product was marketed. A British company – the famous Rank Hovis McDougall (RHM) made the decision to develop fungi as a food substitute during the 1960s. This was followed by more than 15 years of research, development and toxicity testing before the approval for the sale of the new fungal food product was obtained from the British Ministry of Agriculture, Fisheries and Food (MAFF) (Edelman, Fewell, & Solomons, 1983; Angold, Beech, & Taggart, 1989).

Fungi in Food Security

Fungi as a food are a good source of protein as they contain essential and non-essential amino acids. Archaeological evidence reveals traces associated with edible wild mushrooms by the inhabitants of Chile approximately 13,000 years ago (Rojas & Mansur, 1995). In China, the consumption of wild fungi dates back several hundred years before the birth of Jesus Christ (Aaronson, 2000). Fermented food was used in Sumeria and Babylon (Elander & Lowe, 1994). However, the utilization of fungi in food production and their full potential were not explored until the second half of the 20th century when these were boosted by the advent of the golden age of industrial microbiology. Since then the diverse community of fungi has been exploited in a myriad of food products for human consumption.

More than 3000 fungal isolates around the world were analysed in research and development for protein content. The strain ATCC PTA-2684 (named *F. Graminearum* and re-identified as *F. venenatum*, O'Donnell, Cigelnik, & Casper, 1998) was recognized as the best organism for myco-protein production. Once the production organism is chosen for the process, fermentation and product development can proceed along with the safety testing. Approximately 12 years of research and various demonstrations of the myco-protein took place before it was given to be consumed by volunteers without harmful effects (Solomons, 1987). The nutritional value of myco-protein was tested and it was also found that it is comparable to eggs in amino acid composition (Miller & Dwyer, 2001), does not contained cholesterol and has a substantial fibre content. After receiving approval by MAFF, the toxicity and allergen testing of QuornTM continued (Miller & Dwyer, 2001; Tee, Gordon, Welch, & Taylor, 1993). QuornTM has high fibre content, was found to help decrease blood cholesterol levels (Turnbull, Leeds, & Edwards, 1992) and also may encourage reduced energy intake (Turnbull, Walton, & Leeds, 1993; Burley, Paul, & Blundell, 1993). It is now a popular meat alternative product sold in supermarkets across the globe.

From a biological point of view, a significant aspect of myco-protein production is the contribution of the morphology of hyphae to the final product and the effect that cultivation in long-term, submerged, continuous flow cultures has on the morphology of hyphae. The filamentous fungus was chosen for the production of this meat substitute as it was believed that the mycelia would add a fibrous texture and high protein content comparable to that of meat (Edelman, Fewell, & Solomons, 1983). An optimal branch length was determined for the process of production. As several other filamentous fungi (Forss, Gadd, Lundell, & Williamson, 1974; Righelato, 1976) grow when *F. venenatum* A3/5 is cultivated in a continuous flow system over long time periods, mutants which alter the branching patterns arise and eventually displace the parental strain. Such types of mutants are considered unsuitable for the formation of the final product (Trinci, 1992).

Table 4 provides information about some edible macrofungal species. Macrofungi, such as mushrooms, brackets and truffles, have spore-bearing structures visible to the naked eye. This is a selection from a much longer list which contains thousands of species.

Fungi, such as *Monascus purpureus*, have been traditionally used for the production of red wine (Went, 1895). Produced by the fungus group mucorales, β-carotene is an additive agent in a variety of foods (van den Ende & Stegwee, 1971). The toxic and allergic characteristics of artificial colours have led to the production of food colour by the process of fermentation of *Monascus purpureus* on rice to prepare 'koji' or 'ang-kak' (red rice), which has been used as a traditional Chinese food and medicine since 800 A.D. (Li, Zhu, Wang, Zhu, Chang, & Kritchevsky, 1998). The reaction of rubropunctatin and the orange pigments monascorubrin with amino acids present in the fermentation fungus media produces red pigments monascorubramine which are water soluble and rubropunctamine. Fungi are also used in the production of soybean cheese, wine, red rice and authentic food in many Asian and European countries. The accumulation of β-carotene by mating of different fungal strains is strongly linked to sexual interaction which results in the production of a hormone-like substance. A major component of this is trisporic acid, which stimulates pigment production and enhances the food quality. Tables 5 and 6 list some foods obtained directly from fungi and the nutrient composition of various fungi.

ALGAE

Algae are chlorophyll bearing thallophytes characterized by very simple structural organization which uses light energy, carbon dioxide and ions dissolved in the water for the synthesis of complex molecules and producing biomass. Microalgae are components of phytoplankton and benthic communities which can live anywhere in fresh or transitional waters as well as in marine water. The microalgae can also

Table 4. Nutritional properties of edible macrofungi

Fungi	Application	References
Flammulina velutipes	Mannofucogalactan is a heterogalactan which is derived from its mycelia and has great nutritional value.	Ko et al. (1995), Carbonero et al. (2008) and Smiderle et al. (2008)
Lentimula edodes	It contains high protein with all essential amino acids and also has high content of vitamin D.	Murata et al. (2002) and Rossi et al. (1993)
Pleurotus ostreaus	It is considered to be rich in fiber, carbohydrates and protein. Its cell also has vitamins and minerals.	Cohen et al. (2002)
Tuber melanosporum	It has tantalizing taste and aroma with great economic value in the food industry.	Breene (1990)
Auricularia polytricha	It is rich in potassium (K), phosphorus (P), magnesium (Mg) and strontium (Sr).	Sheu et al. (2004)
Ganoderma lucidum	Long history of use for promoting health and longevity, in the cellular structure protein contains only 7.3% of dry weight whereas glucose is 11% and metals 10.2% of the dry mass.	Bao et al. (2002) and Zhang et al. (2002)
Morchella semilibera	It has spongy texture of young morels which makes delicious dishes.	Carbonero et al. (2008)
Agaricus bisporus	It is rich in vitamins like vitamin B and minerals like phosphorus, sodium, potassium and selenium.	Beelman et al. (2003)

Table 5. Direct plant-based foods obtained from macrofungi

	Macrofungi	Application
1	*Oriental food fermentation*	*Fungus species*
	• Ang-kak • Miso • Hamanatto • Ontjom • Shoyu (soy sauce) • Tempeh	• *Monascus purpurea* • *Aspergillus oryzae, A. Sojae* • *Aspergillus oryzae* • *Neurospora intermedia* • *Aspergillus oryzae, A. Sojae* • *Rhizopus oligosporus*
2	*Edible macro fungi*	*Fungus species*
	• Oyster mushroom • Truffle • Button mushroom and Champignon • Enoki and Winter mushroom • Shiitake	• *Pleurotus* sp. • *Tuber melanosporum* • *Agaricus bisporus, A. Bitorquis* • *Flammulina velutipes* • *Lentinula edodes*

Source: (Wainwright,1992)

Table 6. Nutrient composition of different fungi [% of dry matter]

Fungus Name	Protein	Carbohydrates	Fat
Amanita loosii	20	-	
Amanita caesarea	15	-	14
Boletu Boletus	38	47	9
Boletus erythropus	15	57	1
Boletus loyo	22	50	1
Lactarius phlebophyllum	30	51	9
Lactarius deliciosus	27	-	7
Lactarius indigo	13	-	-
Ramaria flava	14	-	
Russula cyanoxantha	17	-	8
Suillus luteus	20	57	4
Termitomyces microcarpus	49	49	10
Tirmania nívea	14	21	-
Tricholoma populinum	13	70	9
Tricholoma saponaceum	5	-	7
Terfezia claveryi	8	17	-
Suillus granulatus	14	70	2
Rsuuula sp.	29	55	6
Russula delica	17	-	-

Source: Modified from (Kiger, 1959; Degreef, 1990; Leon-Guzman, Silva, & Lopez, 1997; Boa, 2004; Çaglarirmak, Ünal, & Ötles, 2002; Turner, Kuhnlein, & Egger, 1987)

colonize on mountain ice, sand, polar surfaces and on rocks. Many of the microalgae live in a symbiotic association with fungi and lichens. The term microalgae in phycology (a branch of botany which deals with the study of algae) includes microscopic algae like *Sensu stricto* and some of the photosynthetic bacteria like cyanobacteria formerly known as Cyanophyceae (Milledge, 2012).

Use of Algae

Microalgae can be seen as promising candidates for a broad range of applications in the white biotechnology which "uses living cells – from yeast, moulds, bacteria and plants—and enzymes to synthesize products that are easily degradable, require less energy and create less waste during their production" (Frazzetto, 2003, p. 835). This includes the production of food and the creation of different green energy carriers. All these unicellular microorganisms contain a versatile polyphyletic group which has common capability of photosynthetic fixation of carbon dioxide for generating various algal cell components and energy (Craggs, 1996). The microalgae have eukaryotic organization of the cell (Hoffmann, 1998; Lau, 1997) with the most important characteristic feature of the algal cell being that it maintains peculiar biological features which vary from species to species.

In different situations, such as under limited light, the cells of algae can change their nutrient regime and assimilate organic carbon (Ugwu, Ogbonna, & Tanaka, 2002) as well as different inorganic nutrients such as phosphorus and nitrogen (Hase, Oikawa, Sasao, Morita, & Watanabe, 2000). They can use contaminated water for their growth without an aerobic situation being created and maintained (Hall, 2003; Carlozzi, 2003). Microalgae produce oxygen (O_2) during the process of photosynthesis and provide a safe and cheap substitute to mechanical aeration while also contributing to carbon dioxide (CO_2) mitigation (Vega-Estrada, 2005).

Algae in Food Security

The use of microalgae by indigenous populations has been occurring for centuries but their cultivation is only a few decades old. It is believed that approximately 30,000 species of microalgae exist at present in which few thousand strains are kept in collections and few hundred have been investigated for their different chemical composition (Borowitzka,1999; Chaumont, 1993; Radmer & Parker, 1994; Olaizola,2003).

Commercial scale production of microalgae was realized in Taiwan based on *Chlorella spp.* Countries such as USA, China and India are using the blue-green algae *Arthrospira* and *Spirulina* as food. Also, genera that are predominantly cultivated for their high quality of protein content are *Scenedesmus, Anabaena*, or *Synechococcus* (Olaizola, 2003). Further advantages from products produced by these microalgae are that they contain sugars, fatty acids, pigments, enzymes, vitamins (B-vitamins, namely B1, B2, B3, B5, B6, B7, B9, B12l; vitamins A, C, and E) and other important bioactive compounds which display anti-cancer, anti-inflammatory and antibiotic (Chlorellin) effects (Olaizola, 2003; Pratt, Daniels, Eiler, Gunnison, Kumler, Oneto,... & Smith, 1944; Pasquet, Chérouvrier, Farhat, Thiéry, Piot, Bérard, ... & Picot, 2011). The most common microphages which are used commercially as food are *Dunaliella salina, Chlorella vulgaris, Spirulina maxima, Haematococcus pluvialis*. For example, the algae-like *Chlorella vulgaris* is widely produced and marketed as a food supplement in countries, such as US, Japan, China and other Asian countries and also across Europe. *Chlorella spp.* is being considered and widely used in the food market because of its wide range of nutrients, including vitamins, minerals as well as carotenoids. According to Spolaore et al. (2006), *Chlorella* have beta– 1, 3–glucan which is an immunostimulator, a free-radical scavenger and a reducer of blood lipids. It is a crucial health–promoting species of microphages that helps for many conditions and disorders, including wounds, anaemia, hypertension, diabetes as well as in malnutrition of infants (Yamaguchi, 1997). *Chlorella* also has other health properties which act against atherosclerosis and hypercholesterolemia through the phospholipids, glycolipids and antitumor actions by glycoprotein, peptides and nucleotides (Yamaguchi, 1997).

Another approach of obtaining food from microalgae is the application of Dunaliella salina which is generally a halotolerant, naturally occurring algae in salted lakes. *Dunaliella salina* can accumulate β-carotene which is used as natural food colouring agent and contains provitamin A (retinol). The variety from Pink Lake, Victoria contains 14% carotenoids in dry weight. Some of the strains may also contain 10% β-carotene under nutrient-stressed, high-salt and high-light conditions (Aasen, Eimhjellen, & Liaaen- Jensen, 1969; Ben-Amotz & Avron,1980; Oren, 2005). Except β-carotene, it also produces glycerol. *Haematococcus pluvialis* is another alga which can accumulate the highest level of astaxanthin in nature (1.5–3.0% dry weight).

Another species of microalgae is *Spirullina spp* which grows profusely in marine lakes in Africa and Mexico, and the local communities have been using it as a primary staple food in their daily diets since ancient times (Yamaguchi,1997). Spirullina nowadays is processed into powder and health supplements as superfood for its pure and high nutritional qualities. It has high protein content as well as high alpha-linolenic acid level due to which its production around the world is approximately 3,000 tonnes/year (Yamaguchi, 1997). *Spirulina spp.* is also the main source of the natural pigment called 'phycocyanin', used as a natural food, cosmetic colouring and biochemical tracer in immunoassays (Shimamatsu, 2004; Reinehr & Costa, 2006). Other potential microalgal species that can be used for future food production are *Nannochloris spp., Chlorella spp., Dunaliella spp., Parietochloris incisa, Neochloris oleoabundans* and *B. braunii* because of their special capacity to accumulate large quantities of lipids contents (Li, 2008).

The high nutritional content and the role microalgae play as a human food source are recognised throughout history. This is documented in the ancient Chinese literature of two and a half millennia ago (Tseng, 2004). Because of their valuable protein content *Spirulina spp.* and *Nostoc spp.* are used in East-Asia, South Africa and Mexico for nutritional purposes. Other algae share similar high protein properties. *Spirulina platensis* and *Arthrospira maxima* contain approximately 60% of crude protein. *Chlorella vulgaris* harbours more than 50% of proteins in its cell mass (Becker, 2007).

More recently, attention has been drawn on the marine microalgae (whose habitats are salty water) *Diacronema vlkianum* and *Isochrysis galbana* because of their potential ability to produce long chain polyunsaturated fatty acids (LC-PUFA). These algae mainly produce docosahexaenoic acid (DHA, 22:63) and eicosapentaenoic acid (EPA, 20:53), which are accumulated as oil droplets in prominent lipid bodies in the cellular structure (Liu & Lin, 2001). Microalgae like *Galbana* were shown as the most suitable source of nutrition for rapid growth, while the other species *D. vlkianum* resulted in high growth rates and low mortality for the pacific oyster *Crassostrea gigas* larvae (Wikfors, Ferris, & Smith, 1992; Ponis Probert, Veron, Mathieu, & Robert, 2006). These algae are also potentially promising for the food industry as a valuable source of proteins, lipids and LC-PUFAs, and for supplying tocopherols, sterols, colouring pigments and other nutraceuticals (Donato, Vilela, & Bandarra, 2003). The nutritional composition of different algae is presented in Table 7.

Table 7. Nutrients composition of different algae [% of dry matter]

Name of Algae	Carbohydrates	Proteins	Lipids
Spirulina platensis	8–14	46–63	4–9
Synechococcus sp.	15	63	11
Anabaena cylindrica	25–30	43–56	4–7
Chlamydomonas rheinhardii	17	48	21
Aphanizomenon flos-aquae	23	62	3
Chlorella pyrenoidosa	26	57	2
Dunaliella salina	32	57	6
Porphyridium cruentum	40–57	28–39	9–14
Euglena gracilis	14–18	39–61	14–20
Spirogyra sp.	33–64	6–20	11–21
Scenedesmus obliquus	10–17	50–56	12–14
Arthrospira maxima	13–16	60–71	6–7
Spirulina maxima	13–16	60–71	6–7
Chlorella vulgaris	12–17	51–58	14–22

Source: Modified from (Becker, 2007; Sialve Bernet, & Bernard, 2009)

YEAST

Yeasts are eukaryotic microorganisms and are most commonly referred to as unicellular fungi, although unicellular growth occurs within several fungal taxonomic orders and many types of yeast can grow by forming pseudo-hyphae just as is the case of fungi. The history of yeast's association with human society is synonymous with the evolution of beer, wine and bread, as global food and beverage commodities which originated 5,000 years ago.

Use of Yeast

Understanding the microbiology of all yeast-based food products commenced in the mid-1600s when the first observations of the cells of yeast were reported by the great microbiologist Antonie van Leeuwenhoek. His findings laid dormant until the classic studies of another great microbiologists Louis Pasteur in France and Emil Christian Hansen in Denmark during 1850–1900, which heralded the beginnings of the disciplines of biochemistry and microbiology. Further studies by other microbiologists – Alexandre Guilliermond in France and Albert Kluyver in the Netherlands, in the early 1900s revealed that yeasts are a unique group of microorganisms with a major role in food and beverage production.

Yeast species are found mainly in association with plants or animals in nature but their numbers are also abundant in aquatic environments. They can colonize a wide range of habitats both naturally or in connection with human activities, mainly because of their ability to grow and to survive in different and stressful environments. Different species of yeast have simple nutritional requirements as they have the ability to colonize dry surfaces for a long time. They can grow rapidly on inexpensive substrates in bioreactors with particularly relevant yeast in the selection of biocontrol agents. Another important feature yeasts possess is that they do not produce any allergenic spores or mycotoxins, as is the case of many mycelial fungi or antibiotics which might be produced by bacterial antagonists.

However, understanding the ecological fitness of the potential yeast biocontrol agents and the different developing strategies to enhance their stress tolerance are necessary for their efficacy and for commercial application. Not many other microorganisms match yeast in terms of economic, historical and scientific significance as the spontaneous fermentation of beer, wine, and cereal doughs is the oldest biopreservation technology, which has been practically applied since ancient times. In terms of their potentials among different microorganisms, yeasts have been extensively studied because they possess many significant features that make them suitable as bio control agents. Table 8 presents a list of fermented food products obtained from yeast and the application of yeast in beverage and food processing is displayed in Table 9.

Yeast in Food Security

Yeast has long been considered the organism of choice in the production of alcoholic beverages, bread and many other industrial products. The products obtained from yeast are currently making a large impact in agriculture as well as in the food industry. Yeasts have been a very important tool in the food production industry and nowadays it would be almost impossible to imagine a world devoid of all fermented products, such as bread, wine and beer. Compared to other microorganisms, e.g. bacteria, yeasts are slow growing. For example, if bacteria and yeast are placed in the optimum environment to grow, it is most likely that the faster growing bacteria would quickly outgrow and outcompete the slower growing yeast, becoming the dominant flora. However, if the medium is changed from one providing the optimum growth conditions of most bacteria, to an environment-which-is acidic, or of low water activity, but high in sugar content, then the yeast would have an advantage and would rapidly exceed and overtake the growth of the bacteria.

Yeast-based products are mainly derived from *Saccaromyces cerevisiae*, which have been used for many years as additives in food processing and in different ingredients. The products include antioxidants, lavourants, vitamins, colourants and polysaccharides, enzymes etc. (Dawson, 2002; Abbas, 2006; De Vuyst & Neysens, 2005). Other species of yeast are *C. Pyralidae, D. Hansenii, D. marasmus, Filobasidium floriforme,* C. zeylanoides, C. famata, H. Burtonii, *Kluyveromyces wickerhamii, P. Membranaefaciens, Tetrapisispora phaffii, Torulaspora globosa, Ustilago maydis, W. Anomalus* and *Williopsis mrakii.*

Table 8. Yeast involved in fermented food products

Yeast Species	Applications
Saccharomyces cerevisiae and Schizosaccharomyces sps.	Alcoholic beverages
S. uvarum, S. cerevisiae	Beers
S. exigus, S. Cerevisiae, Candida krusei, Hansenula anomala, Pichia and Pichia and *Hansenula anomala* Beers *S. cerevisiae*	Cakes and breads
C. holmii; C. krusei, C. famata	Cocoa
S. cerevisiae,	Wines

Source: Modified from (Cristiani & Monnet, 2001)

Table 9. Yeast species in food and beverage applications

Yeast Species	Strains	Application and Antagonist	References
Debaryomyces hansenii	1021	It is involved in olive fermentation against *Candida boidinii* and *Kluyveromyces lactis*	Llorente et al. (1997)
C. pyralidae	IWBT Y1140, IWBT Y1057	Involved in grape juice against *Brettanomyces bruxellensis*	Mehlomakulu et al. (2014)
Filobasidium floriforme	NRRL Y7454	Involved in apple against *Brettanomyces cinerea*	Filonow et al. (1999)
S. cerevisiae	Cf8, M12	Involved in winemaking apple *Pichia membranifaciens* *Brettanomyces bruxellensis*, and *Dekkera anomala*,	Ullivarri et al. (2014)
Torulaspora globosa	1S100, 1S111, 1S112, 2S01, 2S04	Involved in sorghum and maize against *Colletotrichum sublineolum* and *Colletotrichum graminicola*	Rosa-Magri et al. (2011)
Ustilago maydis	possibly causing the leakage of K+ or NH4+ from cells	Involved in grape juice against *B. bruxellensis* strains	Santos et al. (2011)
Williopsis mrakii	169 = NCYC (National Collection of Yeast Culture) 251	Involved in yogurt and maize silage against *Candida krusei* and *Saccharomyces cerevisiae*	Lowes et al. (2000)
Aureobasidium pullulans	PI1	On grape berries against *B. cinerea*	Parafati et al. (2015)
A. pullulans	L47	Involved in strawberries grown under plastic tunnels against *B. cinerea* and *R. stolonifer*; on apple against *B. cinerea* and *P. expansum*	Lima et al. (1997) and Ippolito et al. (2011)

The action of yeasts helps in many processes for high-value cash crops, such as cocoa beans and coffee beans (Schwan & Wheals, 2003). For example, cocoa beans must be fermented to generate the precursors of chocolate flavour, and various species of *Hanseniaspora, Saccharomyces, Pichia, Candida, and Issatchenkia* contribute in this phenomenon (Nielsen, Hønholt, Tano-Debrah, & Jespersen, 2005; Ardhana & Fleet, 2003).

Traditional fermented foods and beverages are produced in many countries in the world across Africa, Asia and South America from raw materials such as wheat, cassava, rice, soy beans, maize and fruit. Fermentation is an essential process contributing to the safety, quality and nutritional value of these products. Various aspects of the microbial ecology are just starting to emerge, demonstrating important contributions from numerous yeast species (Aidoo, Rob Nout, & Sarkar, 2005; Nout & Boekhout, 2003). Collectively, the ecological studies of yeasts in various products other than wine, beer and bread are providing good knowledge base for developing a new generation of yeast starter cultures, beyond the *S. cerevisiae*.

FUTURE PROSPECTIVE

Microbial protein is a valuable component in increasing food security (Suman, Nupur, Anuradha, & Pradeep, 2015). The microbial protein producing microbes grow fast and produce large quantitative of protein from a small area of land and in a limited time which makes this option attractive for future research and development. In addition to the high nutritional values of microbial protein, it can be produced throughout the year as it is independent of seasonal restrictions or other climatic conditions. Being an excellent source of nutritive proteins, its other cellular components can also have increasing importance driving new uses of microbial-based by-products. Containing essential nutrients, such as nitrogen, phosphorus and nutritive resources, the production of these microbes as food is gaining renewed industrial interest, especially in the context of the circular economy. The transition towards widespread uses as well as the appreciation of the microbial protein as a major supply route for food needs in the near future needs to be prepared with care and foresight, especially in terms of quality and regulatory issues.

CONCLUSION

The rapidly increasing population of the world generates challenges and issues related to providing the necessary foodstuff to sustain humanity. One of the best and convenient solutions to this problem may be generation of microbial mediated single-cell protein. Algae, yeast and fungi are the main components of the single-cell protein as they comprise vitamins, proteins, minerals, and essential fatty acids. Another important feature of this protein is the rate of digestibility which is generally high compared to animal protein.

Therefore, in the light of protein shortage, microbes may offer significant possibilities in protein production. They can be used partially or totally to replace conventional animal protein, and also, because of their ability to grow on waste substances, they can help in the reduction of environment pollution and recycling of materials. Microbial protein may be a sustainable alternative future protein source to fulfil the food requirements of humans. The processing of this protein is also important because of susceptibility to contamination and needs to be monitored. In the coming years, scientists are looking at single-cell protein as an important source with more durability and resistance to outer contamination.

REFERENCES

Aaronson, S. (2000). Fungi. In K. F. Kiple & K. C. Ornelas (Eds.), *The Cambridge world history of food* (pp. 313–336). Cambridge, UK: Cambridge University Press.

Aasen, A. J., Eimhjellen, K. E., Liaaen-Jensen, S., Heinegård, D., Balaban, A. T., & Craig, J. C. (1969). An Extreme Source of beta-Carotene. *Acta Chemica Scandinavica*, *23*, 2544–2545. doi:10.3891/acta.chem.scand.23-2544 PMID:5368546

Abbas, C. A. (2006). Production of antioxidants, aromas, colours, flavours, and vitamins by yeasts. In A. Querol & G. H. Fleet (Eds.), *Yeasts in food and beverages* (pp. 285–334). Berlin, Germany: Springer. doi:10.1007/978-3-540-28398-0_10

Adedayo, M. R., Ajiboye, E. A., Akintunde, J. K., & Odaibo, A. (2011). Single cell proteins: As nutritional enhancer. *Advances in Applied Science Research*, *2*(5), 396–409.

Aidoo, K. E., Rob Nout, M. J., & Sarkar, P. K. (2005). Occurrence and function of yeasts in Asian indigenous fermented foods. *FEMS Yeast Research*, *6*(1), 30–39. doi:10.1111/j.1567-1364.2005.00015.x PMID:16423068

Angold, R., Beech, G., & Taggart, J. (1989). *Food biotechnology*. Cambridge, UK: Cambridge University Press.

Anupama, & Ravindra, P. (2000). Value-added food: Single cell protein. *Biotechnology Advances, 18*(6), 459–479. doi:10.1016/S0734-9750(00)00045-8

Ardhana, M. M., & Fleet, G. H. (2003). The microbial ecology of cocoa bean fermentations in Indonesia. *International Journal of Food Microbiology*, *86*(1), 87–99. doi:10.1016/S0168-1605(03)00081-3 PMID:12892924

Azzam, A. M. (1992). Production of metabolites, industrial enzymes, amino acids. *Journal of Environmental Science & Engineering*, *56*, 67–99.

Bao, X. F., Wang, X. S., Dong, Q., Fang, J. N., & Li, X. Y. (2002). Structural features of immunologically active polysaccharides from Ganoderma lucidum. *Phytochemistry*, *59*(2), 175–181. doi:10.1016/S0031-9422(01)00450-2 PMID:11809453

Becker, E. W. (2007). Micro-algae as a source of protein. *Biotechnology Advances*, *25*(2), 207–210. doi:10.1016/j.biotechadv.2006.11.002 PMID:17196357

Becker, E. W., & Venktaraman, L. V. (1982). *Biotechnology and exploitation of algae: The Indian approach. A comprehensive report on the cultivation and utilization of microalgae performed at the Central Food Technological Research Institute, Mysore, India.* Eschborn, Germany: German Agency for Technical Cooperation.

Beelman, R. B., Royse, D. J., & Chikthimmah, N. (2003). Bioactive components in button mushroom Agaricus bisporus (J. Lge) Imbach (Agaricomycetideae) of nutritional, medicinal, and biological importance. *International Journal of Medicinal Mushrooms*, *5*(4), 321–338. doi:10.1615/InterJMedicMush.v5.i4.10

Beer, E. (2015). *"No success model" for new protein ingredients*. Retrieved from http://www.foodnavigator.com/Market-Trends/No-success-model-for-new-protein-ingredients

Bekatorou, A., Psarianos, C., & Koutinas, A. A. (2006). Production of food grade yeasts. *Food Technology and Biotechnology*, *44*(3), 407–415.

Ben-Amotz, A., & Avron, M. (1980). Glycerul β-carotene and dry al gal meal production by commercial cultivation of Dunaltella. In G. Shelef & C. J. Soeder (Eds.), *Algal biomass: Production and use* (pp. 603–610). Amsterdam: Elsevier.

Boa, E. (2004). *Wild edible fungi: A global overview of their use and importance to people*. Rome, Italy: Food and Agriculture Organization of the United Nations. Retrieved from http://www.fao.org/docrep/007/y5489e/y5489e00.htm#TopOfPage

Borowitzka, M. A. (1999). Commercial production of microalgae: Ponds, tanks, tubes and fermenters. *Journal of Biotechnology*, *70*(1–3), 313–321. doi:10.1016/S0168-1656(99)00083-8

Bozakouk, A. H. (2002). *Acid hydrolysis of Phragmites austral is powder for production of single cell protein by* Candida utilis (Master thesis). Garyounis University, Benghazi, Libya.

Boze, H., Moulin, G., & Galzy, P. (1992). Production of food and fodder yeasts. *Critical Reviews in Biotechnology*, *12*(1–2), 65–86. doi:10.3109/07388559209069188 PMID:1733522

Braude, R., Hosking, Z. D., Mitchell, K. G., Plonka, S., & Sambrook, I. E. (1977). Pruteen, a new source of protein for growing pigs. I. Metabolic experiment: Utilization of nitrogen. *Livestock Production Science*, *4*(1), 79–89. doi:10.1016/0301-6226(77)90022-7

Breene, W. M. (1990). Nutritional and medicinal value of specialty mushrooms. *Journal of Food Protection*, *53*(10), 883–894. doi:10.4315/0362-028X-53.10.883

Burley, V. J., Paul, A. W., & Blundell, J. E. (1993). Influence of a high-fibre food (myco-protein) on appetite: Effects on satiation (within meals) and satiety (following meals). *European Journal of Clinical Nutrition*, *47*, 409–409. PMID:8396025

Çaglarirmak, N., Ünal, K., & Ötles, S. (2002). Nutritional value of wild edible mushrooms collected from the Black Sea region of Turkey. *Micologa Aplicada International*, *14*(1), 1–5.

Caplice, E., & Fitzgerald, G. F. (1999). Food fermentations: Role of microorganisms in food production and preservation. *International Journal of Food Microbiology*, *50*(1-2), 131–149. doi:10.1016/S0168-1605(99)00082-3 PMID:10488849

Carbonero, E. R., Gorin, P. A. J., Iacomini, M., Sassaki, G. L., & Smiderle, F. R. (2008). Characterization of a heterogalactan: Some nutritional values of the edible mushroom Flammulina velutipes. *Food Chemistry*, *108*(1), 329–333. doi:10.1016/j.foodchem.2007.10.029

Carlozzi, P. (2003). Dilution of solar radiation through "culture" lamination in photobioreactor rows facing south–north: A way to improve the efficiency of light utilization by cyanobacteria (*Arthrospira platensis*). *Biotechnology and Bioengineering*, *81*(3), 305–315. doi:10.1002/bit.10478 PMID:12474253

Chaumont, D. (1993). Biotechnology of algal biomass production: A review of systems for outdoor mass culture. *Journal of Applied Phycology*, *5*(6), 593–604. doi:10.1007/BF02184638

Cohen, R., Persky, L., & Hadar, Y. (2002). Biotechnological applications and potential of wood-degrading mushrooms of the genus Pleurotus. *Applied Microbiology and Biotechnology*, *58*(5), 582–594. doi:10.100700253-002-0930-y PMID:11956739

Committee on World Food Security. (2012). *Global strategic framework for food security and nutrition*. Rome, Italy: Food and Agriculture Organization of the United Nations. Retrieved from http://www.fao.org/docrep/meeting/026/ME498E.pdf

Craggs, R. J., Adey, W. H., Jessup, B. K., & Oswald, W. J. (1996). A controlled stream mesocosm for tertiary treatment of sewage. *Ecological Engineering*, *6*(1–3), 149–169. doi:10.1016/0925-8574(95)00056-9

Cristiani, G., & Monnet, V. (2001). Food micro-organisms and aromatic ester synthesis. *Sciences des Aliments*, *21*(3), 211–230. doi:10.3166da.21.211-230

Dawson, K. A. (2002). Not just bread and beer: New applications for yeast and yeast products in human health and nutrition. In T. P. Lyons & F. A. Jaques (Eds.), *Nutritional biotechnology in the feed and food industry* (pp. 225–232). Nottingham, UK: Nottingham University Press.

De Vuyst, L., & Neysens, P. (2005). The sourdough microflora: Biodiversity and metabolic interactions. *Trends in Food Science & Technology*, *16*(1–3), 43–56. doi:10.1016/j.tifs.2004.02.012

Degreef, J. (1990). Inventaire, valeur alimentaire et culture de champignons da Shaba. Memoire fin d'études, Chaire d'Ecologie et Phytosociologie, Faculté des Sciences Agrononiques, de Gembloux, Belgium.

Donato, M., Vilela, M. H., & Bandarra, N. M. (2003). Fatty acids, sterols, α-tocopherol and total carotenoids composition of *Diacronema vlkianum*. *Journal of Food Lipids*, *10*(4), 267–276. doi:10.1111/j.1745-4522.2003.tb00020.x

Draaiisma, R. B., Wijffels, R. H., Slegers, P. M., Brentner, L. B., Roy, A., & Barbosa, M. J. (2013). Food commodities from microalgae. *Current Opinion in Biotechnology*, *24*(2), 169–177. doi:10.1016/j.copbio.2012.09.012 PMID:23084075

Edelman, J., Fewell, A., & Solomons, G. L. (1983). Myco-protein – a new food. *Nutrition Abstracts and Reviews*, *53*(6), 471–480.

Elander, R. P., & Lowe, D. A. (1994). Fungal biotechnology: An overview. In K. Arora, L. Ajello, & K. G. Mukerji (Eds.), *Handbook of applied mycology* (pp. 1–34). New York, NY: Marcel Dekker.

Enzing, C., Ploeg, M., Barbosa, M., & Sijtsma, L. (2014). *Microalgae-based products for the food and feed sector: An outlook for Europe*. JRC Scientific and Policy Reports. Institute for Prospective Technological Studies. Joint Research Centre. European Commission. Retrieved from ftp://ftp.jrc.es/pub/EURdoc/JRC85709.pdf

Eshel, G., Shepon, A., Makov, T., & Milo, R. (2014). Land, irrigation water, greenhouse gas, and reactive nitrogen burdens of meat, eggs, and dairy production in the United States. *Proceedings of the National Academy of Sciences of the United States of America*, *111*(33), 11996–12001. doi:10.1073/pnas.1402183111 PMID:25049416

Ezeh, A. C., Bongaarts, J., & Mberu, B. (2012). Global population trends and policy options. *Lancet*, *380*(9837), 142–148. doi:10.1016/S0140-6736(12)60696-5 PMID:22784532

Filonow, A. B., Vishniac, H. S., Anderson, J. A., & Janisiewicz, W. J. (1996). Biological control of *Botrytis cinerea* in apple by yeasts from various habitats and their putative mechanisms of antagonism. *Biological Control*, *7*(2), 212–220. doi:10.1006/bcon.1996.0086

Food and Agriculture Organization of the United Nations (FAO), International Fund for Agricultural Development (IFAD), United Nations' Children's FUND (UNICEF), World Food Programme (WFP), & World Health Organization (WHO). (2017). *The state of food security and nutrition in the world 2017: Building resilience for peace and food security*. Rome, Italy: FAO. Retrieved from www.fao.org/3/a-I7695e.pdf

Forss, K. G., Gadd, G. O., Lundell, R. O., & Williamson, H. W. (1974). *Process for the manufacture of protein-containing substances for fodder, foodstuffs and technical applications*. US Patent Office, Patent No. 3809614.

Frazzetto, G. (2003). White biotechnology. *European Molecular Biology Organization (EMBO). Reports*, *4*(9), 835–837. doi:10.1038j.embor.embor928

Frey, C. N. (1930). History and development of the modern yeast industry. *Industrial & Engineering Chemistry*, *22*(11), 1154–1162. doi:10.1021/ie50251a012

Gabriel, A., Victor, N., & du Preez James, C. (2014). Cactus pear biomass, a potential lignocellulose raw material for single cell protein production (SCP): A review. *International Journal of Current Microbiology and Applied Sciences*, *3*(7), 171–197.

Garcia-Garibay, M., Gómez-Ruiz, L., Cruz-Guerrero, A. E., & Bárzana, E. (2014). Single-cell protein/The algae. In C. Batt & C. A. Batt (Eds.), *Encyclopedia of food microbiology* (2nd ed.; pp. 425–430). Cambridge, MA: Academic Press; doi:10.1016/B978-0-12-384730-0.00309-8

Goldberg, I. (1985). *Single cell protein*. Berlin, Germany: Springer. doi:10.1007/978-3-642-46540-6

Gour, S., Mathur, N., Singh, A., & Pradeep, B. (2015). Single cell protein production: A review. *International Journal of Current Microbiology and Applied Sciences*, *4*(9), 251–262.

Grand View Research. (2015). *Global beef market size is expected to reach $2151 billion by 2020*. Retrieved from https://www.grandviewresearch.com/press-release/global-beef-market

Greenpeace. (2018). *Less is more: Reducing meat and dairy for a healthier life and planet*. Scientific background of the Greenpeace vision of the meat and dairy system towards 2050. Retrieved from https://storage.googleapis.com/p4-production-content/international/wp-content/uploads/2018/03/6942c0e6-longer-scientific-background.pdf

Grewal, H., Kalra, K., & Kahlom, S. (1990). Improvement of lipase production at different stirring speeds and oxygen levels. *Journal of Research*, *1*, 90–96.

Haider, M., & El-Hassy, M. (2000). Biodegradation of orange and pineapple peels for production of single cell protein. *Garyounis Journal of Science*, *1*, 7–23.

Hall, D. O., Fernández, F. G., Guerrero, E. C., Rao, K. K., & Grima, E. M. (2003). Outdoor helical tubular photobioreactors for microalgal production: Modeling of fluid-dynamics and mass transfer and assessment of biomass productivity. *Biotechnology and Bioengineering*, *82*(1), 62–73. doi:10.1002/bit.10543 PMID:12569625

Harun, R., Singh, M., Forde, G. M., & Danquah, M. K. (2010). Bioprocess engineering of microalgae to produce a variety of consumer products. *Renewable & Sustainable Energy Reviews*, *14*(3), 1037–1047. doi:10.1016/j.rser.2009.11.004

Hase, R., Oikawa, H., Sasao, C., Morita, M., & Watanabe, Y. (2000). Photosynthetic production of microalgal biomass in a raceway system under greenhouse conditions in Sendai city. *Journal of Bioscience and Bioengineering*, *89*(2), 157–163. doi:10.1016/S1389-1723(00)88730-7 PMID:16232719

Hoffmann, J. P. (1998). Wastewater treatment with suspended and nonsuspended algae. *Journal of Phycology*, *34*(5), 757–763. doi:10.1046/j.1529-8817.1998.340757.x

Ippolito, A., El Ghaouth, A., Wilson, C. L., & Wisniewski, M. (2011). Control of postharvest decay of apple fruit by *Aureobasidium pullulans* and induction of defense responses. *Postharvest Biology and Technology*, *19*(3), 265–272. doi:10.1016/S0925-5214(00)00104-6

Kellershohn, J., & Russell, I. (2015). Yeast biotechnology. In R. Rai V (Ed.), Advances in food biotechnology. Chichester, UK: Wiley Blackwell. doi:10.1002/9781118864463.ch18

Khachatourians, G. G., & Arora, D. K. (2002). *Applied mycology and biotechnology* (Vol. 2). Amsterdam: Elsevier.

Kiger, C. J. (1959). Etude de la composition chimique et de la valeur alimentaire de 57 espèces de champignons supérieurs. *Révue de Mycologie*, *24*, 161–170.

Kirsop, B. (n.d.). Food from micro-organisms. *Proceedings of the Nutrition Society*, *44*(1), 13–17.

Ko, J. L., Hsu, C. I., Lin, R. H., Kao, C. L., & Lin, J. Y. (1995). A new fungal immunomodulatory protein, FIP-fve isolated from the edible mushroom, Flammulina velutipes and its complete amino acid sequence. *European Journal of Biochemistry*, *228*(2), 244–249. doi:10.1111/j.1432-1033.1995.tb20256.x PMID:7705335

Kuhad, R. C., Singh, A., Tripathi, K. K., Saxena, R. K., & Eriksson, K. E. L. (1997). Microorganisms as an alternative source of protein. *Nutrition Reviews*, *55*(3), 65–75. doi:10.1111/j.1753-4887.1997.tb01599.x PMID:9170892

Kuzmanova, S., Vandeska, E., Dimitrovski, A., & Doneva, D. (1989). Microbial procedures from utilization of food industry wastes. In *DECHEMA biotechnology conferences*. Retrieved from https://www.cabdirect.org/cabdirect/abstract/19911368140

Lau, P. S., Tam, N. F. Y., & Wong, Y. S. (1997). Wastewater nutrients (N and P) removal by carrageenan and alginate immobilized Chlorella vulgaris. *Environmental Technology*, *18*(9), 945–951. doi:10.1080/09593331808616614

Leon-Guzman, M. F., Silva, I., & Lopez, M. G. (1997). Proximate chemical composition, free amino acid contents and free fatty acid contents of some edible mushrooms from Querétaro, Mexico. *Journal of Agricultural and Food Chemistry*, *45*(11), 4329–4332. doi:10.1021/jf970640u

Li, C., Zhu, Y., Wang, Y., Zhu, J. S., Chang, J., & Kritchevsky, D. (1998). Monascus purpureus-fermented rice (red yeast rice): A natural food product that lowers blood cholesterol in animal models of hypercholesterolemia. *Nutrition Research (New York, N.Y.)*, *18*(1), 71–81. doi:10.1016/S0271-5317(97)00201-7

Li, Y., Horsman, M., Wang, B., Wu, N., & Lan, C. Q. (2008). Effects of nitrogen sources on cell growth and lipid accumulation of green alga Neochloris oleoabundans. *Applied Microbiology and Biotechnology*, *81*(4), 629–636. doi:10.100700253-008-1681-1 PMID:18795284

Lima, G., Ippolito, A., Nigro, F., & Salerno, M. (1997). Effectiveness of *Aureobasidium pullulans* and *Candida oleophila* against postharvest strawberry rots. *Postharvest Biology and Technology*, *10*(2), 169–178. doi:10.1016/S0925-5214(96)01302-6

Litchfield, J. H. (1983). Single-cell proteins. *Science*, *219*(4585), 740–746. doi:10.1126cience.219.4585.740 PMID:17814036

Liu, C. P., & Lin, L. P. (2001). Ultrastructural study and lipid formation of Isochrysis sp. CCMP1324. *Botanical Bulletin of Academia Sinica*, *42*, 207–214.

Llorente, P., Marquina, D., Santos, A., Peinado, J. M., & Spencer-Martins, I. (1997). Effect of salt on the killer phenotype of yeasts from olive brines. *Applied and Environmental Microbiology*, *63*, 1165–1167. PMID:9055432

Lowes, K. F., Shearman, C. A., Payne, J., MacKenzie, D., Archer, D. B., Merry, R. J., & Gasson, M. J. (2000). Prevention of yeast spoilage in feed and food by the yeast mycocin HMK. *Applied and Environmental Microbiology*, *66*(3), 1066–1076. doi:10.1128/AEM.66.3.1066-1076.2000 PMID:10698773

Matassa, S., Boon, N., Pikaar, I., & Verstraete, W. (2014). Microbial protein: Future sustainable food supply route with low environmental footprint. *Microbial Biotechnology*, *9*(5), 568–575. doi:10.1111/1751-7915.12369 PMID:27389856

Mehlomakulu, N. N., Setati, M. E., & Divol, B. (2014). Characterization of novel killer toxins secreted by wine-related non-*Saccharomyces* yeasts and their action on *Brettanomyces* spp. *International Journal of Food Microbiology*, *188*, 83–91. doi:10.1016/j.ijfoodmicro.2014.07.015 PMID:25087208

Milledge, J. J. (2012). Microalgae-commercial potential for fuel, food and feed. *Food Science and Technology (Campinas)*, *26*(1), 28–30.

Miller, G. T. Jr. (1985). *Living in the environment* (4th ed.). Belmont, CA: Wadswork Publishing.

Miller, S. A., & Dwyer, J. T. (2001). Evaluating the safety and nutritional value of mycoprotein. *Food Technology*, *55*(7), 42–47.

Murata, Y., Shimamura, T., Tagami, T., Takatsuki, F., & Hamuro, J. (2002). The skewing to Th1 induced by lentinan is directed through the distinctive cytokine production by macrophages with elevated intracellular glutathione content. *International Immunopharmacology*, *2*(5), 673–689. doi:10.1016/S1567-5769(01)00212-0 PMID:12013506

Nielsen, D. S., Hønholt, S., Tano-Debrah, K., & Jespersen, L. (2005). Yeast populations associated with Ghanaian cocoa fermentations analysed using denaturing gradient gel electrophoresis (DGGE). *Yeast (Chichester, England)*, *22*(4), 271–284. doi:10.1002/yea.1207 PMID:15704234

Nout, M. R., & Boekhout, T. (2003). Traditional fermented products from Africa, Latin America and Asia. In T. Boekhout & V. Robert (Eds.), *Yeasts in food: Beneficial and detrimental aspects* (pp. 451–473). Hamburg, Germany: Behr's Verlag. doi:10.1533/9781845698485.451

O'Donnell, K., Cigelnik, E., & Casper, H. H. (1998). Molecular phylogenetic, morphological, and mycotoxin data support reidentification of the quorn mycoprotein fungus as *Fusarium venenatum*. *Fungal Genetics and Biology*, *23*(1), 57–67. doi:10.1006/fgbi.1997.1018 PMID:9501477

Olaizola, M. (2003). Commercial development of microalgal biotechnology: From the test tube to the marketplace. *Biomolecular Engineering*, *20*(4–6), 459–466. doi:10.1016/S1389-0344(03)00076-5 PMID:12919832

Oren, A. (2005). A hundred years of Dunaliella research: 1905–2005. *Saline Systems*, *1*(1), 2. doi:10.1186/1746-1448-1-2 PMID:16176593

Osho, A. (1995). Production of metabolites, industrial enzymes, amino acids, vitamins, single cell protein. *Journal of Research*, *6*, 521–529.

Parafati, L., Vitale, A., Restuccia, C., & Cirvilleri, G. (2015). Biocontrol ability and action mechanism of food-isolated yeast strains against *Botrytis cinerea* causing postharvest bunch rot of table grape. *Food Microbiology*, *47*, 85–92. doi:10.1016/j.fm.2014.11.013 PMID:25583341

Pasquet, V., Chérouvrier, J. R., Farhat, F., Thiéry, V., Piot, J. M., Bérard, J. B., ... Picot, L. (2011). Study on the microalgal pigments extraction process: Performance of microwave assisted extraction. *Process Biochemistry*, *46*(1), 59–67. doi:10.1016/j.procbio.2010.07.009

Pessoa, J. A., Mancilha, I. M., & Sato, S. (1996). Cultivation of Candida tropicalis in sugarcane hemicellulosic hydrolysate for microbial protein production. *Journal of Biotechnology*, *51*(1), 83–88. doi:10.1016/0168-1656(96)01572-6

Ponis, E., Probert, I., Veron, B., Mathieu, M., & Robert, R. (2006). New microalgae for the Pacific oyster Crassostrea gigas larvae. *Aquaculture (Amsterdam, Netherlands)*, *253*(1–4), 618–627. doi:10.1016/j.aquaculture.2005.09.011

Prabhat, S. (2017). *Difference between yeast and fungus*. Retrieved from http://www.differencebetween.net/science/nature/difference-between-yeast-and-fungus/

Pratt, R., Daniels, T. C., Eiler, J. J., Gunnison, J. B., Kumler, W. D., Oneto, J. F., ... Smith, J. H. C. (1944). Chlorellin, an antibacterial substance from chlorella. *Science*, *99*(2574), 351–352. doi:10.1126cience.99.2574.351 PMID:17750208

Radmer, R. J., & Parker, B. C. (1994). Commercial applications of algae: Opportunities and constraints. *Journal of Applied Phycology*, *6*(2), 93–98. doi:10.1007/BF02186062

Ranganathan, J. (2013). *The global food challenge explained in 18 Graphics*. World Resources Institute. Retrieved from https://www.wri.org/blog/2013/12/global-food-challenge-explained-18-graphics

Reinehr, C. O., & Costa, J. A. V. (2006). Repeated batch cultivation of the microalga Spirulina platensis. *World Journal of Microbiology & Biotechnology*, *22*(9), 937–943. doi:10.100711274-006-9138-0

Righelato, R. C. (1976). Selection of strains of Penicillium chrysogenum with reduced penicillin yields in continuous cultures. *Journal of Chemical Technology and Biotechnology (Oxford, Oxfordshire)*, *26*(1), 153–159. doi:10.1002/jctb.5020260125

Riviere, J. (1977). Microbial proteins. In M. O. Moss & J. E. Smith (Eds.), *Industrial applications of microbiology* (pp. 105–149). London, UK: Surry University Press.

Rojas, C., & Mansur, E. (1995). Ecuador: Informaciones generals sobre productos non madereros in Ecuador. In *Memoria, consulta de expertos sobre productor forestales no madereros para America Latina y el Caribe* (pp. 208–223). Santiago, Chile: Regional Office for Latin America and the Caribbean, Food and Agriculture Organization of the United Nations.

Rosa-Magri, M. M., Tauk-Tornisielo, S. M., & Ceccato-Antonini, S. R. (2011). Bioprospection of yeasts as biocontrol agents against phytopathogenic molds. *Brazilian Archives of Biology and Technology*, *54*(1), 1–5. doi:10.1590/S1516-89132011000100001

Roser, M., & Ritchie, H. (2018). *Hunger and undernourishment*. Our World in Data. Retrieved from https://ourworldindata.org/hunger-and-undernourishment

Rossi, V., Jovicevic, L., Nistico, V., Orticelli, G., Troiani, M. P., & Marini, S. (1993). In vitro antitumor activity of Lentinus edodes. *Pharmacological Research*, *27*, 109–110. doi:10.1006/phrs.1993.1089

Saima, M. A., Khan, M., Anjum, M., Ahmed, S., Rizwan, M., & Ijaz, M. (2008). Investigation on the availability of amino acids from different animal protein sources in golden cockerels. *Journal of Animal and Plant Sciences*, *18*(2-3), 53–55.

Santos, A., Navascués, E., Bravo, E., & Marquina, D. (2011). *Ustilago maydis* killer toxin as a new tool for the biocontrol of the wine spoilage yeast *Brettanomyces bruxellensis*. *International Journal of Food Microbiology*, *145*(1), 147–154. doi:10.1016/j.ijfoodmicro.2010.12.005 PMID:21195497

Schwan, R. F., & Wheals, A. E. (2003). Mixed microbial fermentations of chocolate and coffee. In T. Boekhout & V. Robert (Eds.), *Yeasts in food: Beneficial and detrimental aspects* (pp. 429–449). Hamburg, Germany: Behr's Verlag. doi:10.1533/9781845698485.429

Sheu, F., Chien, P. J., Chien, A. L., Chen, Y. F., & Chin, K. L. (2004). Isolation and characterization of an immunomodulatory protein (APP) from the Jew's Ear mushroom Auricularia polytricha. *Food Chemistry*, *87*(4), 593–600. doi:10.1016/j.foodchem.2004.01.015

Shimamatsu, H. (2004). Mass production of Spirulina, an edible microalga. *Hydrobiologia*, *512*(1–3), 39–44. doi:10.1023/B:HYDR.0000020364.23796.04

Sialve, B., Bernet, N., & Bernard, O. (2009). Anaerobic digestion of microalgae as a necessary step to make microalgal biodiesel sustainable. *Biotechnology Advances*, *2*(4), 409–416. doi:10.1016/j.biotechadv.2009.03.001 PMID:19289163

Smiderle, F. R., Carbonero, E. R., Sassaki, G. L., Gorin, P. A., & Iacomini, M. (2008). Characterization of a heterogalactan: Some nutritional values of the edible mushroom Flammulina velutipes. *Food Chemistry*, *108*(1), 329–333. doi:10.1016/j.foodchem.2007.10.029

Solomons, G. L. (1987). Myco-protein: Safety evaluation of a novel food. In C. M. Chambers, P. L. Chambers, & D. S. Davies (Eds.), *Mechanisms and models in toxicology. Archives of Toxicology* (Vol. 11, pp. 191–193). Berlin, Germany: Springer. doi:10.1007/978-3-642-72558-6_30

Spolaore, P., Joannis-Cassan, C., Duran, E., & Isambert, A. (2006). Commercial applications of microalgae. *Journal of Bioscience and Bioengineering*, *101*(2), 87–96. doi:10.1263/jbb.101.87 PMID:16569602

Suman, G., Nupur, M., Anuradha, S., & Pradeep, B. (2015). Single cell protein production: A review. *International Journal of Current Microbiology and Applied Sciences*, *4*(9), 251–262.

Tee, R. D., Gordon, D. J., Welch, J. A., & Taylor, A. N. (1993). Investigation of possible adverse allergic reactions to mycoprotein ('Quorn'). *Clinical and Experimental Allergy*, *23*(4), 257–260. doi:10.1111/j.1365-2222.1993.tb00319.x PMID:8319120

Thrane, U., Andersen, B., Frisvad, J. C., & Smedsgaard, J. (2007). The exo-metabolome in filamentous fungi. In J. Nielsen & J. Hewitt (Eds.), *Metabolomics: A powerful tool in systems biology* (pp. 235–252). Berlin, Germany: Springer. doi:10.1007/4735_2007_0230

Tovar, D., Zambonino, J., Cahu, C., Gatesoupe, F. J., Vázquez-Juárez, R., & Lésel, R. (2002). Effect of live yeast incorporation in compound diet on digestive enzyme activity in sea bass (*Dicentrarchus labrax*) larvae. *Aquaculture (Amsterdam, Netherlands)*, *204*(1-2), 113–123. doi:10.1016/S0044-8486(01)00650-0

Trinci, A. P. J. (1992). Myco-protein: A twenty-year overnight success story. *Mycological Research*, *96*(1), 1–13. doi:10.1016/S0953-7562(09)80989-1

Tseng, C. K. (2004). The past, present and future of phycology in China. *Hydrobiologia*, *512*(1-3), 11–20. doi:10.1023/B:HYDR.0000020363.37807.ec

Turnbull, W. H., Leeds, A. R., & Edwards, D. G. (1992). Mycoprotein reduces blood lipids in free-living subjects. *The American Journal of Clinical Nutrition*, *55*(2), 415–419. doi:10.1093/ajcn/55.2.415 PMID:1734679

Turnbull, W. H., Walton, J., & Leeds, A. R. (1993). Acute effects of mycoprotein on subsequent energy intake and appetite variables. *The American Journal of Clinical Nutrition*, *58*(4), 507–512. doi:10.1093/ajcn/58.4.507 PMID:8257542

Turner, N. J., Kuhnlein, H. V., & Egger, K. N. (1987). The cottonwood mushroom (*Tricholoma populinum*): A food resource of the Interior Salish Indian peoples of British Columbia. *Canadian Journal of Botany*, *65*(5), 921–927. doi:10.1139/b87-127

Ugwu, C., Ogbonna, J., & Tanaka, H. (2002). Improvement of mass transfer characteristics and productivities of inclined tubular photobioreactors by installation of internal static mixers. *Applied Microbiology and Biotechnology*, *58*(5), 600–607. doi:10.100700253-002-0940-9 PMID:11956741

Ullivarri, M. F., Mendoza, L. M., & Raya, R. R. (2014). Killer activity of *Saccharomyces cerevisiae* strains: Partial characterization and strategies to improve the biocontrol efficacy in winemaking. *Antonie van Leeuwenhoek*, *106*(5), 865–878. doi:10.100710482-014-0256-7 PMID:25145824

van den Ende, H., & Stegwee, D. (1971). Physiology of sex in Mucorales. *Botanical Review*, *37*(1), 22–36. doi:10.1007/BF02860299

Vega-Estrada, J., Montes-Horcasitas, M. C., Domínguez-Bocanegra, A. R., & Canizares-Villanueva, R. O. (2005). Haematococcus pluvialis cultivation in split-cylinder internal-loop airlift photobioreactor under aeration conditions avoiding cell damage. *Applied Microbiology and Biotechnology*, *68*(1), 31–35. doi:10.100700253-004-1863-4 PMID:15657723

Vincent, W. A. (1969). Algae for food and feed. *Process Biochemistry*, *4*(6), 45–47.

Wainwright, M. (1992). *An introduction to fungal biotechnology*. Chichester, UK: John Wiley and Sons.

Went, F. A. F. C. (1895). Monascus purpureus, le champignon de l'âng-quac. Une nouvelle thélébolée. *Annales des Sciences Naturelles. Botanique et Biologie*, *8*, 1–18.

Wikfors, G. H., Ferris, G. E., & Smith, B. C. (1992). The relationship between gross biochemical composition of cultured algal foods and growth of the hard clam, Mercenaria mercenaria (L.). *Aquaculture (Amsterdam, Netherlands)*, *108*(1–2), 135–154. doi:10.1016/0044-8486(92)90324-E

Yamaguchi, K. (1997). Recent advances in microalgal bioscience in Japan, with special reference to utilization of biomass and metabolites: A review. *Journal of Applied Phycology*, *8*(6), 487–502. doi:10.1007/BF02186327

Zhang, J., Tang, Q., Zimmerman-Kordmann, M., Reutter, W., & Fan, H. (2002). Activation of B lymphocytes by GLIS, a bioactive proteoglycan from Ganoderma lucidum. *Life Sciences*, *71*(6), 623–638. doi:10.1016/S0024-3205(02)01690-9 PMID:12072151

Zubi, W. (2005). *Production of single cell protein from base hydrolyzed of date extract by product by the fungus Fusarium graminearum* (Master thesis). Garyounis University, Benghazi, Libya.

ADDITIONAL READING

Aaronson, S. (2000). Fungi. In K. F. Kiple & K. C. Ornelas (Eds.), *The Cambridge world history of food* (pp. 313–336). Cambridge, UK: Cambridge University Press.

Adedayo, M. R., Ajiboye, E. A., Akintunde, J. K., & Odaibo, A. (2011). Single cell proteins: As nutritional enhancer. *Advances in Applied Science Research*, *2*(5), 396–409.

Anupama, & Ravindra, P. (2000). Value-added food: Single cell protein. *Biotechnology Advances*, *18*(6), 459–479. doi:10.1016/S0734-9750(00)00045-8

Beer, E. (2015). "No success model" for new protein ingredients. Retrieved from http://www.foodnavigator.com/Market-Trends/No-success-model-for-new-protein-ingredients

Boa, E. (2004). *Wild edible fungi: A global overview of their use and importance to people*. Rome, Italy: Food and Agriculture Organization of the United Nations; Retrieved from http://www.fao.org/docrep/007/y5489e/y5489e00.htm#TopOfPage

Caplice, E., & Fitzgerald, G. F. (1999). Food fermentations: Role of microorganisms in food production and preservation. *International Journal of Food Microbiology*, *50*(1-2), 131–149. doi:10.1016/S0168-1605(99)00082-3 PMID:10488849

Gabriel, A., Victor, N., & du Preez James, C. (2014). Cactus pear biomass, a potential lignocellulose raw material for single cell protein production (SCP): A review. *International Journal of Current Microbiology and Applied Sciences*, *3*(7), 171–197.

Greenpeace (2018). *Less is more: Reducing meat and dairy for a healthier life and planet*. Scientific background of the Greenpeace vision of the meat and dairy system towards 2050. Retrieved from https://storage.googleapis.com/p4-production-content/international/wp-content/uploads/2018/03/6942c0e6-longer-scientific-background.pdf

Querol, A., & Fleet, G. H. (Eds.). (2006). *Yeasts in food and beverages*. Berlin, Germany: Springer. doi:10.1007/978-3-540-28398-0

Suman, G., Nupur, M., Anuradha, S., & Pradeep, B. (2015). Single cell protein production: A review. *International Journal of Current Microbiology and Applied Sciences*, *4*(9), 251–262.

KEY TERMS AND DEFINITIONS

Actinomycetes: Gram-positive bacteria (that is, retain the crystal violet stain from the Gram test) used in antibiotics and food.

Algae: A large and diverse group of aquatic organisms; they range from unicellular microalgae (e.g., chlorella) to multicellular formations (e.g., giant kelp); seaweed is the most complex marine algae.

Bacteria: Single-cell organisms; they can live outside (on land and water) or inside the human body.

Fermentation: A process of chemical breakdown of food substances by bacteria, yeasts or other microorganisms which improves the taste and preserves the products.

Food Supplement: A concentrated source of nutrients, such as amino acids, fatty acids, fiber, minerals, and vitamins that is taken to correct food deficiencies; within Europe food supplements are regulated as foods and this is the approach taken in this chapter.

Fungi: A large group of eukaryotic organisms whose cells consist of a nucleus enclosed within membrane; the fungi kingdom covers 80,000 species, including the most commonly known mushrooms, yeasts, and molds.

Microbial Protein: A form of dehydrated microbial culture of cells, obtained from microorganisms.

Protein: An organic substance – polymer chains of amino acids, considered an essential nutrient for the human body; there are 20 types of amino acids representing the building blocks for the human proteins – 11 are non-essential which can be synthesized by the human organism and 9 are essential which need to be provided by food.

Single-Cell Protein: Protein obtained from microorganisms consisting of one cell (also referred as unicellular).

Yeast: Single-cell fungi used in the preparation of food (e.g., bread) and drinks (e.g., beer).

ENDNOTE

1 Potential use of microbial protein as animal feed is outside the scope of this book.

Chapter 7
Soybeans Consumption and Production in China:
Sustainability Perspective

Xiumei Guo
Curtin University, Australia

Xiaoling Shao
Nanjing Audit University, China

Shagufta M. Trishna
Curtin University, Australia

Dora Marinova
Curtin University, Australia

Amzad Hossain
Curtin University, Australia & Rajshahi University, Bangladesh

ABSTRACT

China is the world's top consumer and largest importer of soybeans used as human food and livestock feed. Since the 1980s, China's meat consumption has been growing despite this being an inefficient way of feeding the world's largest population. It diverts resources which can be used directly for human consumption. If the Chinese people were to maintain or expand their high consumption of soybean-based foods instead of switching to a meat-rich diet, greenhouse gas emissions would be reduced, and natural resource use improved. This chapter examines the trends in soy consumption and production in China and explores people's dietary preferences for soybeans, including concerns about the import of genetically modified soybeans. Without diverting soybeans to animal feed, the demand for them will decrease and will make China more self-sufficient. This study also provides educational guidance about the health benefits of plant-based foods and environmental damage associated with high consumption of animal-based products.

DOI: 10.4018/978-1-5225-7350-0.ch007

INTRODUCTION

It is well known that China is not able to produce enough soybeans for processing to meet the growing demands for human soybean-based food and animal feed for livestock. With the increasing household incomes of the Chinese people, their demand for meat and other animal-based foods is also growing. Foods that were once considered unaffordable or foreign are now part of the transition to more western style dietary habits (Ma Verkuil, Reinbach, & Meinert, 2017).

Although pork continues to be the dominate animal protein, there is surging interest in beef and poultry with China's total and per capita meat consumption on the rise since the 1980s (Nam, Jo, & Lee, 2010). Meat production reached 86.45 million tonnes in 2014 and annual meat consumption was 61.82 kg per person per year in 2013 (Ritchie & Roser, 2018). In 2014, the number of livestock animals raised for human consumption in China included 480 million pigs, 114 million cattle and 5.58 billion poultry compared to respectively 326 million, 52 million and 1.18 billion in 1980 (Ritchie & Roser, 2018).

Since the discovery by animal nutritionists that combined with grain, soybean can be used very efficiently as feed for livestock and poultry to boost the production of animal protein, soybeans have been consistently given to farm animals (Brown, 2011). As China's appetite for animal-based products, such as meat and milk grew, so did the conversion of soybeans to animal meal (Brown, 2011). According to Brown from the Earth Policy Institute (2011, p. 95), "since half of the world's pigs are in China, the lion's share of soy use is in pig feed. Its fast-growing poultry industry is also dependent on soybean meal".

This is in sharp contrast with the traditional use of soy which was domesticated as a garden plant by Chinese farmers around 1100 BC (NC Soybean Producers Association, 2014). The legume plant was named "miracle crop" because of its versatile properties and its ability to produce oil and other byproducts suitable for human consumption, such as tofu and soy drinks (U.S. Soybean Export Council, 2006). More recently, soybeans have been grown commercially all around the world for animal feed. In this day and age, "[s]oybean oil is the most widely used edible oil in the world and soybean meal is the leading protein and energy source for animal feeds" (U.S. Soybean Export Council, 2006, p. 4). Soy is also used in cosmetics, pharmaceutical, manufacturing and other industries, as a lubricant, in inks, paints and varnishes as well as biofuel.

The list of applications is long, but nowhere is soy as wanted as it has been as animal feed. This has led to land clearing and conversion to grow soy in some of the most important from a biodiversity point of view places, such as the rainforests of the Amazon (Brügger, Marinova, & Raphaely, 2016). The conflict between the use of soybeans as food and feed on a limited planet has escalated to enormous proportions and China (together with all other high-meat consuming countries) is contributing to large scale deforestation, greenhouse gas emissions, biodiversity loss and inefficient ways for feeding the human population (Schmidinger, Bogueva, & Marinova, 2018). Instead of being used for feeding people directly, soybeans are prepared as animal meal and fed to livestock. In the case of pork – the most popular meat choice in China, 11 calories are fed to the animal to produce 1 calorie for human consumption (Eshel, Shepon, Makov, & Milo, 2014). The respective figures for beef are 38 and for poultry 9 (Eshel, Shepon, Makov, & Milo, 2014).

A solution to the global demand for soy has been through genetic engineering and the development of genetically modifies (GM) versions of soybeans. More than 90% of the soy planted in the US is genetically engineered with the assertions that this helps increase yields and reduce the use of pesticides (Brookes & Barfoot, 2017). However, some disagree with such a view (e.g. Satheesh, 2012) and are of the opinion that GM seeds have not delivered better performance than conventional soy. There are also serious concerns raised about the ethics, risks and impacts (often unknown) on human wellbeing and the health of the planet from GM crops. Such concerns are widely spread across the globe (Bawa & Anilakumar, 2013; Bodnar, 2018) and this is an area with a large gap between public perception and scientific position – 88% of scientists believe that it is safe to eat GM crops while only 37% of the general public are of the same opinion (Funk & Rainie, 2016). These concerns are particularly valid for China as the country currently imports large amounts of soybeans to be used as animal feed and human food, including from USA and Brazil where GM seeds are allowed.

Although in the last few decades China has experienced a growing human consumption of soybeans, the direct use for food products, such as tofu, soy drinks and soy sauce, remains low. It has increased from 10% of the crop used directly as food products in 2010 (Brown, 2011) to 14% in 2017 (China Industrial Information Network, 2018). The remaining 86% are used as pressed oil and animal feed. It is well established that plants and soy in particular are more environmentally friendly than meat production and generate lower greenhouse gas emissions, less pollution and have less requirements for land (Raphaely & Marinova, 2016). Research also shows that soy food products are recognized as the best protein alternatives to meat as they contain a complete set of essential amino acids (Marsall & Marinova, 2019). Therefore, promoting soybeans-based food consumption in China – the country with the world's largest population, becomes an urgent task when tackling environmental issues locally and globally.

This chapter elaborates on the issue about soy production and consumption in China making the argument that if soy is used for direct human consumption with a shift to nutritious and healthy plant-based food options, there will be much less need for this versatile crop to grown in the current excessive amounts. Hence, there will also be no need for genetic engineering and genetic modifications to artificially increase the already healthy yield capacities of this miracle crop. The chapter examines first China's soybean production and trade. Then it analyses the actual soybean consumption in the country, including presenting an overview of some soy-based meat alternatives. It finally makes policy recommendations to encourage behavior changes away from high meat consumption which will be beneficial for China from a human health as well as planetary wellbeing point of view, and given this country's population size such a transition will be advantageous for the entire globe.

The study aims to improve Chinese consumers' awareness of considering the environment when making decisions about what to choose to eat and encourages soybean consumption, with less meat intake, in order to achieve the decoupling between food consumption and production and the environmental damage, simultaneously maintaining better health. In addition, this study highlights the importance of social marketing and policy interventions in encouraging healthy food consumption to maintain the sustainability of modern society.

SOYBEANS PRODUCTION AND TRADE IN CHINA

China has a long history of soybeans production, which has been playing an essential role in poverty reduction by providing plant protein resources and a healthy edible vegetable oil for Chinese people. Planting soybeans domestically in rural areas of the country has contributed significantly to agricultural sustainability. With soybean being a legume, it transfers the atmospheric nitrogen into the soil and does not require as much fertiliser while improving the fertility of the land. The Rhizobia bacteria which infects the roots of all legumes, supplies enough nitrogen to the plant from the air and also helps with nitrogen fixing of the soil (Mosaic Company, 2018). Farmers learn to rotate the crops, maintain the nitrogen cycle and reduce the need to use fertilisers with soybeans fixing naturally and biologically the soil's fertility. In temperate and tropical climates, atmospheric nitrogen transfer to the soil through the symbiotic association between the Rhizobia bacteria and the legume plants, including soybeans, represents a renewable source for fixing soil fertility even for arid and semi-arid lands (Zahran, 1999). The Chinese government is keen to promote sustainable soybeans production for optimizing the planting structure and balancing the supply and demands of soybeans with other agricultural crops (Ministry of Agriculture and Rural Affairs of the People's Republic of China, 2016).

Due to population growth, industrialisation and fast urbanisation, China's planting areas for soybeans have been generally decreasing in the past 20 years (see Figure 1). The amount of land for soybean planting peaked in the years 2004 and 2005 to over 9.5 million hectares but in recent years has been lower at around 6.5 million hectares. Although total soybeans output per hectare has remained relatively stable since 2011 (see Figure 2) due to improved agricultural technology, planting innovation or favourable weather conditions (FAO, 2016), the overall production has decreased because of the shrinking amount of available land.

In 2016, the planting area for soybeans started to increase after six years of decrease. It is expected that China will continue to restoratively increase its soybeans planting areas in 2018 at an 1.1% annual growth rate towards achieving a total output of 15 million tonnes (Xinhua, 2018). In 2017, the total soybeans production was already 14.3 million tonnes (Wang, 2018). The domestic sources of soybeans outputs were mainly from the traditional agricultural provinces – Heilongjiang (41%), Anhui (11%), Inner Mongolia, Henan and others. As a matter of priority, the Chinese Government is encouraging farmers to continue to increase the land areas available to grow soy in order to reduce the country's current reliance on imports (Wang, 2018).

This however will not be enough for the current demand for soy products. Since 1994, China has been a net importer of soybeans and due to the large trade volumes, it has been the top soybeans importer in

Figure 1. Soybeans planting areas in China and annual change (1000 hectare), 1997–2016
Source: National Bureau of Statistics of China (1997–2016)

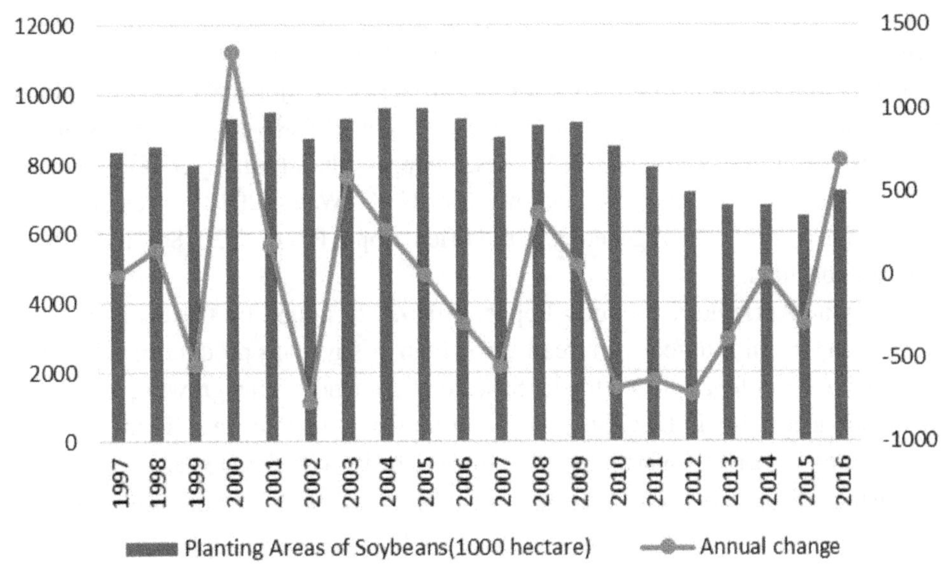

Figure 2. Total production of soybeans (10 thousand tonnes) and soybeans output per hectare (kg) in China, 1997–2016
Source: National Bureau of Statistics of China (1997-2016)

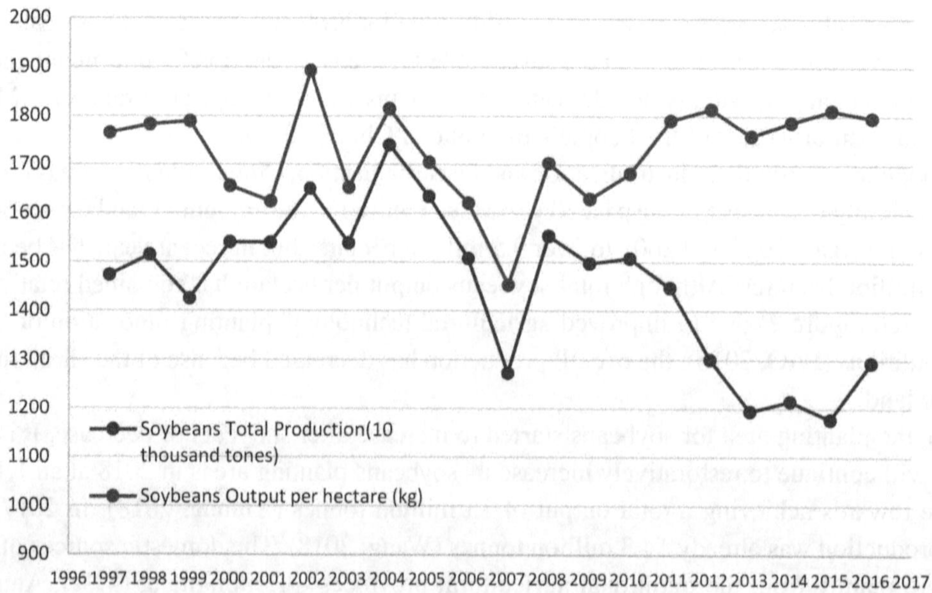

**For a more accurate representation see the electronic version.*

the world since 2013 (The Statistics Portal, 2018). China's gradually growing imports of soybeans (see Figure 3) reached over 83 million tonnes in 2016 followed by a further drastic 14.8% increase to 95.5 million tonnes in 2018 (Wang, 2018).

Due to this continuous demand for soybeans products, including for human consumption and as animal feed, the soybean import is likely to be maintained at a high level (Central People's Government of the People's Republic of China, 2018b). The shortage of land availability for soybeans planting means that with the current consumption patterns China will continue to depend on imports from foreign countries for the supply of this valuable commodity for domestic use. In 2017, 87% of all domestically consumed soybeans were imported (Wang, 2018) and China in fact accounts for two-thirds of the global imports (Sheldon, 2018). A lot of the imported soy comes from the two major players on the global market – USA and Brazil (Sheldon, 2018). The two countries are the world's largest producers of soybeans and essentially dominate the global market. In 2017, Brazil supplied half (namely 53%) of China's soybeans imports whilst another third (namely 34%) came from USA (Wang, 2018). These two countries are also the world's largest producers of genetically modified crops (Reuters, 2018a), including a lot of the soybeans exported to China.

In 2018, the China Agriculture Outlook Report 2016-2025 issued by the Ministry of Agriculture emphasized the need for China to focus on steady growth in its soybeans production. However, given the limited area available for cultivation and the large scale of demand from a growing population and fast-expanding livestock sector, it will be very challenging for China to achieve self-sufficiency in soybean production. Therefore, imported soybeans will continue to be in large demand as a supply channel in the future (Food Business Net, 2016). Together with this, the concerns about GM crops will also remain.

Figure 3. Soybeans export and import and the net import in China, 1980–2016
Source: National Bureau of Statistics of China (1997-2016)

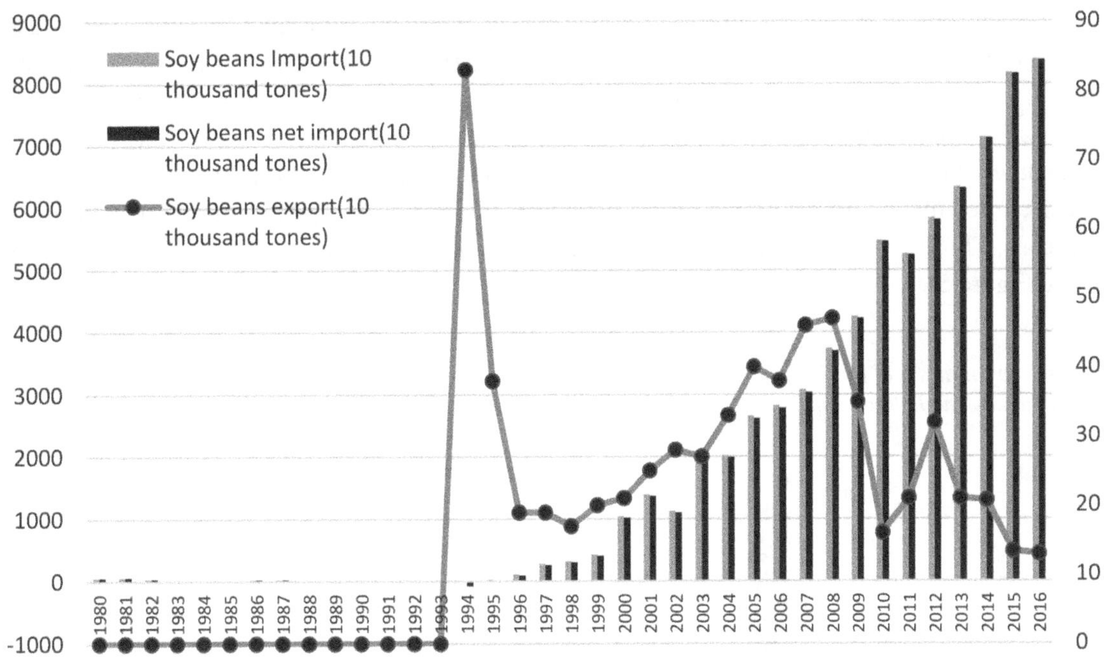

**For a more accurate representation see the electronic version.*

SOYBEANS CONSUMPTION IN CHINA

Until mid-20th century China was the largest soybean producer in the world (Jamet & Chaumet, 2016). Since then, China has become the world's largest consumer of soybeans and this consumption is continuously increasing. According to the National Agricultural Market Research Centre of the Agricultural University of China (2018), China's soybeans consumption grew over 4% in 2017. The Chinese customs data report even higher growth of more than 14% for both domestic supply and imports of soybeans (Wang, 2018). These recent changes are driven by growing domestic demand which is shifting not only in volumes but also in nature.

Overview

China has a long history of plantation and consumption of soybeans. The consumption volume was relatively stable before 1984 at around 800 tonnes from which 70% were consumed as human food products, 20% as pressed products, such as soy oil and animal meals, and the remaining 10% were used directly as animal feed, seeds and loss in the agricultural system. After 1985, the amount of pressed products, including animal meals, increased significantly between 28% and 45%, while the use for human food products decreased to 50%–60%. This trend continued in the 1990s and 2000s. At the moment, the pressed products, including animal meals, dominate China's domestic consumption at 83% while direct human food consumption is at 14% and direct animal feed at 2% (China Industrial Information Network, 2018). Seeds for re-planting represent only 0.8% of the soybean consumption and the loss in the system is low at 0.2%.

Soybeans for Animal Consumption

It is very important to unpack the pressed oil category as it combines products destined for industrial use and animal feed. In 2017, 80% of the pressed soybeans (or 66.4% of the total consumed soybeans) were used to manufacture animal soybean meals (China Industrial Information Network, 2018). Together with the 2% directly used as feed, this means that more than two-thirds of the soybeans consumed in China, namely 68.4%, are used to feed livestock animals (see Figure 4). China's pig, cattle and poultry livestock are fed mainly from soybean meals, which has intensified both, the demand and consumption of soybeans. This is not surprising given the high numbers of these animals raised for human consumption (see Figure 5).

Figure 4. Soybeans use in China [%], 2017
Source of data: China Industries Information Net (2018)

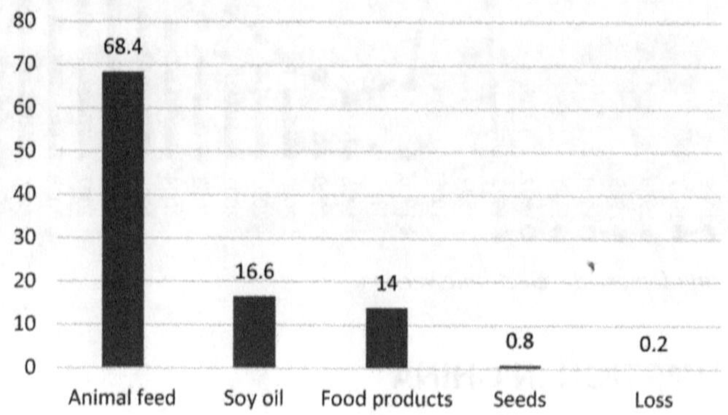

Figure 5a. Cattle livestock counts (heads) in China, 1961–2014
Source: Ritchie & Roser (2018)

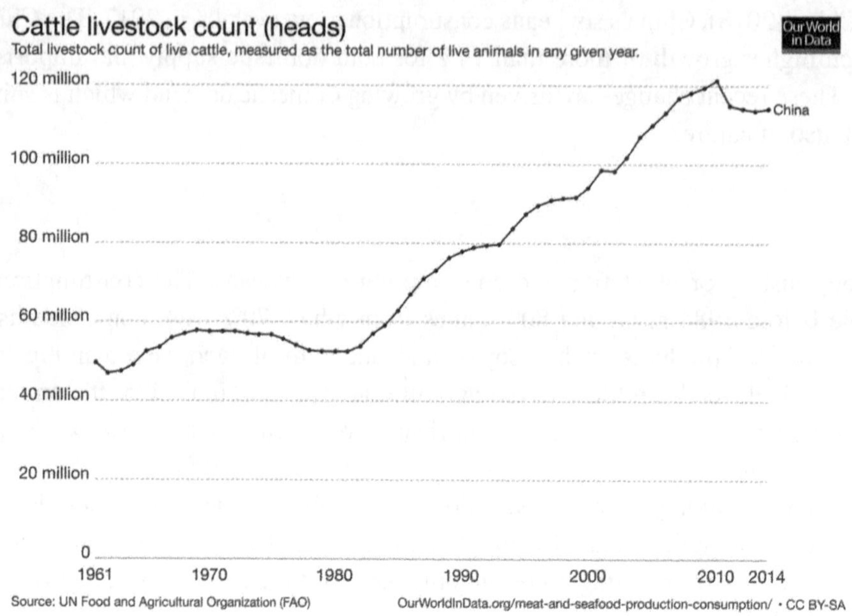

Figure 5b. Pig livestock counts (heads) in China, 1961–2014
Source: Ritchie & Roser (2018)

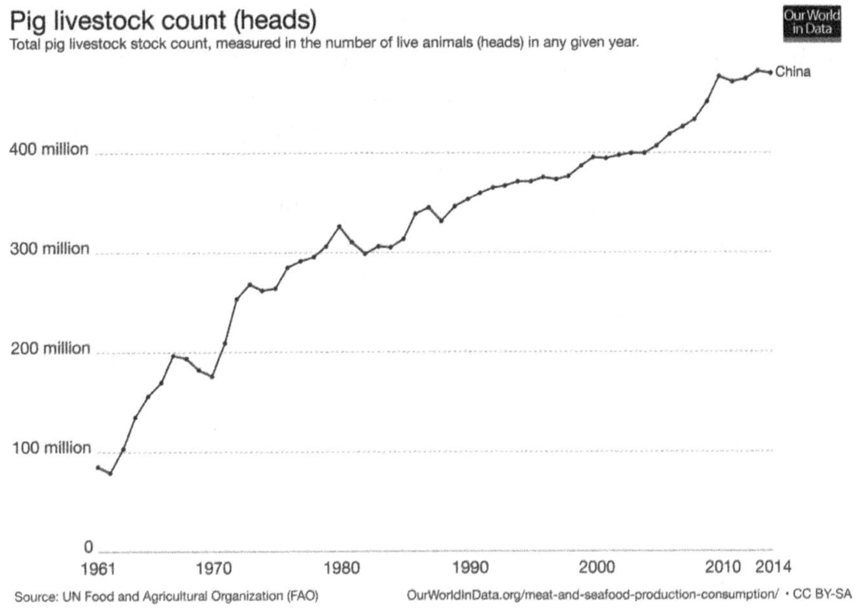

Figure 5c. Poultry livestock counts (heads) in China, 1961–2014
Source: Ritchie & Roser (2018)

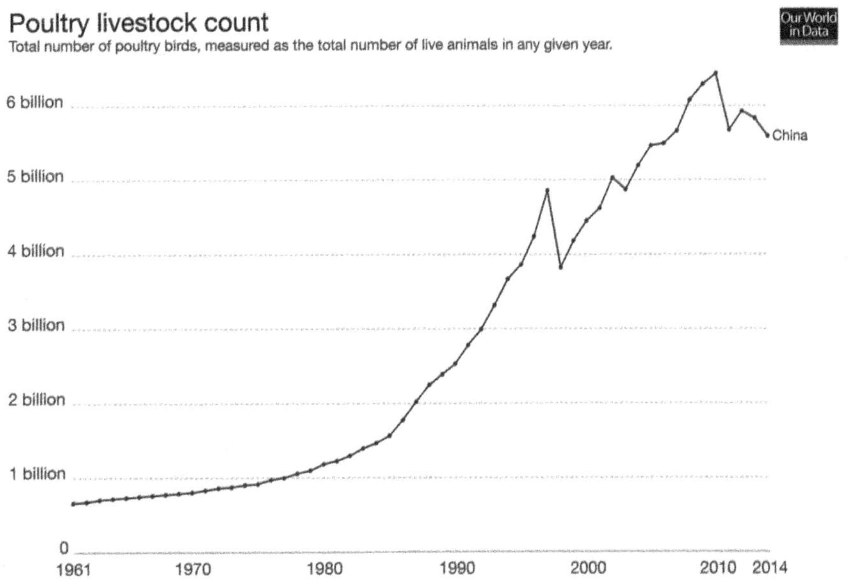

Soybeans for Human Consumption

Figure 6 presents the versatility of uses of the soybeans. Direct food products and soy oil represent less than a third (namely 30.6%) of the total soybean consumption despite the myriad of uses (see Figure 4). Of particular interest because of their nutritional value are the bean products (the middle column of Figure 6) for direct human intake which in 2017 were only 14% of the total soybean consumption in China. From the category of fat products, soybean oil can be used for cooking and is sold at a reasonable price.

The other soybean products have many specialised applications within and outside the food industry. For example, from the group of deeply processed soybeans products, soy lecithin which supports the development of the human nervous system is widely used in the confectionery food industry as well as for medicines, paper and in the leather industries.

The soybean products include non-fermented varieties, such as tofu (including water tofu and dried tofu), yuba (referred also as tofu skin or bean curd sheet), soymilk and others, and fermented varieties, such as fermented bean curd, stinky tofu, bean paste, soy sauce etc. Other bean products cover bean sprouts, fried soy products, smoked soy products, frozen soy products, soy flour etc. The soy flour can be used to make a variety of foods, including baby foods. All bean products contain the nine essential amino acids and are complete high-content proteins which can replace meat and other animal-based products.

In the past, the bean products were widely used in traditional recipes and typical Chinese dishes; however, in more recent years with the country's fast economic growth leading to increasing household incomes, there has also been a dietary shift towards increasing demand and intake of animal products. The liberation of trade since the 1978 open door economic policy, allowed imports of soybeans which fuelled their use as animal feed. Previously, China did not have enough feed to support large livestock production, but the soybean imports allowed this sector to expand (Jamet & Chaumet, 2016). Increased supply of domestic meat – coupled also with meat imports, also as a result from trade liberalisation (Guo, Raphaely, & Marinova, 2016), resulted in a fast expansion of the place of animal proteins in the Chinese dietary preferences (see Figure 7). The humble soybean, the miracle crop, caused a dietary revolution that affects not only the Chinese population but also all populations on this planet.

Figure 6. Use of soybeans in China
Source: The Authors.

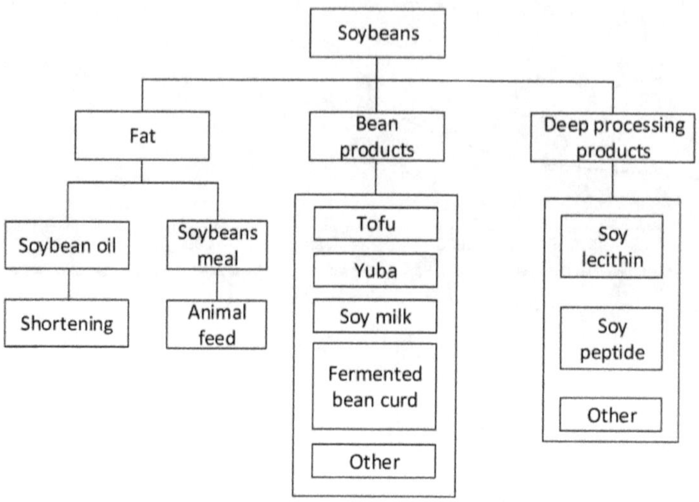

Figure 7. Per capita meat supply in China [kg], 1961–2013
Source: Ritchie & Roser (2018)

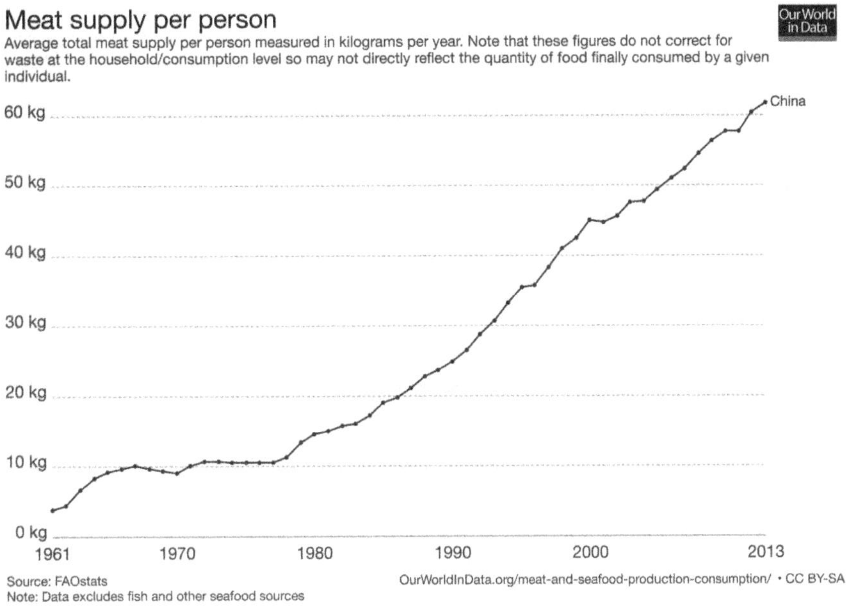

Dietary Changes

There is ample research evidence demonstrating the positive role of plant-based foods (e.g. Raphaely & Marinova, 2016; Craig, 2018), and particularly soybeans intake (Barret, 2006), in the human diet. Furthermore, there is convincing evidence about the negative impacts of excessive meat consumption on human health, including in the case of China (Campbell & Campbell, 2006). However, China has been on a dietary trend of increased meat consumption (see Figure 7) with the westernisation of diets spreading not just in urban centres but also to rural people (see Figure 8). Although the direct consumption of soybeans products has increased recently, this was accompanied with higher increases in the consumption of meat-based foods. This is fuelling the demand for soybeans, establishing a strong dependence on imports and with it, creating concerns about GM products.

The response from the Chinese Government has been two-pronged. On the one hand, farmers are encouraged to grow more soybeans which should play the leading role in domestic food consumption (China Economy Net, 2018). The aim is to make sure that at least the domestic human consumption is self-sufficient. On the other hand, China's latest dietary guidelines are recommending significant cut in the current levels of meat consumption in line with other reputable international health bodies. They are calling for an intake of 40 g to 75 g of meat per day and overall halving of the country's current meat consumption (Froggatt & Wellesley, 2016). This will also reduce the environmental and climate change pressure caused by the country's large livestock sector – a move, strongly supported by the global community (Milman & Leavenworth, 2016).

With China having the largest meat consumption in the world, a replacement of animal-based proteins with soybeans will reduce greenhouse gas emissions, improve land use, reduce competition for water and contribute towards better health outcomes (Aleksandrowicz, Green, Joy, Smith, & Haines, 2016). Soybean products are essentially meat replacement alternatives with a much smaller environmental impact.

Figure 8. Per capita consumptions of soybeans, meat and poultry by rural residents [kg], China, 2007–2016
Source of data: National Bureau of Statistics (2018)

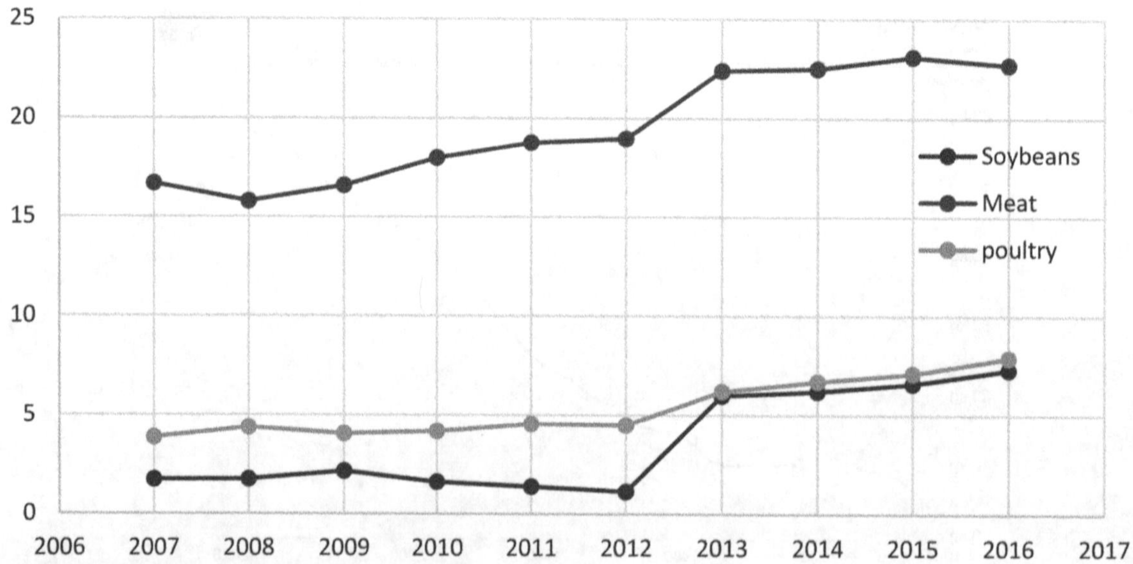

**For a more accurate representation see the electronic version.*

Halving meat consumption in China will also pave the way to restoring the traditional beauty of soy-based products as a healthy dietary choice. It will allow China to lead a global transformation along the lines of those initiated by international non-government organisations, such as Greenpeace (2018). This will further reduce the pressure on imports and the global demand for ever-increasing yields of soybeans. Even if GM soy is to be shown to be harmless in the long run, the opportunity to take a more precautionary approach in preserving and cultivating a diversity of local strains will be beneficial for biodiversity conservation.

POLICY IMPLICATIONS

There is always need for future research to find out people's reaction to the strong messages about reducing meat consumption in China. Dietary behavioural changes are easy as an avenue to pursue because they do not require large investment in technology. They are however very difficult to implement as they often go against the zones of comfort or habit that we all have. Smart strategies are required to shift people's perceptions and encourage them to implement personal changes with multiple individual and societal co-benefits. Below is a list of suggestions for policy interventions which can target more sustainable food consumption in China.

Promote Consumption of Local Soybean Products

With Chinese people being concerned about the imported genetically modified soybeans, particularly those from the US and Brazil, it is important to promote the consumption of local soybean products as healthy and tasty meat alternatives. Given that local soybeans are believed not to be genetically modified, this will reduce the anxiety about soybean foods. Indirectly it will also send the message that human health is a priority for the government while global scientists cooperate to solve the puzzle around GM foods.

Promote the Health Benefits of Soybeans Consumption

At this moment in time, the Western world is rediscovering the health benefits of soybeans and their nutritional values associated with high content of complete proteins and low cholesterol (The Dutch Soy Coalition, 2008). The Chinese doctors often recommend patients with obesity and high risk of type 2 diabetes, heart disease and high blood pressure, to eat tofu. Such recommendations should be given to everybody as a preventative measure.

Barriers of people to consume soybeans include unclear information, particularly the source of the soybeans beans (Wong & Chan, 2016), nutritional values and benefits. Possible interventions can include getting people familiar with soy-based products, such as soy milk. According to Ahenkora et al. (2012, p. 188), "acceptability is influenced by the degree of familiarity with soymilk and high satisfaction derived from exposure influenced consumer intention to purchase". There should also be ways to communicate the latest research results about the effects of soybeans consumptions on humans, for example, in reducing cancer incidents and mortality (Lu, Pan, Ye, Duan, Xu, Yin … Zhang, 2017), improving metabolic diseases and cardiovascular health (Liu, Ho, Hao, Chen, Woo, Wong, … Ling, 2016).

Subsidies to Soybean Farmers

China implemented an urgent campaign to increase soybeans output (Reuters, 2018b). The government encourages Chinese citizens to consume soybeans as food through implementing a structural reform on the agricultural supply side and accelerating soybeans planting (Soybeans Association of Heilongjiang Province, 2018). To maximize growers' economic profits, the government also provides subsidies to the soybeans planting farmers and it guarantees higher floor prices for the government purchasing of soybeans (Yoon, 2018). Through agricultural innovation and technology, the outputs can be improved and the quality of soybeans increased which is beneficial to the oil yield. For example, farmers in China's north-eastern provinces were offered higher subsidies for growing soybean than for corn directing their choice of crops (Reuters, 2018c).

Media Influence

The media can be used to encourage the consumption of soybeans products (AIC Technology, 2017) by providing educational materials and well-balanced perspectives on the qualities of soy products, including baby food. According to the medium and long-term planning outline for China's food and nutrition development 2011–2020, by 2020, China's per capita soybean consumption should reach 13 kg. The media can influence the uptake of soybean products by offering a platform for advertising quality and brand competition to provide excellent products, services and consumer experiences.

Social Marking

According to Firestone et al. (2017), social marketing is playing an important role in improving health on a global scale where the benefits are for the greater good. Bogueva et al. (2018) specifically examined social marketing opportunities for reduction in meat consumption. For example, Sun et al. (2007) described the success social marketing has already had in Guizhou, China for the use of iron-fortified soy sauce in enhancing women's knowledge, attitudes, perceptions about benefits and barriers, willingness to buy, and consumption.

Soybeans products and technology exhibitions are held frequently and are an ideal opportunity for social marketing. For example, the Eighth China International Exhibition of Soyfood Processing Technology and Equipment (2018) was held in Shanghai in 2018. During the event, there were many activities conducted, such as the annual soybeans enterprises conference, soybeans food festival, soybeans food tasting, local soybeans products exhibition and soybeans processing technology and innovation exhibits.

Investment in Research and Development

There is a need to provide more variety and new products on the market. Investment in research and development at a company level can deliver such innovative quality products in response to the needs of different customer groups. China's soybean food processing technology and equipment also need to adapt to the increasing consumer demands for soy products with good and strong technical support for the soybean food production enterprises.

CONCLUSION

Irrespective as to where one sits within the current debate surrounding GM soybeans and their impacts on the ecosystems of the planet, public health needs to be protected and environmental harm avoided (Maghari & Ardekani, 2011). An easier alternative to achieve this is to divert the inefficiently used feed calories into direct human nutrition. The current and increasing consumption of soybeans as animal feed can be circumvented by influencing human diets towards decreased intake of animal products and taking advantage of the numerous benefits plant-based diets have.

In 2017, China's domestic soybeans consumption was over 1.1 billion tonnes with 68.4% of this amount used as animal feed. This is an inefficient use of resources as well as a pathway which leads to economic dependence on imports. Encouraging people to consume directly more soy-based products, such as tofu, yuba and soy milk, is beneficial for their personal health, offers environmental co-benefits and allows to build a self-reliant domestic market for human consumption. Rather than succumb to the westernisation of the Chinese diet, China should focus its efforts on achieving the goal of halving the current levels of meat production. The miracle soy crop has a special place in this process.

REFERENCES

Ahenkora, K., Arthur, K., & Banahene, S. (2012). Marketing strategy for soybean products: Familiarity breeds content. *International Journal of Marketing Studies*, *4*(3), 186–189. doi:10.5539/ijms.v4n3p186

Aleksandrowicz, L., Green, R., Joy, E. J. M., Smith, P., & Haines, A. (2016). The impacts of dietary change on greenhouse gas emissions, land use, water use, and health: A systematic review. *PLoS One*, *11*(11), e0165797. doi:10.1371/journal.pone.0165797 PMID:27812156

Barret, J. R. (2006). The Science of soy: What do we really know? *Environmental Health Perspectives, 114*(6), A352–A358. Retrieved from https://ehp.niehs.nih.gov/doi/pdf/10.1289/ehp.114-a352 PMID:16759972

Bawa, A. S., & Anilakumar, K. R. (2013). Genetically modified foods: Safety, risks and public concerns – a review. *Journal of Food Science and Technology, 50*(6), 1035–1046. doi:10.100713197-012-0899-1 PMID:24426015

Bodnar, A. (2018). The scary truth behind fear of GMOs. *Biology Fortified*. Retrieved September 12, 2018 from https://www.biofortified.org/2018/02/scary-truth-gmo-fear/

Bogueva, D., Marinova, D., & Raphaely, T. (Eds.). (2018). *Handbook of research on social marketing and its influence on animal origin food product consumption*. Hershey, PA: IGI Global. doi:10.4018/978-1-5225-4757-0

Brookes, G., & Barfoot, P. (2017). Farm income and production impacts of using GM crop technology 1996–2015. *Biotechnology in Agriculture and the Food Chain, 8*(3), 156–193. doi:10.1080/21645698.2017.1317919 PMID:28481684

Brown, L. R. (2011). *Full planet, empty plates: The new geopolitics of food scarcity. Earth Policy Institute*. New York, NY: W. W. Norton & Company.

Brügger, P., Marinova, D., & Raphaely, T. (2016). Animal production and consumption: an ethical educational approach. In T. Raphaely & D. Marinova (Eds.), *Impact of meat consumption on health and environmental sustainability* (pp. 295–312). Hershey, PA: IGI Global. doi:10.4018/978-1-4666-9553-5.ch017

Campbell, T. C., & Campbell, T. M. (2006). *The China Study: The most comprehensive study of nutrition ever conducted and the startling implications for diet, weight loss and long-term health*. Dallas, TX: Benbella Books.

Central People's Government of the People's Republic of China. (2018b). *Domestic soybeans should give full play to food consumption advantages*. Retrieved September 13, 2018 from http://www.gov.cn/zhengce/2018-05/21/content_5292362.htm

Central People's Government of the People's Republic of China. (2018b). *Figure: The report about China's agricultural forecast shows that China's soybeans planting areas will be restoratively increasing in 2018*. Retrieved September 13, 2018 from http://www.gov.cn/xinwen/2018-04/20/content_5284607.htm

China Economy Net. (2018) *Domestic soybeans should play a leading role in food consumption. This year's output may hit a new high in 10 years*. Retrieved September 13, 2018 from http://www.ce.cn/cysc/sp/info/201805/21/t20180521_29192313.shtml

China Industrial Information Network. (2018). *Analysis on China's soybeans' production and consumption in 2018 including export and import*. Retrieved September 13, 2018 from http://www.chyxx.com/industry/201805/638760.html

Craig. (Ed.). (2018). *Vegetarian nutrition and wellness*. Boca Raton, FL: CRC Press.

Eighth China International Exhibition of Soyfood Processing Technology and Equipment. (2018). *Exhibition network*. Soy Products Professional Association of the National Food Industry Association of China. Retrieved September 13, 2018 from http://www.onezh.com/web/index_54336.html

Eshel, G., Shepon, A., Makov, T., & Milo, R. (2014). Land, irrigation water, greenhouse gas, and reactive nitrogen burdens of meat, eggs, and dairy production in the United States. *Proceedings of the National Academy of Sciences of the United States of America, 111*(33), 11996–12001. doi:10.1073/pnas.1402183111 PMID:25049416

Firestone, R., Rowe, C. J., Modi, S. N., & Sievers, D. (2017). The effectiveness of social marketing in global health: A systematic review. *Health Policy and Planning, 32*(1), 110–124. doi:10.1093/heapol/czw088 PMID:27476502

Food and Agriculture Organization of the United Nations (FAO). (2016). *The state of food and agriculture: Climate change, agriculture and food security*. Rome, Italy: FAO. Retrieved September 13, 2018 from http://www.fao.org/3/a-i6030e.pdf

Food Business Net. (2016). *China's soybeans consumption reached to about 100 million tones, with 80% imported, what are they use for?* Retrieved September 13, 2018 from http://news.21food.cn/33/2772440.htmlhttp://news.21food.cn/33/2772440.html

Froggatt, A., & Wellesley, L. (2016). *China shows way with new diet guidelines on meat*. Chatham House. Retrieved September 13, 2018 from https://www.chathamhouse.org/expert/comment/china-shows-way-new-diet-guidelines-meat

Funk, C., & Rainie, L. (2016). *Public and scientists' views on science and society*. Pew Research Centre Internet & Technology. Retrieved September 13, 2018 from http://www.pewinternet.org/2015/01/29/public-and-scientists-views-on-science-and-society/

Greenpeace. (2018). *Less is more: Reducing meat and dairy for a healthier life and planet*. Scientific background of the Greenpeace vision of the meat and dairy system towards 2050. Retrieved from https://storage.googleapis.com/p4-production-content/international/wp-content/uploads/2018/03/6942c0e6-longer-scientific-background.pdf

Guo, X., Raphaely, T., & Marinova, D. (2016). China's growing meat demands, implications for sustainability. In T. Raphaely & D. Marinova (Eds.), *Impact of meat consumption on health and sustainability* (pp. 221–233). Hershey, PA: IGI Global. doi:10.4018/978-1-4666-9553-5.ch011

Jamet, J. P., & Chaumet, J. M. (2016). Soybean in China: Adapting to the liberalization. *OCL (Oilseeds & Fats, Crops and Lipids), 23*(6), D604. doi:10.1051/ocl/2016044

Liu, Z., Ho, S., Hao, Y., Chen, Y., Woo, J., Wong, S. Y., ... Ling, W. (2016). Randomised controlled trial of effect of whole soy replacement diet on features of metabolic syndrome in postmenopausal women: Study protocol. *BMJ Open, 6*(9), e012741. doi:10.1136/bmjopen-2016-012741 PMID:27678545

Lu, D., Pan, C., Ye, C., Duan, H., Xu, F., Yin, L., … Zhang, S. (2017). Meta-analysis of soy consumption and gastrointestinal cancer risk. *Nature Scientific Reports, 7*, 4048. Retrieved September 13, 2018 from https://www.nature.com/articles/s41598-017-03692-y

Ma, X. Q., Verkuil, J. M., Reinbach, H. C., & Meinert, L. (2017). Which product characteristics are preferred by Chinese consumers when choosing pork? A conjoint analysis on perceived quality of selected pork attributes. *Food Science & Nutrition*, *5*(3), 770–775. doi:10.1002/fsn3.457 PMID:28572967

Maghari, B. M., & Ardekani, A. M. (2011). Genetically modified foods and social concerns. *Avicenna Journal of Medical Biotechnology*, *3*(3), 109–117. PMID:23408723

Marsall, P., & Marinova, D. (2019). Health benefits of eating more plant foods and less meat. In D. Bogueva, D. Marinova, T. Raphaely, & K. Schmidinger (Eds.), Environmental, health and business opportunities in the new meat alternatives market. Hershey, PA: IGI Global.

Milman, O., & Leavenworth, S. (2016). China's plan to cut meat consumption by 50% cheered by climate campaigners: New dietary guidelines could reduce greenhouse gas emissions by 1bn tonnes by 2030, and could lessen country's problems with obesity and diabetes. *The Guardian*. Retrieved September 13, 2018 https://www.theguardian.com/world/2016/jun/20/chinas-meat-consumption-climate-change

Ministry of Agriculture and Rural Affairs of the People's Republic of China. (2016). *Ministry of Agriculture's guide on promotion of soybeans' production*. Retrieved September 13, 2018 from http://www.moa.gov.cn/govpublic/ZZYGLS/201604/t20160412_5091357.htm

Mosaic Company. (2018). *Crop nutrition: Nitrogen*. Retrieved September 12, 2018 from https://www.cropnutrition.com/efu-nitrogen

Nam, K. C., Jo, C., & Lee, M. (2010). Meat products and consumption culture in the East. *Meat Science*, *86*(1), 95–102. doi:10.1016/j.meatsci.2010.04.026 PMID:20510536

National Agricultural Market Research Centre of the Agricultural University of China. (2018). *China's soybeans consumption grew over 4%*. Retrieved September 13, 2018 from http://finance.sina.com.cn/money/future/agri/2018-02-02/doc-ifyremfz3770325.shtml

National Bureau of Statistics. (2018). *Rural residents' per capita consumptions of soybeans, meat and poultry*. Retrieved September 13, 2018 from http://www.stats.gov.cn

National Bureau of Statistics of China. (1997–2016). *China statistical yearbook*. Retrieved September 13, 2018 from http://www.stats.gov.cn/english/statisticaldata/annualdata/

North Carolina (NC) Soybean Producers Association. (2014). *History of soybean*. Retrieved September 12, 2018 from http://ncsoy.org/media-resources/history-of-soybeans/

Raphaely, T., & Marinova, D. (Eds.). (2016). *Impact of meat consumption on health and environmental sustainability*. Hershey, PA: IGI Global. doi:10.4018/978-1-4666-9553-5

Reuters. (2018a). *Brazil boasts world's second largest genetically modified crop area: ISAAA. Environment*. Retrieved September 13, 2018 from https://www.reuters.com/article/us-brazil-gmo/brazil-boasts-worlds-second-largest-genetically-modified-crop-area-isaaa-idUSKBN1JN1KW

Reuters. (2018b). *CORRECTED-China launches "emergency" campaign to boost soy output*. Retrieved September 13, 2018 from https://www.reuters.com/article/china-soybeans/china-launches-emergency-campaign-to-boost-soy-output-idUSL3N1SA32I

Reuters. (2018c). *China grants more subsidies to soy farmers as it cuts corn stocks*. Retrieved September 13, 2018 from https://www.reuters.com/article/us-china-agriculture-subsidies/china-grants-more-subsidies-to-soy-farmers-as-it-cuts-corn-stocks-idUSKCN1HA176

Ritchie, H., & Roser, M. (2018). *Meat and seafood production & consumption*. Our World in Data. Retrieved September 12, 2018 from https://ourworldindata.org/meat-and-seafood-production-consumption

Satheesh, P. V. (2012). Genetically engineered promises & farming realities. In D. Bollier & S. Helfrich (Eds.), *The wealth of the commons: A world beyond market and state* (pp. 141–146). Amherst, MA: The Common Strategies Group.

Schmidinger, K., Bogueva, D., & Marinova, D. (2018). New meat without livestock. In D. Bogueva, D. Marinova, & T. Raphaely (Eds.), *Handbook of research on social marketing and its influence on animal origin food product consumption* (pp. 344–361). Hershey, PA: IGI Global. doi:10.4018/978-1-5225-4757-0.ch023

Sheldon, I. (2018). Why China's soybean tariffs matter. *The Conversation*. Retrieved September 13, 2018 from http://theconversation.com/why-chinas-soybean-tariffs-matter-94476

Soybeans Association of Heilongjiang Province. (2018). Document No. 1 of the Central Committee in 2017: Several suggestions on deepening the structural reform of the supply side of agriculture and accelerating the cultivation of new momentum of agricultural and rural development. *Economy Daily*. Retrieved September 13, 2018 from http://www.chinasoy.com.cn/Page_Views/AllView.aspx?ID=2716&cateorgy=%e7%89%b9%e5%88%ab%e5%85%b3%e6%b3%a8

Sun, X., Guo, Y., Wang, S., & Sun, J. (2007). Social marketing improved the consumption of iron-fortified soy sauce among women in China. *Journal of Nutrition Education and Behavior*, *39*(6), 302–310. doi:10.1016/j.jneb.2007.03.090 PMID:17996625

Technology, A. I. C. (2017). *How to create a powerful social media strategy for food products*. Retrieved September 13, 2018 from https://aictechnologies.com.au/2017/05/23/selling-food-products-social-media/

The Dutch Soy Coalition. (2006). *Soy: Big business, big responsibility. Addressing the social- and environmental impact of the soy value chain*. Retrieved September 13, 2018 from http://www.bothends.org/uploaded_files/document/2006_Soy_big_business.pdf

The Statistics Portal. (2018). *Import volume of soybeans worldwide from 2013/14 to 2017/18, by country (in million metric tons)*. Statista. Retrieved September 13, 2018 from https://www.statista.com/statistics/612422/soybeans-import-volume-worldwide-by-country/

U.S. Soybean Export Council. (2006). *U.S. Soy: International Buyers' Guide*. Retrieved September 12, 2018 from https://ussec.org/wp-content/uploads/2015/10/buyers-guide.pdf

Wang, O. (2018). *China orders farmers to grow more soybeans despite deal to buy more produce from US: Officials have ordered farmers to grow more soybeans even though Donald Trump has claimed China will buy 'as much as our farmers can produce'*. US–China Trade Wars. South China Morning Post. Retrieved September 13, 2018 from https://www.scmp.com/news/china/economy/article/2147296/china-orders-farmers-grow-more-soybeans-despite-deal-buy-more

Wong, A. Y., & Chan, A. W. (2016). Genetically modified foods in China and the United States: A primer of regulation and intellectual property protection. *Food Science and Human Wellness*, *5*(3), 124–140. doi:10.1016/j.fshw.2016.03.002

Xinhua. (2018). China soybean production forecast to recover. *China Daily*. Retrieved September 13, 2018 fromhttp://www.chinadaily.com.cn/a/201804/23/WS5add59d9a3105cdcf6519e50.html

Yoon, E. (2018). *How China is working hard to be less reliant on US farmers*. Retrieved September 13, 2018 from https://www.cnbc.com/2018/05/11/china-looking-to-rely-less-on-us-for-soybeans.html

Zahran, H. H. (1999). *Rhizobium*-legume symbiosis and nitrogen fixation under severe conditions and in an arid climate. *Microbiology and Molecular Biology Reviews*, *63*(4), 968–989. PMID:10585971

ADDITIONAL READING

Aleksandrowicz, L., Green, R., Joy, E. J. M., Smith, P., & Haines, A. (2016). The impacts of dietary change on greenhouse gas emissions, land use, water use, and health: A systematic review. *PLoS One*, *11*(11), e0165797. doi:10.1371/journal.pone.0165797 PMID:27812156

Bawa, A. S., & Anilakumar, K. R. (2013). Genetically modified foods: Safety, risks and public concerns – a review. *Journal of Food Science and Technology*, *50*(6), 1035–1046. doi:10.100713197-012-0899-1 PMID:24426015

Funk, C., & Rainie, L. (2016). *Public and scientists' views on science and society*. Pew Research Centre Internet & Technology. Retrieved September 13, 2018 from http://www.pewinternet.org/2015/01/29/public-and-scientists-views-on-science-and-society/

Greenpeace (2018). *Less is more: Reducing meat and dairy for a healthier life and planet*. Scientific background of the Greenpeace vision of the meat and dairy system towards 2050. Retrieved from https://storage.googleapis.com/p4-production-content/international/wp-content/uploads/2018/03/6942c0e6-longer-scientific-background.pdf

Jamet, J. P., & Chaumet, J. M. (2016). Soybean in China: Adapting to the liberalization, *OCL (Oilseeds & Fats, Crops and Lipids)*, *23*(6), D604. doi:10.1051/ocl/2016044

Maghari, B. M., & Ardekani, A. M. (2011). Genetically modified foods and social concerns. *Avicenna Journal of Medical Biotechnology*, *3*(3), 109–117. PMID:23408723

Ritchie, H., & Roser, M. (2018). *Meat and seafood production & consumption*. Our World in Data. Retrieved September 12, 2018 from https://ourworldindata.org/meat-and-seafood-production-consumption

Schmidinger, K., Bogueva, D., & Marinova, D. (2018). New meat without livestock. In D. Bogueva, D. Marinova, & T. Raphaely (Eds.), *Handbook of research on social marketing and its influence on animal origin food product consumption* (pp. 344–361). Hershey, PA: IGI Global. doi:10.4018/978-1-5225-4757-0.ch023

Sheldon, I. (2018). *Why China's soybean tariffs matter*. The Conversation. Retrieved September 13, 2018 from http://theconversation.com/why-chinas-soybean-tariffs-matter-94476

Zahran, H. H. (1999). *Rhizobium*-legume symbiosis and nitrogen fixation under severe conditions and in an arid climate. *Microbiology and Molecular Biology Reviews, 63*(4), 968–989. PMID:10585971

KEY TERMS AND DEFINITIONS

Fermented: A food which has been through the process of fermentation, that is, chemical breakdown of its substance by bacteria, yeast, or other microorganisms.

Genetically Modified (GM): (Also genetically engineered). Applied to food crops and organisms whose genetic material, namely deoxyribonucleic acid (DNA) has been altered in a way that does not occur in nature.

Import: Goods or commodities brought into a country across its borders.

Protein: An organic substance – polymer chains of amino acids, considered an essential nutrient for the human body; there are 20 types of amino acids representing the building blocks for the human proteins – 11 are non-essential which can be synthesized by the human organism and 9 are essential which need to be provided by food.

Social Marketing: Marketing which aims at inducing a behavioral change and maintaining such behavior for the greater social good, including benefits for the individual and society as a whole.

Soy: (Also soybean and soya). A legume plant native to East Asia with a high content of complete protein and beneficial nutritional value.

Soybean Meal: A product prepared from soybeans to be used in animal feed as a source of protein; very often soybean meal is made from the residue after the oil from the soybean has been extracted.

Trade Liberalization: Removal of barriers or other restrictions to the free movement of goods and commodities between countries.

Tofu: Soy bean curd often used as a meat alternative; it is very popular in Asian countries, such as China and Japan, and more recently has started to also be included in a western type of diet.

Yuba: (Also tofu skin, bean curd skin, bean curd sheet or bean curd robe). A food product made from soybeans during the boiling of soy milk; the thin skin formed at the top of the boiling pan is collected and dried in sheets which can be used as wraps.

Chapter 8
Nutritional Properties of Edible Insects

Anna K. Żołnierczyk
Wrocław University of Environmental and Life Sciences, Poland

ABSTRACT

Insects are the biggest animal group on earth. They constitute as much as 80% of the animal kingdom. Over 2000 species of insects are consumed in Central and South America, Africa, Asia, Australia, and New Zealand. Currently almost 1 billion people on this planet suffer from hunger, and we must strive to increase the efficiency of food production. One of the possible solutions is to use insects as a source of food. An important advantage of insect production is the high environmental safety compared to conventional livestock. Conventional animal husbandry is responsible for at least 18% of total greenhouse gas emissions and large consumption of drinking water. A much smaller amount of water is used to produce insect meat and insects require far less feed. Production of insect protein requires much less land and energy than the more widely consumed forms of animal protein. The nutritional usefulness of edible insects varies depending on the species, on the stage of development of the insect and the method of breeding and feeding. Insects have a high nutritional value. They are a rich source of protein which includes all eight essential amino acids (phenylalanine, isoleucine, leucine, lysine, methionine, threonine, tryptophan, and valine). Edible insects contain on average 10-30% of fat in dry matter and they are good source of edible oil which contains more than 50% of polyunsaturated fatty acids (PUFA) desirable for nutritional and health reasons. The average energy value of edible insects is about 400-500 kcal/100g of dry matter. Insects also contain a variety of water soluble or lipophilic vitamins and minerals. Their consumption can build a well-balanced diet. Insects can be regarded as safe, if properly managed and consumed, but international food regulations are needed.

INTRODUCTION

From the beginning of human existence on the Earth, most of the protein supplied with food was taken by hunting or fishing, but in many places collection of insects was necessary to allow to supplement nutritional deficiencies (Tosi & Daccordi, 1983). Insects are the biggest animal group on earth, they constitute as much as 80% of the animal kingdom. It is estimated that over 2000 species of insects are

DOI: 10.4018/978-1-5225-7350-0.ch008

consumed in almost 80% of the countries in the world (Wageningen University and Research, 2017). Eggs, larvae and adult forms of insects (Figure 1) are eaten as food in Central and South America, Africa, Asia, Australia and New Zealand.

The taste and flavour of insects are very diverse (Payne, 2018). We can compare them to the ingredients we know and the ways of cooking are no different from the traditional ones. Also, insects can absorb the taste of the chosen seasoning with which they are fed. The total number of ethnic groups practicing entomophagy (from the Greek words ἔντομον éntomon meaning "insect" and φᾰγεῖν phagein meaning "to eat") exceeds 3000 (Ramos-Elorduy, 1998; MacEvilly, 2000). Entomophagy is not well accepted in western European populations but it is common in the world.

It is considered that eating insects may reduce the environmental risks (FAO, 2016). Insect breeding compared to livestock farms (pigs, cattle and poultry) releases six to ten times less ammonia (Oonincx, Kgomotso, & Letswiti, 2010). Conventional animal husbandry is responsible for at least 18% of greenhouse gas emissions and massive consumption of drinking water. Much smaller amounts of water are used to produce insect meat. Insects are able to derive their moisture demand from food. Also they require far less feed. For example, the production of 1 kg of live animal weight of crickets requires as little as 1.7 kg of feed (Collavo, Glew, Huang, Chuang, Bosse, & Paoletti, 2005). Typically, 1 kg of live animal weight in a conventional production system demands 2.5 kg of feed for chicken, 5 kg for pork and 10 kg for beef (Smil, 2002). Moreover, the production of insect protein takes much less land and energy than the more widely consumed forms of animal protein (Halloran, Hanboonsong, Roos, & Bruun, 2017; Oonincx & de Boer, 2012; Premalatha, Abbasi, Abbasi, & Abbasi, 2011). Edible insects can be grown at home, on small farms or large industrial facilities anywhere in the world. The interest in using insects for nutritional purposes is justified because (apart from nutritional qualities) insects are characterized by high survival capacity in various ecological conditions, short life cycle and high reproductive ability (DeFoliart, 1999; Illgner & Nel, 2000; Renault, Laparie, McCauley, & Bonte, 2018). However, it is first necessary to establish international food regulations regarding the safety of insect food products (Rumpold & Schluter, 2013a). Also, in countries where there is no tradition of eating insects, it takes time for people to get used to new possibilities.

The science of edible insects is a relatively new field of scientific research. Large-scale breeding is also a small percentage of the sources from which edible insects are obtained – for the most part they

Figure 1. Larvae of the mealworms Tenebrio molitor (left) and the adult form of the Jamaican field cricket Gryllus assimilis (right)
Photo credit: Tomasz Lewandowski

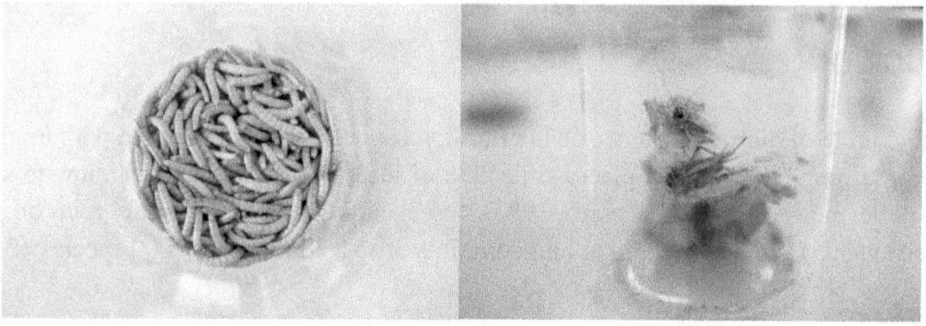

are collected in a natural environment. Insects are mainly material for animal feed. However, in recent years, there has been an increased interest in the subject of insects as a source of food, both among the scientific community and consumers. The global human population is growing by around 70 million people each year. If the growth rate continues, by 2050 the population will probably reach as much as 9 billion. To feed all these people, we will have to produce almost twice as much food as at present. This may be difficult to implement, because we are already using 70% of the agricultural land for cattle farming. Furthermore, we cause pollution of the environment and our activities lead to rapid climate changes, which adversely affect agricultural production. Considering the fact that currently almost 1 billion people on earth suffer from hunger, we must strive to increase the efficiency of food production (FAO, 2009). One of the possible solutions is to use insects as a source of food. Probably in the future, populations from developed countries will need to adapt to other sources of animal proteins because the traditional breeding of beef, poultry or pork will become insufficient.

NUTRITIONAL VALUE OF EDIBLE INSECTS

The nutritional usefulness of edible insects varies depending on the species and method of breeding and feeding. Even within the same species of insects, the composition of nutritional compounds changes depending on the stage of development of the insect.

As shown in Figure 2, the main component of the nutrient composition of insects represents protein. The average amount of protein contents in edible insects varies between 42% for beetles and grubs and 63% for crickets, grasshoppers and locusts (see Figure 2). Despite such a large difference, many species of insects can cover human demand for energy, protein or minerals.

Energy

The energy value of raw insects (see Table 1) is in the range of 89 kcal/100g for grasshopper (Cyrtacanthacris tatarica) to 1272 kcal/100g for green ant (Oecophylla smaragdina) (van Huis, 2013). The average energy value of edible insects is about 400–500 kcal/100g of dry matter (see Figure 3).

The energy value of edible insects is generally subject to a large variation which comes primarily from differences between developmental stages and depends mainly on the fat content. Insects in the early stages of development (larvae, pupae, maggots and grubs) are usually richer in energy compared to adults because they generally contain more fat whereas insects containing more protein have lower energy content. Such an information can be very useful for creating specific diets, for example for people who want to reduce the amount of fats in their food.

Protein

From a nutritional point of view, apart from the energy value, the most important aspect is the protein content in the diet. We are constantly hearing messages that protein should be the basis of our diet, because it is the building block of all cells and participates in important life processes. Due to the increasing cost of animal proteins, population growth, and increasing need for protein-rich options in the developed and less developed countries, alternative food sources are highly needed. Hence, insect consumption can help with food and feed insecurity and thus replace the conventional animal source in the future.

Figure 2. Average nutrient contents [%] (based on dry matter) of edible insects groups belonging to the same order

NFE – nitrogen-free extract, the fraction containing sugars and starches plus small amounts of other materials (Rumpold & Schluter, 2013a and b).

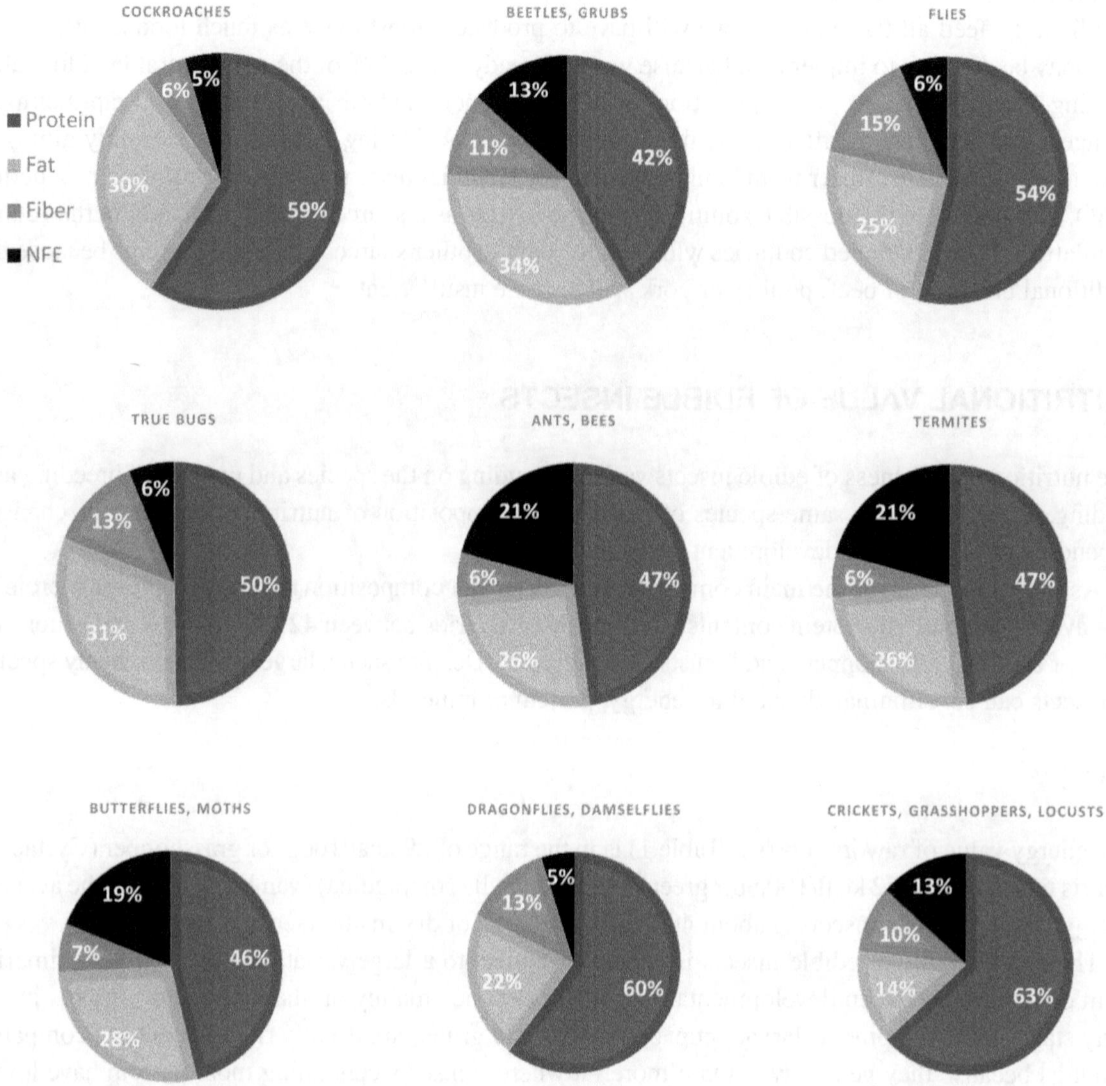

The findings from scientific research (see Table 2) show that insects are rich in protein (5–70%), while the protein content of boiled beef meat varies within the range of 11–27% (raw 19–26%), reptiles 11–27% and seafood 13–28%. The protein content of insects varies strongly by species. Also, the digestibility of protein from individual food products is diverse. For example, the digestibility of protein from egg white is 95–100%, dairy products 70–80%, vegetables and fruit 90–100% and meat about 65% (beef – 98%) while for insect protein it is from 76 to 96% (Ramos-Elorduy, Moreno, Prado, Perez, Otero, & de Guevara, 1997). For example, for dried, traditionally prepared mopane worms (caterpillars of the moth *Gonimbrasia belina*) the protein digestibility is 85.8% (Dreyer & Wehmeyer, 1982). Removal of chitin improves further the digestibility of insect protein (Finke, 2007). In wheat flour, there is about 10% of

Figure 3. Average energy value [kcal/100g of dry matter] for selected orders of insects (Rumpold & Schluter, 2013b)

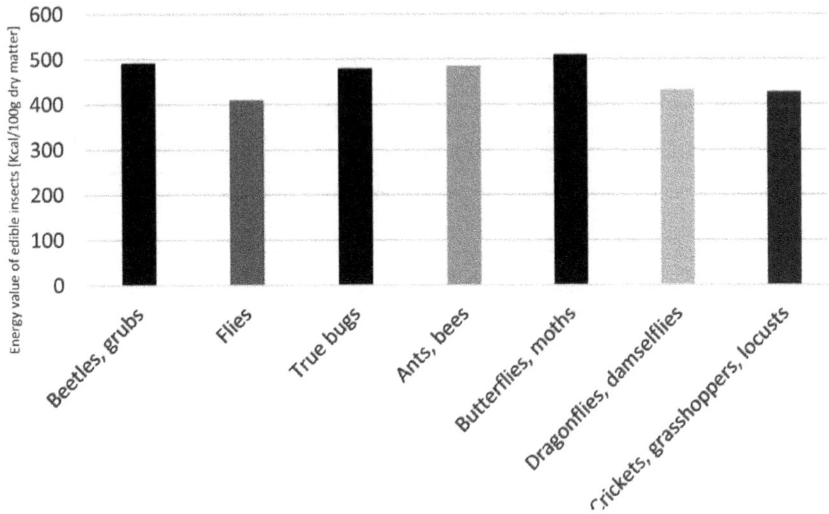

Table 1. Energy content [kcal/100g] based on dry matter or fresh weight for selected edible insects

Edible Insects		Stage	Energy Content [kcal/100g]	
Scientific Name	**Common Name**		**Based on Dry Matter**	**Based on Fresh Weight**
Tenebrio molitor	Yellow mealworm	Adult	380	139
Tenebrio molitor	Yellow mealworm	Pupae	550	-
Tenebrio molitor	Yellow mealworm	Larva	539.63–577.44	206
Zophobas morio	Zophobas	Larva	575	-
Bombyx mori	Domesticated silkworm	Larva	390	94
Macrotermes bellicosus	Termite	Adult	28–46	
Macrotermes subhyalinus	Termite/dried, flour	Adult	535	-
Locusta migratoria	Migratory locust	Adult	-	179
Cytracanthacris tatarica	Grasshopper	Adult	-	89
Acheta domesticus	House cricket	Adult	455	-
Melanoplus femurrubrum	Red-legged grasshopper	Adult	361	160
Oecophylla smaragdina	Green ant	Adult	-	1272
Atta mexicana	Leaf-cutter ant	Adult	555	404

Source of data: (Rumpold & Schluter, 2013b)

protein, which due to its vegetable origin, belongs to a deficient source of essential amino acids. In pasta, the protein content is about 5%. A flour made from crickets contains about 70% full-value protein and pasta made with just 10% cricket flour contains 14% of protein (which is more than the 13% protein in eggs). As protein sources, the nutritive value of edible insects is as good as that of other animals and plants or even better.

Table 2. Protein content in selected insect species and traditional food sources

Animal Group	Species and Common Name	Edible Product	Protein Content [g/100g fresh weight]
Insects	Locusts and grasshopppers (*Locusta migratoria, Acridium melanorhodon, Ruspolia differens*)	Larva	14–18
	Locusts and grasshopppers (*Locusta migratoria, Acridium melanorhodon, Ruspolia differens*)	Adult	13–28
	Chapulines – Mexico (*Sphenarium purpurascens*)	Adult	35–48
	Silkworm (*Bombyx mori*)	Larva	54–70
	Yellow mealworm (*Tenebrio molitor*)	Larva	14–25
	Crickets (*Gryllidae*)	Adult	8–25
	Termites (*Isoptera*)	Adult	13–28
	Cockroaches (*Blattodea*)	Adult	44–66
	Beetles (*Coleoptera*)	Adult	9–70
	Beetles (*Coleoptera*)	Larva/pupae	12–53
	Flies (*Diptera*)	Adult	36–56
	Flies (*Diptera*)	Larva/pupae	63–64
	Ants, bees (*Hymenoptera*)	Adult	5–66
	Ants, bees (*Hymenoptera*)	Larva/pupae	40–61
	High quality Cricket Flour (*Acheta domestica*)[a]	Adult	67.8
	Cricket protein pasta (*Acheta domestica*)[b]	Adult	14
Cattle		Beef (raw)	19–26
Reptiles (cooked)	Turtles (*Chelodina rugosa, Chelonia depressa*)	Flesh	25–27
		Intestine	18
		Heart	17–23
		Liver	11–27
Fish and seafoods (raw)	Finfish	Mackerel	16–28
		Tilapia	16–19
	Crustaceans (*Crustacea*)	Shrimp	13–27
		Lobster	17–19
	Molluscs (*Mollusca*)	Cuttlefish, squid	15–18

[a]http://www.bizarrefood.com/insect-bug-flour-powder (Accessed 13.04.2018)
[b]http://nutribug.com/product/cricket-protein-pasta/ (Accessed 13.04.2018)
Source of data: (Rumpold & Schluter, 2013b; Chen Feng, Zhang, & Chen, 2010)

Protein content also depends on the feed. For example, reared grasshoppers that are fed with bran, have almost double the protein content of those fed on maize. Furthermore, the protein content of insects also depends on their metamorphosis stage (Ademolu, Idowu, & Olatunde, 2010): adults usually have higher protein content than instars (see Table 2).

However, it is not the amount of protein but the quality – the amino acid composition – that determines whether the food ration will be wholesome. A standard protein is one that contains all amino acids in quantities and proportions corresponding to human needs. Particularly important is the content of essential amino acids. The reference protein for adults is egg white but the key sources of proteins in human diets are often cereal proteins. They are low in lysine, tryptophan and threonine. In some insect species, these amino acids are very well represented (Bukkens, 2005). Insects offer a complete animal protein that includes all 8 essential amino acids (phenylalanine, isoleucine, leucine, lysine, methionine, threonine, tryptophan and valine) (see Table 3).

Comparing the average content of amino acids in the larvae of mealworm and beef (see Figure 4), tryptophan is found in insects and is not present in beef, four of the exogenous amino acids (isoleucine, leucine, lysine, and valine) are found in insects in a larger amount then in beef, threonine is in a comparable amount, and only two of the amino acids – phenylalanine and methionine, are present in insect larvae in lower amounts than in beef (about half as much) (USDA, 2012). The average amount of amino acids in edible insects is higher than that in beef (see Figure 4).

Eating foods which do not contain all essential amino acids requires a balanced diet and if not done properly can lead to health disorders. Based to their origin, we distinguish animal proteins and vegetable proteins. Animal protein which occurs in meat, fish, seafood, eggs, milk and dairy products (cheese, yoghurt and buttermilk) is a complete protein. Vegetable protein derived from vegetables, fruits, legumes, nuts, seeds and cereals is a source of incomplete protein (it does not contain all essential amino acids) with only a few exceptions. Animal protein has a higher nutritional value and through eating meat, fish, eggs and dairy products, it is easier to provide the body with the necessary dose of protein. Plant proteins contain less lysine, methionine, tryptophan and valine and people on a vegetarian or vegan diet need to balance their intake of a variety of foods. Although animal proteins are complete, the excess of animal products in the diet is not beneficial for human health. Animal products contain cholesterol and a lot of fat, so they can increase the risk of cardiovascular disease and obesity. There is growing interest in alternative protein sources to feed the increasing world population and insects represent one of the potential sources to exploit (Janssen, Vincken, van den Broek, Fogliano, & Lakemond, 2017).

Table 3. Average amino acids content of selected edible insects

Group of Insects	Amino Acids																		
	His	Ile	Leu	Lys	Met	Cys	Met+ Cys	Phe	Tyr	Phe+ Tyr	Thr	Trp	Val	Arg	Ser	Pro	Ala	Gly	Glu
Cockroaches	19.4	29.9	56.4	48.0	29.8	11.6	41.4	30.6	62.3	92.9	34.6	6.0	53.8	41.5	41.9	65.0	56.6	58.7	99.7
Beetles, grubs	26.3	45.6	74.2	50.6	16.2	14.6	31.9	47.1	55.7	98.6	35.2	10.1	51.9	53.9	42.6	64.1	69.5	55.2	123.7
Flies	22.3	32.6	57.4	62.9	27.2	5.3	36.6	50.6	56.7	107.3	38.8	28.3	46.9	49.6	60	27.8	58.9	45.1	98.6
True bugs	15.7	31.5	49.8	28.0	21.7	12.9	32.2	34.4	38.7	63.8	29.9	10.3	44.3	24.9	10.3	-	26.4	16.4	23.7
Ants, bees	27.0	47.8	78.4	53.8	23.8	12.9	30.5	47.5	55.3	104.3	41.7	10.3	60.5	43.5	38.2	66.7	72.3	81.3	134.3
Termites	51.4	51.1	78.3	54.2	7.5	18.7	26.2	43.8	30.2	74.0	27.5	14.3	73.3	69.4	-	-	-	-	-
Butterflies, moths	23.7	40.4	62.7	57.7	22.1	12.2	34.7	46.3	49.1	95.8	40	11.2	54.1	46.9	48.34	44.9	48.9	43.8	103.4
Dragonflies, damselflies	21.2	39.6	74.8	53.9	19.3	12.8	29.8	46.6	61.5	100.3	35.8	8.1	50.3	53.6	41.9	53.9	77.4	54.0	94.5
Crickets, grasshoppers, locusts	15.0	30.0	59.0	45.0	16.0	6.0	22.0	-	-	30.0	23.0	6.0	39.0	-	-	-	-	-	-

His – histidine; Ile – isoleucine; Leu – leucine; Lys – lysine; Met – methionine; Cys – cysteine; Phe – phenylalanine; Tyr – tyrosine; Thr – threonine; Trp – tryptophan; Val – valine; Arg – arginine; Ser – serine; Pro – proline; Ala – alanine; Gly – glycine; Glu – glutamic acid.

Source of data: (Rumpold & Schluter, 2013b).

Figure 4. Average content of amino acids in selected edible insects and beef [mg/g dry matter protein] (Finke, 2002; Oonincx & Dierenfeld, 2012; Rumpold & Schluter, 2013b)

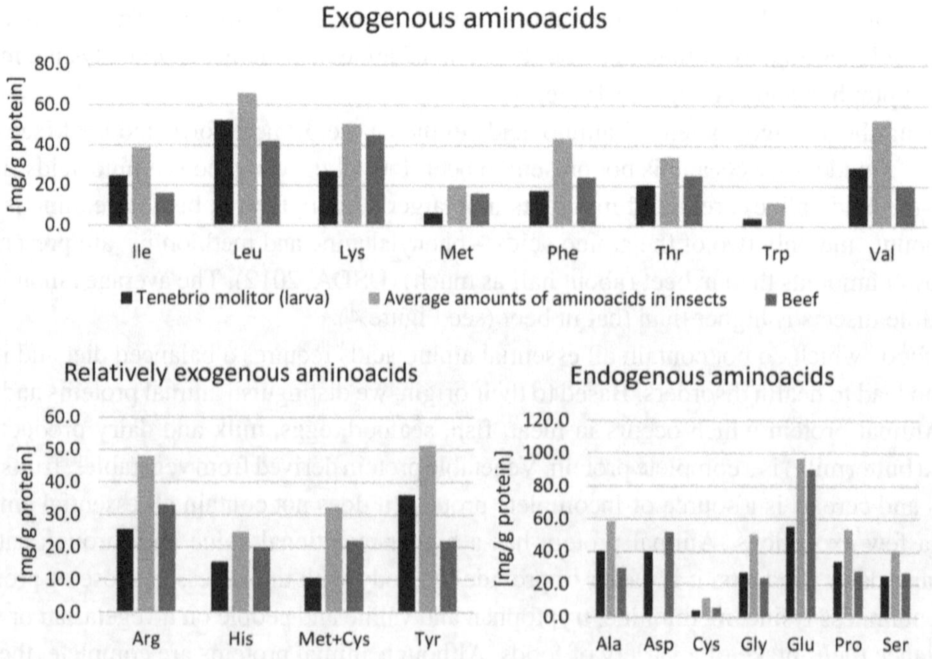

Fiber

Edible insects contain a variable but significant amounts of fiber which ranges from several to several dozen percent (see Table 4). The exoskeleton of insects is made of chitin (the most common form of fiber in the body of insects). Fiber content is measured by crude fiber (CF), acid detergent fiber (ADF), and neutral detergent fiber (NDF) (Finke, 2002, 2007; Pennino, Dierenfeld, & Behler, 1991; Barker, Fitzpatrick, & Dierenfeld, 1998). For plant-based foods, the compositing of the various components of these fibers is well established: ADF is composed usually of cellulose and lignins while NDF is composed of cellulose, hemicellulose and lignin (Van Soest & Robertson, 1977). Insects contain significant amounts of both ADF and NDF; however the components that make up these fibers are unknown. Some authors have suggested that the fiber in insects represents chitin because chitin – linear polymer of b-(1-4) N-acetyl-D-glucosamine units, is similar structurally to cellulose – linear polymer of b-(1-4)-D-glucopyranose units (Barker, Fitzpatrick, & Dierenfeld, 1998). Chitin from insect exoskeletons acts in the human body like cellulose and because of this effect it is often called "animal fiber". It does not have a nutritional role in the human body as it is not digested like cellulose. The enzyme chitinase is found in human gastric juices (Paoletti, Norberto, Damini, & Musumeci, 2007), but it has been found that it may be inactive. Active chitinase response in the body dominates among people from tropical countries where the consumption of insects has a long-term tradition (Lee, Simpson, & Wilson, 2008). Chitin affects the work of the digestive system and the regulation of fat metabolism in the body. It reduces the appetite and inhibits the absorption of fats and sugars from the gastrointestinal tract, thus it lowers the calorie content of the diet. Chitin is used mainly in dietetics for the production of dietary supplements supporting slimming.

Table 4. Fiber content in selected insect species

Edible Insects		Stage	Fiber Content [% in Dry Matter]
Scientific Name	**Common Name**		
Locusta migratoria	African migratory locust	Nymph	27
Acheta domesticus	House cricket	Nymph	14.9–15.7
Acheta domesticus	House cricket	Adult	16.3–22
Gryllus assimilis	Jamaican field cricket	Nymph	8
Sphenarium purpurascens	Chapulines	Adult	4–11
Bombyx mori	Silkworm	Larva	5.9–6.4
Tenebrio molitor	Yellow mealworm	Larva	5–15
Tenebrio molitor	Yellow mealworm	Pupae	5.1
Tenebrio molitor	Yellow mealworm	Adult	20.2
Apis mellifera	Honey bee	Larva	1–1.3
Apis mellifera	Honey bee	Pupae	2.7–3
Apis mellifera	Honey bee	Adult	2–11

Source of data: (Chen, Feng, Zhang, & Chen, 2010; Rumpold & Schluter, 2013b; Kouřimská & Adámková, 2016)

Lipids

Just like source of protein, insects can also be a rich source of fatty acids. Fats and carbohydrates are important nutritive elements in the human body. They are the main energy sources and fat is the most energetic ingredient of human food. Carbohydrates in insects are formed mainly by chitin. Insect fat is composed of 80% triacylglycerols and about 20% phospholipids. Fatty acid in triacylglycerols can either be saturated, unsaturated, or essential. The intake of fatty acid plays a key role in human health. Limiting the amount of saturated fatty acids (SFA) consumed in the diet may reduce the risk of cardiovascular disease. Edible insects contain on average 10 to 30% of fat in the dry matter (refer to Figure 2). Usually this is higher in the larval stages than in adults (Rumpold & Schluter, 2013a; Chen, Feng, Zhang, & Chen, 2010). Grubs and beetle larvae belong to the insects with the highest fat content. For example, the African palm weevil larvae in the early stage (*Rhynhophorus phoenicis*) contains about 70% fat, and adult rhinoceros beetles (*Oryctes rhinoceros*) contain only 0.7% fat (Rumpold & Schluter, 2013b). Termites are also a good source of fat with a fat content more than 30% in their bodies. The fat content of grasshoppers, crickets, locusts, flies and dragonflies is lower.

Furthermore, insects are a good source of edible oil which contains more than 50% of polyunsaturated fatty acids (PUFA), mainly consisting of arachidonic, linolenic and linoleic acids desirable for nutritional and health reasons. The PUFA content of selected groups of edible insects is as high as about 30% (Figure 5). Oils extracted from several species of insects are rich in PUFA and often they contain the essential linoleic acid and α-linolenic acids which are very important for the proper development of children and infants (Womeni, Linder, Tiencheu, Mbiapo, Villeneuve, Fanni, & Parmentier, 2009). Adding essential fatty acids (EFA) to our daily diet is a great way to support human health. They support cardiovascular health, brain function and development, skin health, and offer many other benefits to our body. The composition of fatty acids in insects is promising – the content of linoleic acid, α-linolenic acids, omega-3

Figure 5. Average content [%] of the main groups of fatty acids in selected groups of edible insects (Bukkens, 2005)
SFA – saturated fatty acids; MUFA – monounsaturated fatty acids; PUFA – polyunsaturated fatty acids

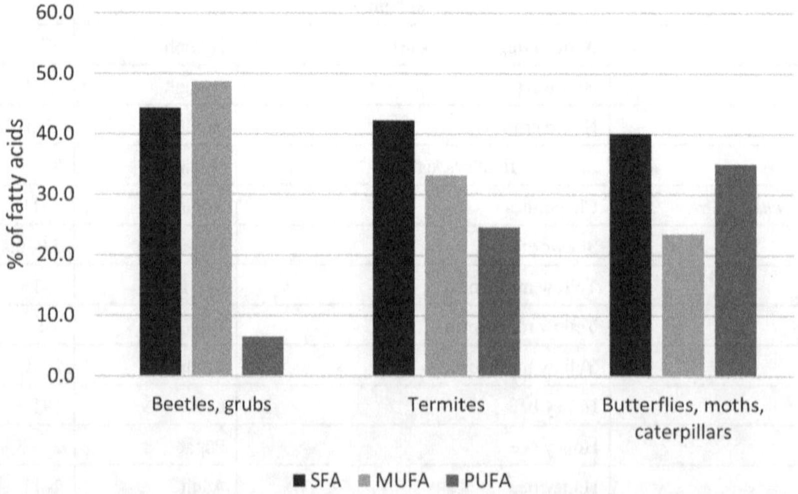

and omega-6 confirms that they can be successfully used in the diet as a source of unsaturated fatty acids (Womeni, Linder, Tiencheu, Mbiapo, Villeneuve, Fanni, & Parmentier, 2009).

Minerals

Vitamins and minerals are essential nutrients that our body needs in small amounts to work properly. Two kinds of minerals exist: macrominerals which the body needs in relatively larger amounts, namely calcium, phosphorus, magnesium, sodium, potassium, sulfur and chloride, and trace minerals required in small quantities, namely iron, zinc, manganese, copper, iodine, fluoride, cobalt and selenium. Most people should be able to get all the nutrients they need by eating a varied and balanced diet. We can find a variety of these minerals in fruits, vegetables, meat, and other foods. Minerals are necessary for three main reasons: building strong bones and teeth, controlling body fluids inside and outside the cells and turning the food we eat into energy. Essential minerals include calcium, iron, magnesium and potassium. It is difficult to refer to the content of micro- and macrominerals in edible insects in the context of the recommended daily intake because these values are set at different levels for various food products. Insects are rich in minerals and their content is higher than the content in the meat of slaughtered animals (Bukkens, 2005). For example, the concentration of iron in most insect species – 31–77 mg/100 g dry weight, is higher than in beef – 6mg/100g dry weight (Bukkens, 2005). The required amount of iron depends on its bioavailability, the consumer's age and sex. For example, this value is 9.1 mg per day for an adult male and 19.6 mg per day for an adult female. The content of iron, zinc and calcium are presented in Table 5. In the selected insects, the calcium content ranges from 32 in ants (*Onyoso mammon*) to 2010 in housefly larvae (*Musca domestica*) mg/100 g dry matter. The best absorbed calcium comes from milk and dairy products (the content of calcium in fat milk is about 120 and in various species of yellow cheese on average of 600 mg per 100 g of dry matter); however, many fruit and vegetables are also a good source of this macromineral.

Table 5. Content of calcium, iron and zinc in selected insects

Edible Insects			Content of Elements [mg/100g Dry Matter]		
Scientific Name	**Common Name**	**Stage**	**Ca**	**Fe**	**Zn**
Rhyncophorus phoenicis	African Palm Weevil	Larva	131	22.8	21.1
Tenebrio molitor	Yellow mealworm	Larva	45.8	5.46	12.5
Musca domestica	Housefly	Larva	2010	60.4	23.7
Bombyx mori	Silkworm	Larva	102.3	9.5	17.8
Onyoso mammon	Ant	Adult	32.6	17.7	11.1
Oyala	Termite	Adult	84.7	332	11.9
Ogawo	Termite	Adult	83	93.9	8.1
Agoro	Termite	Adult	132	161	14.3
Onjiri mammon	Cricket	Adult	341	1562	25.1
Average content of minerals [mg/g dry matter] in meat					
Beef			4-27	6	12.5
Pork			5-28	1.5	-
Poultry			5-14	1.2	-
Recommended nutrient intake [mg/day]					
Adult/male			750	9.1	4.2
Adult/female			750	19.6	3

Source of data: (Bukkens, 1997; FAO/WHO, 2001; Finke, 2002; Christensen, Orech, Mungai, Larsen, Friis, & Aagaard-Hansen, 2006; Hwangbo, 2009; Elemo, Elemo, Makinde, & Erukainure, 2011; Zielińska, Baraniak, Karaś, Rybczyńska, & Jakubczyk, 2015)

Edible insects have the potential to provide specific micronutrients such as copper, iron, magnesium, manganese, phosphorous, selenium, and zinc. There are even suggestions that the consumption of insects could decrease some trace minerals deficiency in developing countries (Christensen, Orech, Mungai, Larsen, Friis, & Aagaard-Hansen, 2006).

Vitamins

Insects contain a variety of water soluble or lipophilic vitamins (Finke, 2002; Chen, Feng, Zhang, & Chen, 2010; Oonincx & Dierenfeld, 2012). Many species that have been analyzed contain only negligible amounts of these vitamins (Bukkens, 2005). The content of vitamins and minerals in wild edible insects is seasonal and in the case of farm bred species it can be controlled by properly selected feed.

Organoleptic Properties

Regardless of the content of nutrients, the culinary treatment determines the original value of food. Insects can be consumed immediately after catching, or after 1–3 days of starvation (Ramos-Elorduy, 1998). Depending on the species of insect and taste preferences, the insects can be fried, baked, grilled, cooked, marinated, smoked or dried. Organoleptic properties are an important criterion for the selection of the consumed insect and the process of its culinary treatment. The taste of insects is very diverse and can usually be compared with the taste of well-known dishes (see Table 6). In addition, if insects are fed with an intensely flavored feed, for example with cinnamon, they would have the taste of the feed used. The skeleton of adult forms of some insects (for example beetles) has a very big influence on the texture – the insects are crispy and fragile like crackers or crisps. Insects are more often consumed in the early stages of maturity as larvae as they contain less chitin, which can irritate the digestive system and therefore, they are a better digestible food (Pino-Moreno & Ramos-Elorduy, 2006).

RISK FROM EATING INSECTS

Insects are considered as suitable alternatives to mainstream animal sources of food such as chicken, pork, beef and fish as well as an alternative feed. Because they are available in nature periodically (depending on the season and access to food), insect breeding can overcome their shortage (Yen, 2009, 2010; Sileshi & Kenis. 2010). Around the world, including in Europe, there are farms where insects are bred for animal feed and for human food. One of the important advantages of insect production is the high environmental safety compared to conventional livestock (Nakagaki & Defoliart, 1991).

Also, there are special devices on the market for home insect cultivation, for example mealworm larvae. The mealworms can be fed sustainably on vegetable scraps from the kitchen (Figure 6), because such waste can be used safely as animal feed (Fontenot, 1999).

Eating insects however could pose certain risks that must be taken into account. Such risk is expected to be comparable to other animal production systems (EFSA, 2015). There is need for further research to better assess the microbiological and chemical risks from insects destined as food and animal feed. Another risk associated with the consumption of insects is that some of them can contain naturally present toxic substances, such as cyanogenic glycosides (Zagrobelny, Dreon, Gomiero, Marcazzan, Glaring, MøLler, & Paoletti, 2009). Moreover, eating insects can also cause allergies (EFSA, 2015). In China, there were over 1000 patients who suffered anaphylactic reactions after consuming silkworm pupa (Ji, Zhan, Chen, & Liu, 2008). A number of instances of allergic reactions to cochineal–biologically derived colorant obtained from the dried bodies of female cochineal insects (*Dactylopius coccus Costa/ Coccus cacti* L.), including anaphylaxis, have also been reported (Kagi, Wuthrich, & Johansson, 1994;

Table 6. Taste and smell of selected edible insects

Insect	Taste and Smell
Ants, termites	Sweet, nutty
Beetle larvae	Whole grain bread
Woodworm larvae	Fat meat
Dragonflies larvae, moths larvae	Fish
Cockroaches	Mushrooms
Bedbugs	Apples
Wasps	Pine
Maggots	Fried potatoes

Source of data: (Ramos-Elorduy, 1998)

Figure 6. Madagascar's cockroach (Gromphadorhina portentosa) grown at home fed with leftover meals
Photo: Tomasz Lewandowski

DiCello, Myc, Baker, & Baldwin, 1999). Only a few cases of anaphylactic shock have been described following consumption of Mopane caterpillar (Kung, Fenemore, & Potter, 2011; Okezie, Kgomotso, & Letswiti, 2010).

The risk of food allergy after insect consumption needs further investigation and greater attention. Similar to other animal products, insects contain their gut microflora (Klunder, Wolkers-Rooijackers, Korpela, & Nout, 2012). This makes them susceptible to microbiological hazards if proper heat treatment or storage conditions are not applied. Edible insects need to be processed and stored with care to preserve microbiological safety, that is, to control the microflora inhabiting the body of insects (Cerritos & Cano-Santana, 2008). Therefore, further studies are needed to improve our knowledge of the microbiota thriving in insects with potential uses for food and feed production. More studies concerning the influence of the rearing conditions and processing on the associated microbiota of edible insects are also necessary (Garofalo, Osimani, Milanovic, Taccari, Cardinali, Aquilanti, … Clementi, 2017).

Parasites represent another potential hazard in relation to insect consumption. They could be present in edible insects and should be considered in the case of insects' consumption as food (Chai, Shin, Lee, & Rim, 2009). For example, particular attention should be paid to *Cryptosporidium parvum*, which is an important lethal agent for immunocompromised individuals (Graczyk, Knight, & Tamang, 2005). Chemical hazards in insects depend on their habitat and plant feed contamination and can be controlled by selected farming and dietary conditions. For example, crickets are able to introduce contaminants from solid waste into the food web by preying on discarded consumer products (Gaylor, Harvey, & Hale, 2012). Insects can be regarded as safe, if properly managed and consumed. Furthermore, international food regulations need to be established for food safety of insect products.

INTEREST IN EDIBLE INSECTS ON THE FOOD MARKET

Insect breeding carried out under controlled feeding conditions has resulted in the creation of new food products responding to market needs (Figure 7). The most popular are protein bars from mixtures of various species of insects as well as flour and pasta with the addition of ground crickets (University of California, Riverside, 2016). Currently, a wide range of protein rich powders/flours from insects is being produced. Insect flour can be used in a number of ways, for example, for energy bars, bread, pasta and much more. It can be used to increase the protein and nutrient content of any food. For example, 100 g of pasta with crickets delivers 14 g of protein and is also a good source of essential vitamins and minerals. Probably with the increase of consumer acceptance, the availability will increase not only in online stores but people will be able to buy insects in every supermarket. Some supermarkets are already announcing this (Borkhataria, 2017).

Nowadays an increasing number of restaurants around the world are introducing insects into their menu. An example is chocolate dessert with roasted ants (Figure 8). It was created as part of the realization of the educational project "Between Meadow and Forest" (Food Think Tank, 2014). A group of chefs, scientists and students from the Faculty of Biotechnology and Food Science of Wroclaw University of Environmental and Life Sciences, local farmers, photographers, filmmakers, sound engineers, artisans, architects, bartenders and baristas participated in the process. The essence of the project was to explore the richness of meadows and forest environments in terms of finding new sources of food and unfamiliar flavors and aromas.

Some of the world's best chefs, like Rene Redzepi from Noma (Copenhagen, Denmark) – ranked four times as the Best Restaurant in the World by Restaurant magazine, serve ants and promote insect food. The "Cook it Raw" project (Cook it Raw, 2015) created in 2009 at the climate summit in Copenhagen, invited chefs to cook dishes with minimal energy consumption, from what was available nearby. The episode filmed in Poland in 2012 ended with a scene in which Rene Redzepi stood on a meadow with a grasshopper in his hand saying that this is our future, but we are not ready for it.

Although entomophagy is common in the world it is not yet accepted by western European populations. Popularization and information spreading about the advantages of eating insects could facilitate the integration of entomophagy in our feeding habits and behaviors (Megido, Sablon, Geuens, Brostaux, Alabi, Blecker, … Francis, 2014).

Figure 7. Chocolate with roasted larvae of the mealworms (Tenebrio molitor) made by members of the Student Scientific Circle of Molecular Cuisine, Faculty of Biotechnology and Food Science, Wroclaw University of Environmental and Life Sciences
Photo: Tomasz Lewandowski

Figure 8. Chocolate dessert with roasted ants
Photo: Jędrzej Stelmaszek/Food Think Tank

CONCLUSION

The existing research shows that insects have a high nutritional value, they are a rich source of protein and microelements (Banjo, Lawal, & Sangonuga, 2006). One of the important advantages of insect production is the reduced environmental impact compared to conventional livestock (Nakagaki & Defoliart, 1991). In addition, species such as crickets or mealworms contain many unsaturated fatty acids and vitamins. Their consumption can allow to build a well-balanced diet. Although some insects contain less protein than typical meat animal products, most of them also contain less fat, food products prepared from insects will have a lower energy value with the same nutritional value. The traditional use of insects as food is common in tropical countries but westerners should become more aware of the fact that their bias against insects as food has an adverse impact on the environmental health of the planet.

REFERENCES

Ademolu, K. O., Idowu, A. B., & Olatunde, G. O. (2010). Nutritional value assessment of variegated grasshopper, Zonocerus variegatus (L.) (Acridoidea: Pygomorphidae) during post-embryonic development. *African Entomology*, *18*(2), 360–364. doi:10.4001/003.018.0201

Banjo, A. D., Lawal, O. A., & Sangonuga, E. A. (2006). The nutritional value of fourteen species of edible insects in southern western Nigeria. *African Journal of Biotechnology*, *5*(3), 201–298. doi:10.5897/AJB05.250

Barker, D., Fitzpatrick, M. P., & Dierenfeld, E. S. (1998). Nutrient composition of selected whole invertebrates. *Zoo Biology*, *17*(2), 123–134. doi:10.1002/(SICI)1098-2361(1998)17:2<123::AID-ZOO7>3.0.CO;2-B

Borkhataria, C. (2017). Would you eat a burger made from insects? Mealworm-based food line set to hit grocery stores in Switzerland next week. *Daily Mail Australia*. Retrieved from https://www.dailymail.co.uk/sciencetech/article-4790414/Food-insects-sold-Swiss-supermarket.html

Bukkens, S. G. F. (1997). The nutritional value of edible insects. *Ecology of Food and Nutrition*, *36*(2-4), 287–319. doi:10.1080/03670244.1997.9991521

Bukkens, S. G. F. (2005). Insects in the human diet: Nutritional aspects. In M. G. Paoletti (Ed.), *Ecological implications of minilivestock: potential of insects, rodents, frogs and snails* (pp. 545–577). Enfield, NH: Science Publishers.

Cerritos, R., & Cano-Santana, Z. (2008). Harvesting grasshoppers Sphenarium purpurascens in Mexico for human consumption: A comparison with insecticidal control for managing pest outbreaks. *Crop Protection Journal*, *27*(3–5), 473–480. doi:10.1016/j.cropro.2007.08.001

Chai, J. Y., Shin, E. H., Lee, S. H., & Rim, H. J. (2009). Foodborne intestinal flukes in Southeast Asia. *Korean Journal of Parasitology*, *47*(Suppl), S69–S102. doi:10.3347/kjp.2009.47.S.S69 PMID:19885337

Chen, X., Feng, Y., Zhang, H., & Chen, Z. (2010). Review of the nutritive value of edible insects. In P. B. Durst, D. V. Johnson, R. N. Leslie & K. Shono (Eds), *Forest insects as food: Humans bite back* (pp. 85–92). Bangkok, Thailand: Food and Agriculture Organization of the United Nations. Retrieved from http://www.fao.org/docrep/012/i1380e/i1380e00.pdf

Christensen, D. L., Orech, F. O., Mungai, M. N., Larsen, T., Friis, H., & Aagaard-Hansen, J. (2006). Entomophagy among the Luo of Kenya: A potential mineral source? *International Journal of Food Sciences and Nutrition*, *57*(3–4), 198–203. doi:10.1080/09637480600738252 PMID:17127470

Collavo, A., Glew, R. H., Huang, Y.-S., Chuang, L. T., Bosse, R., & Paoletti, M. G. (2005). House cricets small-scale farming. In M. G. Paoletti (Ed.), *Ecological implications of minilivestock: potential of insects, rodents, frogs and snails* (pp. 519–544). Enfield, NH: Science Publishers.

Cook it raw. (2015). *Menu*. Retrieved from http://www.cookitraw.org/home/

DeFoliart, G. R. (1999). Insects as food: Why the Western attitude is important. *Annual Review of Entomology*, *44*(1), 21–50. doi:10.1146/annurev.ento.44.1.21 PMID:9990715

DiCello, M. C., Myc, A., Baker, J. R. Jr, & Baldwin, J. L. (1999). Anaphylaxis after ingestion of carmine colored foods: Two case reports and a review of the literature. *Allergy and Asthma Proceedings*, *20*(6), 377–382. doi:10.2500/108854199778251816 PMID:10624494

Dreyer, J. J., & Wehmeyer, A. S. (1982). On the nutritive value of Mopanie worms. *South African Journal of Science*, *78*, 33–35.

Elemo, B. O., Elemo, G. N., Makinde, M. A., & Erukainure, O. L. (2011). Chemical evaluation of African palm weevil, Rhychophorus phoenicis, larvae as a food source. *Journal of Insect Science*, *11*(146), 1–6. doi:10.1673/031.011.14601 PMID:22236060

European Food Safety Authority (EFSA). (2015). Risk profile related to production and consumption of insects as food and feed. *EFSA Journal*, *13*(10), 4257. doi:10.2903/j.efsa.2015.4257

Finke, M. D. (2002). Complete nutrient composition of selected invertebrates commonly fed to insectivores. *Zoo Biology*, *21*, 269–285. doi:10.1002/zoo.10031

Finke, M. D. (2007). Estimate of chitin in raw whole insects. *Zoo Biology*, *26*(2), 105–115. doi:10.1002/zoo.20123 PMID:19360565

Fontenot, J. P. (1999). Nutrient recycling: The North American experience — review. *Asian-Australasian Journal of Animal Sciences*, *12*(4), 642–650. doi:10.5713/ajas.1999.642

Food and Agriculture Organization of the United Nations (FAO). (2009). *How to feed the world in 2050?* Rome, Italy: FAO. Retrieved from http://www.fao.org/fileadmin/templates/wsfs/docs/expert_paper/How_to_Feed_the_World_in_2050.pdf

Food and Agriculture Organization of the United Nations (FAO). (2016). *The contribution of insects to food security, livelihoods and the enviroment.* Rome, Italy: FAO. Retrieved from http://www.fao.org/docrep/018/i3264e/i3264e00.pdf

Food and Agriculture Organization of the United Nations/World Health Organization (FAO/WHO). (2001). *Human vitamin and mineral requirements.* Rome, Italy: Food and Nutrition Division, FAO. Retrieved from http://www.fao.org/3/a-y2809e.pdf

Food Think Tank. (2014). *Between Meadow and Forest.* Retrieved from http://foodthinktank.pl/miedzylakaalasem (in Polish)

Garofalo, C., Osimani, A., Milanovic, V., Taccari, M., Cardinali, F., Aquilanti, L., ... Clementi, F. (2017). The microbiota of marketed processed edible insects as revealed by high-throughput sequencing. *Food Microbiology*, *62*, 15–22. doi:10.1016/j.fm.2016.09.012 PMID:27889142

Gaylor, M. O., Harvey, E., & Hale, R. C. (2012). House crickets can accumulate polybrominated diphenyl ethers (PBDEs) directly from polyurethane foam common in consumer products. *Chemosphere*, *86*(5), 500–505. doi:10.1016/j.chemosphere.2011.10.014 PMID:22071374

Graczyk, T. K., Knight, R., & Tamang, L. (2005). Mechanical transmission of human protozoan parasites by insects. *Clinical Microbiology Reviews*, *18*(1), 128–132. doi:10.1128/CMR.18.1.128-132.2005 PMID:15653822

Halloran, A., Hanboonsong, Y., Roos, N., & Bruun, S. (2017). Life cycle assessment of cricket farming in north-eastern Thailand. *Journal of Cleaner Production*, *156*, 83–94. doi:10.1016/j.jclepro.2017.04.017

Hwangbo, J., Hong, E. C., Jang, A., Kang, H. K., Oh, J. S., Kim, B. W., & Park, B. S. (2009). Utilization of house fly-maggots, a feed supplement in the production of broiler chickens. *Journal of Environmental Biology*, *30*, 609–614. PMID:20120505

Illgner, P., & Nel, E. (2000). The geography of edible insects in Sub-Saharan Africa: A study of the Mopane Caterpillar. *The Geographical Journal*, *166*(4), 336–351. doi:10.1111/j.1475-4959.2000.tb00035.x

Janssen, R. H., Vincken, J.-P., van den Broek, L. A. M., Fogliano, V., & Lakemond, C. M. M. (2017). Nitrogen-to-protein conversion factors for three edible insects: Tenebrio molitor, Alphitobius diaperinus, and Hermetia illucens. *Journal of Agricultural and Food Chemistry*, *65*(11), 2275–2278. doi:10.1021/acs.jafc.7b00471 PMID:28252948

Ji, K. M., Zhan, Z. K., Chen, J. J., & Liu, Z. G. (2008). Anaphylactic shock caused by silkworm pupa consumption in China. *Allergy, 63*(10), 1405–1410. doi:10.1111/j.1398-9995.2008.01838.x PMID:18782121

Kagi, M. K., Wuthrich, B., & Johansson, S. G. (1994). Campari-orange anaphylaxis due to carmine allergy. *Lancet, 344*(8914), 60–61. doi:10.1016/S0140-6736(94)91083-9 PMID:7912327

Klunder, H. C., Wolkers-Rooijackers, J., Korpela, J. M., & Nout, M. J. R. (2012). Microbiological aspects of processing and storage of edible insects. *Food Control, 26*(2), 628–631. doi:10.1016/j.foodcont.2012.02.013

Kouřimská, L., & Adámková, A. (2016). Nutritional and sensory quality of edible insects. *NFS Journal, 4*, 22–26. doi:10.1016/j.nfs.2016.07.001

Kung, S. J., Fenemore, B., & Potter, P. C. (2011). Anaphylaxis to mopane worms (Imbrasia belina). *Annals of Allergy, Asthma & Immunology, 106*(6), 538–540. doi:10.1016/j.anai.2011.02.003 PMID:21624756

Lee, K. P., Simpson, S. J., & Wilson, K. (2008). Dietary protein-quality influences melanization and immune function in an insect. *Functional Ecology, 22*(6), 1052–1061. doi:10.1111/j.1365-2435.2008.01459.x

Lefevre, M., Kris-Etherton, P. M., Zhao, G., & Tracy, R. P. (2004). Dietary fatty acids, hemostasis, and cardiovascular disease risk. *Journal of the Academy of Nutrition and Dietetics, 104*(3), 410–419. doi:10.1016/j.jada.2003.12.022 PMID:14993864

MacEvilly, C. (2000). Bugs in the system. *Nutrition Bulletin, 25*(4), 267–268. doi:10.1046/j.1467-3010.2000.00068.x

Megido, R. C., Sablon, L., Geuens, M., Brostaux, Y., Alabi, T., Blecker, C., ... Francis, F. (2014). Promising attitude for entomophagy development. *Journal of Sensory Studies, 29*(1), 14–20. doi:10.1111/joss.12077

Nakagaki, B. J., & Defoliart, G. R. (1991). Comparison of diets for mass-rearing Acheta domesticus (Orthoptera: Gryllidae) as a novelty food, and comparison of food conversion efficiency with values reported for livestock. *Journal of Economic Entomology, 84*(3), 891–896. doi:10.1093/jee/84.3.891

Okezie, O. A., Kgomotso, K. K., & Letswiti, M. M. (2010). Mopane worm allergy in a 36-year-old woman: A case report. *Journal of Medical Case Reports, 4*(1), 42. doi:10.1186/1752-1947-4-42 PMID:20205892

Oonincx, D. G., & Dierenfeld, E. S. (2012). An investigation into the chemical composition of alternative invertebrate prey. *Zoo Biology, 31*(1), 40–54. doi:10.1002/zoo.20382 PMID:21442652

Oonincx, D. G. A. B., & de Boer, I. J. M. (2012). Environmental impact of the production of mealworms as a protein source for humans – a life cycle assessment. *PLoS One, 7*(12), 51145. doi:10.1371/journal.pone.0051145 PMID:23284661

Oonincx, D. G. A. B., van Itterbeeck, J., Heetkamp, M. J. W., van den Brand, H., van Loon, J. J. A., & van Huis, A. (2010). An exploration on greenhouse gas and ammonia production by insect species suitable for animal or human comsumption. *PLoS One, 5*(12), e14445. doi:10.1371/journal.pone.0014445 PMID:21206900

Paoletti, M. G., Norberto, L., Damini, R., & Musumeci, S. (2007). Human gastric juice contains chitinase that can degrade chitin. *Annals of Nutrition & Metabolism*, *51*(3), 244–251. doi:10.1159/000104144 PMID:17587796

Payne, C. (2018, January 18). Edible insects: Do insects actually taste any good? *BBC News*. Retrieved from: https://www.bbc.co.uk/news/world-42639877

Pennino, M., Dierenfeld, E. S., & Behler, J. L. (1991). Retinol, alpha-tocopherol and proximate nutrient composition of invertebrates used as feed. *International Zoo Yearbook*, *30*(1), 143–149. doi:10.1111/j.1748-1090.1991.tb03477.x

Pino-Moreno, J. M., & Ramos-Elorduy, J. J. (2006). Algunos ejemplos de aprovechamiento comercial de varios insectos comestibles y medicinales. In E. Estrada, J. N. Romero, A. M. Equihua, C. L. Luna L., & J. L. A. Rosas (Eds.), Entomología mexicana (pp. 524-533). Mexico City, Mexico: Sociedad Mexicana de Entomología.

Premalatha, M., Abbasi, T., Abbasi, T., & Abbasi, S. A. (2011). Energy-efficient food production to reduce global warming and ecodegradation: The use of edible insects. *Renewable & Sustainable Energy Reviews*, *15*(9), 4357–4360. doi:10.1016/j.rser.2011.07.115

Ramos-Elorduy, J. (1998). *Creepy crawly cuisine: The gourmet guide to edible insects*. Paris, France: Park Street Press.

Ramos-Elorduy, J., Moreno, J. M. P., Prado, E. E., Perez, M. A., Otero, J. L., & de Guevara, O. L. (1997). Nutritional value of edible insects from the state of Oaxaca, Mexico. *Journal of Food Composition Analisys*, *10*(2), 142–157. doi:10.1006/jfca.1997.0530

Renault, D., Laparie, M., McCauley, S. J., & Bonte, D. (2018). Environmental adaptations, ecological filtering, and dispersal central to insect invasions. *Annual Review of Entomology*, *63*(1), 345–368. doi:10.1146/annurev-ento-020117-043315 PMID:29029589

Rumpold, B. A., & Schluter, O. K. (2013a). Nutritional composition and safety aspects of edible insects. *Molecular Nutrition & Food Research*, *57*(5), 802–823. doi:10.1002/mnfr.201200735 PMID:23471778

Rumpold, B. A., & Schluter, O. K. (2013b). Potential and challenges of insects as an innovative source for food and feed production Innovative. *Food Science and Emerging Technologies*, *17*, 1–11. doi:10.1016/j.ifset.2012.11.005

Sileshi, G. W., & Kenis, M. (2010). Food security: Farming insects. *Science*, *328*(5978), 568–568. doi:10.1126cience.328.5978.568-a PMID:20430996

Smil, V. (2002). Worldwide transformation of diets, burdens of meat production and opportunities for novel food proteins. *Enzyme and Microbial Technology*, *30*(3), 305–311. doi:10.1016/S0141-0229(01)00504-X

Tosi, E., & Daccordi, M. (1983). Insetti commestibili. Slurp! *Nigrizia*, *101*(6), 49–52.

United States Department of Agriculture (USDA). (2012). *The Pherobase: Database of pheromones and semiochemicals*. Retrieved from http://www.pherobase.com/

University of California, Riverside. (2016). *Entomophagy (Eating insects).* San Diego, CA: Ceter for Invasive Species Research. Retrieved from http://cisr.ucr.edu/entomophagy.html

van Huis, A., Dicke, M., & van Loon, J. J. A. (2015). Insects to feed the world. *Journal of Insects as Food and Feed, 1*(1), 3–5. doi:10.3920/JIFF2015.x002

van Huis, A., Itterbeeck, J. V., Klunder, H., Mertens, E., Halloran, A., Muir, G., & Vantomme, P. (2013). *Edible insects: Future prospects for food and feed security.* Rome, Italy: Food and Agriculture Organisation of the United Nations. Retrieved from http://www.fao.org/docrep/018/i3253e/i3253e.pdf

Van Soest, P. J., & Robertson, J. B. (1977). What is fibre and fibre in food? *Nutrition Reviews, 35*(3), 12–22. doi:10.1111/j.1753-4887.1977.tb06532.x PMID:840446

Wageningen University and Research. (2017). *List of edible insects of the world (April 1, 2017).* Retrieved from https://www.wur.nl/en/Expertise-Services/Chair-groups/Plant-Sciences/Laboratory-of-Entomology/Edible-insects/Worldwide-species-list.htm

Womeni, H. M., Linder, M., Tiencheu, B., Mbiapo, F. T., Villeneuve, P., Fanni, J., & Parmentier, M. (2009). Oils of insects and larvae consumed in Africa: Potential sources of polyunsaturated fatty acids. *OCL – Oleagineux, Corps Gras, Lipides, 16*(4), 230-235. Retrieved from https://www.ocl-journal.org/articles/ocl/abs/2009/04/ocl2009164p230/ocl2009164p230.html

Yen, A. L. (2009). Edible insects: Traditional knowledge or western phobia? *Entomological Research, 39*(5), 289–298. doi:10.1111/j.1748-5967.2009.00239.x

Yen, A. L. (2010). Edible insects and other invertebrates in Australia: future prospects. In P. B. Durst, D. V. Johnson, R. N. Leslie & K. Shono (Eds), *Forest insects as food: Humans bite back* (pp. 65–84). Bangkok, Thailand: Food and Agriculture Organization of the United Nations. Retrieved from http://www.fao.org/docrep/012/i1380e/i1380e00.pdf

Zagrobelny, M., Dreon, A. L., Gomiero, T., Marcazzan, G. L., Glaring, M. A., MøLler, B. L., & Paoletti, M. G. (2009). Toxic moths: Source of a truly safe delicacy. *Journal of Ethnobiology, 29*(1), 64–76. doi:10.2993/0278-0771-29.1.64

Zielińska, E., Baraniak, B., Karaś, M., Rybczyńska, K., & Jakubczyk, A. (2015). Selected species of edible insects as a source of nutrient composition. *Food Research International, 77*, 460–466. doi:10.1016/j.foodres.2015.09.008

ADDITIONAL READING

Borkhataria, C. (2017). *Would you eat a burger made from insects? Mealworm-based food line set to hit grocery stores in Switzerland next week.* Daily Mail Australia. Retrieved from https://www.dailymail.co.uk/sciencetech/article-4790414/Food-insects-sold-Swiss-supermarket.html

Bukkens, S. G. F. (1997). The nutritional value of edible insects. *Ecology of Food and Nutrition, 36*(2-4), 287–319. doi:10.1080/03670244.1997.9991521

Durst, P. B., Johnson, D. V., Leslie, R. N., & Shono, K. (Eds.). (2010). *Forest insects as food: Humans bite back*. Proceedings of a workshop on Asia-Pacific resources and their potential for development, Chiang Mai, Thailand, 19-21 February, 2008. Bangkok, Thailand: Food and Agriculture Organization of the United Nations. Retrieved from http://www.fao.org/docrep/012/i1380e/i1380e00.pdf

Food and Agriculture Organization of the United Nations (FAO). (2016). *The contribution of insects to food security, livelihoods and the enviroment*. Rome, Italy: FAO. Retrieved from http://www.fao.org/docrep/018/i3264e/i3264e00.pdf

Oonincx, D. G. A. B., van Itterbeeck, J., Heetkamp, M. J. W., van den Brand, H., van Loon, J. J. A., & van Huis, A. (2010). An exploration on greenhouse gas and ammonia production by insect species suitable for animal or human comsumption. *PLoS One*, *5*(12), e14445. doi:10.1371/journal.pone.0014445 PMID:21206900

Paoletti, M. G. (Ed.). (2005). *Ecological implications of minilivestock: Potential of insects, rodents, frogs and snails*. Enfield, NH: Science Publishers. doi:10.1201/9781482294439

Ramos-Elorduy, J. (1998). *Creepy crawly cuisine: The gourmet guide to edible insects*. Paris, France: Park Street Press.

Rumpold, B. A., & Schluter, O. K. (2013b). Potential and challenges of insects as an innovative source for food and feed production Innovative. *Food Science and Emerging Technologies*, *17*, 1–11. doi:10.1016/j.ifset.2012.11.005

Wageningen University and Research. (2017). List of edible insects of the world (April 1, 2017). Retrieved from https://www.wur.nl/en/Expertise-Services/Chair-groups/Plant-Sciences/Laboratory-of-Entomology/Edible-insects/Worldwide-species-list.htm

Zielińska, E., Baraniak, B., Karaś, M., Rybczyńska, K., & Jakubczyk, A. (2015). Selected species of edible insects as a source of nutrient composition. *Food Research International*, *77*, 460–466. doi:10.1016/j.foodres.2015.09.008

KEY TERMS AND DEFINITIONS

Animal Protein: Protein is built from building blocks (amino acids); our bodies make amino acids from scratch, or by modifying others but a few amino acids (known as the essential amino acids) must come from food. Animal sources of protein tend to deliver all amino acids we need. Other protein sources, such as fruits, vegetables, grains, nuts and seeds, may lack one or more essential amino acids.

Conventional Livestock: Domesticated animals raised in an agricultural setting to produce meat, eggs, milk, leather, wool, and other products.

Diet: The sum of the food consumed by a person or another organism.

Edible Insects: Insects which can be consumed by humans.

Essential Fatty Acid (EFA): An unsaturated fatty acid that is essential to human health, but cannot be manufactured in the body; supplementation with EFAs could be useful as a treatment for certain neurological disorders.

European Food Safety Agency (EFSA): An organization which provides scientific advice and communications about existing and emerging risks associated with the food chain.

Energy Value/Content: The amount of energy available from an item of food when digested, mostly from carbohydrates and fats.

Entomophagy: From the Greek words ἔντομον *éntomon* – insect, and φᾰγεῖν *phagein* – to eat; the human use of insects as food.

Essential Amino Acid: An amino acid which is required for normal health and growth but cannot be synthesized *de novo* (from scratch) by the organism, and thus must be supplied in the diet.

Essential Nutrient: A nutrient required for normal body functioning which cannot be synthesized by the organism and must be provided by the diet.

Fatty Acids Profile: Percentage of fatty acids in food.

Fiber: Dietary material containing substances such as cellulose, lignin, and pectin, that are resistant to the action of digestive enzymes.

Food Allergy: An immune system reaction that occurs soon after eating a certain food; causes digestive problems, hives or swollen airways, in some people, a food allergy can cause severe symptoms or even a life-threatening reaction known as anaphylaxis.

Food and Agriculture Organization of the United Nations (FAO): A specialized agency of the United Nations which leads international efforts to defeat hunger.

Greenhouse Gas: Gas that contributes to the greenhouse effect by absorbing infrared radiation (for example carbon dioxide and chlorofluorocarbons).

Insect Farming: The practice of raising insects as livestock. Insect farming in a closed or indoor environment is an important means for making food available continuously year-round.

Insect Protein: A new source for animal feed and food; as protein sources, the nutritive value of edible insects is as good as other animals (or plants) or even better.

Macrominerals: A number of minerals, such as calcium, phosphorus, magnesium, sodium, potassium, chloride, and sulfurase, which are needed in large amounts to maintain the proper functioning of an organism.

Monounsaturated Fatty Acids (MUFAs): Acids with one double bond in the fatty acid chain, the remaining carbon atoms are bound by single bond.

Nutrient Content: A source of nourishment, especially a nourishing ingredient in a food.

Nutritive Value: The contribution of a food to the nutrient content of the diet. This value depends on the quantity of the food which is digested and absorbed and the amounts of the essential nutrients (protein, fat, carbohydrate, minerals, vitamins) which it contains.

Organoleptic Properties: The aspects of food that an individual experiences via the senses—including taste, sight (color), smell, and touch (texture).

Polyunsaturated Fatty Acids (PUFAs): Fatty acids with two or more double bonds between the carbon atoms.

Protein Bars: Lower in carb, vitamins, and dietary minerals and significantly higher in protein than other bars; they are mainly used by athletes for muscle building.

Saturated Fatty Acids (SFAs): Fatty acids in which all carbon atoms in the hydrocarbon chain are joined by single bonds. They exist mostly as components of fats (triglycerides) or other lipids of animal origin; a diet high in saturated fatty acids may contribute to a high blood cholesterol level.

Trace Minerals: Essential minerals, such as iron, zinc, selenium, fluoride, chromium, copper, iodine, manganese, and molybdenum, which help the body perform regulatory and structural functions.

Vitamins: A group of organic compounds which are essential for normal growth and nutrition and are required in small quantities in the diet because they cannot be synthesized by the body; they have diverse biochemical functions.

World Health Organization (WHO): An international organization whose primary role is to direct international health within the United Nations' system and to lead partners in global health responses.

Chapter 9
Understanding Edible Insects as Food in Western and Eastern Societies

Giovanni Sogari
University of Parma, Italy

Aijun Liu
Nanjing Agricultural University, China

Jie Li
Cornell University, USA

ABSTRACT

In the forthcoming decades, insects might become an important alternative protein source for human consumption. However, what do consumers think about eating insects? The answer is still not very clear, and large differences exist between Western and Eastern societies. While the former has never really experienced edible insects as food, (some) Eastern countries have already practiced entomophagy for a long time. To better understand consumers' perception in both types of societies, a literature review was carried out. The results show that in the Western countries, the consumption of edible insects will depend primarily on availability in the market (i.e., regulatory framework and industry), product category (i.e., processed or unprocessed, familiar or unfamiliar), communication, and marketing. Nonetheless, more research studies are needed to explore Eastern consumers and the development of the edible insect market and industry in Asian countries.

INTRODUCTION

The Food and Agriculture Organization (FAO) of the United Nations (UN) forecasts a 2.7% annual increase in meat production in the coming years, primarily driven by population growth in developing countries (FAO, 2017). This increase, along with the limited land area available, may pose a challenge for the meat industry worldwide, and alternative protein sources are most likely to be needed to feed a

DOI: 10.4018/978-1-5225-7350-0.ch009

growing population (van Huis, 2016). The heavy ecological footprint of the livestock industry is another reason for the search for alternatives (FAO, 2006). Over the past several years, the FAO, through its publications, in particular *Edible insects, future prospects for food and feed security*, (FAO, 2013), work programs, and field projects, has been promoting the idea of insects as food to help increase consumer acceptability in its member countries (Vantomme, 2015).

In addition to their possible consumption by humans, insects are also being studied as an innovative ingredient for animal feeds, including feed for conventional livestock (e.g. poultry and swine), fish, and pets. This is because insect-based animal feeds are high in protein content, especially when processed and transformed into insect flour (van Huis, Van Itterbeeck, Klunder, Mertens, Halloran, Muir, & Vantomme, 2013). Although insect-based feeds appear to be a viable option, further studies are needed regarding the nutritional, environmental and economic benefits of using them to reduce and replace conventional animal feeds (Rumpold & Schlüter, 2013a; Gasco, Finke, & van Huis, 2018). Particularly, it is not known what additional health benefits exist for animals that consume insect-based feeds, relative to traditional feed formulae, such as fishmeal and soybeans (Gasco, Finke, & van Huis, 2018).

Also for human consumption, insect-based food appears to be a healthy choice because it is an excellent source of protein, fat, energy and fiber (see Table 1). Nutrient composition varies greatly depending on species (Belluco, Losasso, Maggioletti, Alonzi, Paoletti, & Ricci, 2013; Rumpold & Schlüter, 2013b). Additionally, insect production is generally considered to be more sustainable from an environmental point of view (Hartmann & Siegrist, 2017). Therefore, insect proteins are seen as an alternative sustainable source for both human and animals.

INSECTS AS FOOD

Insect species are numerous and heterogeneous. Their class belongs to the arthropods and includes more than a million species (Rumpold & Schlüter, 2013a). Insects are utilized in the sericulture industry (e.g. silkworms), as a coloring agent (e.g. cochineal red), and in the apiculture sector (e.g. honey bees). They are also edible and therapeutic (e.g. high protein foods and pharmaceuticals) with great ecological

Table 1. Average content of protein, fat and energy of specific insect orders

Insect Orders	Protein (% Dry Matter)	Fat (% Dry Matter)	Fiber (%)	Energy (kcal/100 g)
Blattodea (cockroaches)	57.30	29.90	5.31	-
Coleoptera (adult beetles, larvae)	40.69	33.40	10.74	490.30
Hemiptera (true bugs)	48.33	30.26	12.40	478.99
Hymenoptera (ants, bees)	46.47	25.09	5.71	484.45
Isoptera (termites)	35.34	32.74	5.06	
Lepidoptera (butterflies, moths)	45.38	27.66	6.60	508.89
Odonata (dragonflies, damselflies)	55.23	19.83	11.79	431.33
Orthoptera (crickets, grasshoppers, locusts)	61.32	13.41	9.55	426.25

Source of data: (Rumpold & Schluter, 2013b)

importance (e.g. biomass recycling and indicator of water pollution) and have various other uses (e.g. forensic investigation) (Lokeshwari & Shantibala, 2010).

"Entomophagy", meaning consuming insects (Evans, Alemu, Flore, Frøst, Halloran, Jensen, … Eilenberg, 2015), is derived as a word from the Greek *éntomon* (insect) and *phagein* (to eat). It is practised in many different societies around the world as shown in Figure 1 (van Huis, 2013), and has been a part of human eating practices since prehistory (Sogari & Vantomme, 2014; van Huis, Van Itterbeeck, Klunder, Mertens, Halloran, Muir, & Vantomme, 2013). Insects can be prepared and eaten raw, fried, boiled, roasted or ground and in various life stages. Jongema (2017) estimates that around 2,111 known edible insect species are currently consumed by humans, and these tend to be species that are easier to access (e.g. wingless and slow moving) and abundant (Raubenheimer & Rothman, 2013; Sogari & Vantomme, 2014).

The most commonly consumed species (see Figure 2) are within the Coleoptera group, including but not limited to beetles (659 species), Lepidoptera represented by caterpillars (362 species), Hymenoptera represented by bees, wasps and ants (321 species), Orthoptera represented by grasshoppers, locusts and crickets (278 species), Hemiptera represented by true bugs (237 species), Odonata represented by dragonflies (61 species), Isoptera represented by termites (59 species), and Diptera represented by flies (37 species) (Jongema, 2017). A large proportion of these insects are captured in the wild (Rumpold & Schlüter, 2013a) using trap technology (e.g. lights, nets, fire or water-filled bins) (Durst & Johnson, 2010). However, there are many domesticated species (e.g. bees and silkworms) that have been farmed for specific purposes (e.g. honey and silk) (Raubenheimer & Rothman, 2013).

Figure 1. Recorded edible insect species, by country
Source: https://www.wur.nl/en/Expertise-Services/Chair-groups/Plant-Sciences/Laboratory-of-Entomology/Edible-insects/Worldwide-species-list.htm

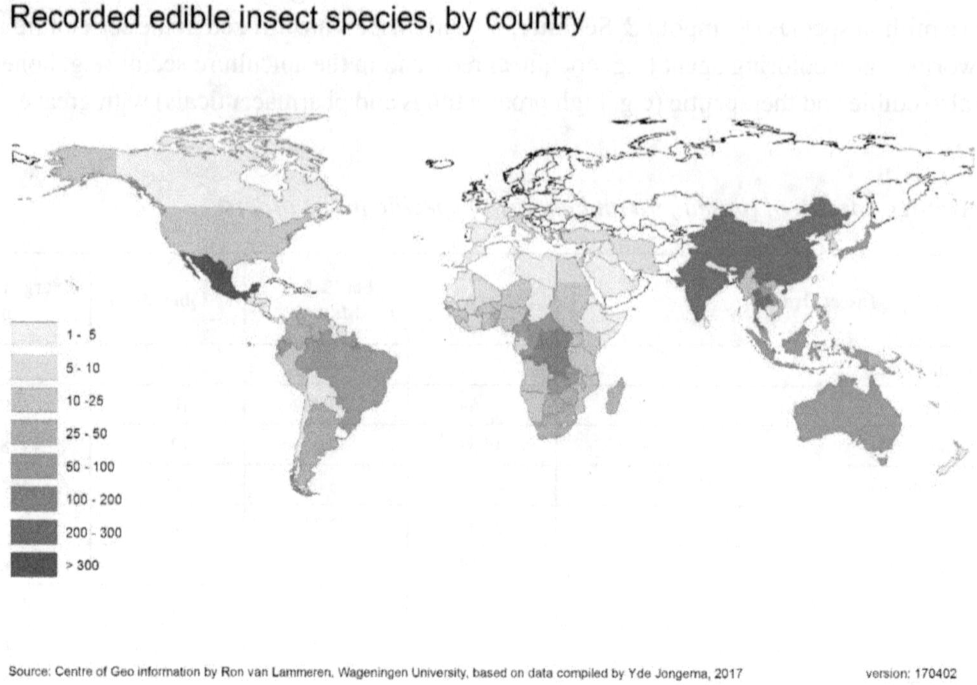

Figure 2. Number of recorded edible insect species per group in the world (n=2,111)
Source: https://www.wur.nl/en/Expertise-Services/Chair-groups/Plant-Sciences/Laboratory-of-Entomology/Edible-insects/Worldwide-species-list.htm

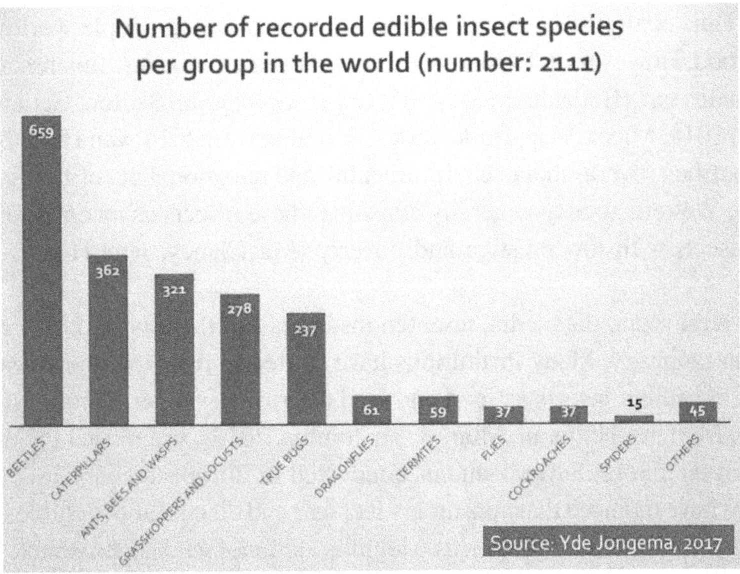

In the last few decades, insect consumption has declined, especially in countries where entomophagy was very common in the past. This is mostly due to the introduction and adoption of new food habits closer to those of Western cultures, particularly among younger and urban populations (Looy, Dunkel, & Wood, 2014; Obopile & Seeletso, 2013; van Huis, Dicke, & van Loon, 2015; Yen, 2015). On the other hand, there has recently been an increasing interest in insect food products in Western society, which traditionally considers eating insects as "rural" and "barbarian" (Caparros Megido, Sablon, Geuens, Brostaux, Alabi, Blecker, … Francis, 2014; van Huis, 2013).

Insects might become an important alternative protein source for both human consumption and animal feed. However, what do consumers think about this idea? The answer is unknown, and it is thus crucial to explore the differences between two types of societies: the Western countries, where edible insects are considered a novel food, and (some) Eastern countries, where entomophagy has long been a normal practice. To better understand consumers from both types of societies, a literature review is performed, drawing principally from the Web of Science database and secondarily from other databases which have provided high-quality, peer-reviewed journal articles and grey literature (e.g. contributions to scientific conferences and reports). The structure of this literature review includes: (1) entomophagy in Western and Asian societies; (2) consumers' perception of edible insects as food, and; (3) the market perspective.

ENTOMOPHAGY AND WESTERN SOCIETY

In some Western countries, especially Europe, the issue of rearing edible insects as food and feed has become a hot topic among scientific communities, companies and policy makers, as well as among consumers who look at entomophagy with either curiosity or disgust. From a historical point of view, entomophagy is not a common eating practice and is quite new to Western societies. However, the ancient Greeks and Romans considered some types of insects, especially grasshoppers and beetle larvae, as a delicacy (DeFoliart, 1995 and 1999).

A study by van Huis et al. (2013) estimated that about two billion people worldwide consider consuming insects as food. However, in Europe and other Western countries, this practice has never been widespread and remains rare (Bodenheimers, 1951; Caparros Megido, Sablon, Geuens, Brostaux, Alabi, Blecker, … Francis, 2014; Mlcek, Rop, Borkovcova, & Bednarova, 2014; van Huis, 2013). Although the FAO (2013) has identified several social, environmental, and nutritional benefits associated with human insect consumption, Western society generally considers these insects as an emergent food source, and associates eating insects with low prestige and poverty (MacClancy, Jeya Henry, & Macbeth, 2007; Sogari, 2015).

Over the past several years, the media, research institutes and the food industry have paid considerable attention to entomophagy. Many institutions have started to research topics such as the costs and benefits of introducing edible insects as food and feed (Deroy, Reade, & Spence, 2015; van Huis, Van Itterbeeck, Klunder, Mertens, Halloran, Muir, & Vantomme, 2013), suggesting the emergence of a possible niche market in the future. Several studies conducted in European countries (e.g. Belgium, Italy, and the Netherlands) have reported that consumers feel fear and disgust about edible insects, but are also curious to try these novel products (Caparros Megido, Sablon, Geuens, Brostaux, Alabi, Blecker, … Francis, 2014; Materia & Cavallo, 2015; Pascucci & De-Magistris, 2013; Menozzi, Sogari, Veneziani, Simoni, & Mora, 2017b; Sogari, Menozzi, & Mora, 2017).

ENTOMOPHAGY AND EASTERN SOCIETY

Surprisingly, consumption of edible insects is decreasing in Eastern countries where insects have long been a part of the traditional diet, and this decrease has been even more marked in urban areas. This is primarily due to globalization, which has made available new food varieties and improvements in food technology and has allowed Western culture's contributions to gastronomy to diffuse worldwide (Yen, 2009; Hartmann, Shi, Giusto, & Siegrist, 2015). However, entomophagy, was and is still an important dietary behavior in many parts of the world, especially in Asia (van Huis, Van Itterbeeck, Klunder, Mertens, Halloran, Muir, & Vantomme,, 2013). China has a long history of consuming edible insects dating back over 2000 years. They are raised and bred mostly in rural China for human consumption, medicine and animal feed (Feng, Chen, Zhao, He, Sun, Wang, & Ding, 2017). The earliest document on insect use by humans concerned the large scale breeding of silk worms around 5000 years ago (Luo, 1997). It is estimated that 178 insect species from 96 genera, 53 families and 11 orders are commonly consumed in China in the present day (Dobermann, Swift, & Field, 2017). Eggs and adults are mostly processed and prepared for snacks, while larvae and pupae are mostly consumed as a main course in restaurants (Chen, Feng, & Chen, 2009). Preparation of edible insects includes deep frying, braising, stewing, stewing after frying, boiling, and roasting. There are 20 to 30 popular species used in restaurants year-round, including grasshoppers, silkworm pupae, wasps, bamboo insects and stink bugs (Chen, Feng, & Chen, 2009). The Chinese Ministry of Health has recently promoted silkworm pupae as a new food source, creating great scientific interest in this topic (Raheem, Carrascosa, Oluwole, Nieuwland, Saraiva, Millán, & Raposo, 2018). Most of Chinese consumers are familiar with edible insects, and thus their acceptance of insects as food is generally high (Hartmann, Shi, Giusto, & Siegrist, 2015; Feng, Chen, Zhao, He, Sun, Wang, & Ding, 2017).

Human insect consumption is also popular in other Asian countries including Thailand, Japan, Laos, and Borneo. In Thailand, about 150 mostly wild-caught insect species are consumed, and they constitute a critical part of Thai people's diets (Yhoung–Aree, Puwastien, & Attig, 2010). In particular this country has one of the most advanced cricket farming systems in the world with a total number of farms around 20,000 and an averaged production of 7,500 tonnes per year (Hanboonsong, Jamjanya, & Durst, 2013).

In Japan, insect food consumption has generally declined in most places. People who live in mountainous areas are the primary consumers of insects in Japan, and they eat species such as the long-horned beetle caterpillar and wasps (Nonaka, 2010). Eating wild-caught insects is very common among the Laotian people and the percentage of the Laotian population that consumes insects on a regular basis is the highest in the world (Hanboonsong & Durst, 2014). Moreover, it seems their willingness to eat insects more frequently will increase if some barriers are addressed, such as seasonality and cost issues (Barennes, Phimmasane, & Rajaonarivo, 2015). In Borneo, a country with 80 commonly consumed insect species, the local population collects insects at various life stages, including eggs, larvae or nymphs, and pupae or adults (Chung, 2010).

CONSUMER ATTITUDE TOWARD INSECT CONSUMPTION LITERATURES

In order to explore consumer attitude toward insect consumption, a study was conducted using the Web of Science database to extract documents (such as articles, reviews, books, chapters, conference proceedings, etc.) published over the past 20 years. Several keywords were used in the title, abstract or keywords of the publications for the database query, including *perception OR attitude OR acceptance OR acceptability OR behavior OR receptiveness OR sensory OR disgust OR preference OR neophobia* AND *entomophagy OR edible insect OR insects as food OR eating insects OR insect product OR insect-based product OR insect-based product OR insect consumption*. The total output was manually screened by evaluating the title and/or the abstract for relevance.

A total of 139 works was identified including articles focusing on the Western world and the Eastern world (see Figure 3). We analyze the data separately for the two types of societies.

In the Western World

Over the past several years, Schösler, De Boer and Boersema (2012) were some of the first authors to investigate people's interest in insects as food in Europe. Their results indicate that consumers have a positive attitude to familiar food products processed with insect protein, especially among younger generations. They also found that consumers' responses rely heavily on the insect's shape and appearance, that is the visibility of the insects. Sogari, Menozzi and Mora (2018) in a study investigating the role of sensory-liking expectations before and after tasting a visible and processed house cricket in a familiar food product, suggested that for some "consumer groups" the presence of the whole insect increases the pleasure of the taste, although the aspect is still repellent.

De Boer, Schösler and Boersema (2013) found that proteins derived from insects (for example, snacks made with locusts) have a lower preference rating compared to other environmentally-friendly protein sources such as lentils, seaweed or hybrid meat. In Belgium, Caparros Megido et al. (2014) carried out

Figure 3. Publication trend of articles on "consumers and edible insects" indexed by Web of Science (1997-2017)

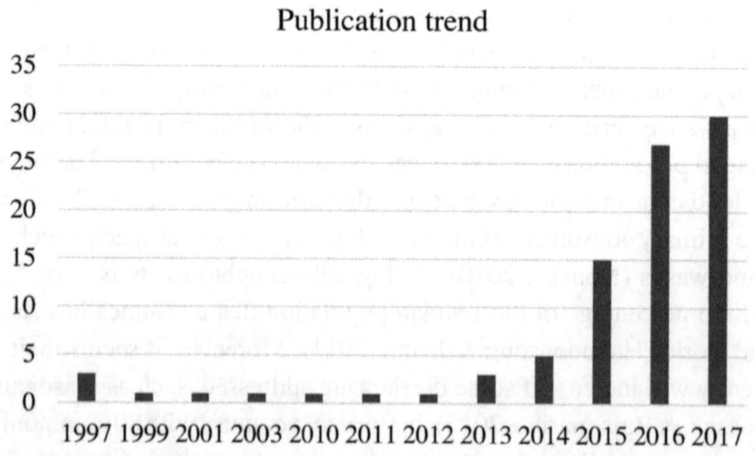

an unstructured hedonic test with consumers (sample n=189) who positively accepted the tasting of two insect products (a mealworm, *Tenebrio molitor L.* and a house cricket, *Acheta domesticus L.*). They found that familiar flavors, crispy textures and being younger are the factors that lead to increased approval of the sampled products.

In a study in Italy (Sogari, 2015), a group of people were given the possibility to taste an insect within a bug banquet. Most of the participants indicated curiosity as the first reason for trying such products, while feelings of disgust and negative opinions from family and friends regarding entomophagy were the main barriers. More recently, Hartmann and Siegrist (2017) conducted a literature review to explore the potential of introducing alternative protein sources to European consumers. They found that consumers are general unwilling to change their dietary habits (i.e. lower meat consumption) to a more sustainable protein source (e.g. insects).

In a study on the barriers and drivers of the insects' acceptance as food for the Italian Millennial generation, Cavallo and Materia (2018) found that product-specific features are crucial for acceptance. For example, the invisibility of the insect as ingredient (e.g. used as powder) is an important way to approach for the first time such a novel food.

Consumer attitude toward insect food is also influenced by cultural background and the availability of insects and insect-based products in the market. For example, House (2016) suggests that in the Netherlands, a country where such products are commercially available, some people are already starting to eat insects on a regular basis. Interestingly, his results suggest that repeated consumption behavior is mainly affected by the same factors influencing general food choices, such as price, taste and other practical reasons (e.g. availability). A comparative study between Australia and the Netherlands (Lensvelt & Steenbekkers, 2014) highlights the importance of education when it comes to consuming edible insects. Despite long tradition among Aboriginal Australians to consume insects, the dominant Western culture is still in need of understanding the suitability of such a food.

Socio-demographic characteristics such as education, background (Cicatiello, De Rosa, Franco, & Lacetera, 2016; De Boer, Schösler, & Boersema, 2013; Sogari & Vantomme, 2017), age (Schösler, De Boer, & Boersema, 2012), and gender (Menozzi, Sogari, Veneziani, Simoni, & Mora, 2017a; Verneau,

La Barbera, Kolle, Amato, Del Giudice, & Grunert, 2016) might play a crucial role in the acceptance of edible insects as food. According to Verbeke (2015), the early adopters of insects in Western society are younger males who dislike meat and are open to trying novel foods.

In the Eastern World

Based on the previous search on consumer attitude toward insects as food, the results indicated that there is a limited number of studies in Asian countries although entomophagy has a long history in most Southeast Asian countries. Most of the previous research focuses on general overviews of edible insect species and the market for them, as well as the production situation in each country. Limited research is available about consumer attitude toward insect consumption with only a few exceptions. Hartmann et al. (2015) conducted a survey to compare cultural differences between Chinese and German consumer preferences for insect-based food. They found that Chinese people rated insect taste, nutritional value, familiarity and social acceptance more highly than Germans. Chinese people are also more willing to try various processed (e.g. cricket flour cookies) and unprocessed (e.g. crickets) foods than Germans. Studies also indicate that Chinese consumers are more familiar with the idea of consuming insects compared to other consumers because eating insects has long been a part of Chinese culture (Luo, 1997; Hartmann, Shi, Giusto, & Siegrist, 2015). Barennes, Phimmasane and Rajaonarivo (2015) conducted a national consumer survey in Laos (n=1059). They find that most consumers have a positive attitude toward insect-based foods. However, availability and seasonality issues result in a decreasing consumption trend in Laos, where commercial insect farming is still in an introductory stage.

Anecdotal evidence (Chen, Wongsiri, Jamyanya, Rinderer, Vongsamanode, Matsuka, … Oldroyd, 1998) suggests that women in Thailand consume more insects when they are pregnant, particularly bee nests, because they believe their consumption is healthy for the baby. Bamboo caterpillar has also been traditionally consumed in Thailand (Leksawasdi, 2001). Additionally, although an earlier report (Mitsuhashi, 1997) indicated that most Japanese people have no desire to eat insects and even dislike them, the growing import of fresh insects for consumption in Japan may imply that Japanese consumers have an increasingly positive attitude toward fresh insects (Chen, Wongsiri, Jamyanya, Rinderer, Vongsamanode, Matsuka, … Oldroyd, 1998). In the case of South Korea, where insects-based food is a part of the standard diets and government actively encourages the adoption of edible insects in the Korean cuisine, the younger generation still appears to have a negative attitude toward using insects as food (Kim, 2014). A very recent consumer study in South Korea conducted by Ryu, Shin, Kim and Kim (2017) used a conjoint analysis and choice simulation with 203 respondents to analyse consumer preference for cookies made using edible insects. They find that consumers are willing to choose the edible insects' cookie if it is combined with other optimal product features (e.g. medium width, thin, medium price, and butter).

MARKET FOR EDIBLE INSECTS

Currently, few data are available on the market of edible insects for food and animal feed in the world. In 2016, a Global Market Insights report indicated the estimated 2015 value of the insect market to be US$ 33 million covering U.S., Belgium, France, UK, the Netherlands, China, Thailand, Vietnam, Brazil and Mexico (Global Market Insights Inc., 2016). This figure is likely to grow to US% 522 million by 2023.

The report predicts a growth of the market in the global community, with both consumer awareness as well as acceptance increasing (Han, 2017; Global Market Insights Inc., 2016). The UK, Netherlands and France topped the edible insects market among the European countries, which are likely to have a promising growth of the edible insects market in the next few years. The edible insect market in the U.S. is estimated to be worth US$ 50 million by 2023.

Regarding specific industry products, insect-based snacks are the primary outlets for edible insects (Global Market Insights Inc., 2016). The value of the industry exceeded US$ 11 million in 2015, and it is expected that the market for insect-based protein bars will grow over 42% by 2023. Additionally, the use of insect-based flour is estimated to increase by 42% and its value will exceed US$ 165 million by 2023.

Concerning international trade of edible insects, import and export of insect-based foods are also an important part of economic growth throughout Southeast Asia. Most of this quantity of edible insects (800 tonnes) comes from Cambodia, Myanmar, Lao PDR and China. These countries sell both cooked and fresh edible bugs in wholesale and local markets. For example, the import value in Thailand alone is 40 million Thai Baht annually (approximately US$ 1.14 million) (Hanboonsong & Durst, 2013).

Western countries currently face many challenges in mass rearing of insects. One of the challenges is the high cost of the production process, which includes rearing, harvesting and processing (Rumpold & Schlüter, 2013a). It is also difficult to manage small-scale farms and integrate recycled organic waste in the insect supply chain (van Huis, Van Itterbeeck, Klunder, Mertens, Halloran, Muir, & Vantomme, 2013).

CONCLUSION AND FUTURE IMPLICATIONS

Will it be an impossible task to convince consumers to eat insects as an alternative protein source to meat? This question addressed by van Huis (2017), editor of the Journal of Insects as Food and Feed, is a key issue for the whole insect food industry both in the Western and Eastern worlds.

First, in European and other Western countries, the consumption of edible insects will primarily depend on availability in the market (including regulatory framework and industry), product category (e.g. processed or unprocessed, familiar or unfamiliar), communication and marketing (e.g. taste experience appeals to pleasant taste) (van Huis, 2016; Menozzi, Sogari, Veneziani, Simoni, & Mora, 2017b; Sogari, Menozzi, & Mora, 2017).

Sogari (2015) showed that emphasizing the environmental and health benefits of a diet that includes entomophagy can increase curiosity and willingness to try insects for the first time. However, willingness to try processed and unprocessed insect products does not necessarily mean that consumers are ready to embrace entomophagy in their diet (Tan, Fischer, Tinchan, Stieger, Steenbekkers & Trijp, 2015). Moreover, stressing sustainability benefits does not change the hedonic evaluation (tastiness) of insect products (Tan, Fischer, Tinchan, Stieger, Steenbekkers & Trijp, 2015; van Huis, 2017). Therefore, changing negative perceptions about taste experience is crucial (Sogari, Menozzi, & Mora, 2018). Moreover, in this first stage, insect-based products should not contain any visible insects and should be sold in appealing packaging in order to be successful (Cavallo & Materia, 2018). Shelomi (2015) suggests that stakeholders interested in developing entomophagy should be aware that better production and marketing (e.g. packaging) strategies may work better than convincing consumers that eating bugs is healthy.

Few contributions have been published which explore the link between the legislative issues – e.g. food safety and production standards (see for example, Helble & Wind, 2017) and the business prospective (i.e. potential market trend). There might be opportunities to create profitable businesses in the edible insect sector, but growth of the industry will mostly depend on how and whether the new regulation of the European Union about novel food (EU Regulation 2015/2283) will facilitate the authorization procedure for commercializing such new products.

On the other hand, some Eastern countries have a long history of entomophagy and many people are familiar with collecting, farming and rearing edible insects. Particularly, some Southeast Asia countries such as China, Thailand, South Korea have already had an edible insect supply chain and have well-established farms as well as trade channels (Dobermann, Swift, & Field, 2017). However, despite the fact that Asian countries play an important role in producing and consuming edible insects, reputable studies focusing on edible insects markets, consumers, and the supply chain in Asia are very limited. Looking toward the future, more research needs to be done about consumers and the development of the edible insects market and industry in the Asian countries.

REFERENCES

Barennes, H., Phimmasane, M., & Rajaonarivo, C. (2015). Insect consumption to address undernutrition, a national survey on the prevalence of insect consumption among adults and vendors in Laos. *PLoS One*, *10*(8), 1–16. doi:10.1371/journal.pone.0136458 PMID:26317772

Belluco, S., Losasso, C., Maggioletti, M., Alonzi, C. C., Paoletti, M. G., & Ricci, A. (2013). Edible insects in a food safety and nutritional perspective: A critical review. *Comprehensive Reviews in Food Science and Food Safety*, *12*(3), 296–313. doi:10.1111/1541-4337.12014

Bodenheimers, F. S. (1951). Insets as human food. The Huage, The Netherlands: Springer-Science+Business Media. doi:10.1007/978-94-017-6159-8

Caparros Megido, R., Sablon, L., Geuens, M., Brostaux, Y., Alabi, T., Blecker, C., ... Francis, F. (2014). Edible insects acceptance by belgian consumers: Promising attitude for entomophagy development. *Journal of Sensory Studies*, *29*(1), 14–20. doi:10.1111/joss.12077

Cavallo, C., & Materia, V. C. (2018). Insects or not insects? Dilemmas or attraction for young generations: A case in Italy. *International Journal on Food System Dynamics*, *9*(3), 226–239. doi:10.18461/ijfsd.v9i3.932

Chen, P. P., Wongsiri, S., Jamyanya, T., Rinderer, T. E., Vongsamanode, S., Matsuka, M., ... Oldroyd, B. P. (1998). Honey bees and other edible insects used as human food in Thailand. *American Entomologist*, *44*(1), 24–29. doi:10.1093/ae/44.1.24

Chen, X., Feng, Y., & Chen, Z. (2009). Common edible insects and their utilization in China. *Entomological Research*, *39*(5), 299–303. doi:10.1111/j.1748-5967.2009.00237.x

Chung, Y. C. A. (2010). Edible insects and entomophagy in Borneo. In P. B. Durst, D. V. Johnson, R. N. Leslie & K. Shono (Eds), *Forest insects as food: Humans bite back* (pp. 141–150). Bangkok, Thailand: Food and Agriculture Organization of the United Nations. Retrieved September 12, 2018 from http://www.fao.org/docrep/012/i1380e/i1380e00.pdf

Cicatiello, C., De Rosa, B., Franco, S., & Lacetera, N. (2016). Consumer approach to insects as food: Barriers and potential for consumption in Italy. *British Food Journal, 118*(9), 2271–2286. doi:10.1108/BFJ-01-2016-0015

De Boer, J., Schösler, H., & Boersema, J. J. (2013). Motivational differences in food orientation and the choice of snacks made from lentils, locusts, seaweed or "hybrid" meat. *Food Quality and Preference, 28*(1), 32–35. doi:10.1016/j.foodqual.2012.07.008

Defoliart, G. R. (1995). Edible insects as minilivestock. *Biodiversity and Conservation, 4*(3), 306–321. doi:10.1007/BF00055976

DeFoliart, G. R. (1999). Insects as food: Why the western attitude is important. *Annual Review of Entomology, 44*(80), 21–50. doi:10.1146/annurev.ento.44.1.21 PMID:9990715

Deroy, O., Reade, B., & Spence, C. (2015). The insectivore's dilemma, and how to take the West out of it. *Food Quality and Preference, 44*, 44–55. doi:10.1016/j.foodqual.2015.02.007

Dobermann, D., Swift, J. A., & Field, L. M. (2017). Opportunities and hurdles of edible insects for food and feed. *British Nutrition Foundation Nutrition Bulletin, 42*(4), 293–308. doi:10.1111/nbu.12291

Durst, P., & Johnson, D. (2010). Edible forest insects: Exploring new horizons and traditional practices. In P. B. Durst, D. V. Johnson, R. N. Leslie & K. Shono (Eds), *Forest insects as food: Humans bite back* (pp. 1–4). Bangkok, Thailand: Food and Agriculture Organization of the United Nations. Retrieved September 12, 2018 from http://www.fao.org/docrep/012/i1380e/i1380e00.pdf

European Parliament and Council. (2015). Regulation (EU) 2015/2283 of the European Parliament and of the Council of 25 November 2015 on novel foods, amending Regulation (EU) No 1169/2011 of the European Parliament and of the Council and repealing Regulation (EC) No 258/97 of the European Parliament and of the Council and Commission Regulation (EC) No 1852/2001.

Evans, J., Alemu, M. H., Flore, R., Frøst, M. B., Halloran, A., Jensen, A. B., ... Eilenberg, J. (2015). "Entomophagy": An evolving terminology in need of review. *Journal of Insects as Food and Feed, 1*(4), 293–305. doi:10.3920/JIFF2015.0074

Feng, Y., Chen, X., Zhao, M., He, Z., Sun, L., Wang, C., & Ding, W. (2017). Edible insects in China: Utilization and prospects. *Insect Science, 25*(2), 184–198. doi:10.1111/1744-7917.12449 PMID:28225201

Food and Agriculture Organization of the United Nations (FAO). (2006). *Livestock's long shadow: Environmental issues and options.* Rome, Italy: Food and Agriculture Organization of the United Nations (FAO). Retrieved September 12, 2018 from http://www.fao.org/docrep/010/a0701e/a0701e.pdf

Food and Agriculture Organization of the United Nations (FAO). (2013). *Edible insects: Future prospects for food and fed security*. Rome, Italy: FAO. Retrieved September 12, 2018 from http://www.fao.org/docrep/018/i3253e/i3253e.pdf

Food and Agriculture Organization of the United Nations (FAO). (2017). *The state of food and agriculture. Leveraging food systems for inclusive rural transition*. Rome, Italy: FAO. Retrieved September 12, 2018 from http://www.fao.org/3/a-i7658e.pdf

Gasco, L., Finke, M., & van Huis, A. (2018). Can diets containing insects promote animal health? *Journal of Insects as Food and Feed*, *4*(1), 1–4. doi:10.3920/JIFF2018.x001

Global Market Insights Inc. (2016). *Edible insects market size by product (beetles, caterpillars, grasshoppers, bees, wasps, ants, scale insects & true bugs), by application (flour, protein bars, snacks), industry analysis report, regional outlook (U.S., Belgium, Netherlands, UK, France, China, Thailand, Vietnam, Brazil, Mexico): Application potential, price trends, competitive market share & forecast, 2016 – 2023.* Retrieved April, 12, 2018 from https:// www. gminsights. com/ industry-analysis/edible-insects-market

Han, R., Shin, J. T., Kim, J., Choi, S. Y., & Kim, W. Y. (2017). An overview of the South Korean edible insect food industry: Challenges and future pricing/promotion strategies. *Entomological Research*, *47*(3), 141–151. doi:10.1111/1748-5967.12230

Hanboonsong, Y., & Durst, B. P. (2014). *Edible insects in Lao PDR: Building on tradition to enhance food security*. RAP Publication 2014/12, Food and Agriculture Organization of the United Nations (FAO), Regional Office for Asia and the Pacific, Bangkok, Thailand, 56. Retrieved September 12, 2018 from http://www.fao.org/3/a-i3749e.pdf

Hanboonsong, Y., Jamjanya, T., & Durst, P. B. (2013). *Six-legged livestock: Edible insect farming, collection and marketing in Thailand.* Bangkok, Thailand: Food and Agriculture Organization of the United Nations, Regional Office for Asia and the Pacific. Retrieved September 12, 2018 from http://www.fao.org/3/a-i3246e.pdf

Hartmann, C., Shi, J., Giusto, A., & Siegrist, M. (2015). The psychology of eating insects: A cross-cultural comparison between Germany and China. *Food Quality and Preference*, *44*, 148–156. doi:10.1016/j.foodqual.2015.04.013

Hartmann, C., & Siegrist, M. (2017). Consumer perception and behaviour regarding sustainable protein consumption: A systematic review. *Trends in Food Science & Technology*, *61*, 11–25. doi:10.1016/j.tifs.2016.12.006

Helble, S., & Wind, C. (2017). Food based on edible insects is a developing new market in Europe and also a challenge for the official food control authorities. *Tierärztliche Umschau*, *72*(4), 135–139.

House, J. (2016). Consumer acceptance of insect-based foods in the Netherlands: Academic and commercial implications. *Appetite*, *107*, 47–58. doi:10.1016/j.appet.2016.07.023 PMID:27444958

Jongema, Y. (2017). *List of edible insects of the world (April 1, 2017).* Wageningen University. Retrieved March 30, 2018 from http://www.wageningenur.nl/en/Expertise-Services/Chair-groups/Plant-Sciences/Laboratory-of-Entomology/Edible-insects/Worldwide-species-list.htm

Kim, Y. M. (2014). *The 50 ways to cook edible insects*. Seoul, South Korea: Bumwoo.

Leksawasdi, P. (2001). Bamboo caterpillar in Thailand. *Khon Kaen Agriculture Journal*, *29*(1), 15–21. (in Thai)

Lensvelt, E. J. S., & Steenbekkers, L. P. (2014). Exploring consumer acceptance of entomophagy: A survey and experiment in Australia and the Netherlands. *Ecology of Food and Nutrition*, *53*(5), 543–561. 865 doi:10.1080/03670244.2013.879

Lokeshwari, R. K., & Shantibala, T. (2010). A review on the fascinating world of insect resources: Reason for thoughts. *Psyche, 11*, 207570. doi:10.1155/2010/207570

Looy, H., Dunkel, F. V., & Wood, J. R. (2014). How then shall we eat? Insect-eating attitudes and sustainable foodways. *Agriculture and Human Values, 31*(1), 1–11. doi:10.100710460-013-9450-x

Luo, Z. Y. (1997). Insects as food in China. *Ecology of Food and Nutrition, 36*(2–4), 201–207. doi:10.1080/03670244.1997.9991515

MacClancy, J., Jeya Henry, C., & Macbeth, H. (2007). *Consuming the inedible: Neglected dimensions of food choice*. New York, NY: Berghahn Books.

Materia, V. C., & Cavallo, C. (2015). Insetti per l'alimentazione umana: Barriere e drivers per l'accettazione da parte dei consumatori. *Rivista di Economia Agraria, LXX*(2), 139–161. doi:10.13128/REA-18004

Menozzi, D., Sogari, G., Veneziani, M., Simoni, E., & Mora, C. (2017a). Eating novel foods: An application of the theory of planned behaviour to predict the consumption of an insect-based product. *Food Quality and Preference, 59*, 27–34. doi:10.1016/j.foodqual.2017.02.001

Menozzi, D., Sogari, G., Veneziani, M., Simoni, E., & Mora, C. (2017b). Explaining the intention to consume an insect-based product: A cross-cultural comparison. In A. H. Seyal & M. N. A. Rahman (Eds.), *Theory of planned behavior: New research* (pp. 201–215). New York, NY: Nova Publishers.

Mitsuhashi, J. (2010). Insects as traditional foods in Japan. *Ecology of Food and Nutrition, 36*(2–4), 187–189. doi:10.1080/03670244.1997.9991514

Mlcek, J., Rop, O., Borkovcova, M., & Bednarova, M. (2014). A comprehensive look at the possibilities of edible insects as food in Europe – a review. *Polish Journal of Food and Nutrition Sciences, 64*(3), 147–157. doi:10.2478/v10222-012-0099-8

Nonaka, K. (2010). Cultural and commercial roles of edible wasps in Japan. In P. B. Durst, D. V. Johnson, R. N. Leslie & K. Shono (Eds), *Forest insects as food: Humans bite back* (pp. 123–130). Bangkok, Thailand: Food and Agriculture Organization of the United Nations. Retrieved September 12, 2018 from http://www.fao.org/docrep/012/i1380e/i1380e00.pdf

Obopile, M., & Seeletso, T. G. (2013). Eat or not eat: An analysis of the status of entomophagy in Botswana. *Food Security, 5*(6), 817–824. doi:10.100712571-013-0310-8

Pascucci, S., & De-Magistris, T. (2013). Information bias condemning radical food innovators? The case of insect-based products in the Netherlands. *The International Food and Agribusiness Management Review, 16*(3), 1–16. Retrieved from https://ageconsearch.umn.edu/bitstream/156421/2/20130022.pdf

Raheem, D., Carrascosa, C., Oluwole, B. O., Nieuwland, M., Saraiva, A., Millán, R., & Raposo, A. (2018). Traditional consumption of and rearing edible insects in Africa, Asia and Europe. *Critical Reviews in Food Science and Nutrition*, 1–20. doi:10.1080/10408398.2018.1440191 PMID:29446643

Raubenheimer, D., & Rothman, J. M. (2013). Nutritional ecology of entomophagy in humans and other primates. *Annual Review of Entomology, 58*(1), 141–160. doi:10.1146/annurev-ento-120710-100713 PMID:23039342

Rumpold, B. A., & Schlüter, O. K. (2013a). Potential and challenges of insects as an innovative source for food and feed production. *Innovative Food Science & Emerging Technologies, 17*, 1–11. doi:10.1016/j.ifset.2012.11.005

Rumpold, B. A., & Schlüter, O. K. (2013b). Nutritional composition and safety aspects of edible insects. *Molecular Nutrition & Food Research, 57*(5), 802–823. doi:10.1002/mnfr.201200735 PMID:23471778

Ryu, J. P., Shin, J. T., Kim, J., & Kim, Y. W. (2017). Consumer preference for edible insect-containing cookies determined by conjoint analysis: An exploratory study of Korean consumers. *Entomological Research, 47*(2), 74–83. doi:10.1111/1748-5967.12203

Schösler, H., De Boer, J., & Boersema, J. J. (2012). Can we cut out the meat of the dish? Constructing consumer-oriented pathways towards meat substitution. *Appetite, 58*(1), 39–47. doi:10.1016/j.appet.2011.09.009 PMID:21983048

Shelomi, M. (2015). Why we still don't eat insects: Assessing entomophagy promotion through a diffusion of innovations framework. *Trends in Food Science & Technology, 45*(2), 311–318. doi:10.1016/j.tifs.2015.06.008

Sogari, G. (2015). Entomophagy and Italian consumers: An exploratory analysis. *Progress in Nutrition, 17*(4), 311–316.

Sogari, G., Menozzi, D., & Mora, C. (2017). Exploring young foodies' knowledge and attitude regarding entomophagy: A qualitative study in Italy. *International Journal of Gastronomy and Food Science, 7*, 16–19. doi:10.1016/j.ijgfs.2016.12.002

Sogari, G., Menozzi, D., & Mora, C. (2018). Sensory-liking expectations and perceptions of processed and unprocessed insect products. *International Journal on Food System Dynamics, 9*(3), 226–239. doi:10.18461/ijfsd.v9i4.942

Sogari, G., & Vantomme, P. (2014). *A tavola con gli insetti*. Fidenza, Italy: Mattioli 1885.

Tan, H. S. G., Fischer, A. R. H., Tinchan, P., Stieger, M., Steenbekkers, L. P. A., & Trijp, H. C. M. (2015). Insects as food: Exploring cultural exposure and individual experience as determinants of acceptance. *Food Quality and Preference, 42*, 78–89. doi:10.1016/j.foodqual.2015.01.013

van Huis, A. (2013). Potential of insects as food and feed in assuring food security. *Annual Review of Entomology, 58*(1), 563–583. doi:10.1146/annurev-ento-120811-153704 PMID:23020616

van Huis, A. (2016). Edible insects are the future? *The Proceedings of the Nutrition Society, 75*(03), 294–305. doi:10.1017/S0029665116000069 PMID:26908196

van Huis, A. (2017). Edible insects: Marketing the impossible? *Journal of Insects as Food and Feed, 3*(2), 67–68. doi:10.3920/JIFF2017.x003

van Huis, A., Dicke, M., & van Loon, J. J. A. (2015). Insects to feed the world. *1st International Conference Insects to Feed the World - Summary Report, 1*, 3–5. Retrieved September 12, 2018 from http://www.fao.org/edible-insects/86385/en/

van Huis, A., Van Itterbeeck, J., Klunder, H., Mertens, E., Halloran, A., Muir, G., & Vantomme, P. (2013). *Edible insects. Future prospects for food and feed security. Food and Agriculture Organization of the United Nations* (Vol. 171). Retrieved September 12, 2018 from http://www.fao.org/docrep/018/i3253e/i3253e.pdf

Vantomme, P. (2015). Way forward to bring insects in the human food chain. *Journal of Insects as Food and Feed*, *1*(2), 121–129. doi:10.3920/JIFF2014.0014

Verbeke, W. (2015). Profiling consumers who are ready to adopt insects as a meat substitute in a Western society. *Food Quality and Preference*, *39*, 147–155. doi:10.1016/j.foodqual.2014.07.008

Verneau, F., La Barbera, F., Kolle, S., Amato, M., Del Giudice, T., & Grunert, K. (2016). The effect of communication and implicit associations on consuming insects: An experiment in Denmark and Italy. *Appetite*, *106*, 30–36. doi:10.1016/j.appet.2016.02.006 PMID:26855371

Yen, A. L. (2009). Edible insects: Traditional knowledge or western phobia? *Entomological Research*, *39*(5), 289–298. doi:10.1111/j.1748-5967.2009.00239.x

Yen, A. L. (2015). Insects as food and feed in the Asia Pacific region: Current perspectives and future directions. *Journal of Insects as Food and Feed*, *1*(1), 33–55. doi:10.3920/JIFF2014.0017

Yhoung–Aree, J., Puwastien, P., & Attig, A. G. (2010). Edible insects in Thailand: An unconventional protein source? *Ecology of Food and Nutrition*, *36*(2-4), 133–149. doi:10.1080/03670244.1997.9991511

ADDITIONAL READING

Belluco, S., Losasso, C., Maggioletti, M., Alonzi, C. C., Paoletti, M. G., & Ricci, A. (2013). Edible insects in a food safety and nutritional perspective: A critical review. *Comprehensive Reviews in Food Science and Food Safety*, *12*(3), 296–313. doi:10.1111/1541-4337.12014

Evans, J., Alemu, M. H., Flore, R., Frøst, M. B., Halloran, A., Jensen, A. B., ... Eilenberg, J. (2015). "Entomophagy": An evolving terminology in need of review. *Journal of Insects as Food and Feed*, *1*(4), 293–305. doi:10.3920/JIFF2015.0074

Food and Agriculture Organization of the United Nations (FAO). (2013). *Edible insects: Future prospects for food and fed security*. Rome, Italy: FAO. Retrieved September 12, 2018 from http://www.fao.org/docrep/018/i3253e/i3253e.pdf

Gasco, L., Finke, M., & van Huis, A. (2018). Can diets containing insects promote animal health? *Journal of Insects as Food and Feed*, *4*(1), 1–4. doi:10.3920/JIFF2018.x001

Lokeshwari, R. K., & Shantibala, T. (2010). A review on the fascinating world of insect resources: Reason for thoughts. *Psyche*, *11*, 207570. doi:10.1155/2010/207570

Rumpold, B. A., & Schlüter, O. K. (2013). Potential and challenges of insects as an innovative source for food and feed production. *Innovative Food Science & Emerging Technologies*, *17*, 1–11. doi:10.1016/j.ifset.2012.11.005

Rumpold, B. A., & Schlüter, O. K. (2013). Nutritional composition and safety aspects of edible insects. *Molecular Nutrition & Food Research*, *57*(5), 802–823. doi:10.1002/mnfr.201200735 PMID:23471778

Sogari, G. (2015). Entomophagy and Italian consumers: An exploratory analysis. *Progress in Nutrition*, *17*(4), 311–316.

Sogari, G., Menozzi, D., & Mora, C. (2018). Sensory-liking expectations and perceptions of processed and unprocessed insect products. *International Journal on Food System Dynamics*, *9*(3), 226–239. doi:10.18461/ijfsd.v9i4.942

KEY TERMS AND DEFINITIONS

Consumer Attitude: A settled way of thinking or feeling about something of consumer value.

Eastern Culture: A term used to refer to a heritage of social norms, ethical values, traditional customs, belief systems, political systems, and specific artefacts and technologies that have some origin or association with Asia.

Eastern Society: A group of people living in or originating from the East, in particular Asia.

Edible Insects: All types of insects which are considered edible for human consumption (around 2,000 species today).

Entomophagy: The eating practice of consuming insects, which is derived from the Greek words *éntomon* (insect) and *phagein* (to eat).

Insect-Based Food: All kinds of food products which use insect ingredients in the preparation (e.g., insect powder).

Market Perspective: A term which covers broad issues, including the global economy, market environment, and/or sector overviews.

Western Culture: A term used to refer to a heritage of social norms, ethical values, traditional customs, belief systems, political systems, and specific artefacts and technologies that have some origin or association with Europe.

Western Society: A group of people living in or originating from the West, in particular Europe, the United States, Canada, Australia, and New Zealand.

Section 3
Business Aspects of New Meat Alternatives

Chapter 10
Building a Market for New Meat Alternatives:
Business Activity and Consumer Appetite in the Netherlands

Hans Dagevos
Wageningen University and Research, The Netherlands

Ella Tolonen
The Good Food Institute, Finland

Jaco Quist
Delft University of Technology, The Netherlands

ABSTRACT

This chapter provides an overview of developments in the Netherlands on new meat alternatives with a focus on plant-based meat substitutes and lab-grown meat. It devotes attention to both the supply side of the market (business activity) and the demand side (consumer appetite). The first concerns developments in the meat substitutes' innovation system since the 1990s until now. It concludes that the Netherlands has become a major player. The latter concerns the supportive purchasing power of consumers regarding the building of a viable and strong market for new meat alternatives. It is concluded that available consumer studies provide evidence for being cautiously optimistic. The closing parts of this chapter, however, bring to the fore that a transition from the current high-meat diets to more sustainable and healthier diets with more non-meat sources of proteins is anything but self-evident. However encouraging and energetic modern developments in the Netherlands are, much progress is needed as it comes to consumer acceptance of new meat alternatives, producer capacity to innovate, concentrate strengths, and capture market share, as well as governmental support for reducing the adverse effects of today's meat consumption and production levels in accordance with Sustainable Development Goal 12 concerning responsible consumption and production.

DOI: 10.4018/978-1-5225-7350-0.ch010

Der Gedanke, dass alles so ist, wie es ist, weil es nicht anders sein könnte, lässt alles Nachdenken über Alternativen erstarren. – Philipp Blom, Was auf dem Spiel steht (2017: 20)[1]

INTRODUCTION

Finding new solutions to relieve and reduce the massive and multiple (i.e., environmental, human health, animal welfare and food security) problems associated with the excessive global production and consumption of meat, is a matter of growing urgency. After a few decades of research that has produced mounting and compelling evidence about the adverse effects of animal agriculture and overconsumption of meat, one of the new shoots on this tree of knowledge is a paper in *Science* by Poore and Nemecek (2018). This distinguished study also clearly corroborates the need for dietary change from current highly animal-based diets to more sustainable plant-based diets.

In such a broad dietary transition, the reduction and replacement of farmed meat products in our food consumption patterns, is essential. One promising and emerging avenue are the so-called "new meat alternatives" as an option that can provide protein foods with considerable lower environmental impact. In the remainder of this chapter we refer to new meat alternatives, particularly to plant-based and cellular alternatives, i.e., meat substitutes and cultured meat. More generally, however, the term meat alternatives or alternative protein sources could refer to all alternative protein product categories to current animal products (meat, dairy, eggs or fish) ranging from algae, seaweed, duckweed, rape seed, to pulses, molds and mushrooms, soy-based products, nuts, and to insects (see Figure 1 for an overview). As Figure 1 shows, some product categories have already been on the Dutch market for a long time, for instance nuts, mushrooms, legumes, and texturized vegetable products (e.g., soy-based or seitan), while cultured meat has not hit the market yet. Relatively new on the market are newly-advanced meat substitutes, and insects. Current meat substitutes and cultured meat have in common that both aim directly at imitating and therefore replacing meat in a meal. These two product categories both belong within the avenue of new meat alternatives due to their technological novelty, whereas insects and algae belong to it due to their novelty as a source for food in the Dutch context. Both these pairs of categories relate symbiotically in the sense that the novel food sources can make use of the new meat substitution technologies, for example, to increase familiarity. As indicated, our main focus with respect to new meat alternatives will be on plant-based meat substitutes and animal-based cultured meat.

In the following section, "Where a small country can be great," we give an impression of the progressive business activities in the Netherlands, as it appears that this tiny country is quite prominent and innovative when it comes to new meat alternatives. Several Dutch companies are pioneering and leading in the field of developing new meat alternatives. For example, the Dutch company Vivera launched the first plant-based steak on the UK-market in May 2018, followed by market introductions in the Netherlands and other European countries in the summer of 2018. Another prominent Dutch company manufacturing alternative plant-based meat substitutes, i.e., the Vegetarian Butcher, will follow in Vivera's track when their plant-based steak has its market introduction in due course. Meatless is another front-running company in the Netherlands which produces lupine-based material for meat substitutes. Also in the emerging field of cultured meat, a world's first originates from the Netherlands: the Dutchman Mark Post presented the first cultured meat hamburger in August 2013. All in all, quite a lot of business activity and innovation are to be found in the Dutch context, which is important to evaluate in the light of building a market for new meat alternatives.

In the subsequent section, "Consumer appetite for meat alternatives," we switch attention from the supply side to the demand side of the market, and investigate what we can expect from food consumers as supporters of making more environmentally-friendly and healthy dietary choices in general, and more specifically, making the new meat alternative choice. Our exploration leads us to academic consumer studies and their results. Do consumers appreciate new meat alternatives and do they contribute to the acceleration of plant foods as an attractive and ordinary alternative to replace meat?

The closing section, "Toward a market for new meat alternatives," gives an impression of the obstacles and drivers accompanying a dietary shift in the pursuit of stimulating a new balance between animal protein and plant protein. While supportive arguments and sense of urgency are abundantly available to make the dietary transition from less meat-centred to more plant-oriented meals, mainstream production practices and consumer culture are obstinate. In practice, many uncertainties and hurdles are on the road to a strong and viable market for new meat alternatives.

WHERE A SMALL COUNTRY CAN BE GREAT

When thinking about product innovation in anticipation of a meatless future, one may probably point to Silicon Valley as the place to be. This would be quite right though, because various progressive enterprises and start-ups in the field of new meat alternatives – particularly meat substitutes and cultured meat ("cellular agriculture," or abbreviated: "cell-ag") – are based in that region. The Netherlands, however, has proven to also be a highly innovative hub.

Figure 1. A brief categorization of different protein sources relevant in the Netherlands. Products within the side of novelty are regarded as new meat alternatives.

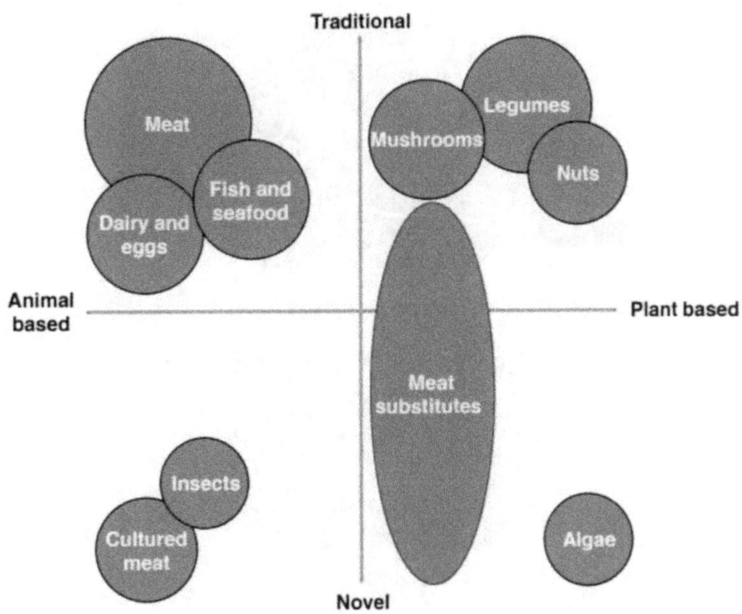

Business Developments in the Dutch Delta

Perhaps it is surprising to hear that next to the well-known Californian tech-hub, the "Dutch delta" is also an important place on the map of the meat alternatives world. As it happens, in the Netherlands, a small country in the north-west of Europe, various pioneering companies are based and much business activity can be registered in the realm of new meat alternatives. We start with a sketch of developments in the meat substitutes business.

Several companies (e.g. Ojah and Meatless) belong to the global front runners which for decades now have been developing plant-based raw materials for the production of meat substitutes. Other early companies active in the Dutch market of vegetable meat substitutes are Garden Gourmet (originally an Israeli company called Tivall which entered the Dutch market in 1986, and was renamed after being taken over by the food giant Nestlé in 2015), Goodbite (since 2002) and Vivera (since 1990). The Vegetarian Butcher is a newer business, established in 2010. Vivera and the Vegetarian Butcher have both developed a broad range of texturalized plant-based meat replacements that continues to grow, and are both pioneering in bringing plant-based beef to the market. Both companies belong to the biggest players in the field of meat analogues in the Netherlands.

Vivera's surprising presentation of the world's first plant-based steak on the UK-market in May 2018 gave this somewhat introvert company a unique opportunity to generate media exposure (Figure 2). Until then, the Vegetarian Butcher and its amusing founder Jaap Korteweg (Figure 3) have been frequently

Figure 2. The veggie steak product of the Dutch company Vivera

Figure 3. The founder of the Vegetarian Butcher Jaap Korteweg
Source: Bart Homburg

ahead of Vivera in getting free publicity and social media coverage. This strategy contributes to the strong reputation of the quality and tastiness of the Vegetarian Butcher's meat analogues, which are predominantly sold under the Vegetarian Butcher brand and named after their animal-based equivalent with a deliberate spelling error (e.g., the Dutch word for minced meat is "gehakt" but is spelled by the Vegetarian Butcher as "gehackt", which could be translated as "hacked meat"[2]) (Ingenbleek & Zhao, 2018). The extravert strategy of the Vegetarian Butcher resonated also in the autumn of 2015 when the company needed money for a new factory. Banks proved to be reluctant to finance this, whereupon the Vegetarian Butcher decided to raise money through crowdfunding with the help of an environmental interest group that mobilized its members to invest in the company. In less than three weeks the Vegetarian Butcher reached its goal and generated 2.5 million euro. In fact, this could have been more if the collected amount of money were not the legal maximum for a crowdfunding campaign in the Netherlands. After the successful crowdfunding campaign, the Vegetarian Butcher managed to find a bank that was most willing to loan the rest of the money needed to build and equip the factory that is going to produce 50 million meat analogues a year for the Dutch market and the growing amount of countries of export (Ingenbleek & Zhao, 2018).

Current business activity and actual market penetration of plant-based meat substitutes in the Netherlands are less self-evident than one probably would suppose given today's state of affairs. In fact, meat alternatives were a rather marginal business activity in the Netherlands for a long period in which several small and medium-sized enterprises (SMEs) produced traditional meat alternatives mainly for reform and organic food shops. Supermarkets, in their turn, limited themselves to selling tofu and tempeh, and alternative protein sources were not on the Dutch agricultural and food research and development (R&D) agenda before the mid-1990s. All this changed both gradually and considerably in the following decades (see Quist, 2007). The Dutch government started to be a participant in funding research projects on new meat alternatives – the Novel Protein Foods project in the second half of the 1990s was the first one, followed by the Profetas programme in 1999–2004, and more recently (since 2017), the Plant Meat Matters research project. They have been helpful and still help several producers and manufactures to improve new meat alternatives and enable them to provide the market with products that have increasing chances to invite supermarkets to supply and consumers to purchase these new meat substitutes. By now, producers, supermarkets, non-governmental organizations (NGOs) and scientists are joining forces in the Green Protein Alliance to collectively increase the market share of plant-based meat substitutes and the share of plant-based foods in people's diets. By now, developments in the field of manufacturing alternative plant-based meat substitutes in the past decades have resulted in the current availability of meat-substitute foods which imitate meat on an unprecedented quality level. Thanks to mission-driven, innovative companies and research efforts made by groups in Utrecht University and Wageningen University, the Netherlands belongs to the leading countries in the world with respect to meat substitution technology (Tolonen, 2018, p. 58).

Other Entrepreneurial Activities

The Dutch meat alternatives market is dominated by various types of plant-based meat substitutes. Efforts to improve and scale up production methods as well as the product quality (texture, taste and nutritional value) have borne fruit in the last decade. Today, all sorts of meat-free products are on the Dutch market based on, for instance, legumes, lupine, tofu, rice, or beetroot. In the past years a lot of efforts are directed especially toward creating meat substitution products from local ingredients, such as lupine or wheat, and toward avoiding over-representation of soy-based products.

In addition to such plant-based meat substitutes are alternative products based on dairy. On the Dutch market, dairy-based meat replacements are rather prominent mainly due to Valess. In 2005 the Dutch dairy giant Friesland Campina launched its new milk-based meat substitute Valess with a large marketing campaign – which turned out to be both beneficial to Valess and meat substitutes that were already on the supermarket shelves at the time.

Meat substitutes based on seaplants are a more recent and less prominent phenomenon. Noteworthy with respect to seaweed as seaplant-based meat alternative in the Dutch context is the seaweed farm "Zeewaar" on the supply side as well as other (research) initiatives in algae cultivation. Notable on the demand side are the pioneering business activities of the Dutch Weed Burger and the company Olijck that have managed to develop seaweed burgers which can be found, for instance, in the supermarkets of the largest retailer in the Netherlands, Albert Heijn. Fungi-based substitutes are available in the form of tempeh and Quorn, and also belong to the small fractions of the meat alternatives market in the Netherlands.

For the supply and demand of insects as alternative to current animal products, in spite of the growing research and societal interest in the potential of this source of protein in the Dutch diet in the last decade, these products have not managed to sneak their way into the Dutch food market. Not very helpful for a possible breakthrough was that the introduction of an insect-based burger in the conventional supermarkets of Jumbo in 2014 failed after having been on the shelves for several months.

Hitherto, an even smaller market – to be more precise: no market – exists for cultured meat, which is based on animal cells grown in a laboratory. Worldwide, several teams and start-ups are working at the commercialization and upscaling of this technology. The relationship between cultured meat and the Netherlands takes back to the second half of the previous century when the Dutchman Willem van Eelen was ahead of time daydreaming about growing muscles outside of an animal. Although his own early and enduring endeavours to grow animal muscle tissue had only limited results, van Eelen is certainly one of the two pioneering Dutchmen on this innovation. The other one is Mark Post of Maastricht University, and co-founder of the Dutch-based Mosa Meat in 2016, who succeeded in producing the first lab-grown burger patty offering it to the world in 2013. This burger, with a price tag of 250,000 euro, developed by using stem-cell technology and made out of thousands of individual muscle strips, was brought to London by Post personally, and prepared and eaten in a live television broadcast in front of many international journalists. The news of this technological breakthrough quickly spread all over the world, making its way to headlines of prominent newspapers. It was a fascinating expression of the possibilities of cultured meat, and generated much enthusiasm for cultured meat (Shapiro, 2018, p. 68–74; Tolonen, 2018, p. 53).

With respect to the relationship between the Netherlands and cultured meat, two other milestones are worthy of mention. First, the Dutch government was an early adopter considering it was the first national government in 2005 to fund a research project on cultured meat that was led by Utrecht University. Second, and more recently, the American high-tech food start-up Just flew a parcel of its cultured meat products to the Netherlands: a sausage ended in a science museum and the duck chorizo ground meat was planned to be introduced in a Dutch restaurant in the spring of 2018 – by way of tribute to Van Eelen. Notwithstanding that this introduction was prohibited for food safety reasons (Novel Food Law), this example indicates that the moment of cautious market introduction of cultured meat might be not too far away. In addition, Post anticipates that cultured meat will be available and affordable for many consumers in the coming decade. More specifically, his current planning is to have cultured meat on the menu of a number of restaurants in the beginning of the next decade and thereafter to bring cultured meat to the market in the course of the 2020s.

Comparing Dutch Development With Other Countries

Entrepreneurial activity in the Netherlands shows some differences in comparison to business developments in other countries. One of the differences is that traditional meat processors and meat-substitute manufacturers seem to have relatively little interest in each other. That is, interference and interaction appear limited. For example, the largest meat producing company in the Netherlands, Vion, has never tried to bring meat substitutes to the market or to incorporate a meat substitute manufacturer. However, during a short period in 2012-2013 Vion developed and marketed a range of so-called hybrid meat products, i.e., meat analogues in which part of the ground meat is mixed with (30-40%) plant-based ingredients, under the brand name Hackplus. However, Vion's attempts and ambitions to broaden its product portfolio with hybrid sausages, burgers and mince products were short-lived, stating that the consumer demand for these products failed. Vion's decision seems not to represent some sort of actual resistance or hostility by the Dutch meat sector towards meat substitutes, but rather a calculating estimation that this business activity is not a profitable one for them at the moment. More generally, it appears that in Dutch meat business circles one is perhaps more uncomfortable with the idea of becoming a general protein producer than with the idea of an emerging meat alternatives market.

This quick disappearance of enthusiasm contrasts with the business strategy of one of Vion's German counterparts: Rügenwalder Mühle. Since 2014 this meat enterprise has shifted toward more plant-oriented directions and invests substantially in innovation and marketing of vegetarian and vegan meat analogues. One of the largest Canadian meat processors, Maple Leaf Foods, sees similar opportunities and has comparable interest in plant-based meat substitutes as a way to diversify its product portfolio. Other examples of a "hybridization" between traditional meat companies and the plant foods market, are presented by the investments of one of the world's largest meat processors, Tyson Foods, in the plant-based meat producer start-up companies Beyond Meat and Memphis Meats, or the French multinational and dairy stronghold Danone completing its acquisition of the American plant-based foods and beverage company Whitewave Foods in April 2017. In the Netherlands, however, the mainstream Dutch meat company the Enkco Food Group detached the brand of Vivera in 2014 and made it semi-independent from the meat department of the company. In other words, the suggestion by Hicks et al. (2018, p. 8–9) that the meat industry should increase its involvement in meat alternatives and "flexitarian foods", is not very much heard in the Netherlands so far.

Exception to the rule is the meat processor Zwanenburg Food Group which lately seeks market growth via non-meat products such as plant-based snacks, vegetarian soups and sauces. Perhaps the collaboration with Meyn, one of the world's largest manufacturers of chicken processing technology, in a Wageningen University project – to be more precise, the abovementioned Plant Meat Matters research project – aimed at finding new technologies and applications in plant-based meat substitutes, also adds to development within the Netherlands resembling tendencies beyond the Dutch borders.

Another difference is that the Dutch "green protein" industry has not yet attracted much venture capital and investors' interest. For the past years the investment by Google co-founder Sergey Brin in Post's cultured meat hamburger was an isolated example comparable with some of the investments one can point to with respect to meat substitute manufacturers or cultured animal meat companies outside the Netherlands. More recently, however, it is possible to add another telling example: in July 2018, the start-up Mosa Meat has raised 7.5 million euro, and one of the main investors is Bell Food Group, a main Swiss meat processor. To date these are the first examples of European investments in a Dutch meat alternatives company, and do not change the overall picture that the R&D activities in the Netherlands as yet have been primarily funded by the meat substitute producing companies themselves with occasional financial support by the Dutch government.

CONSUMER APPETITE FOR MEAT ALTERNATIVES

The role of the consumers in the Dutch meat alternatives market is the other side of the marketing equation. They are still finding their way into the emerging new choices.

The State of the Field

Let's start with the bright side. The turnover of the Dutch market for plant-based meat substitutes is on the rise. This market has grown from 27 million euro in the beginning of this century to 60 million euro in the beginning of the current decade to considerably over 80 million euro nowadays. Also, it is expected that it will not take many years to pass the boundary of 100 million euro. Market figures quickly moving into the upward direction should not be trivialized, but at the same time we should also realize that it is a niche market and will remain so in the foreseeable future. Compared with the Dutch meat market the new market for meat alternatives is about 2% of the "old" meat market's size. Moreover, the current meat market is hardly shrinking considering recent meat consumption figures. After an annual decrease in meat consumption of about half a kilo per person per year (which happens to be the equivalent of the maximum recommended meat intake per week according to the national dietary guidelines in the Netherlands) since 2010, the consumption figures of 2016 show that this declining trend has stagnated for the moment.

More generally, only a small minority of the Dutch food consumers meet the dietary guidelines which point overall to a diet that is less animal-based and in favor of plant foods. Despite the fact that consumer interest in sustainable and healthy eating exists – rising consumer expenditures for sustainable foods in recent years enforce this as well as the gradual normalization of flexitarianism in the Netherlands, and seems to grow due to consumer awareness and concern about relationships between food and sustainability (examples include climate change and loss of biodiversity) topics[3] and health (body weight, cardiovascular diseases) issues[4], an undeniable break with dietary habits as we know it still has to be realized.

Do modern consumer studies on consumer appetite for new meat alternatives give reason to be optimistic about consumer acceptance and appreciation of alternative protein products? This is explored below.

What Consumer Studies Envision: Meat Substitutes

Having a look at several of the recent studies available, a first thing that can be noticed is that scholarly attention for "new" proteins from a consumer perspective is growing. Consumer studies with a focus on both meat substitutes and cultured meat have made their appearance in the scientific journals.

As already mentioned in the introductory section, plant-based meat substitutes in the sense of including textured vegetable proteins and excluding unprocessed nuts, mushrooms or beans, are the most common meat alternative in today's world of food. Therefore, we start with this more established product category. The first eye-catching development is that particularly in the last decade the product quality and variety of plant-based meat analogues have made much improvement far beyond the basic level of tofu and tempeh. Moreover, the broad range of vegan meat substitutes is supplied in an increasing number of mainstream supermarkets (the two largest Dutch supermarket chains, Albert Heijn and Jumbo, also offer their own inhouse brand of different meat substitutes), online outlets as well as various kinds of bars, restaurants and food or other cultural festivals. These developments facilitate consumers to try and taste such meat-free products as well as contribute to the normalization of the meat-free choice. Next to vegetarians and vegans whose consumer life could become a bit easier and varied with those alternate meat products, meat eaters are an important target group of the new meat alternatives market, and among those, more particularly the wide target group of flexitarians (Verain, Dagevos, & Antonides, 2015; Dagevos, 2016; Dagevos & Reinders, 2018). While for a group of vegetarians and vegans it is unnecessary or unappealing to wholeheartedly embrace meat substitutes because they feel aversion to the idea of imitating meat – as meat analogues, by definition, do – flexitarians often take up a different position. As meat eating semi-vegetarians who regularly choose to have a low- or non-meat dish, flexitarians' interest in plant foods is, in principle, supportive to the market growth of meat replacements. Since a flexitarian foodstyle is characterized by meat moderation, flexitarians are consequently searching for meat alternatives to practice their reduced meat foodstyle.

Closer scrutiny, however, reveals that the incorporation of meat replacements into the habitual food pattern of many consumers is not self-executing. Consumers' reluctance to purchase, prepare and eat meat substitutes regularly originates from various sources: product unfamiliarity, insecurity about how to cook plant-based meat replacements, food neophobia, image of artificialness, perceived negative (sensory) quality, (high) price, appearance or (lack of information on) healthiness of meat-free alternatives (see e.g., Apostolidis & McLeay, 2016; Hartmann & Siegrist, 2017; Hoek, Luning, Weijzen, Engels, Kok, & De Graaf, 2011).

Such results obtained in contemporary consumer studies are confirmed by Weinrich (2018) who finds in her exploratory study based on focus groups discussions that the power of habit is influential to Dutch consumers' inclination to hold on to eating meat. A major reason why many consumers are not eager to consume meat substitutes is that they adore the taste of meat and do not want to miss it. At the same time many consumers feel insecure about eating meat and the consumption of meat substitutes in the light of a healthy diet. As much as consumer awareness about meat and sustainability issues often appears to be relatively low, as multiple consumer studies have shown in the past few years (see Dagevos & Reinders, 2018, p. 108 for several references; see e.g., also Garnett, Mathewson, Angelides, & Borthwick, 2015, p. 27), the same seems to hold for consumer confidence with respect to the position of animal-based products in a healthy diet. Apparently, it is neither common knowledge that a reduction of the consumption of meat does not imply negative health effects nor that current meat substitutes have been improved by and large so much that they provide sufficient nutrients and vitamins. Low food literacy as well as lack of cooking skills are thus important practical factors that restrain present-day Dutch consumers from opting for meat substitutes – irrespective of whether Weinrich's comparative study finds some evidence that meat substitutes are more established in the Netherlands in comparison to Germany and France. This indicates that at this early stage of the dietary transition toward more sustainable plant-based diets – a phase that is first and foremost about gradually changing consumer preferences among flexitarians and omnivores – clear information on both environmental and health effects of people's diets is still a basic necessity.

By way of link between this section on "classic" plant-based meat alternatives in the context of meat, and the following section that pays some attention to cultured meat as an emerging new meat alternative, a recent study by Slade (2018) is useful. Slade (2018) conducts a hypothetical choice experiment in which Canadian consumers were given a choice for ordering a burger made from beef, plant-based protein or cultured meat, with equal prices and the message that all burgers have the same taste as well as nutritional profile. Survey respondents were asked to imagine that they were in a fastfood restaurant offering three types of burgers: a beef-based one, a plant-based one and a cultured meat burger. This results in a majority of two-thirds of the respondents opting for the conventional beef burger, while one-third of the respondents would choose the plant-based burger (21%) or the cultured meat burger (11%) respectively. Other findings are: (1) most participants did not believe the statement that all burgers taste similar; (2) vegetarians prefer plant-based burgers rather than cultured meat burgers; (3) women are more inclined to choose a plant-based burger than a cultured meat burger and vice versa for men, and; (4) frequent meat eaters are less likely to opt for plant-based burgers than for cultured meat burgers. The latter suggests that "meat lovers" are perhaps easier to persuade to purchase cultured meat alternatives than plant-based ones – which opens a window of opportunity for developing a market for cultured meat alternatives.

What Consumer Studies Envision: Cultured Meat

The mass production of cultured meat – also known as in vitro meat and artificial meat (Hocquette, 2016) or clean meat (Shapiro, 2018) – would mean a big change with the present. Although the technology of growing meat in-vitro from animal stem cells is at an early stage, it is frequently taken as a possible and promising pathway toward a future in which the problematic effects of conventional meat production are solved by and large. Now the market introduction of simulated meat-like products is coming closer, it is time to raise the question what we know about consumer acceptance of cultured meat. Bryant and Barnett (2018) try to find answers to this question by analyzing the scholarly papers (N=14) that have been published in peer-reviewed journals since 2014.

In advance, it is possible to formulate a couple of consumer concerns with respect to cultured meat. First, and even stronger than in the case of meat substitutes, consumers could distance themselves from cultured meat because it is perceived as unnatural, artificial, high-tech food – even with the disqualification that it is disgusting. In line with this consumer perception, a second possible issue of concern for consumers may be that they guess that cultured meat favors industrial concentration and big corporations on the one hand and endangers small and local food production systems on the other. An objection that has been made recently (June 2018) by several Dutch scholars (Jan van der Valk from Utrecht University, Martijn Katan from VU University Amsterdam) could potentially evolve into a third consumer concern when it will not be solved properly or remain unclear. It is about serum from unborn calves that current cultured meat technology demands. Apart from ethical and animal welfare considerations, this present dependence of cultured meat production from material that is part of livestock farming is quite inconvenient and a difficult message to communicate to consumers by a sector that tries to replace conventional meat production.[5]

The review paper by Bryant and Barnett (2018) shows that particularly the first-mentioned objection to cultured meat has been found in consumer studies. Unnaturalness appears to be a common consumer characterization or association. The link between unnaturalness and disgust finds also some empirical evidence in current research. Consumer doubts about cultured meat seem not to be much different than frequently found in consumer studies on conventional foods: safety, healthiness, taste and price appear to concern consumers. Next to personal concerns are also societal concerns expressed about distrust of companies producing cultured meat and their impact on traditional agriculture. Such anxieties are close to the abovementioned second concern.

Besides consumer uncertainty and doubts, Bryant and Barnett (2018) find enough reason in their synthesis of recent studies about consumer perceptions of cultured meat to be optimistic about consumer willingness to try cultured meat when available. In comparison to conventional meat and meat substitutes, much lack of information and unfamiliarity surround cultured meat at present. However, the promise of cultured meat technology to produce meat in ways which mean substantial improvement in terms of environmental friendliness and animal welfare conditions, is a key benefit with much consumer appeal. Less unambiguous are findings regarding perceived healthiness of cultured meat, but positive consumer perceptions are found with respect to public and personal health benefits. Also, global food security as a problematic issue in which cultured meat may have a positive impact as "protein producer for the poor", is addressed sometimes, Bryant and Barnett note (2018, p. 14). Such positive associations with cultured meat regarding environmental benefits, food security merits and animal welfare improvements, are also found among a small group (N=10) of Dutch participants in a study by Bekker and colleagues (2017, p. 90).

All in all, research findings so far are preliminary and whereas cultured meat is not fictitious anymore it is a novel food technology with a clean record of commercial market performance, though. Until further notice it seems wise to remember that whatever high the expectations are about cultured meat as the meat of the future, it will take quite a while before the cultured-meat market has reached such a stage of maturity that a large portion of food consumers are genuine clean-meat enthusiasts – let alone, have evolved into "neomnivores", who only eat cell-ag-produced meat products.

TOWARDS A MARKET FOR NEW MEAT ALTERNATIVES

However encouraging and energetic modern developments in the Netherlands are, much progress is needed in this country as well as across the globe for the consumer to accept the new meat alternatives. Despite the hindrances and food traditions to overcome, there is enough reason to suggest that the market for new meat alternatives will continue to grow and expand in the future.

Degrees of Freedom

Bryant and Barnett (2018) treat cultured animal products as a regular food product category in foreseeing that it is more likely that such usual behavioral determinants like price, taste or food safety will remain decisive rather than that environmental and animal welfare motives will be central to the buying decisions of tomorrow's consumers. Of course, time will tell, but it is also possible to think differently and more sophisticatedly about dietary choices.

First, portraying food consumers as always and automatically interested in price and convenience is rather caricatural. Both conceptual and empirical consumer studies show that food choices are also influenced by citizen virtues devoted to social justice, animal welfare or environmental friendliness. In other words, some prudence is in order as it comes to "a 'rational choice' vision on the consumer" as a self-oriented, calculating and status quo-biased creature (Middlemiss, 2018, p. 166, see also pp. 76–88).

Second, food choice and consumer acceptance are not only guided by a complex of interacting variables related to the person and the product, but also by the context. Contemporary research pays increasingly attention to the latter by focusing on the food environment. It is acknowledged that product popularity or acceptability as well as people's preferences and prejudices, practices and principles are substantially shaped and steered, constrained and cultivated by the socio-cultural and material conditions. What is being encouraged and endorsed rather than disapproved, excluded, or, for instance, considered out of date, is better not ignored when our aim is to improve our understanding of the chances to develop a market for new meat alternatives and realize new balances in animal and plant foods. Consumer choice is not free as such but surrounded by degrees of freedom.

From this viewpoint, it is not difficult to see that many barriers exist that prevent or frustrate consumers to change their high-meat diet (see also Stubbs, Scott, & Duarte, 2018, p. 130). One of the most evident factors in our persistent desire to eat meat is that meat eating has deep roots in food habits and food culture. Consuming meat is part of personal and cultural identity and abstaining from eating meat implies often in everyday life at present that food consumers have the nerve to deviate from social and cultural norms to a greater or smaller extent. Other obstacles frequently pointed to are lack of consumer interest as well as lack of knowledge of modern practices in animal farming, of links between meat consumption and environmental issues, or of the nutritional necessity of meat in a healthy diet. Also, economic reasons could interfere with making meat replacement or meat reduction acceptable and attractive alternatives: contributing as consumer to a profitable livestock sector in which producers can earn a living, and another obvious condition is that meat is often offered at low prices and special offers as well as prominently displayed in retail outlets.

In sum, socio-cultural, material and economic-institutional aspects influence consumer mentality and modi operandi, and these conditions are anything but automatically supportive to a change in dietary habits in which new meat alternatives are accepted and preferred options by food consumers. All this also holds for the Netherlands that still can be typified as a high meat-consuming country. It also reminds us of the fact that consumers can be potential change agents of alternative markets, but at the same time, awareness is raised that more is required than relying on individual consumers having unrestricted freedom of choice and adequate capacity to take full responsibility. To quote Mason and Lang (2017, p. 193): "If there are limits to how much we really choose our diets, (...) we need to think more seriously about how to engineer systematic change, and not leave the transition to sustainable diets to individual choice. Consumers need help in this cultural transition."

Alternative Paradigms

Building a market for new meat alternatives means to build an alternative that challenges the current world of food and animal agriculture. For sure, reasonable doubts are possible with respect to the possibilities of and for alternatives that resist the mainstream food market (see e.g., Ritzer, 2017).

However, Lang and Heasman's thought-provoking *Food Wars* (2004) provides inspiration. Lang and Heasman envision that two opposing paradigms are challenging the mainstream food system. The prevailing productionist paradigm is going to bifurcate into the so-called life sciences integrated paradigm and the ecologically integrated paradigm respectively. Briefly, and to put it in our own words, the first relies heavily on technological solutions to solve societal problems and to innovate the food system. The life sciences paradigm generates a world of food and agriculture that is full of smart agriculture, smart refrigerators, smart product packaging, big data, digitalization, robotization, as well as personal health monitoring and personalized nutrition. Hence, the problems of the world and the food system should be resolved first and foremost by changing production processes and product innovation.

The alternate perspective offered by the ecological paradigm contrasts fundamentally with the engineering approach of the "old" productionist paradigm and the life sciences paradigm as its "successor" by addressing attitude and behavioral change of consumers as key in solving problems of the food system and consumer society at large. Hope is cherished on progressive food movements such as slow food, community supported agriculture (CSA), urban agriculture, agro-ecology, flexitarianism and vegetarianism, freeganism, or locavorism. Shared objectives and values could build bridges to other groups like public health advocates, animal rights activists, anti-globalists, genetically modified organisms (GMO) opponents, or voluntary simplifiers. Roughly put, while the life sciences paradigm would argue that all societal efforts, attention and sympathy for ecologically sustainable and healthy diets have not translated yet into major changes in meat eating habits among mainstream food consumers, the logic of the ecological paradigm would be that technological progress contributes more to the increasing global demand for meat as one of the main problems the world of food is facing than to its solution.

At first sight, cultured meat seems to fit better in the life sciences paradigm and meat substitutes in the eco paradigm. However, perhaps both meat alternatives should not be presented as competitive but rather as complementary, and sharing the goal of moving – via other avenues – toward lower animal meat-based diets. Besides, possibly cultured meat is going to find its entrance to the market when combined with plant-based foods: hybrid products of cultured and plant-based meat. Anyhow, Lang and Heasman's (2004) *Food Wars* is above all a useful reminder that the chances of new meat alternatives to succeed and develop beyond niche markets to more viable and stronger markets, depend very much on the dominant production practices and consumer culture. Similar to the reflections on consumer choice in the previous subsection, it also applies to (niche) markets that an enabling environment is vital for further development.

The Clock Is Ticking

The observed entrepreneurial activity and examined consumer appetite give reason to believe that on both sides of the market drivers exist and untapped potential is present that are going to contribute in the coming years to the emerging markets of meat analogues and cultured meat. Of course, developments in the new meat alternatives domain are at its early stages. Much is uncertain and circumstances are fragile. Moreover, the state of art of plant-based meat substitutes differs considerably from cultured meat. Meat substitutes are much ahead of the latter in establishing themselves in the minds and meals of many Dutch consumers as well as in the introduction of products on the Dutch market that go beyond patties and mince: the plant-based steak is already available and affordable. The cultured meat sector is far away from the stage of developing whole meat items like a steak, lamb chop or chicken breast at this moment.

After concentrating on the supply and demand side of the market, we bring this chapter to a close by saying a few words on governmental support. Albeit that the Dutch government has funded several research projects on new proteins in the course of time, meat reduction as policy goal is strictly avoided so far. Nevertheless, since 2011 several government-related councils and advisory boards in the Netherlands have addressed "the meat problem". The latest example is given in March 2018 by the plain advice of the Council for Environment and Infrastructure (2018) to adopt a food policy that aims to reduce animal protein consumption to no more than 40% of total protein consumption by 2030. This actually means a major switch in comparison to the current consumption of animal-sourced foods and it demands a serious acceleration in the replacement of these products with plant-based proteins (the normal Dutch diet at present is the opposite of what the Council aims for in 2030 and contains around 60% animal-based proteins and about 40% plant-based proteins).

In order to act upon such advices, an active role for a government or other public bodies in stimulating and facilitating the dietary shift away from farmed meat could be given interpretation through: (1) subsidizing new meats; (2) taking away needless regulation; (3) imposing meat taxation; (4) supporting information campaigns to raise consumer awareness of the protein transition; (5) contributing to meatless as default option in choice architectures, cultural frames of reference and social practices; (6) taking "divestment" initiatives, i.e., remove or reduce agricultural subsidies that encourage unsustainable production practices; or (7) appointing oneself as a launching customer of vegetable-based proteins by public procurement (see also Schmidinger, Bogueva, & Marinova, 2018, pp. 354–355; Stubbs, Scott, & Duarte, 2018, p. 132).

Additionally, governments may be expected to be sensitive to the Sustainability Development Goals (SDGs) currently. With respect to new meat alternatives, especially SDG number 12 "Responsible consumption and production" is relevant. Moving into the direction of sustainable diets also helps to solve some of the other pressing world problems addressed by the SDGs, but most directly SDG 12 (Figure 4). The extent to which new meat alternatives will become the meat of the future will contribute to reach the goal of responsible consumption and production. A major reduction in the (over)consumption of animal meat products and a major correction in the course of increasing global meat production are critically

Figure 4. United Nations Sustainable Development Goal 12

important to keep faith in a bright future for humankind and the planet that hosts us. Scientific opinion is clear about this. In following, business activity, consumer appetite, and governmental attention should fuel the drastic departure we have to make from conventional animal proteins as the highly-praised and absolutely normal component of our diets. To borrow Shapiro's words (2018, p. 222): "The clock is ticking to find better ways to feed ourselves".

REFERENCES

Alexander, P., Brown, C., Arneth, A., Dias, C., Finnigan, J., Moran, D., & Rounsevell, M. D. A. (2017). Could consumption of insects, cultured meat or imitation meat reduce global agricultural land use? *Global Food Security*, *15*, 22–32. doi:10.1016/j.gfs.2017.04.001

Apostolidis, C., & McLeay, F. (2016). Should we stop meating like this? Reducing meat consumption through substitution. *Food Policy*, *65*, 74–89. doi:10.1016/j.foodpol.2016.11.002

Bekker, G. A., Tobi, H., & Fischer, A. R. H. (2017). Meet meat: An explorative study on meat and cultured meat as seen by Chinese, Ethiopians and Dutch. *Appetite*, *114*, 82–92. doi:10.1016/j.appet.2017.03.009 PMID:28323057

Bryant, C., & Barnett, J. (2018). Consumer acceptance of cultured meat: A systematic review. *Meat Science*, *143*, 8–17. doi:10.1016/j.meatsci.2018.04.008 PMID:29684844

Council for Environment and Infrastructure (Rli). (2018). *Sustainable and Healthy: Working Together towards a Sustainable Food System.* The Hague, the Netherlands: Council for Environment and Infrastructure.

Dagevos, H. (2016). Exploring flexitarianism: Meat reduction in a meat-centred food culture. In T. Raphaely & D. Marinova (Eds.), *Impact of meat consumption on health and environmental sustainability* (pp. 233–243). Hershey, PA: IGI Global. doi:10.4018/978-1-4666-9553-5.ch012

Dagevos, H., & Reinders, M. J. (2018). Flexitarianism and social marketing: Reflections on eating meat in moderation. In D. Bogueva, D. Marinova, & T. Raphaely (Eds.), *Handbook of research on social marketing and its influence on animal origin food product consumption* (pp. 105–120). Hershey, PA: IGI Global. doi:10.4018/978-1-5225-4757-0.ch007

Derbyshire, E. J. (2017). Flexitarian diets and health: A review of the evidence-based literature. *Frontiers in Nutrition, 3*, 55. doi:10.3389/fnut.2016.00055 PMID:28111625

Garnett, T., Mathewson, S., Angelides, P., & Borthwick, F. (2015). *Policies and actions to shift eating patterns: What works? A review of the evidence of the effectiveness of interventions aimed at shifting diets in more sustainable and healthy directions.* Oxford, UK: Food Climate Research Network and Chatham House.

Hartmann, C., & Siegrist, M. (2017). Consumer perception and behaviour regarding sustainable protein consumption: A systematic review. *Trends in Food Science & Technology, 61*, 11–25. doi:10.1016/j.tifs.2016.12.006

Hicks, T. M., Knowles, S. O., & Farouk, M. M. (2018). Global provision of red meat for flexitarian diets. *Frontiers in Nutrition, 5*, 50. doi:10.3389/fnut.2018.00050 PMID:29963555

Hocquette, J.-F. (2016). Is *in vitro* meat the solution for the future? *Meat Science, 102*, 167–176. doi:10.1016/j.meatsci.2016.04.036 PMID:27211873

Hoek, A. C., Luning, P. A., Weijzen, P., Engels, W., Kok, F. J., & De Graaf, C. (2011). Replacement of meat by meat substitutes: A survey on person- and product-related factors in consumer acceptance. *Appetite, 42*, 265–272. doi:10.1016/j.appet.2003.12.003 PMID:21315123

Ingenbleek, P., & Zhao, Y. (2018). *The Vegetarian Butcher: On its way to become the world's biggest "meat" producer?* [unpublished manuscript]. Wageningen: Marketing and Consumer Behaviour Group, Wageningen University.

Lang, T., & Heasman, M. (2004). *Food wars: The global battle for mouths, minds and markets*. London: Earthscan.

Mason, P., & Lang, T. (2017). *Sustainable diets: How ecological nutrition can transform consumption and the food system*. Abingdon, UK: Routledge.

Middlemiss, L. (2018). *Sustainable consumption: Key issues*. Abingdon, UK: Routledge. doi:10.4324/9781315628035

Poore, J., & Nemecek, T. (2018). Reducing food's environmental impacts through producers and consumers. *Science, 360*(6392), 987–992. doi:10.1126cience.aaq0216 PMID:29853680

Quist, J. (2007). *Backcasting for a sustainable future: The impact after ten years*. Delft, The Netherlands: Eburon Publishers.

Reynolds, M. (2018). The clean meat industry is racing to ditch its reliance on foetal blood. *Wired.* Retrieved from http://www.wired.co.uk/article/scaling-clean-meat-serum-just-finless-foods-mosa-meat

Ritzer, G. (2017). Can there really be "true" alternatives within the food and drink markets? If so, can they survive as alternative forms? *Journal of Marketing Management*, *33*(7-8), 652–661. doi:10.1080/0267257X.2017.1312843

Schmidinger, K., Bogueva, D., & Marinova, D. (2018). New meat without livestock. In D. Bogueva, D. Marinova, & T. Raphaely (Eds.), *Handbook of research on social marketing and its influence on animal origin food product consumption* (pp. 344–361). Hershey, PA: IGI Global. doi:10.4018/978-1-5225-4757-0.ch023

Shapiro, P. (2018). *Clean meat: How growing meat without animals will revolutionize dinner and the world*. New York, NY: Gallery Books.

Slade, P. (2018). If you build it, will they eat it? Consumer preferences for plant-based and cultured meat burgers. *Appetite*, *125*, 428–437. doi:10.1016/j.appet.2018.02.030 PMID:29501683

Stubbs, R. J., Scott, S. E., & Duarte, C. (2018). Responding to food, environment and health challenges by changing meat consumption behaviours in consumers. *Nutrition Bulletin*, *43*(2), 125–134. doi:10.1111/nbu.12318

Tolonen, E. (2018). Meat substitution in the Netherlands: How to further accelerate the Green Protein Transition from an innovation perspective (MSc thesis). Delft University of Technology & Leiden University, The Netherlands.

Tuomisto, H. L., Ellis, M. J., & Haastrup, P. (2014). Environmental impacts of cultured meat: alternative production scenarios. In R. Schenck, & D. Huizenga (Eds.), *Proceedings of the 9th International Conference on Life Cycle Assessment in the Agri-Food Sector (LCA Food 2014), 8-10 October 2014, San Francisco, USA* (pp. 1360–1366). Vashon, WA: ACLCA. Retrieved from https://core.ac.uk/download/pdf/38629617.pdf

Tuomisto, H. L., & Teixeira de Mattos, M. J. (2011). Environmental impacts of cultured meat production. *Environmental Science & Technology*, *45*(14), 6117–6123. doi:10.1021/es200130u PMID:21682287

Verain, M., Dagevos, H., & Antonides, G. (2015). Flexitarianism: A range of sustainable food styles. In L. A. Reisch & J. Thøgersen (Eds.), *Handbook of research on sustainable consumption* (pp. 209–233). Cheltenham, UK: Edward Elgar Publishing.

Weinrich, R. (2018). Cross-cultural comparison between German, French, and Dutch consumer preferences for meat substitutes. *Sustainability*, *10*(6), 1819. doi:10.3390u10061819

ADDITIONAL READING

Alexander, P., Brown, C., Arneth, A., Dias, C., Finnigan, J., Moran, D., & Rounsevell, M. D. A. (2017). Could consumption of insects, cultured meat or imitation meat reduce global agricultural land use? *Global Food Security*, *15*, 22–32. doi:10.1016/j.gfs.2017.04.001

Apostolidis, C., & McLeay, F. (2016). Should we stop meating like this? Reducing meat consumption through substitution. *Food Policy*, *65*, 74–89. doi:10.1016/j.foodpol.2016.11.002

Bryant, C., & Barnett, J. (2018). Consumer acceptance of cultured meat: A systematic review. *Meat Science, 143*, 8–17. doi:10.1016/j.meatsci.2018.04.008 PMID:29684844

Garnett, T., Mathewson, S., Angelides, P., & Borthwick, F. (2015). *Policies and actions to shift eating patterns: What works? A review of the evidence of the effectiveness of interventions aimed at shifting diets in more sustainable and healthy directions.* Oxford, United Kingdom: Food Climate Research Network and Chatham House.

Mason, P., & Lang, T. (2017). *Sustainable diets: How ecological nutrition can transform consumption and the food system.* Abingdon, United Kingdom: Routledge.

Middlemiss, L. (2018). *Sustainable consumption: Key issues.* Abingdon, United Kingdom: Routledge. doi:10.4324/9781315628035

Raphaely, T., & Marinova, D. (Eds.). (2016). *(release date October 2015) Impact of Meat Consumption on Health and Environmental Sustainability.* Hershey, PA: IGI Global. doi:10.4018/978-1-4666-9553-5

Schmidinger, K., Bogueva, D., & Marinova, D. (2018). New meat without livestock. In D. Bogueva, D. Marinova, & T. Raphaely (Eds.), *Handbook of research on social marketing and its influence on animal origin food product consumption* (pp. 344–361). Hershey, PA: IGI Global. doi:10.4018/978-1-5225-4757-0.ch023

Shapiro, P. (2018). *Clean meat: How growing meat without animals will revolutionize dinner and the world.* New York, NY: Gallery Books.

Stubbs, R. J., Scott, S. E., & Duarte, C. (2018). Responding to food, environment and health challenges by changing meat consumption behaviours in consumers. *Nutrition Bulletin, 43*(2), 125–134. doi:10.1111/nbu.12318

KEY TERMS AND DEFINITIONS

Alternative Protein Sources: Alternatives to current animal products (such as meat, dairy, eggs, and fish) range from algae, seaweed, duckweed, rape seed, to pulses, molds and mushrooms, soy-based products, nuts, and to insects.

Cultured Meat: Meat based on animal cells grown in a laboratory, also known as cellular or in vitro meat, to use the term that is becoming obsolete, or clean meat, to the use the term that is currently gaining popularity and is probably going to replace the term cultured meat.

Flexitarian: A human diet which aims at reducing or abstaining from the consumption of all kinds of meat for several days per week.

Food Wars: Cultural concept describing the clashes and conflicts between the opposing two paradigms in today's and tomorrow's world of food (i.e., the so-called life sciences integrated paradigm and the ecologically integrated paradigm, respectively).

New Meat Alternatives: Meat analogues or meat substitutes which are plant-based, lab-grown, or use ingredients other than livestock, such as insects.

Sustainable Development Goals (SDGs): The 17 broad and interdependent Sustainable Development Goals came into effect in the beginning of 2016, as developed by the United Nations Development Programme and aiming for a more sustainable world in many aspects including poverty, hunger, health, education, gender equity, clean water and sanitation, clean and affordable energy, good jobs and economic growth, innovation and infrastructure, reduced inequalities, sustainable cities and communities, sustainable consumption and production, climate action, life below water, life on land, peace and justice, and partnerships for the goals. Various SDGs are more or less food related, but from this chapter's perspective SDG 12 Sustainable Production and Consumption Patterns in particular is important.

ENDNOTES

1 The idea that everything is as it is, because it could not be otherwise, makes all thinking about alternatives freeze [translated from German].

2 Vivera has copied this way of labeling by the Vegetarian Butcher with a variation on the theme. In recent years (and similar to examples in Germany and the US), in the Netherlands some politicians and authorities have complained against the marketing strategy of manufacturing companies of plant-based meat substitutes to give names of conventional meat products to their products. Most recently this so-called "Schnitzelgate" flashed up in 2017.

3 For a recent study indicating that sustainability benefits – particularly in terms of land use footprint of production method – of reducing meat consumption appear more substantial than meat analogues or cultured meat (the latter finding contrasts with previous research by Hanna Tuomisto and colleagues, 2011 and 2014), see Alexander et al., 2017.

4 For a review study in which a first line of evidence is presented between flexitarian diets and (potential) health benefits, see Derbyshire, 2017.

5 Shapiro (2018, p. 62) claims that "several cellular ag companies have already completely done away with such serum, typically by using plant-based or synthetic serums or by simply figuring out ways of going serum-free". However, solving this problem appears to be less simple than expressed here and therefore premature to conclude that this serum problem has been fully solved now for the cultured meat sector in general. See also Reynolds (2018).

Chapter 11
What's New?
A History of Meat Alternatives in the UK

Malte B. Rödl
The University of Manchester, UK

ABSTRACT

The "new market" for meat alternatives promises meaty profits and attracts enormous interest by consumers and investors alike. In this chapter, the historical development of meat alternatives is reviewed in an attempt to identify what is "new" about this particular market in the United Kingdom as an example of a Western country. Beginning in Victorian England, through the Wars into the 21st century, the societal background and developments leading into various episodes of markets for meat alternatives are discussed. Together with a description of the "new" market, historical continuities and current opportunities are outlined. It is concluded that health, environment, and business opportunities have played an important role throughout the history of the market, but the significance of this market in the commercial world is new.

INTRODUCTION

The 'new meat alternatives market' promises a healthy, ethical, and sustainable product that is made from plant ingredients and fabricated to replicate meat. With an increasing societal awareness for the ethical, environmental, and individual health impacts of eating animals, more and more consumers decide to either cut down or avoid the consumption of meat and on other animal products altogether (Mintel, 2017). Also, with that new market, there are new products, new companies, and large investments into meat-free meats (The Economist, 2015). In contrast to the rather recent mainstream business expectations and investments, British consumers have been able to buy a diverse range of meat alternatives in most major supermarkets for over two decades, and for many decades prior outside of the mainstream market.

Most accounts of the history of meat alternatives start with soybeans in ancient China, where the highly proteinaceous crop has been used and cultivated for over three millennia; according to Shurtleff and Aoyagi (2014), tofu (coagulated soy protein) was first mentioned in a document from 965 CE, in which the consumption of tofu as an alternative to meat is advocated. Cooked wheat gluten (today known

DOI: 10.4018/978-1-5225-7350-0.ch011

as seitan) has also been used in China for many centuries; the creation of fibrous, meat-like foods from it was first mentioned in 1301. Another 'traditional' meat alternative available today is tempeh (fermented soybean cake) which is likely to exist since the early 1600s, in Java, Indonesia (Shurtleff & Aoyagi, 2011, 2014, p. 5). In ancient China, plant-based meat-like products were especially popular in the country's Buddhist periods, as meat was then forbidden for religious reasons (Shurtleff & Aoyagi, 2014).

The first references to replacing meat in the UK originate in Victorian England, when a vegetarian community of considerable size emerged. Vegetarian advocates promoted a purist diet based on wholesomeness, taste, price, and simplicity. However, it was also recognised that popularisation of the diet required more appealing foods, hence it was not uncommon to create cutlets, sirloins, etc. from vegetables (Gregory, 2007, p. 129). Vegetarian recipe books featured alternatives to meat dishes such as sausages, steaks, or cutlets. Towards the end of the 19th century, people called for more diversity in the vegetarian diet and replacements for animal ingredients. In consequence, a large variety of nut meats and other protein-rich products were created and sold (Gregory, 2007).

These two examples of ancient China and Victorian England show that the idea to recreate meat is likely to be as old as people turning away from meat; either due to religious, ethical, environmental, or health reasons. Acknowledging that with £ 292 million market value, the UK market for meat alternatives is the second biggest in the world after the US (Euromonitor International, 2017a), the question arises what is 'new' about this 'new market'. To address this issue, this review aims to give an insight into the origins and development of commercial meat alternatives, and contrasts these with current market developments. In the first section, the history of commercial meat alternatives in the UK is described from the late 19th century to the early 21st century. Particular emphasis is placed on societal developments that are interrelated with the history of meat alternatives. The second section describes the 'new' market on the basis of recent market research. The insights from the past and present are combined in order to discuss historical continuities, as well as possible opportunities arising in the current market for meat alternatives. It is summarised that neither health, environmental, or business opportunities are unique to the 'new market'—but rather represent an unprecedented optimism and success of the market.

This review aims to provide an in-depth background on meat alternatives and their societal and technological roots to contextualise the market for the reader. Further, by outlining the development of the UK market, path-dependent opportunities and difficulties are highlighted.

HISTORY OF COMMERCIAL MEAT ALTERNATIVES IN THE UK

Historically, the 'new meat alternatives market can be seen as going through three different periods in the UK during recent times. Its origins are at the end of the 19th century; the search for cheap affordable proteins marks the second phase and the most recent period is associated with the demand for healthier diet from the point of view of individual health but also for planetary wellbeing. They are discussed in turn below.

The Origins: Religious Substitute, Health Food, and War Nutrition

As the thought of replacing meat with meat-like products became more common towards the end of the 19th century, meat alternatives came to the market. An important role is ascribed to the American health foods reformer and Seventh Day Adventist John Harvey Kellogg, who was director at a health centre called the Battle Creek Sanatorium in Michigan, US (Shprintzen, 2012; Shurtleff & Aoyagi, 2014). Advocating a vegetarian diet for reasons of health and religious beliefs, Kellogg made the institution entirely vegetarian. By developing tasty and nutritious vegetarian food products, he aided a changing narrative from vegetarianism as a political choice focusing on injustice, towards a focus on physical and medical benefits of vegetarianism. The attention his products gained led to a correspondence with the US Department of Agriculture in 1896, where solutions to potentially rising meat prices or animal diseases were sought. Motivated by this, Kellogg developed a range of products from ground nuts, grain, and/or gluten, which were mixed and cooked to varying degrees. Acknowledging the benefits of the high protein content of meat and its sensory qualities, he created products that had "even higher nutritional value than meat, but exhibited a similar taste and consistency as flesh foods" (Shprintzen, 2012, p. 114). Together with his brother he set up a company to sell the products via mail order and authorised health food stores throughout the country. A patent for meat alternatives was published in 1901 in the United States and the United Kingdom, where the products were described as "a vegetable substitute for meat", emphasising "equal or greater nutritive value", easier digestion, nutritional similarity, a similar flavour, and a texture similar to that of tender meat (Shurtleff & Aoyagi, 2014, p. 6). Meat alternatives became widely acknowledged and popular in the US in the early 20th century, as they promised a healthy lifestyle with an "experience that approximated the desirable components of carnivorous living" (Shprintzen, 2012, p. 124).

The health food movement was similarly set up in the UK, and so around the same time, meat alternatives appeared across the Atlantic. In 1899, *The International Health Association Limited* was founded by Seventh Day Adventists, produced the Sanitarium foods in licence, and sold them in health food outlets across the UK, while previously having imported them (Gregory, 2007; Shurtleff & Aoyagi, 2014, e.g. pp. 44, 48, 394). An increasing number of companies produced a large variety of vegetarian foods including meat alternatives, which contributed to a vast extension of the vegetarian diet at the turn of the century. Brands such "Nut Cream, Meatose, Vejola, Nut-vego, Savoury Nut Meat, [or] 'Nutton'" were used in a range of dishes, including curries, stews, and cold meats (Gregory, 2007, p. 132). The main ingredients for these products were nuts (such as peanuts and hazelnuts), seeds (such as pine nuts), or grains (such as whole wheat and gluten).

Despite success in the health reform movement in the UK, meat alternatives remained out of sight from the mainstream market (Shprintzen, 2013), until during the First World War when food needed to be rationed. Consumption of less meat was encouraged from 1915, and rationing was mandatory from 1918, while the government endorsed nutritional alternatives to meat "as a matter of national duty" (Owens, 2016, p. 6). Some of the nut meat companies even advertised in big national newspapers. The scarcity led all sorts of 'health' foods, such as wholegrain bread, being advertised as a nutritious, healthier, and cheap alternative to meat. This wide adoption of meat-less and less-meat diets desisted after the war, partially because it was advertised as a wartime solution (Owens, 2016).

Soy beans became widely used in the UK in the early 20th century as a replacement for other oilseeds in the production of oils, soaps, or glycerine. By the 1930s soy beans had many uses in the chemical industry, while the by-product (soy bean cake) was fed to cattle (Johnson et al., 1992a). Throughout the 1930s and 1940s processing and possible uses of soy beans extended, and full-fat soy bean flour was recognised as a nutritious and proteinaceous ingredient to fortify bread and other foods (Johnson, Meyers, & Burden, 1992b; Shurtleff & Aoyagi, 2015). However, the soy bean 'by-product' was largely unpalatable, and never reached the consumer in pure form. Again, in a situation of scarcity during the Second World War, soy flour was added to many different food products, and "became the main ingredient in British sausages . . . [while people] pretended they are the real thing" (Collingham, 2012, p. 85). The Indian army, fighting as part of the British armed forces, received a lot of "self-heating tins of soya chunks" from the US as the soldiers would not eat various meats for religious reasons. Intended as other animal-based alternatives to meat, the UK government also imported marinated whale meat, canned snoek, and tinned barracuda, which was not greatly appreciated by the public (Collingham, 2012). While a lot of products could be substituted or fortified using soy, their quality was poor (Learmonth, 1963 as cited in Shurtleff and Aoyagi, 2015, p. 2795). These years were deemed important for technological learning, but negatively affected the image of soy as a food, as it was subsequently only considered a functional food ingredient for some time. The first half of the 20th century paved the way for meat-free protein products—meat alternatives—to be recognised as a healthy alternative to meat by some, and a religious alternative by others. The rest at least accepted them as a nutritious necessity in lack of animal protein.

Building Market Foundations: In Search for a Cheap and Reliable Protein Supply

It can be inferred from newspaper reports that meat alternatives gained little traction in the UK market, and the more popular 'synthetic meat' industry in the US was viewed with scepticism (e.g. Raphael, 1969). Growing awareness for the issues of protein malnutrition in the global South, and increasing worldwide meat consumption led to a lot of research concerned with the so called 'new protein foods' (Altschul, 1974). Related to this research, a US Department of Agriculture (USDA) report projected that soy products would replace around 10 to 21% of all processed meats by 1980 (cited in Horan, 1974). While some research was concerned with the refinement of plant-protein, another strand of studies dealt with the so-called single-cell protein (SCP). These are unicellular microorganisms (e.g. bacteria, algae, yeasts, or fungi) which grow on byproducts of various industries, and were largely intended to be used as a protein-rich feed for livestock. Some projects operated on large scales, but most were stopped in the 1980s with the changing political conditions and improved animal farming practices. For instance, the end of the Cold War and the advent of global trade agreements ultimately stabilised prices and 'outperformed' single-cell protein (Ugalde & Castrillo, 2002). One such project in the UK focused on converting starches into protein using fungi, and developing them for human consumption. Whilst adequate cultures were found, obligatory food safety tests delayed the commercialisation of the product by almost a decade until 1985, when the current market leader *Quorn* released its first products in supermarket ready meals (Finnigan, 2011; Trinci, 1992).

Meanwhile, commodity prices for soy beans dropped due to its growing use as feed, and soybean processing methods were improved (Horan, 1974). Soy flour became more refined compared to the one used during the war, and soy protein concentrates and isolates were fabricated on larger scale. New processes were developed which could arrange protein (isolated or as part of emulsions including ingredients like starches, fibres, fats, and salts) into fibrous structures, so that textures similar to meat could be created. These were soon sold, for example, as 'Spun Vegetable Protein', 'Textured Vegetable Protein' (TVP), or 'Textured Soy Protein' in chunks of varying sizes (Johnson, Myers, & Burden, 1992b; Pearson, 1976; Ziemba, 1966). At least from the late 1960s, some of these were sold or even fabricated in England.

These products were mostly in use by food manufacturers in ready meals, or to get more value of processed meats by adding cheaper plant protein (Collins, 1975). In 1969, two vegan and animal rights activists bought some TVP in bulk and repackaged it for the consumption of their friends and acquaintances. After witnessing their popularity, they founded *Direct Foods*, and started to sell products such as a 'burger mix' or 'beef-style' flavoured TVP via mail order (Shurtleff & Aoyagi, 2014, pp. 533, 2686, 2345). In an article published by The Guardian ("A soya point," 1975), their product was pictured alongside other TVP products in a shopping basket, and the article argues that Britain is more of a "meat extenders rather than substitutes" nation.

More entrepreneurs joined the market, specialising for example in Tofu and Tempeh production (e.g., Cauldron in 1981), but also in other fabricated soy-based meat alternatives. By 1985 a variety of meat alternatives were available in health foods shops throughout the country (Shurtleff & Aoyagi, 2014, pp. 2686, 2687). The year 1985 became pivotal for the market, as established food manufacturers with more resources entered. *Quorn* was brought to market in a joint venture between the food manufacturer Rank Hovis McDougall (RHM), and the chemical company Imperial Chemical Industries (ICI), and was subsequently used as an ingredient to various meals in the products of the largest supermarket chains in the UK (Trinci, 1992).

By the mid 1990s, the *Quorn* brand had almost entirely shifted towards producing under their own brand and selling directly to consumers (The Grocer, 1995). Also in 1985, *Direct Foods* was bought by their supplier and major soy processor *British Arkady*, which realised that there was a market opportunity to add value to their fabricated TVP products and sell them directly to consumers. They subsequently bought up various smaller brands that mostly used soya products for various animal-free products and formed the *Haldane Foods Group*, which became a prominent health food manufacturer (Shurtleff & Aoyagi, 2014, p. 2686).

The amount of people not eating meat ranged around 4-5% in the early 1990s (The Grocer, 1994), and 'health foods' sections were established in all major supermarkets (Shurtleff & Aoyagi, 2014, p. 2687). Just as vegetarianism, the market grew significantly (being said to have doubled to £ 25 million in 1991 by The Grocer, 1993a), along with the interest by various businesses. In 1989, the well-known animal rights activist and photographer Linda McCartney published a vegetarian cookbook which involved many home-made or pre-manufactured meat alternatives. Alongside a number of emerging smaller enterprises producing meat alternatives, the established frozen food manufacturer *Ross Young's* recognised a gap in the meat-like and vegetarian ready meals market. They approached Linda McCartney and in 1991 they formed a brand with her name and products based on her recipes (Linda McCartney's Foods, 2016; The Grocer, 1993c). As part of an established food manufacturer and supported by the celebrity, the *Linda McCartney's* brand soon became the market leader in the meat-free ready meals section (The Grocer, 1993b), and is still a prominent brand today. Target groups for this and other brands were young, affluent women, which were also the majority of the increasing vegetarian population (The Grocer, 2001).

The 1990s brought a number of meat scandals, that might have been related to an increase in vegetarianism and meat reduction (Beardsworth & Bryman, 2004). Many brands tapped into this arising mistrust towards meat to promote non-animal protein; part of this was a strong TV presence of many vegetarian brands throughout the 1990s, who fought over predominance in the sector (e.g. The Grocer, 1996, 1999). On the other hand, EU legislation on genetically modified organisms (GMO) in the late 1990s, caused some difficulty for products from, for example, *Linda McCartney's* or *Haldane Foods*, leaving non-soy products such as *Quorn* in advantage. As consumers were generally not in favour of GMO in their food, retailers threatened companies with a boycott if they continued using GM ingredients (especially soy). The public's trust in soy-based meat alternatives dropped, and *Linda McCartney's* announced that they would switch to wheat protein based products. Even some new and promising market entrants using soy, such as *Khero* aiming at health-conscious consumers, could not sustain themselves because of this emerging consumer scepticism (Buckingham, 1999; Dibb & Simkin, 2001). Nonetheless, the reputation recovered over time, and product quality and market size increased throughout the 1990s. As such, by the end of the millennium, "cardboard-like soya sausages" were a thing of the past (Brockes, 2000), as companies offered a wide range of products based on rehydrated and further refined textured soy or wheat protein. Less processed products such as tofu, seitan, and even just vegetables or legumes in burger or sausage shape could be bought.

Apart from unrealised expectations that meat alternatives might solve world hunger, throughout the second half of the 20th century meat alternatives were mostly considered as products for vegetarians and vegans. Meat was considered unhealthy or dangerous due to many meat scandals. While this drove meat reduction, vegetarians were often motivated by ethical reasons (Beardsworth & Bryman, 1999). The market served the portion of these who wanted a convenient protein supply, or appreciated the taste and texture of something meat-like. The mainstream consumer mostly got in touch with them through fortified products or as part of meat extenders.

Targeting the Meat Reducer: Meat Alternatives in the New Millennium

By the late 1990s, demands for a healthier and more diverse diets by mainstream consumers grew, while vegetarianism slightly declined despite food scares such as bovine spongiform encephalopathy (BSE). It was noted that there were "just [not] enough vegetarians to be creating all this demand" (Just Food, 2000). The Grocer (2003) called a distinction between people who liked to eat meat and those who did not 'outdated', as they found that more and more people wanted a "healthier and more varied diet". With health-focused messages, companies thus began to target primarily meat reducers alongside meat-liking vegetarians, in an attempt to mitigate the impeding limits on growth among vegetarian consumers (The Grocer, 2003). Acknowledging that 'suitable for vegetarians' might be off-putting to some consumers, *Quorn*, for example, removed the word 'vegetarian' from their marketing in 2006 (The Grocer, 2006). In that year, half of all British consumers were reported to have bought meat substitutes at least once. Especially women and younger people were reported to like meat substitutes, while affluent males above 45 were the strongest opposed to meat alternatives (Mintel, 2006b). The British population was estimated to have 6% vegetarians and 30% meat reducers, while the market for meat alternatives was worth £ 187 million (Euromonitor International, 2017a; The Grocer, 2006).

In the 1980s, meat alternatives were mostly available in health food outlets, yet by 2006 they were sold in more than 90% of major grocery stores. Here, they were sold within the range of 'meat-free foods', that also comprised meat-free ready meals or delicatessens. *Quorn* was most successful in the big advertising battles in the 1990s, and became the most prominent and largely unchallenged actor in the meat alternatives market. In 2006, around sixty percent of sold meat alternatives were *Quorn*'s mycoprotein, and in Sainsbury's stores the chilled vegetarian section was labelled "Quorn" (The Grocer, 2006). *Haldane Foods* and *Linda McCartney's* were *Quorn*'s largest branded competitors, each with around 2–4% of the market share. Own-label meat alternatives had a considerable market share, but otherwise the competition was negligible, with other companies either selling a small number of products in larger supermarkets, or concentrating on the equally growing health foods and vegetarian market. Those few brands that emerged, did not survive for long or never became mainstream, possibly as the dominance of the market leader in advertising and on the retail shelves set high entrance barriers. This dominance also caused stagnation in the market, as the market largely relied on *Quorn* for its popularisation and growth.

Consumers became increasingly prone to reduce meat intake since the beginning of the 21st century. In 2004, a Mintel study found that 7% of women and 4% of men described themselves as vegetarian, while in an online study in 2017 Mintel found that 9% and 6% respectively followed meat-free and fish-free diets. Other studies are more conservative with their estimates about numbers for vegetarianism, such as a large-scale study by the Vegan Society in 2016, which found that 3.25% of the UK population identified as vegetarian in 2016, while around 1.05% identified as vegan. In all reports over time, vegetarians and vegans are more likely to be young, urban, educated, and female (Mintel, 2004, 2017; The Vegan Society, 2016). Similarly, 17% of women and 8.6% of men in 2004 reported 'they were tending to eliminate meat from their diet', which increased to 30% and 27%, respectively, in 2017. A further 17% planned to reduce on meat intake.

For people avoiding meat, ethics was (and remains) the most prominent reason, whereas for meat reducers, health and weight management were the main motivations. The proportion of people who ate meat alternatives grew from 15% in 2006 to around half in 2017, and were more likely to avoid or reduce their meat consumption. Environmental reasons as well as concerns over antibiotics were not given as an option in earlier Mintel studies, but were prominent reasons in 2017. Other reasons for the substitution of meat with meat alternatives are costs, concerns about food quality and safety (for example through food scandals like BSE, the foot-and-mouth disease, or the use of hormones and antibiotics in industrial animal farming), concerns about the social acceptance of the own diet, and meat alternatives as a transition aid to become vegetarian (Euromonitor International, 2017b; Mintel, 2004, 2006a, 2017, 2017).

THE 'NEW' MARKET

The market for meat alternatives is said to have grown considerably since 2014/15, although various data sources disagree on the extent of this change. Euromonitor International (2017a) estimated the growth in retail value between 2014 and 2017 to be £ 76 million, while Mintel (2017) asserted an £ 11 million growth from 2015 to 2016, and Kantar Worldpanel (2016) estimated more than £ 20 million growth between 2014 and 2016 (see also Figure 1). The 2016 market for meat alternatives was valued at £ 174 million by Kantar Worldpanel (2016), at £ 272 million by Euromonitor International (2017a), and at £ 196 million (£ 272 million with 'meat-free snacks') by Mintel (2017)[1]. Before this current trend, the market flourished less, with Mintel (2017) even reporting a decline in retail value between 2012 and 2015. In the report, this was explained by a general decline in sales of ready-to-cook foods in that time, and therefore may not reflect a change in consumer attitudes.

Market volumes have also increased over time: Kantar Worldpanel (2016) estimated the volume of the market to be 30 million kg, and Mintel (2017) estimated around 34 million kg in 2016. This was attributed to both existing consumers buying more, and new consumers buying meat alternatives. The unit price was reported to have reduced slightly in between 2015 and 2016 and averaged to £ 9.15 per kilo, although the more prominent brands on the market had considerably lower average per kilo prices, going as low as £ 5 per kg for *Linda McCartney's* products at the end of 2015 (Kantar Worldpanel, 2016). Euromonitor International (2017b) estimated an average unit price of around £ 8, with meat being slightly lower. On average, a British household spent £ 6–10 on meat alternatives[2] in 2016, amounting to 101 g per household per year. Kantar Worldpanel (2016) estimated the volume to be 4 kg on aver-

Figure 1. Development of the market for meat alternatives from 2001 to 2017 according to various market research sources.

Notes: Kantar calculates the values from consumption-level data; Mintel and Euromonitor use aggregated retail data. Categorisation differences are likely to be responsible for the remaining differences in the estimates. The price adjustment of Mintel data was reversed.

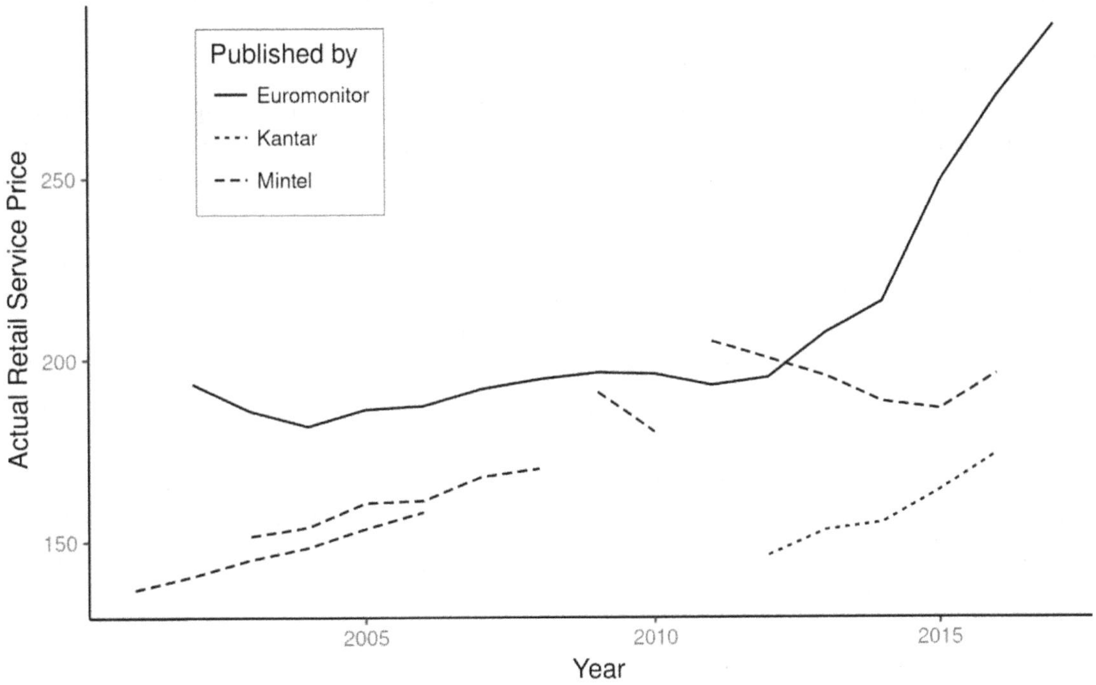

age per household buying meat alternatives regularly. Meat reducers are responsible for around half of meat alternatives sales with a constantly increasing share. However, these numbers are still dwarfed by an average household spend of £ 5.91 a week on an average of 929 grams of meat and meat products, which is 2.43% of the total household expenditure, and around a quarter of the total food expenditure (DEFRA, 2017).

The market is dominated by a few major actors. Firstly and most importantly, *Quorn*, which had a market share of 44.4% in 2016 is now the most prominent meat-free brand in every major supermarket; its market share has been consistently high, ranging from 40 to 46% in between 2007 and 2016. Other brands with a notable market share according to Euromonitor International (2017a) are *Linda McCartney's* (6.1%), *Dalepak* (2.9%), *Cauldron* (part of *Quorn*'s *Marlow Foods*, 3.5%), *Goodlife* (1.7%), and *Birds Eye* (1.7%), which are all sold in one or more of the major supermarkets. Another 17% of the market share is in own-label products, and the remaining 26% are split by around 50-60 smaller brands. Recently, the share of own-label has declined to the benefit of smaller brands (Euromonitor International, 2017a). The last few years have also seen a number of new companies founded which target a diverse range of consumers, from lifestyle wholefood health vegans to meat-loving health-conscious males. Highly successful new and established brands from abroad became known in Britain (e.g. from the US or Netherlands), and invested or announced to invest in the UK market. With more than 40% market share and more than 99% of the total market's advertising expenditure occupied by *Quorn*, market entrance for new brands has not historically been of high success or interest from abroad (Mintel, 2017). *Quorn* products are fungus-based and have changed little beyond product innovation, which may also have had an effect on product quality, diversity, and general innovation activity in the market. While most brands are traded in health foods shops or online, in the major supermarkets they are commonly placed in separate vegetarian categories in the chilled and frozen sections. While this distinction is said to inhibit market growth and occasional purchases from meat reducers, supermarkets might slowly change this practice (see e.g. Glotz, 2018; Smithers, 2018).

Recently, many companies have started to reformulate their products in order to appeal to the small yet vocal vegan market. Being from plant-protein seems important in marketing activity, and meat extenders, or mixed meat- and vegetable-protein products are becoming more popular (Mintel, 2017). With the advent of investment in some markets, the UK market has seen more innovative brands ready to invest in this promising market. Technology will improve so that the texture of meat alternatives even more closely resembles animal flesh, and be considered acceptable even to consumers who have not considered exchanging the beloved meat on their plates.

High Moisture Meat Analogs (HMMA), the next generation of meat alternatives, was developed in the 1980s on laboratory-scale and is slowly becoming commercialised (Zorpette, 2013). Raw ingredients will become more diversified, so that apart from soy, wheat, and occasional fungi, other plant proteins such as pea proteins are used. While so called lab-grown or in-vitro meat, is still far away from coming to the market, the expectations are high. Advancements in ingredients, processing, extrusion, and bio-technology nurture public imagination about meat of the future (e.g. Chiles, 2013). As the trend for healthy, ethical, and environmental alternatives to meat is only going to increase, the consumer demand and associated profit will make it happen.

SOLUTIONS AND RECOMMENDATIONS

Tracing the history of meat alternatives over more than a century exposes patterns across time, and allows to make recommendations for the new market: (1) the ambivalent health credentials of meat are understood as a problem; (2) meat-like plant-based products are framed as a promising solution, and; (3) market success is highly dependent on the social and cultural context. These three points will be elaborated in detail in the following paragraphs.

Firstly, meat eating always was and largely continues to be seen as equally vital and detrimental to health. In the 19th century, when Kellogg recommended not to eat meat for reasons of health, he at the same time acknowledged that protein is essential to the human body and needed to be obtained through foodstuffs other than animal flesh. Later, especially during the two World Wars, meat was scarce and needed to be replaced on nutritional grounds, which made it possible to label any health foods as a suitable alternative to meat. During the Cold War, meat prices were expected to rise and a global protein scarcity was foreseen. As such, research and development efforts were made to find a suitable alternative to the protein needs of the developing world. Since the 1990s, evidence and awareness have increased regarding the detrimental health effects of meat consumption. Replacing meat or having to replace meat for reasons of health, is thus not a recent insight. A lot of consumer education relating to the health impacts of meat consumption has happened in recent years, for example a shift in the framing of dietary guidelines and recommendations. This issue is likely to remain in the public spotlight, as in our contemporary society many people endeavour to live a long and healthy life.

Secondly, the solution to the ambivalent health credentials of meat has always been to provide a tasty, convenient, and healthy slab of protein. This solution is bounded by the human fondness for eating flesh, and the resulting difficulty to alter dietary habits and cultural conditioning. Already in the Victorian period, political vegetarians discussed how to make their diet more attractive and started to commercialise meat lookalikes. Meat alternatives—and more generally alternatives to meat—have been mainly framed as a protein supply, while promising meaty taste and convenience. This was regardless of whether the reasons for not consuming meat were individual, societal, religious, health-related, scarcity-based, or ethical. The challenge for the new market is to establish a meaningful identity as a food choice, rather than a substitute for something else (such as *Quorn* becoming 'yet another protein source' through continuous advertising campaigns). The extension of the market towards more 'authentic' meat alternatives especially for flesh-positive males[3] is a growing opportunity; nonetheless, the market also needs a variety of other convenient and tasty protein products for consumers who neither like to rely on meat-like products, nor on an otherwise 'balanced' diet for their daily protein intake.

Thirdly, establishing meat alternatives as a legitimate foodstuff has always been dependent on the right context and time, as well as on power relations, financial, and social resources. Only in the presence of potential buyers could meat alternatives be reasonable and feasible, but consumers needed to be convinced to opt for these products. What used to be professional acknowledgement for Kellogg, was a network of interested meat avoiders for *Direct Foods*, and an anticipated protein shortage for the development of *Quorn*. Currently, the strong establishment in the UK market, with brands such as *Linda McCartney's* or *Quorn* and a small number of major grocery retailers, poses great challenges for newcomers to gain a market share worth investing in. These resources can be provided either by the meat industry (as has happened in the US or Germany), by organisations established in another market prior to selling in the UK, or by other strong actors in the food industry, who could simplify gain access to the mass market, while providing a safe context for business growth. Beyond being listed and thus available to consumers, businesses must ensure that every consumer can find a healthy protein product—if they desire to find one—in their accustomed routes through the supermarket. This includes plant-based meats next to real meat, but also a separate or attached section of products for different tastes.

Thus far, we have identified three pivotal aspects consistent throughout the history of meat alternatives in the UK. Problems are largely related to health and an efficient protein supply, solutions are about convenience and taste, and implementations require context and power. So what is 'new' about the 'new meat alternatives market'? The new market emerged around 2015 in the UK, when consumption of meat alternatives increased more than ever before. At the same time, an increasing acknowledgement and awareness of the negative health impacts of excessive meat consumption could be observed. In October 2015, an international working group affiliated to the World Health Organization (WHO) declared that processed meat is "carcinogenic to humans" and red meat "probably carcinogenic to humans", which was founded on decades of research inquiry (Bouvard, Loomis, Guyton, Grosse, El Ghissassi, Benbrahim-Tallaa, … Straif, 2015). While not the primary reasons for reducing meat, this is reinforced by increasing awareness of the ecological and ethical issues related to intensive animal farming, as well as repeated food scares related to meat. Finally, the attention and controversy that advancements in technology for non-livestock meats drew, might have played a role. In short, the 'new' market is most likely the result of consumers increasingly aligning their knowledge about meat consumption with their taste buds.

CONCLUSION

The concept of meat alternatives is not new or even recent to the UK. Neither are the ideas that such products are an opportunity in terms of environment, health, or business. In the past, land requirements and global supply chains were identified as an issue such that opportunities to substitute animal produce were most evident in situations of immediate or expected resource scarcity. Environmental opportunities today largely surround the strain of intensive animal agriculture on the planet, particularly regarding carbon emissions. Meat was intermittently acknowledged to be unsuitable to satisfy local or global demands without expanding beyond 'limits to growth'. Occasionally, health opportunities were similarly acknowledged throughout the history of meat alternatives, as the negative effects of a lack of protein had to be balanced with the suggested health impacts of meat consumption. Lastly, business opportunities were responsible for the technological development and for creating niche markets, which enabled the development of a new market in the first place. With a few exceptions, 'new protein foods' research in the 1960s and 1970s did not land directly in consumers' shopping baskets, although they similarly nurtured hope for a better world, and promised success and business opportunities. In contrast, the current market already has a share of 2–3% of the meat market in the UK, and in development activities consumers are the primary target as the optimistic trend suggests that meat-free is the new meat.

REFERENCES

A soya point. (1975, February 14). *The Guardian*, p. 20. Retrieved from the ProQuest database: https://www.proquest.com/libraries/academic/databases/

Altschul, A. (Ed.). (1974). *New protein foods (Vol. 1A Technology)*. New York, NY: Academic Press.

Beardsworth, A., & Bryman, A. (1999). Meat consumption and vegetarianism among young adults in the UK: An empirical study. *British Food Journal*, *101*(4), 289–300. doi:10.1108/00070709910272169

Beardsworth, A., & Bryman, A. (2004). Meat consumption and meat avoidance among young people: An 11-year longitudinal study. *British Food Journal*, *106*(4), 313–327. doi:10.1108/00070700410529573

Bouvard, V., Loomis, D., Guyton, K. Z., Grosse, Y., El Ghissassi, F., & Benbrahim-Tallaa, L., … Straif, K. (2015). Carcinogenicity of consumption of red and processed meat. *The Lancet Oncology*, *16*(16), 1599–1600. doi:10.1016/S1470-2045(15)00444-1

Brockes, E. (2000, December 1). Cheat meat. *The Guardian*, pp. A6–A7. Retrieved from the ProQuest database: https://www.proquest.com/libraries/academic/databases/

Buckingham, L. (1999, September 10). United Biscuits hit by GM fears. *The Guardian*, p. 27. Retrieved from the ProQuest database: https://www.proquest.com/libraries/academic/databases/

Chiles, R. M. (2013). If they come, we will build it: In vitro meat and the discursive struggle over future agrofood expectations. *Agriculture and Human Values*, *30*(4), 511–523. doi:10.100710460-013-9427-9

Collingham, L. (2012). *The taste of war: World War Two and the battle for food*. London, UK: Penguin.

Collins, R. (1975, February 6). Describe a spun protein chop. *The Guardian*, p. 1. Retrieved from the ProQuest database: https://www.proquest.com/libraries/academic/databases/

Department of Environment, Food, and Rural Affairs (DEFRA). (2017). *Family Food 2015*. London, UK: DEFRA. Retrieved February 24, 2018, from https://www.gov.uk/government/uploads/system/uploads/attachment_data/file/597667/Family_Food_2015-09mar17.pdf

Dibb, S., & Simkin, L. (2001). *The marketing casebook: Cases and concepts* (2nd ed.). London, UK: Thomson.

Euromonitor International. (2017a). *Free from Meat: Euromonitor from trade sources/national statistics* (Database). Euromonitor International. Retrieved from https://www.euromonitor.com/

Euromonitor International. (2017b). *Plant-based Protein: Assessing Demand for Sustainable Alternatives* (Briefing). Retrieved from https://www.euromonitor.com/plant-based-protein-assessing-demand-for-sustainable-alternatives/report

Finnigan, T. J. (2011). Mycoprotein: Origins, production and properties. In G. O. Phillips & P. A. Williams (Eds.), *Handbook of Food Proteins* (pp. 335–352). Cambridge, UK: Woodhead Publishing. doi:10.1533/9780857093639.335

Glotz, J. (2018, March 3). Alternative proteins stir up animal passions over meat. *The Grocer*, p. 16. Retrieved from the Factiva database: https://www.dowjones.com/products/factiva/

Gregory, J. (2007). *Of Victorians and vegetarians: The Vegetarian movement in nineteenth-century Britain*. London, UK: Tauris Academic Studies.

Horan, F. (1974). Meat analogs. In A. M. Altschul (Ed.), New protein foods (Vol. 1A Technology, pp. 366–413). New York, NY: Academic Press. doi:10.1016/B978-0-12-054801-9.50014-4

Johnson, L., Myers, D., & Burden, D. (1992a). Early uses of soy protein in Far East, U.S. *INFORM – International News on Fats, Oils and Related Materials*, *3*(3), 282–290.

Johnson, L., Myers, D., & Burden, D. (1992b). Soy protein's history, prospects in food, feed. *INFORM – International News on Fats, Oils and Related Materials*, *3*(4), 429–444.

Just Food. (2000, December 11). *Vegetarian and meat-free foods*. Retrieved April 10, 2018, from https://www.just-food.com/analysis/vegetarian-and-meat-free-foods_id94010.aspx

Kantar Worldpanel. (2016). *The meat reducer*. Custom Market Research.

Learmonth, E. (1963). The future of soya as human food in the United Kingdom. *Soybean Digest*, 14–16.

Leroy, F., & Praet, I. (2015). Meat traditions. The co-evolution of humans and meat. *Appetite*, *90*, 200–211. doi:10.1016/j.appet.2015.03.014 PMID:25794684

Linda McCartney's Foods. (2016). *Linda McCartney's – 25 Years* (Promotional Video). Retrieved July 21, 2017, from http://www.lindamccartneyfoods.co.uk/our-story/our-journey/

Mintel. (2004). *Meat-free foods – UK – December 2004* (Market Research). London, UK: Mintel Group Ltd.

Mintel. (2006a). *Attitudes towards vegetarianism – UK – September 2006* (Market Research). London, UK: Mintel Group Ltd.

Mintel. (2006b). *Meat-free foods – UK - December 2006* (Market Research). London, UK: Mintel Group Ltd.

Mintel. (2017). *Meat-free foods – UK – May 2017* (Market Research). London, UK: Mintel Group Ltd.

Nath, J. (2011). Gendered fare? A qualitative investigation of alternative food and masculinities. *Journal of Sociology (Melbourne, Vic.), 47*(3), 261–278. doi:10.1177/1440783310386828

O'Doherty Jensen, K., & Holm, L. (1999). Preferences, quantities and concerns: Socio-cultural perspectives on the gendered consumption of foods. *European Journal of Clinical Nutrition, 53*(5), 351–359. doi:10.1038j.ejcn.1600767 PMID:10369489

Office for National Statistics. (2016). *Families and households in the UK: 2016*. Retrieved September 10, 2018 from https://www.ons.gov.uk/releases/familiesandhouseholdsintheuk2016

Owens, J. (2016). *Vegetarian propaganda and the Great War* (Project Report, BA History). Lancaster, UK: Lancaster University. Retrieved September 10, 2018 from https://www.academia.edu/31737175/Vegetarian_propaganda_and_the_Great_War

Pearson, A. M. (1976). Meat extenders and substitutes. *Bioscience, 26*(4), 249–256. doi:10.2307/1297347

Raphael, A. (1969, October 27). Meatless meat—synthetic way to the US heart. *The Guardian*, p. 3. Retrieved from the ProQuest database https://www.proquest.com/libraries/academic/databases/

Rothgerber, H. (2013). Real men don't eat (vegetable) quiche: Masculinity and the justification of meat consumption. *Psychology of Men & Masculinity, 14*(4), 363–375. doi:10.1037/a0030379

Shprintzen, A. D. (2012). Looks like meat, smells like meat, tastes like meat: Battle creek, protose and the making of modern American vegetarianism. *Food, Culture, & Society, 15*(1), 113–128. doi:10.2752/175174412X13190510222066

Shprintzen, A. D. (2013). *The Vegetarian Crusade: The Rise of an American Reform Movement, 1817–1921*. Chapel Hill, NC: The University of North Carolina Press.

Shurtleff, W., & Aoyagi, A. (2011). *History of tempeh and tempeh products (1815-2011): Extensively annotated bibliography and sourcebook*. Lafayette, CA: Soyinfo Center.

Shurtleff, W., & Aoyagi, A. (2014). *History of meat alternatives (965 CE to 2014): Extensively annotated bibliography and sourcebook*. Lafayette, CA: Soyinfo Center. Retrieved from http://www.soyinfocenter.com/pdf/179/MAL.pdf

Shurtleff, W., & Aoyagi, A. (2015). *History of Soybeans and Soyfoods in the United Kingdom and Ireland (1613-2015). Includes England, Great Britain, Scotland and Wales*. Lafayette, CA: Soyinfo Center. Retrieved September 10, 2018 from http://www.soyinfocenter.com/pdf/183/UK.pdf

Smithers, R. (2018, June 10). Sainsbury's to sell vegan 'fake meats' next to real thing in trial. *The Guardian*. Retrieved June 24, 2018, from http://www.theguardian.com/lifeandstyle/2018/jun/11/sainsburys-to-sell-vegan-fake-meats-next-to-real-thing-in-trial

The Economist. (2015). Silicon Valley gets a taste for food. Green Food. *The Economist*. Retrieved from https://www.economist.com/technology-quarterly/2015/03/05/silicon-valley-gets-a-taste-for-food

The Grocer. (1993a, April 3). Mince Alternative to Beef up Veggie Sales - Haldane Foods. *The Grocer*, p. 33. Retrieved from the Factiva database: https://www.dowjones.com/products/factiva/

The Grocer. (1993b, August 21). McCartney Revamp Targets Bird's Eye's Pole Position. *The Grocer*, p. 35. Retrieved from the Factiva database: https://www.dowjones.com/products/factiva/

The Grocer. (1993c, June 12). Focus on Healthy Food - Tv Backs McCartney Ready Meals - Ross Young's. *The Grocer*, p. 82. Retrieved from the Factiva database: https://www.dowjones.com/products/factiva/

The Grocer. (1994, July 30). Focus on Vegetarian Foods - Lentil and Nut Roast Image Is History. *The Grocer*, p. 41. Retrieved from the Factiva database: https://www.dowjones.com/products/factiva/

The Grocer. (1995, July 29). Marketing - Marlow Foods. *The Grocer*, p. 41. Retrieved from the Factiva database: https://www.dowjones.com/products/factiva/

The Grocer. (1996, October 5). Mainstream Battle for Meat-Free Brands. *The Grocer*, p. 32. Retrieved from the Factiva database: https://www.dowjones.com/products/factiva/

The Grocer. (1999, March 13). The great pretenders. *The Grocer*, p. 47. Retrieved from the Factiva database: https://www.dowjones.com/products/factiva/

The Grocer. (2001, June 2). Steak-loving Americans will soon have a new word to add to t. *The Grocer*, p. 38. Retrieved from the Factiva database: https://www.dowjones.com/products/factiva/

The Grocer. (2003, April 12). Basking in mainstream approval. *The Grocer*, p. 47. Retrieved from the Factiva database: https://www.dowjones.com/products/factiva/

The Grocer. (2006, February 4). Meat-free not just for vegetarians. *The Grocer*, p. 47. Retrieved from the Factiva database: https://www.dowjones.com/products/factiva/

The Vegan Society. (2016, May 17). Find out how many vegans are in Great Britain. *The Vegan Society*. Retrieved March 8, 2018, from https://www.vegansociety.com/whats-new/news/find-out-how-many-vegans-are-great-britain

Trinci, A. P. (1992). Myco-protein: A twenty-year overnight success story. *Mycological Research*, *96*(1), 1–13. doi:10.1016/S0953-7562(09)80989-1

Ugalde, U. O., & Castrillo, J. I. (2002). Single cell proteins from fungi and yeasts. In G. G. Khachatourians & D. K. Arora (Eds.), *Applied mycology and biotechnology* (Vol. 2, pp. 123–149). Amsterdam: Elsevier.

Ziemba, J. V. (1966). Let soy proteins work wonders for you. *Food Engineering*, *38*(5), 82–84, 87–90, 93.

Zorpette, G. (2013). Muscling out meat. *IEEE Spectrum*, *50*(6), 64–70. doi:10.1109/MSPEC.2013.6521036

ADDITIONAL READING

Beardsworth, A., & Bryman, A. (2004). Meat consumption and meat avoidance among young people: An 11-year longitudinal study. *British Food Journal*, *106*(4), 313–327. doi:10.1108/00070700410529573

Bouvard, V., Loomis, D., Guyton, K. Z., Grosse, Y., El Ghissassi, F., & Benbrahim-Tallaa, L. (1599–1600). … Straif, K. (2015). Carcinogenicity of consumption of red and processed meat. *The Lancet Oncology*, *16*(16), 1599–1600. doi:10.1016/S1470-2045(15)00444-1

Collingham, L. (2012). *The taste of war: World War Two and the battle for food*. London, UK: Penguin.

Gregory, J. (2007). *Of Victorians and vegetarians: The Vegetarian movement in nineteenth-century Britain*. London, UK: Tauris Academic Studies.

Linda McCartney's Foods. (2016). Linda McCartney's – 25 Years (Promotional Video). Retrieved July 21, 2017, from http://www.lindamccartneyfoods.co.uk/our-story/our-journey/

Schmidinger, K., Bogueva, D., & Marinova, D. (2018). New meat without livestock. In D. Bogueva, D. Marinova, & T. Raphaely (Eds.), *Handbook of research on social marketing and its influence on animal origin food product consumption* (pp. 344–361). Hershey, PA: IGI Global. doi:10.4018/978-1-5225-4757-0.ch023

Shurtleff, W., & Aoyagi, A. (2014). *History of meat alternatives (965 CE to 2014): Extensively annotated bibliography and sourcebook*. Lafayette, CA: Soyinfo Center; Retrieved from http://www.soyinfocenter.com/pdf/179/MAL.pdf

Shurtleff, W., & Aoyagi, A. (2015). *History of Soybeans and Soyfoods in the United Kingdom and Ireland (1613-2015). Includes England, Great Britain, Scotland and Wales*. Lafayette, CA: Soyinfo Center. Retrieved September 10, 2018 from http://www.soyinfocenter.com/pdf/183/UK.pdf

Trinci, A. P. (1992). Myco-protein: A twenty-year overnight success story. *Mycological Research*, *96*(1), 1–13. doi:10.1016/S0953-7562(09)80989-1

Zorpette, G. (2013). Muscling out meat. *IEEE Spectrum*, *50*(6), 64–70. doi:10.1109/MSPEC.2013.6521036

KEY TERMS AND DEFINITIONS

High Moisture Meat Analogs (HMMA): Meat alternatives made using high moisture extrusion. Only recently are HMMA products coming to market.

Meat Alternative: Meat analogues or meat substitutes which are plant-based, lab-grown, or use ingredients other than livestock, such as insects.

Meat Reducer: A person who aims to reduce their meat consumption to a limit considered healthy or because of environmental and animal welfare reasons; alternative term for flexitarian.

Mycoprotein: A single-cell protein made of fungus. The strand *fusarium venenatum* is approved as safe for human consumption and used in products of the British brand Quorn.

Protein: An organic substance – polymer chains of amino acids, considered an essential nutrient for the human body; there are 20 types of amino acids representing the building blocks for the human proteins – 11 are non-essential which can be synthesized by the human organism and 9 are essential which need to be provided by food.

Seitan: The Japanese and Western name for cooked wheat gluten; it usually has a fibrous texture, is and can be used as a meat alternative.

Single-Cell Protein (SCP): Refers to microbes converting various raw materials such as oil or paper into proteinaceous biomass mostly used as animal feed.

Tempeh: A soy product traditional for Indonesia. It is the result from a fermentation process which binds soybeans into a cake form.

Textured Vegetable Protein (TVP): Trademark colloquial term for any kind of unflavored, dehydrated, textured product from plant protein.

Tofu: Soy bean curd. Often used as a meat alternative. It is very popular in Asian countries, such as China and Japan, and more recently has started to also be in used a Western type of diet.

ENDNOTES

1 Mintel, whose data is based on Information Resources, Inc. (IRI), estimated a market of £ 559 million for meat free foods; without the categories 'ready meals' and 'pastry-based products' this value is reduced to £ 272 million which is identical to the value estimated by Euromonitor, further excluding the category 'snacks' leads to a value of £ 196 million.

2 Given 27.1 million households in 2016 (Office for National Statistics, 2016, p. 13), market estimates from Kantar Worldpanel (2016) and Euromonitor International (2017a) lead to £ 6.42 and £ 10.04 per household and year respectively.

3 Previous research has shown that historically, culturally, and socially meat consumption tends to be linked to masculinity (e.g., Leroy & Praet, 2015; O'Doherty Jensen & Holm, 1999; Rothgerber, 2013). As such, meat consumption can be understood as a particularly masculine condition that exhibits stronger barriers to change (e.g., Nath, 2011).

Chapter 12
Market for Plant-Based Meat Alternatives

Anusha Thakur
University of Petroleum and Energy Studies, India

ABSTRACT

Shifting consumer preferences towards meat alternatives can be attributed to the factors such as health and ecological benefits, as well as meat adulteration. Increasing consumer demand for better grade of meat alternatives is also expected to boost the market growth in the near future. Protein sources from maize, peas, rice, and chickpeas are anticipated to witness significant growth and new developments. Alternatives such as bean curd or wheat gluten are expected to be the beneficial source of protein and phosphorous. This can be attributed to the fact that 50 g of vital wheat gluten in combination with water produces 2 ounces of gluten in a solid form, which further comprises of nearly 38 g of protein in each serving. However, factors such as the higher cost of meat substitutes inhibit the market growth, particularly in developing economies, wherein the dietary awareness is expected to be lower. Further factors, related to gluten intolerance and soy allergy, are also anticipated to restrain the market growth. This chapter includes a market study of meat alternatives across the world based on analyzing, estimating, and forecasting for the 2015-2025 period. Market determinants of the meat alternatives market are also explored to analyze market drivers, restraints, challenges, opportunities, trends, and developments. The competitive landscape section includes information related to key market players with an overview of product portfolio and strategic initiatives.

INTRODUCTION

Traditional vegetarian and vegan options as well as lab-grown or cultured meat and insect-based food products as discussed in other sections of this book are also part of the emerging new markets for alternatives to livestock meat. This chapter however deals only with plant-based meat alternatives. These meat alternatives, also known as meat substitutes, meat analogues or mock meat, comprise of plant-based proteins, with nearly the same flavors, as well as aesthetic appearance, to resemble beef, poultry or other meat products.

DOI: 10.4018/978-1-5225-7350-0.ch012

In general, a meat based diet uses a substantially larger amount of environmental resources per calorie in comparison to a grain based diet – it takes around 2 kg to 15 kg of plant food products to produce 1 kg of meat (Joshi, 2015). Meat alternative products include textured vegetable protein (TVP), gluten-free vegan meat, seitan, the use of lentils, tempeh and tofu, with some of these foods having been used for a long time. For example, tempeh and tofu have been traditionally used in the Asian countries and are made from soybeans, which are rich in calcium and proteins. Soy based protein meat substitutes are preferred more on account of the facts that they are the cheapest source and also that they have protein digestibility corrected amino acid score equivalent to animal proteins (Hoffman, 2004). Furthermore, lentils and beans are very low in fat content while containing significant amounts of vitamins, minerals, fiber and protein, unlike the meat products. Seitan or wheat gluten has a higher protein content and is very much similar to meat in texture and taste. Such plant-based meat alternatives have low cholesterol as well as low saturated fat content, and are rich sources of proteins, minerals and vitamins.

Consumption of meat products by the group of "flexitarians" or people who are deliberately reducing their meat intake (Dagevos & Voordouw, 2013; Raphaely & Marinova, 2014; Raphaely, Marinova, Crisp, & Panayotov, 2013), is gradually decreasing (Euromonitor International, 2011; Hosie, 2017). This change among the consumers requires equivalent tasty and healthier products which act as substitute to meat products. With the growing health awareness among consumers and their shifting preferences towards plant based proteins, the plant-based meat alternatives are expected to witness surging demand over the coming years (Kenward, 2017). Moreover, factors such as growing demand for religious and wholesome foods, coupled with the rising number of animal diseases such as swine flu and bird flu are expected to increase the consumption of vegetable proteins, and in turn, bolster the market interest in meat alternatives. Increasing consciousness among the consumers for the health benefits of the vegan diet and plant-based protein products is expected to favorably impact the market demand. Continuing consumption of meat products is expected to reinforce concerns related to environmental risks, and animal welfare, which thereby increases the market for substitutes.

Furthermore, there are the health concerns associated with excessive consumption of meat leading to non-communicable diseases, including cardiovascular disease, type 2 diabetes, and certain types of cancers. Unease about cholesterol levels and fat related problems also contributes to the boost in consumer demand towards vegan products. According to the German Nutrition Society, the consumption of meat should be reduced to 2 to 3 servings a week, which is approximately 300 g to 600 g of meat a week (Deutsche Gesellschaft für Ernährung, 2018).

The global meat alternatives market can be segmented on the basis of raw material, product and region. On the basis of raw materials, the meat substitutes can be segregated into wheat-based, mycoprotein, and soy-based products. In relation to soy-based foods product segmentation comprises of tempeh, textured vegetable protein (TVP), tofu, and other soy products. Regional segmentation includes North America, Europe, Asia-Pacific and the rest of the world covering countries from the Middle East, Africa and Latin America.

Investments in research and development are highly important for the players in the global meat alternatives market. Key market players are engaged in the implementation of different entry strategies, such as collaborations, mergers and acquisitions, as well as product portfolio expansion. Further, the market players also emphasize the launch of different plant-based products which are replicas of meat products in taste and smell and also have health benefits. The products of leading companies, such as Beyond Meat (Migala, 2018), offer different health benefits to consumers in comparison to the animal-based beef products. For example, the plant-based food products are significantly richer in protein, they are cholesterol free and their iron content is nearly 25% higher in comparison to the animal-based beef products which are higher in cholesterol and lesser in iron content. Other key market players in the global meat alternatives market include Quorn Foods (Marlow Foods), Morningstar Farms (a division of the Kellogg company), MGP Ingredients Inc., Blue Chip Group (dry food manufacturers), Meatless B.V., Garden Protein International Inc., Vbites Foods Ltd., Amy's Kitchen Inc., Cauldron Foods and Schouten Europe.

The remainder of this chapter first provides a literature review-based description of the various meat alternatives, including market drivers, restrains, opportunities, regulatory and competitive environment. This is followed by a market research study of the factors that are likely to impact (both, positively and negatively) the demand for meat alternatives in the time period up to 2025.

MEAT ALTERNATIVES

Meat is the richest source of protein and also has other nutritional content. Research studies however show that animal agriculture uses disproportionately high amounts of environmental resources – nearly 25% of all freshwater on this planet and nearly 30% of all land on earth (Cowspiracy, 2014). It also leads to significant greenhouse gas emissions and is impacting climate change (FAO, 2016). Meat consumption in the developed regions of the world exceeds the recommended dietary limits (FAO, 2018). For instance, in the European countries, the consumption is 66.2 kg per capita per year in the year 2010 to 2012 (Walker, 2005) which is above the healthy limit of about 500 g per week. With the increasing awareness about negative health and environmental consequences of such meat-rich diets, over the time consumers are expected to shift their preferences towards plant-based food options and look to replace meat. The types of meat alternatives are illustrated below and include soy as well as other plant-based options.

Soy Alternatives

Soy alternatives generally comprise of soy protein and can include wheat gluten, dairy and spices. These products also contain fatty acids, isoflavones, omega 3s, and essential amino acids which are required for the growth of an individual. According to research studies, soy based proteins have the capacity to lower the "bad" low-density lipoprotein (LDL) levels by approximately 4% (Messina, 2016). Soy alternatives further include tempeh and textured soy protein.

Tempeh

Tempeh is a fermented soybean cake derived from cooked grains and soybeans, along with Rhizopus oligoporus culture. This is mainly a soy-derived meat alternative (Malav, 2013).

Textured Soy Protein (TVP)

The textured soy protein consists of nearly 50% protein which is made from soy concentrate and soy flavor. It is also a versatile product. When re-hydrated, TVP becomes very much similar to meat products such as poultry or beef. An additional advantage is that the TVP products also have low sodium content.

Plant Based Alternatives

These products are very similar to the meat products in texture, taste and nutritional values, and are basically derived from plants. They can be in traditional as well non-traditional forms. The non-traditional forms include products which are processed and developed with the help of advancements in science and technology, such as mycoprotein, gluten and conglycinin whereas traditional forms include seitan and tofu. These products provide several key nutrients, including zinc, iron, and protein, along with Vitamin B (niacin, choline, riboflavin, thiamin, cobalamin, and pyridoxine) (Committee on the Scientific Evaluation of Dietary Reference Intakes, 1998).

Few of the meatless products are based on very old recipes of mushrooms, wheat gluten, tempeh, legumes or tofu, to which further additional ingredients are added in order to make them taste like chicken, ham, sausage, beef or generally meat. These products have the ability to retain water and moisture at the time of freezing, thawing, and cooking, which makes them better to be used. Table 1 illustrates the types of protein and their respective sources, such as legumes, wheat or oil seeds.

The plant based protein products which act as substitutes for the meat products are also aimed at the group of flexitarian consumers, who opt for them as one option to reduce their meat consumption. Flexitarians prefer the consumption of meatless products, thereby decreasing their overall meat intake. Several companies, such as Impossible Burger and Beyond Meat, are engaged in innovative research and technologies which can replicate the flavor and feel of the meat products, hence giving the customers a similar experience.

Figure 1 illustrates the annual demand for animal as well as plant-based proteins for the period from 2000 to 2030. The worldwide demand for plant-based proteins in 2010 was estimated at nearly 120 million tonnes/year and is expected to reach nearly 145 million tonnes/year by 2020. Furthermore, the annual demand is expected to be valued at USD 7.5 billion by 2025, with a compound annual growth rate (CAGR) of nearly 7.7% (Prasannan, 2018). This shows that the meat substitutes are anticipated to account for nearly 33% of the total protein market over the forecast period.

In North America, the manufacturers focus more on investing on meatless food products, to thus cope up with the changing consumer needs and requirements. Europe accounted for nearly 39% of the

Table 1. Major non-meat sources which are suitable as meat analogues

Type of Protein	Sources
β-conglycinin	Soybean
• Glycinin, • Vicilin	Legumes
• Legumin, • Albumins, • Globulins • Glutelins	Oil seeds
Gluten • Gliadins • Glutenins	Wheat, rye and barley
Mycoprotein	Fusarium venenatum (Filamentous fungus)

Source of data: (Joshi, 2015)

Figure 1. Annual worldwide demand for plant-based and animal proteins (millions of tonnes/year)
Source: (Thomas, 2015, BIPE based on FAO data)

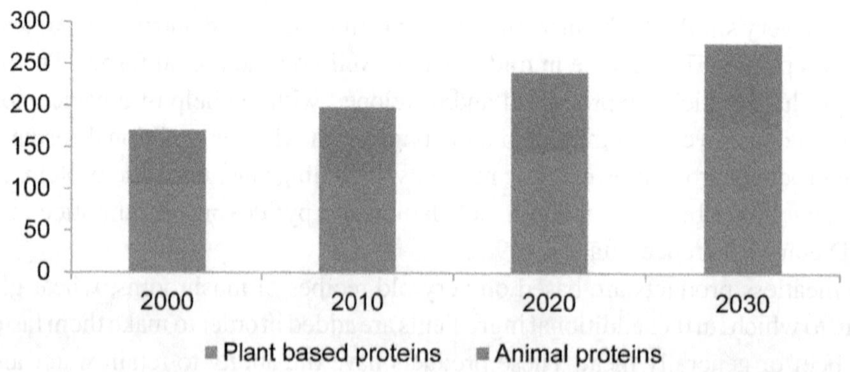

overall market for meat substitute (Prasannan, 2018). The total number of consumers with a completely vegetarian diet in Europe was estimated at approximately 24 million, with the Eurozone witnessing a CAGR of nearly 6.6% over the period from 2011 to 2016, while the growth in the other countries of the European Union was at 2.4% (Halliday, 2018). However, the biggest impact on the demand for meat alternatives is from the flexitarians who are considerably reducing their meat consumption and creating a new market. In UK, Germany, France, Spain and Italy the market for meat alternatives is nearing maturity with an abundance of meat-free food products (Halliday, 2018).

Growing consumer demand for healthy food, rising meat prices, increasing numbers of vegetarians and changing food patterns among the consumers are expected to continue to boost the meat analogue market demand in the future. Various research and development programs combined with marketing initiatives are being taken by the different market players for developing cheap as well as superior substitute products. The market for meat alternatives is expected to witness a surging demand on account of the increasing preferences for vegan options coupled with rising health consciousness among consumers. These alternatives or substitutes resemble the meat products in appearance, flavor and texture and are made up of plant-based ingredients such as tofu, mushrooms, tempeh, and soy.

A contributing factor for the growth of the global meat alternatives market are the health benefits offered by these less cholesterol and high protein content products. The consumption of plant protein helps in treating menopausal problems, diabetes, weight related problems, and polycystic ovary syndrome. In addition to growing health awareness, the increasing demand for vegan food can also be attributed to factors such as religious and ethical views, environmental rights and sustainability.

MARKET DETERMINANTS AND OPPORTUNITIES

This section of the chapter examines different market determinants which illustrate the factors that propel market growth and demand, restraining factors, and various barriers for entry to the meat alternatives market. It also presents the various competitive market players which are engaged in the production or supply of the meat substitute products.

Market Drivers

Three factors are at play in stimulating the market demand for meat alternatives. They are increasing health awareness amongst consumers, growing environmental concerns and risks of disease outbreaks.

Growing Health Awareness Among the Consumers

The most common protein foods consumed in the United States include various animal and meat products which are relatively higher in fat content, in comparison to the plant-based proteins in soy, lentils, and beans (Johns Hopkins Bloomberg School of Public Health, n.d.). Excessive consumption of processed or red meat for a long duration is anticipated to lead to health problems such as risk of colorectal cancer, type 2 diabetes, cardiovascular diseases and increased total mortality. Furthermore, factors such as high energy density, high cholesterol and saturated fat, as well as the presence of various carcinogenic compounds found in the different processed meat products, are expected to pose health risks in their consumption, thereby increasing the demand for meat substitutes in the future.

Increasing Environmental Concerns

Current and future rising consumption of meat products is expected to have negative impact on the planet's climatic as well as environmental conditions. Areas of concern include (Johns Hopkins Bloomberg School of Public Health, n.d.):

- **Animal Waste:** Waste from the Industrial Food Animal Production (IFAP) operations is expected to lead to water pollution, which thereby affects the aquatic and marine life. The manure spill from various swine operations causes outbreak of different toxic microorganisms, affecting the lives of fishes. Additionally, animal wastes are also engaged in releasing more potent greenhouse gases, such as nitrous oxide and methane.
- **Land Usage:** Produced protein and calories from the land are significantly decreased when used in feed crops and then converted into animal products for human consumption.
- **Climatic Change:** Animal agriculture is expected to generate substantial amount of greenhouse gases which affect the climate, thereby leading to extreme weather conditions such as floods, or droughts. The processing, production as well as distribution and retailing of different animal products account for approximately 15 percent of the global greenhouse gas emissions and 9 percent in the USA.

Outbreak of Various Animal Diseases

Nowadays, increasing incidents of disease outbreaks in animals and the potential to spread to people are another factor expected to impact the growth of the meat analogue market. These outbreaks further affect the production capacity for meat products, and their subsequent supply chain processes. In fact, the demand for meat products can also be negatively affected (Pritchett, Thilmany, & Johnson, 2005).

Market Restraints

Several factors such as gluten intolerance as well as soy allergy inhibit the growth of the market for meat alternatives. The phytic acid present in the legume seeds helps in the reduction of the bioavailability of the different essential minerals contained in the plant-based alternatives, by the formation of salts which are excreted (Deshpande, 1984). This results in micronutrient deficiency and mineral depletion in the human body. Gluten is widely available in the cereals (such as barley, rye and wheat) and is also responsible for disorders such as Celiac disease, caused by the consumption of gluten (Sadler, 2004). In this disease, the epithelium of the small intestine gets damaged and this also leads to several other problems, such as early start of osteopenia or osteoporosis, anemia, pancreatic insufficiency, gall bladder malfunction, lactose intolerance and vitamin deficiencies. Such health risk concerns may negatively impact the growth of the market for alternative proteins.

Market Challenges

The list of barriers to entry which need to be overcome as part of the market challenges for these new products includes:

1. Culture is a key factor which might also affect the growth of the market for meat alternatives;
2. The composition of the meat alternatives is expected to pose a challenge to market entry;
3. Low dietary awareness in some sections within society also inhibits the growth of the meat substitutes market. Such low awareness might also have negative health implications;
4. Texture and taste are major hindering factors to producing meatless food products with exactly the same taste as well as texture of that of the meat products.

Market Opportunities

With the growing health consciousness among consumers, people are expected to reduce the consumption of meat products. However, it would be difficult for the customers to find the same taste and texture in the plant-based products. This in turn would provide opportunities for meat substitute market players to develop as well as expand their market for new meat products, so that they can win over the consumers' tastes and feeling of texture, with the help of the meat substitute products. This factor also acts as an opportunity for the meat alternatives market.

Regulatory Scenario

The legislative and regulatory environment can facilitate or slow down this new market. Below are a few examples:

- An amendment was passed to the French Agricultural Bill which prohibits the products based on non-animal ingredients to be labeled with names related to animal-based product, such as bacon flavored strips, soy sausages, vegetable steaks and even "soy yogurt" and "vegan cheese" (Rödl,

2018). Although similar debates have been held in other countries, research evidence shows that despite the opposition from the meat industry, people will continue to consume such meat-free products (Rödl, 2018).

- The Chinese government announced dietary guidelines for meat consumption by the year 2030, which were recommended by the country's Ministry of Health. China currently accounts for nearly 30% of the global meat consumption (Myers, 2016). The recommendations of these guidelines include consumption of around 40–75 g of meat per day. They are expected to alleviate pressure on the water and land resources which is expected to get tense if the consumption of meat products continues to increase. The dietary restrictions are likely to promote interest in traditional and novel meat substitutes.
- The ingredients which are permitted to be used in meat substitutes are required to be applied at a minimum level, wherein, they can be used to improve the syneresis and texture in order to produce characteristics similar to the original meat product. Such regulations can potentially stimulate further research and product development to satisfy the taste preferences of customers.

COMPETITIVE LANDSCAPE

According to the World Wide Fund for Nature (WWF, 2012), livestock production dominates the agricultural sector and is responsible for approximately 70% of the greenhouse gas emissions this industry generates. The United States Department of Agriculture (USDA, 2010; Hyden, 2015) estimates that on a global scale, agriculture is responsible for 30% of all greenhouse gas emissions, including those due to deforestation and land clearing. This makes the livestock sector responsible for 21% of the global greenhouse gas emissions. To mitigate environmental impact and respond to increasing health concerns, various plant based meat-free products were developed. Many players are announcing the introduction of novel innovative meat substitutes which are similar to the meat products not only in texture and taste but also in nutrient content. Below are a few illustrations.

Company 1: Schouten Europe (www.schouteneurope.com)

Company Overview

Schouten Europe is engaged in the production, development and packaging of meat substitutes. It is a Dutch company founded in 1990. The company develops meat substitutes based on soya proteins which can either be frozen or fresh. Schouten Europe works along with the brand manufacturers of meals, salads and snacks, European retailers, and global fast food chains. The company is a member of the Round Table on Responsible Soy (RTRS), which is a global platform wherein the different processing industries, soy trade, social organizations and soy manufacturers work together in collaboration for the sustainable soy production. The company's products are all manufactured at its own specialized production locations and are distributed across nearly 35 countries. Additionally, Schouten Europe has its production facilities at nearly 10 locations in India and Europe.

Product Portfolio

- **Fresh Mince:** This unique vegetarian mince can be used in sausages, bites and hamburgers. Its ingredients are:
    - 21% vegetable proteins (wheat gluten, soy)
    - Starch (wheat, corn, potato)
    - Water, sunflower oil and flavorings
    - Beet root powder
    - Hydrolyzed vegetable protein (soy), spices, Vitamin B12, and salt
    - Egg white.

It can be used for soups and bake-offs.

- **Stir-Fry Strips:** The stir-fry strip pieces consist of threaded texture which is very similar to chicken meat. They are pre-cooked and can be used directly into any existing recipes and are also available in different formats. Their ingredients are:
    - 14% of vegetable proteins (wheat gluten, soy)
    - Starch (wheat, potato, and corn)
    - Water, egg white, and citric acid
    - Sunflower oil, dextrose, spices and herbs, and milk protein (lactose)
    - Ferrous fumarate.

The stir-fry strips are suitable for various applications such as pizzas, main course and other salads, wraps, bake-offs or meals.

Company 2: Beyond Meat (beyondmeat.com)

Company Overview

Beyond Meat is headquartered in the US and was founded in 2009. It is the producer of plant based substitutes which are very much similar to the meat products in terms of texture, and flavor. The company is engaged in offering single serve meals, burgers, beef-free crumbles, beast burgers, and beyond chicken strips. Beyond Meat offers its products through the frozen sections of different stores as well as whole food markets in different regions of the US.

Product Portfolio

Some of the meat substitute products offered by Beyond Meat are illustrated below.

- **Beyond Burger:** Beyond Meat announced the launch of its first plant-based burger, which consists of nearly 20 g of proteins, no gluten or soy and no genetically modified organisms (GMO). Table 2 presents the nutritional data about the burger.

Table 2. Nutrition facts about Beyond Burger

Content	Weight (in g/mg)	Daily Value (in Percent)
Total fat	20 g	31%
a) Saturated fat	5 g	25%
Cholesterol	0 g	0%
Sodium	380 g	16%
Total carbohydrate	5 g	2%
a) Dietary fiber	3 g	13%
b) Sugars	0 g	0%
Protein	20 g	32%

Source of data: beyondmeat.com

- **Beyond Sausage:** The product Beyond Sausage consists of three varieties:
    - Beyond Sausage Brat Original
    - Beyond Sausage Hot Italian
    - Beyond Sausage Sweet Italian.

The nutritional information about the product is presented in Table 3.

- **Beast Burgers 2.0:** These burgers are of two varieties:
    - Beast Burger 2.0
    - Beastly Sliders 2.0

These burgers are an excellent source of B6, iron, vitamin D, B12, and antioxidant vitamins, E, C and A per serving. See Table 4 for their nutritional information.

Table 3. Nutrition facts about Beyond Sausages

Content	Weight (in g/mg)	Daily Value (in Percent)
Total Fat	12 g	18%
a) Saturated fat	5 g	25%
b) Trans fat	0 g	0%
Cholesterol	0 mg	0%
Potassium	230 mg	7%
Total carbohydrate	5 g	2%
a) Dietary fiber	3 g	12%
b) Sugars	0 g	0%
Protein	16 g	25%

Source: beyondmeat.com

Table 4. Nutrition facts about Beyond Beast Burger

Content	Weight (in g/mg)	Daily Value (in percent)
Total Fat	20 g	33%
a) Saturated fat	5 g	25%
b) Trans fat	0 g	
c) Polyunsaturated fat	4 g	
d) Monounsaturated fat	8 g	
Cholesterol	0 mg	0%
Sodium	430 mg	18%
Potassium	360 mg	10%
Total carbohydrate	7 g	2%
a) Dietary fiber	4 g	15%
b) Sugars	0 g	0%
Protein	23 g	37%

Source: beyondmeat.com

Company 3: Impossible Foods (http://impossiblefoods.com)

Company Overview

Impossible Foods – a start-up organization, is headquartered in Redwood City, California. The company develops plant-based dairy and meat products, which are made without animal protein. Impossible Foods is engaged in researching the animal products at a molecular level, selection of the plant based alternative products and then recreating the experience of meat products. The company announced the launch of its first meat analogue product, which has less fat content, more protein content and emits nearly less than 87% of the greenhouse gases.

The company uses a key product known as "heme", which shows the behavioral pattern of the meat products. Heme can be defined as the iron containing molecule in the blood which carries oxygen. Nitrogen fixing plants also contain this product, which is identical to that found in the meat products. Impossible Foods is engaged in utilizing the heme protein with the help of selection of leghemoglobin, found in the soy plants. The company offers a chemical library of various fat and protein contents which are derived from the plants, and conducts further experimentation as well as processing techniques to form a replica of the meat based products.

Product Portfolio

- **Impossible Burger:** The company offers a burger, which uses approximately 87% less greenhouse gas emissions, 74% less water, and 95% less land. Its content is made from coconut oil and potatoes as well as wheat products.

Below is a study which identifies the factors expected to impact on the meat alternatives market both positively as well as negatively in the period until 2025. It takes a global outlook and makes recommendations for strengthening this important economic sector.

RESEARCH METHODOLOGY

The key research objectives of the analysis are:

- To identify recent developments and technological advancements in the market.
- To highlight future prospects and challenges in the meat alternatives market.
- To identify the strategies, and services which are adopted by the different market players.

Covered aspects include market size and estimates of the meat alternatives market in terms of revenue over the forecast period; regulatory and technological scenario as well as regional segmentation covering North America, Europe, Asia Pacific, and the Rest of the World (RoW). This helps determine the size of the market in terms of revenue, understand the factors that favorably impact the market growth and the factors which hinder its development. Identifying the different competitors for the meat analogue market helps to get an idea about the strategies and the products offered by these market players to sustain in the market.

The information is derived from secondary sources collected and gathered through various company annual reports, investor documents, and official statistics published by the various organizations. It is analysed in view of the following factors:

- Market trends and developments.
- Demand and supply estimates.
- Market penetration rates and future opportunities which help understand the regional expansion as well as the product commercialization.
- The analysis is carried out on the basis of the historical market data and their respective trends.

The data is segmented on the basis of different parameters such as product, application or region-wise. Two types of methodologies are used to analyse the collected data:

1. **Top-Down Approach:** The data are collected for the global front and are then segregated into different entities (which include product/region/application)
2. **Bottom-Up Approach:** The data are collected for the different regional entities and are then integrated to obtain the global estimate.

The data are then forecasted on the basis of different market initiatives and trends for over the forecast period of 2015 to 2025. This helps understand as well as gain holistic information about the current market scenario and future trends.

ANALYSIS

The global meat alternatives market is expected to witness substantial growth over the forecast period. This can be attributed to various factors shifting the preferences of the customers towards vegan options. Furthermore, factors such as sustainability, health concerns, animal welfare and traceability are anticipated to additionally change consumer preferences, which in turn, will boost meat analogue demand. Also, the sedentary lifestyle of the consumers plays an essential role in changing their perceptions and choices, thereby favorably impacting demand.

Overview

The market was valued at USD 3.51 billion in 2015, and is expected to reach USD 7.25 billion in 2025, with a CAGR of 7.52%, during the forecast period. Table 5 and Figure 2 illustrate the revenue of the overall market.

Segment Analysis

The segmentation of the market according to raw materials and regions is presented below. The analysis allows for some insights to be drawn.

Raw Material Segment Analysis

Soy accounted for nearly 63% of the total market share (in terms of revenue) in 2017, and dominates the market (see Figure 3). Products based on soy are predominantly found in the regions of North America and South America, thereby favorably impacting the market demand. Theyare rich in protein, low satu-

Table 5. Global meat alternatives market, 2015–2025 (in US$ billion)

2015	2016	2017	2018	2019	2020	2021	2022	2023	2024	2025	CAGR
3.51	3.74	4.00	4.28	4.59	4.93	5.31	5.72	6.18	6.68	7.25	7.52%

Note: CAGR – compounded aggregate growth rate

Figure 2. Global meat alternatives market, 2015–2025 (in US$ billion)

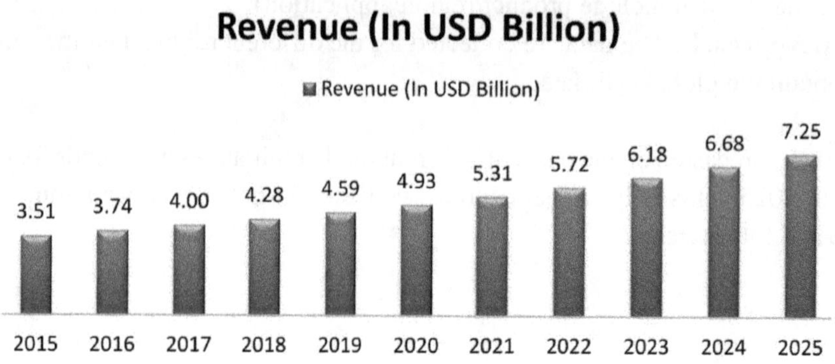

Figure 3. Meat alternatives market by raw material outlook, 2017 (market shares)

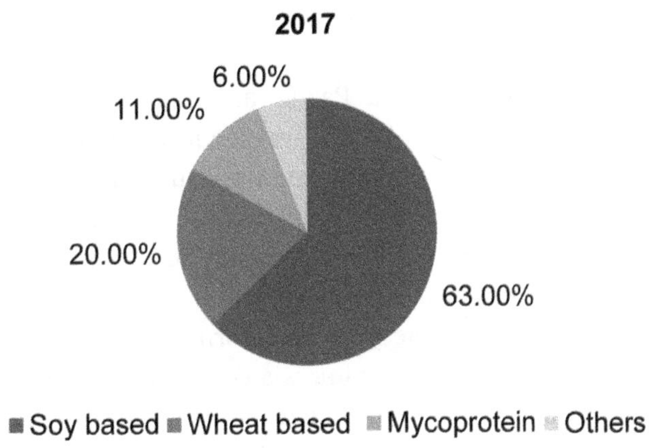

rated fat and are cholesterol free. Moreover, soy has been an individual staple product with a wide range of applications. Major forms of soy protein include textured soy protein (TSP) and tempeh. The wheat based alternatives account for a smaller share in the overall market, which was nearly 20% in 2017.

Regional Segment Analysis

In 2017, Europe accounted for the main share of nearly 42% of the overall market (see Figure 4). Growing demand for the highly nutritious as well as convenient food products is expected to propel market demand. North America accounted for approximately 32% of the total share, thereby witnessing a steady growth along the European region. The United States' market was valued at USD 888.2 million in 2015 and is expected to reach USD 1,085.20 million by 2020. This can be attributed to the negative impacts of meat products and animal wastes on the environmental as well as on climatic conditions. Furthermore, these products also lead to negative impacts on marine life, oceans as well as farmlands, and increase in the greenhouse gas emissions. Additionally, animal rearing is expected to have an impact on natural resources, thereby being one of the reasons for global warming as well. The growing number of consumers with vegetarian or vegan preferences is also expected to pose opportunities for the meat substitutes market. In Germany, the market players' focus is on the diversification of the plant based meat products

Figure 4. Meat alternatives market by regional outlook, 2017 (market shares)

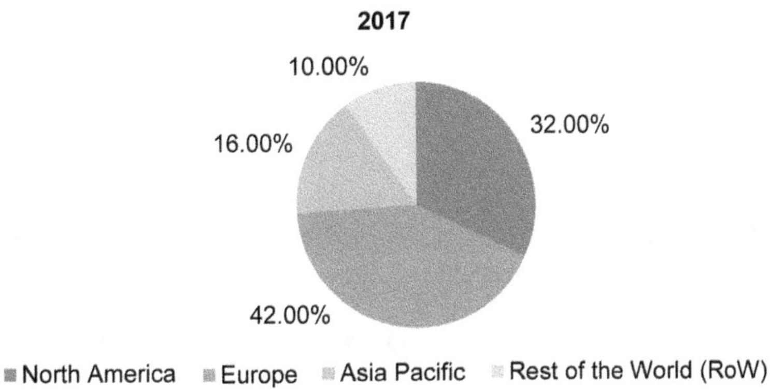

and also making them customer centric. The meat substitutes market in other European countries, such as Italy, Sweden and France, is also expected to become most lucrative during the forecast period.

Factors such as increasing number of vegetarians, vegans and flexitarians also lead to rising spending power for more nutritious food items. The Asia–Pacific accounted for a comparatively smaller share; however, the region witnesses a growth rate of 8.5% over the forecast period. The graph in Figure 5 illustrates the market size of the five major regions of the meat substitutes market.

China was valued at USD 215.1 million in 2015 and is expected to reach USD 351 million by 2020. This can be attributed to the increasing disposable income of the consumers and their changing preferences towards nutritious vegetarian options.

According to the research study conducted by the Informa's Agribusiness Intelligence, the meat alternatives market in the UK is anticipated to witness a considerable growth by nearly 21% by 2021 (News Desk, 2017). This can probably also be attributed to the increasing demand for in-vitro or cellular meat products. The German market was valued at USD 278.7 million in 2015 and is expected to reach USD 383.6 million by 2020. The customers' changing lifestyle and habits coupled with the increasing flexitarian diet propel the meat substitute consumption.

SUGGESTIONS

With the decrease in the consumption of red meat, the amount of carbon emissions can also be reduced. Changing preferences for meat substitute products such as tempeh, tofu, and seitan based foods are gaining significance. The customers are expected to prioritize the consumption of less meat based products in order to reduce the risks of health problems such as obesity, cardiovascular diseases, and diabetes. Furthermore, the impact meat production has on the environmental and climatic conditions is also anticipated to reduce the consumption of red meat among consumers. The factors which are expected to impact the consumption include the following:

- Increasing awareness among the consumers in developed and developing regions for the benefits of vegetarian and vegan options or plant-based food products is expected to favorably impact market demand.

Figure 5. Meat alternatives market by regional outlook, market size (in US$ million)

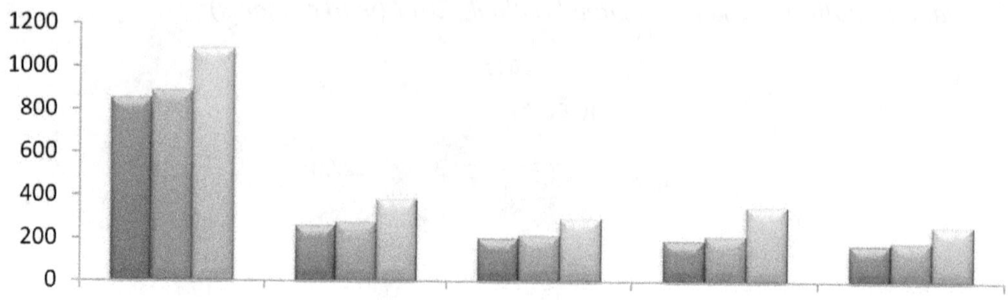

- Additionally, awareness about the benefits of plant-based proteins in all regions across the globe can be improved with the help of health campaigns and programs, which would thereby help them to change their preferences towards vegan products, and adopt a healthy lifestyle.

LIMITATIONS

Performing primary validation of the findings was a major constraint. At this stage, it was difficult to conduct primary interviews to gather market related information from the major market players across the world. However, this is something that can be pursued in the future as the market expands and there is a higher level of familiarity with its products.

CONCLUSION

The global meat analogue market is anticipated to witness substantial growth over the forecast period until 2025, owing to the increasing health consciousness among consumers. Additionally, shifting consumer preferences towards vegetarian and vegan options are expected to bolster the consumption of meat substitutes, thereby favorably impacting the growth of this market across the world. The growing demand for vegan or plant based nutritious food products which are similar to the meat products in their texture, and nutrition content (such as zinc, iron, Vitamin B, and protein) impacts the market demand.

Rising living standards as well as higher disposable incomes are also expected to favorably impact the market, particularly in the Asia–Pacific region. Furthermore, factors such as animal disease outbreaks (swine flu and bird flu in the past) have also led to an increase in the preferences for meat-free food products among consumers. However, factors such as higher costs of the meat substitutes are expected to negatively impact the market growth. Also, soy allergy and gluten intolerance are similarly anticipated to inhibit the market growth over the forecast period.

Key players in the meat substitute products are emphasizing investing in research and development activities, expansion and collaboration for developing innovative products. Manufacturers such as Beyond Burger and Impossible Foods are engaged in developing replicas of meat products, which act as sustainable options for the substitutes. Further, Impossible Foods announced the development of meat-free burgers, keeping in mind the cultural as well as nutritional factors. The company also offers burger products which are healthy, cheap and nutritious with better environmental footprints.

Key takeaway statements from this chapter are:

- Meat substitute products are sustainable, viable as well as profitable.
- In 2017, Europe accounted for the largest share which was nearly 42% of the overall meat substitutes market. This can be attributed to the shifting preferences of consumers towards vegan options and the health benefits offered by them.
- The Asia–Pacific accounted for approximately 16% of the overall market share. However, the Asia–Pacific region was the fastest growing segment because of the increasing living standards of its residents.

- The raw material sources for meat substitutes include wheat-based, soy-based, and mycoprotein, wherein soy accounted for nearly 63% of the overall meat alternatives market. The soy production in the North American region is expected to remain dominant.
- Key market players include Impossible Foods, Beyond Meat, MGP Ingredients Inc., Garden Protein International Inc., Cauldron Foods UK, and Schouten Europe.

REFERENCES

Committee on the Scientific Evaluation of Dietary Reference Intakes. (1998). *Dietary reference intakes: A risk assessment model for establishing upper intake levels for nutrients*. Washington, DC: National Academies Press (USA). Retrieved September 11, 2018 from https://www.ncbi.nlm.nih.gov/books/NBK45189/

Cowspiracy. (2014). *The facts*. Retreived September 11, 2018 from http://www.cowspiracy.com/facts/

Dagevos, H., & Voordouw, J. (2013). Sustainability and meat consumption: Is reduction realistic? Sustainability: Science. *Practice & Policy*, *9*, 60–69. doi:10.1080/15487733.2013.11908115

Deshpande, S. S., & Cheryan, M. (1984). Effects of phytic acid, divalent cations and their interactions on a-amylase activity. *Journal of Food Science*, *49*(2), 516–519. doi:10.1111/j.1365-2621.1984.tb12456.x

Deutsche Gesellschaft für Ernährung. (2018). *10 guidelines of the German Nutrition Society (DGE) for a wholesome diet*. Deutsche Gesellschaft für Ernährung e. V. Retrieved September 11, 2018 from https://www.dge.de/ernaehrungspraxis/vollwertige-ernaehrung/10-regeln-der-dge/10-guidelines-of-the-german-nutrition-society/

Euromonitor International. (2011, August 16). *The war on meat: How low-meat and no-meat diets are impacting the consumer markets*. Retrieved September 11, 2018 from https://blog.euromonitor.com/2011/08/the-war-on-meat-how-low-meat-and-no-meat-diets-are-impacting-consumer-markets.html

Food and Agriculture Organization of the United Nations (FAO). (2016). *The state of food and agriculture: Climate change, agriculture and food security*. Rome, Italy: FAO. Retrieved September 11, 2018 from http://www.fao.org/3/a-i6030e.pdf

Food and Agriculture Organization of the United Nations (FAO). (2018). *Food-based dietary guidelines*. Retrieved September 11, 2018 from http://www.fao.org/nutrition/education/food-dietary-guidelines/regions/en/

Halliday, J. (2018). Flexitarians drive growth for meat substitutes in the EU. *Bites*. Retrieved September 11, 2018 from https://bites.figlobal.com/euromonitor-cultured-meat-flexitarian/

Hoffman, J. R., & Falvo, M. J. (2004). Protein - Which is Best? *Journal of Sports Science & Medicine*, *3*(3), 118–130. PMID:24482589

Hosie, R. (2017). Flexitarianism predicted as key food trend for 2017. *Independent*. Retrieved September 11, 2018 from https://www.independent.co.uk/life-style/food-and-drink/flexitarianism-predicted-as-key-food-trend-for-2017-vegetarian-less-meat-a7465156.html

Hyner, C. (2015). A leading cause of everything: One industry that is destroying our planet and our ability to thrive on it. *Harvard Environmental Law Review*. Retrieved September 19, 2018 from http://harvardelr.com/2015/10/26/elrs-a-leading-cause-of-everything-one-industry-that-is-destroying-our-planet-and-our-ability-to-thrive-on-it/#_ftn1

Johns Hopkins Bloomberg School of Public Health. (n.d.). *Health & environmental implications of U.S. meat consumption & production*. Johns Hopkins University. Retrieved September 11, 2018 from https://www.jhsph.edu/research/centers-and-institutes/johns-hopkins-center-for-a-livable-future/projects/meatless_monday/resources/meat_consumption.html

Joshi, V. K., & Kumar, S. (2015). Meat analogues: Plant based alternatives to meat products – a review. *International Journal of Food and Fermentation Technolology*, *5*(2), 107–119. doi:10.5958/2277-9396.2016.00001.5

Kenward, E. (2017). *Special report: The rise in "flexitarian" lifestyles*. Food Ingredients 1st. Retrieved September 11, 2018 from https://www.foodingredientsfirst.com/news/special-report-the-rise-in-flexitarian-lifestyles.html

Malav, O. P., Talukder, S., Gokulakrishnan, P., & Chand, S. (2013). Meat analog: A review. *Critical Reviews in Food Science and Nutrition*, *55*(9), 1241–1245. doi:10.1080/10408398.2012.689381 PMID:24915320

Messina, M. (2016). Soy and health update: Evaluation of the clinical and epidemiologic literature. *Nutrients*, *8*(12), 754. doi:10.3390/nu8120754 PMID:27886135

Migala, J. (2018). *What is the beyond burger and is it good for you?* Women's Health. Retrieved September 11, 2018 from https://www.womenshealthmag.com/food/a21566428/beyond-meat-burger-ingredients/

Myers, J. (2016). *China consumes more than a quarter of the world's meat. The government wants to change that.* World Economic Forum. Retrieved September 11, 2018 from https://www.weforum.org/agenda/2016/06/china-consumes-more-than-a-quarter-of-the-worlds-meat-the-government-wants-to-change-that/

News Desk. (2017). *UK meat substitutes sales to grow by 25% in four years*. FoodBevMedia. Retrieved September 11, 2018 from https://www.foodbev.com/news/uk-meat-substitutes-sales-grow-25-four-years/

Prasannan, A. (2018). *Meat substitute market by product type (tofu-based, tempeh-based, TVP-based, seitan-based, Quorn-based, and others), source (soy-based, wheat-based, mycoprotein, and other sources), category (frozen, refrigerated, and shelf stable) –global opportunity analysis and industry forecast, 2018-2025.* Allied Market Research. Retrieved September 11, 2018 from https://www.alliedmarketresearch.com/meat-substitute-market

Pritchett, J., Thilmany, D., & Johnson, K. (2005). Animal disease economic impacts: A survey of literature and typology of research approaches. *The International Food and Agribusiness Management Review*, *8*(1), 23–45. Retrieved from https://www.ifama.org/resources/Documents/v8i1/Pritchett-Thilmany-Johnson.pdf

Raphaely, T., & Marinova, D. (2014). Flexitarianism: A more moral dietary option. *International Journal of Sustainable Society*, *6*(1/2), 189–211. doi:10.1504/IJSSOC.2014.057846

Raphaely, T., Marinova, D., Crisp, G., & Panayotov, J. (2013). Flexitarianism (flexible or part-time vegetarianism): A user-based dietary choice for improving personal, population and planetary wellbeing. *International Journal of User-Driven Healthcare*, *3*(3), 34–58. doi:10.4018/ijudh.2013070104

Rödl, M. B. (2018). *Why France banned meat names for vegetarian alternatives*. The Conversation. Retrieved September 11, 2018 from https://theconversation.com/why-france-banned-meat-names-for-vegetarian-alternatives-95549

Sadler, M. J. (2004). Meat alternatives – market developments and health benefits. *Trends in Food Science & Technology*, *15*(5), 250–260. doi:10.1016/j.tifs.2003.09.003

Thomas, C. S. (2015). *Changing diets: Plant protein sustainability crucial for food security*. Retrieved September 19, 2018 from https://www.foodnavigator.com/Article/2015/01/21/Changing-diets-Plant-protein-sustainability-crucial-for-food-security?utm_source=copyright&utm_medium=OnSite&utm_campaign=copyright

United States Department of Agriculture (USDA). (2010). *USDA Climate Change Science Plan*. Retrieved September 19, 2018 from https://www.usda.gov/oce/climate_change/science_plan2010/USDA_CCS-Plan_120810.pdf

Walker, P., Rhubart-Berg, P., McKenzie, S., & Lawrence, R. S. (2005). Public health implications of meat production and consumption. *Public Health Nutrition*, *8*(4), 348–356. doi:10.1079/PHN2005727 PMID:15975179

World Wide Fund for Nature (WWF). (2012). *Climate change on your plate*. WWF Germany. Retrieved September 11, 2018 from https://www.wwf.de/fileadmin/fm-wwf/Publikationen-PDF/Climate_change_on_your_plate.pdf

ADDITIONAL READING

Euromonitor International. (2011, August 16th). The war on meat: How low-meat and no-meat diets are impacting the consumer markets. Retrieved September 11, 2018 from https://blog.euromonitor.com/2011/08/the-war-on-meat-how-low-meat-and-no-meat-diets-are-impacting-consumer-markets.html

Halliday, J. (2018). *Flexitarians drive growth for meat substitutes in the EU*. Bites. Retrieved September 11, 2018 from https://bites.figlobal.com/euromonitor-cultured-meat-flexitarian//

Hosie, R. (2017). Flexitarianism predicted as key food trend for 2017. Independent. Retreived September 11, 2018 from https://www.independent.co.uk/life-style/food-and-drink/flexitarianism-predicted-as-key-food-trend-for-2017-vegetarian-less-meat-a7465156.html

Joshi, V. K., & Kumar, S. (2015). Meat analogues: Plant based alternatives to meat products – a review. *International Journal of Food and Fermentation Technolology*, *5*(2), 107–119. doi:10.5958/2277-9396.2016.00001.5

Migala, J. (2018). *What is the beyond burger and is it good for you?* Women's Health. Retrieved September 11, 2018 from https://www.womenshealthmag.com/food/a21566428/beyond-meat-burger-ingredients/

Raphaely, T., & Marinova, D. (2014). Flexitarianism: A more moral dietary option. *International Journal of Sustainable Society*, *6*(1/2), 189–211. doi:10.1504/IJSSOC.2014.057846

Rödl, M. B. (2018). *Why France banned meat names for vegetarian alternatives.* The Conversation. Retrieved September 11, 2018 from https://theconversation.com/why-france-banned-meat-names-for-vegetarian-alternatives-95549

Sadler, M. J. (2004). Meat alternatives – market developments and health benefits. *Trends in Food Science & Technology*, *15*(5), 250–260. doi:10.1016/j.tifs.2003.09.003

KEY TERMS AND DEFINITIONS

Alternative: (Of one or more things) Available as another possibility or choice.

Market: A location where the buyers and sellers meet to exchange the goods and services at prices determined by the forces of demand and supply.

Meat: Flesh of an animal.

Plant-Based Diet: A diet based on food products derived from plants, which includes vegetables, whole grains, fruits, nuts, legumes, and seeds, with very few or no animal products.

Proteins: Any of the various naturally occurring extremely complex substances that consist of amino-acid residues joined by peptide bonds, contain the elements carbon, hydrogen, nitrogen, oxygen, usually sulfur, and occasionally other elements (such as phosphorous or iron), and include many essential biological compounds (such as enzymes, hormones, or antibodies).

Section 4
Future Directions

Chapter 13
Is There a Future for Cattle Farming?

Clive J. C. Phillips
University of Queensland, Australia

Matti Wilks
University of Queensland, Australia

ABSTRACT

Humans have relied on cattle for production of food and work, as a source of capital, for dung, for fuel, building, and many other uses, for a period of about 10000 years. As a result, cattle biomass is now approximately twice that of humans on the planet. However, in the face of diminishing natural resources for the expanding human population and evidence of livestock pollution, cattle farms are currently criticized widely for their inefficient use of resources, the poor cattle welfare in modern farming systems, and their impact on human health amongst other problems. This chapter explores the reasons why cattle farming may ultimately cease in response to these issues. The replacement of cattle on farms began in the industrial revolution, when traction engines superseded many cattle in field operations. However, the replacement of cattle as food products is only now beginning to accelerate. The acceptability of alternative milks is growing rapidly and that of alternatives to meat products is also increasing. However, the major advance in replacing bovine meat products is under development in the laboratory as cultured meat, grown from a biopsied muscle sample on an edible scaffold in a nutrient media. Significant investment has been made in the process, which is technically feasible but is currently too expensive. This chapter explores current concerns about cattle farming as well as current difficulties in the development of meat alternatives, such as plant-based and clean meat. Through this exploration, the authors examine the potential for cattle farming to survive in the wake of alternatives offered by advanced food technology. Given anticipated success in bringing suitable alternative products to the market, most of the functions of cattle in developed countries are likely to be replaced. The process in developing countries will be much slower. Nonetheless, the authors anticipate that ultimately—perhaps in the far future—food technology developments will end the reliance on traditional cattle farming practices.

DOI: 10.4018/978-1-5225-7350-0.ch013

INTRODUCTION

Humans are currently dependent on breeding and raising cattle for the production of meat, milk, hides, dung and urine, and for agricultural work, transport, sport and capital. Some of these functions have existed for about 10000 years, especially their use as food sources, whereas others, such as the sport of bullfighting, their use in agriculture and for transportation purposes is more recent (Phillips, 2010). This close connection between the human and cattle led to cattle being revered in ancient times and much cattle terminology being adopted into the vernacular languages around the world.

However, although cattle may have served humans well for a relatively brief period in their development, questions are increasingly asked as to whether humans' reliance on cattle is outmoded in the modern era. Human society has evolved at a rapid pace in recent years and the products offered by cattle, which supported, amplified and facilitated our development in former years, may increasingly be seen as being of diminishing importance. This chapter explores the future role of cattle in a rapidly changing world, characterised by the growing human population. It considers the alternatives to farmed cattle, in the form of cattle meat grown *in vitro* and alternative products that offer the same benefits to humans.

BACKGROUND

Some of the concerns surrounding the use of cattle for meat and milk production have been known for many years. Over forty years ago research identified that livestock production systems in the UK were substantially less efficient than crop production systems, in terms of energy input compared to energy output (Leach, 1975, 1976). However, most cattle were fed a diet based on forages at that time, except for brief periods when cereal feeds were cheap. Much of the grazing land in the UK was rough grazing, which did not compete directly with land for producing crops for human consumption.

Nowadays, the extent of cattle production has increased and much land is used to produce feed that could otherwise be utilised to support the growing population of humans (Godfray, Beddington, Crute, Haddad, Lawrence, Muir, ... Toulmin, 2010). The recent growth in world human population and in some regions, most notably Asia, the growing affluence, allow those who were formerly vegetarians for economic reasons to include cattle products in their diet. This, coupled with the growing divide between rich and poor, has led to predictions that there will be mounting shortages of staple foods for the poor, as land is increasingly used for production of high cost, less efficient cattle meat and milk (Phillips, 2015).

CONCERNS ABOUT CATTLE PRODUCTION TODAY

The main concerns about cattle production today include the scale and intensity of the farming operations, their use of natural resources, in particular land, water and energy, pollution from the farms, the use of ingredients in their diet that could be fed to humans, their welfare, the ethics of slaughtering unwanted animals and their impact on human health. They are addressed in turn below.

Concerns Over Scale and Intensity

The concerns about cattle production today are focused on the growth of intensive systems and the increase in the scale of output as a result of growing demand in developing countries. Most of the growth in production systems is in the developing countries themselves in their attempt to become self-sufficient. In the face of growing world population, food security has assumed a new importance. Growing trade in cattle and their products is also possible with the ease of worldwide international trade over the last two decades and the expansion in road and rail freight and shipping opportunities. Many developing countries foresee the looming food shortages that are predicted as an opportunity to develop a lucrative business based on intensive cattle production (Tilman & Clark, 2014).

Use of Resources

Livestock production uses land, water, grains and other forage as well as human resources. On a limited planet, there are concerns as to how such resources are being utilised.

Land

The many benefits to be obtained from cattle and their adaptability, together with our ability to modify their environment, has led to humans keeping about 1.4 billion cattle in all the inhabited continents of the world. Cattle have nearly twice the biomass of humans globally, constituting the largest biomass of any animal on the planet. The greatest concentrations of cattle are in the subtropical regions of Asia, Africa and South America, as well as the temperature regions of north-west Europe. For the poorest sectors of society beef is too expensive to purchase compared with plants; for example, in the USA meat products cost 2-4 times as much per unit of energy compared to plant products – grains, beans, legumes, nuts and seeds (Drewnowski, 2010). In the poorest economies, particularly in South East Asia, the efficiency with which the industry produces edible protein is low for meat production, although for milk it rivals the most efficient meat producers – fish. The ratio between edible protein and feed protein is 0.05 for beef, 0.07 for mutton and goat meat, 0.17 for pork, 0.20 for poultry, 0.25 for cows' milk and 0.25 for trout (Tilman & Clark, 2014). It is not surprising, therefore, that the opportunity exists for vegan and vegetarian diets to reduce land use demand by about 50% (Hallstrom, Carlsson-Kanyama, & Borjesson, 2015).

However, the transport and processing of animal (and plant) products can greatly distort the efficiency of production. Their inefficiency compared to pigs, poultry and fish relates to the need to maintain an active rumen, filled with micro-organisms that conduct an initial digestion of plant matter, which is followed by a secondary digestion of the micro-organisms and any plant material that escaped rumen digestion in the animals' stomach and intestines. This two-stage process is inherently wasteful of energy and protein, but it allows cattle to digest poor quality roughages or byproducts that only micro-organisms can digest. However, this is a slow process, it takes at least 1-2 years to grow a finished beef animal, compared with just 42 days for meat chickens for example. Much energy and protein is therefore used just to maintain the animal.

Water

Water shortages are threatening human survival in heavily populated regions of the world. This coupled with prolonged droughts associated with global warming, means that water is one of the most precious commodities of our modern era. Water use per gram of protein for milk production is 50% more than for pulses and for beef is 20 times greater than for cereals and starchy roots (Mekonnen & Hoekstra, 2012). Beef production typically requires 10,000 litres/kg boneless beef (Legesse, Cordeiro, Ominski, Beauchemin, Kroebel, McGeough, … McAllister, 2018). Industrial enterprises are less efficient in water use than grazing systems because of the water used for crop growth (up to 100 times that required for drinking), waste disposal, cleaning equipment, concrete, processing products from cattle, and in hot conditions, cooling cattle.

Owning cattle creates problems for subsistence farmers to develop water conserving agricultural systems (Wolka, Sterk, Biazin, & Negash, 2018). Cattle may damage the conservation structures, and cattle farmers tend to be focused on animal husbandry and do not tend to have the mentality to optimise water use in crop production.

Cereal Grains and Other Foods for Humans

Modern beef and dairy production systems attempt to speed up the digestive processes of cattle by feeding high quality cereals in feedlots and intensive dairies, respectively, but these could feed more people if used directly. However, it has enabled growth rates of beef cattle to increase and milk production levels to improve. For example, over the last 50 years milk yield per cow per year approximately doubled (Phillips, 2018). This use of cereals also predisposes cattle to digestive disorders, such as acidosis in the rumen (Zebeli, Ghareeb, Humer, Metzler-Zebeli, & Besenfelder, 2015).

Human Resources

Most cattle are kept in the world's poorest regions – the savannah belt of central Africa, the Indian subcontinent and South America. Here a high proportion of people exist on less than US$2 per day. Cattle help to maintain these subsistence farmers, providing food, as well as dung for fuel, housing and land fertilisation. However, they also utilise a human capital resource that could be diverted to more efficient food production purposes. This transition, if it happens in the interest of providing food and fuel more efficiently, will be slow and will need the approval and support of local and national governments. Traditional societies are slow to change voluntarily, particularly if there is a lack of knowledge about alternative occupations.

In developed countries, productivity per person has rapidly increased over the last 50 years. At the same time that milk yield per cow doubled, mean herd size for milk production increased from 30 to 115 cows, with the result that output per farmer increased from 112,500 to 840,000 litres per year. Furthermore, the increasing output per person and specialisation of cattle production enterprises have contributed to rural depopulation and the breakdown of previously sustainable communities.

Cattle Welfare Concerns

The welfare of cattle in large production units that have been intensified to increase output has increasingly been criticised by scientists (e.g. Tucker, 2018). The major concerns are as follows:

- A highly concentrated diet is fed to dairy cows in lactation, which leads to a high prevalence of metabolic diseases, especially acidosis, laminitic lameness, fatty liver disease, and hypocalcaemia. The feeding of a grain-based diet to beef cattle in feedlots has similar effects.
- Poor quality housing of dairy cows in intensively stocked sheds, with small raised beds for them to lie in and little capacity for natural behaviour. High levels of aggression eventuate between cows. In feedlots cattle also have little space, compared with grazing cattle, and respiratory disease is common as a result of dusty conditions. Shade provision for both beef cattle in feedlots and dairy cows at pasture may be inadequate and cause heat stress. Solutions are readily available, in the form of natural or artificial shade, but are rarely required by codes of practice or legislation.
- Diseases are common in dairy cows, including infections such as mastitis, which causes painful conditions in the udder that may persist from one lactation to the next. Lameness, an extremely painful limb disorder, commonly affects about 20% of dairy cows (e.g. Sadiq, Ramanoon, Mansor, Syed-Hussain, & Mossadeq, 2017). Means of treating some of the major diseases, such as mastitis, lameness and respiratory problems, are becoming restricted by the emergence of antibiotic resistance in dairy cows and their wastes (Oliver, 2011; Qian, Sun, Gu, Wang, Sun, Yin, & Duan, 2016).
- The widespread use of antibiotics for disease prevention, rather than cure, is partly held responsible for antimicrobial resistance. Parasites continue to be a problem in intensive cattle operations, in particular stomach and lungworms, with resistance to anthelmintics also becoming evident and a potential problem for the future (Sutherland & Leathwick, 2011). The disease concerns about cattle production are not confined to intensive production systems, though these raise a new spectrum of disorders that were hitherto rare.

Extensive systems, unless they are very well managed, have their own problems of:

- Exposure to drought and flood, and the consequences for food supply;
- Poor conditions for growth of pasture and other herbaceous plants;
- Ectoparasites that are difficult to control in infrequently handled cattle;
- Routine procedures that are painful and are conducted without anaesthetic, in particular castration, dehorning and spaying, and;
- Long distances from market places, necessitating transport with all the associated stress.

The increasing global trade in live cattle worldwide exposes cattle to long distant transport in often overcrowded conditions with inadequate food and water supplies and little or no capacity for natural behaviour (Phillips, 2015). Much of the trade is uncontrolled, sometimes illegal and of unknown impact on the animals' welfare, in particular the trade from India, where cattle cannot legally be slaughtered, to the surrounding countries, where they can. Much of the trade in live cattle is to developing countries, such as Indonesia, where slaughter facilities are rudimentary, offering no facility to efficiently restrain or stun the cattle before slaughter (Tiplady, Walsh, & Phillips, 2012).

The Ethics of Cattle Production

There are ethical concerns about modern production systems that do not necessarily involve poor welfare. Most prominent is the concern that cattle are slaughtered at an unnecessarily young age. Because of the specialisation of the intensive dairy and beef industries in the highly industrialised countries, male calves are of little value. These are usually slaughtered at an age of just one week, after separation from their mother at just a few hours of age.

Public opinion appears against such practices (Weary & von Keyserlingk, 2017). Similarly, but less well known or exposed, is the very short life of the average dairy cow in an intensive production farm, often in the region of just 4-5 years of age, compared with a natural lifespan of about 20-25 years. This is mainly due to the stress caused by high productivity and poor conditions, with most cows killed because they: (1) fail to conceive to produce another calf, and hence more milk, or; (2) get mastitis or lameness.

Pollution

Cattle have been implicated in the growing problem of global warming, as a result of their output of methane, a potent greenhouse gas, and the use of fossil fuels for production of fertilizers and other commodities associated with intensive cattle farming. The carbon emissions, per a gram of edible protein, are much greater for bovine meat – 64 g, compared with milk – 9 g, maize – 1 g and legumes – 0.2 g (Tilman & Clark, 2014). Total greenhouse emissions, which include the highly potent methane that is produced in large quantities during rumen fermentation, are much higher per kg for beef (20 kg) than processed cereals and nuts (3 kg) and milk (1 kg). The opportunity exits for consumers to reduce greenhouse gas emission very substantially, by about 30% of the emissions from diet, by the widespread adoption of a vegan or vegetarian diet, with the reduction in red meat consumption having the most impact (Hallstrom, Carlsson-Kanyama, & Borjesson, 2015). However, this is only possible if the consumption of cheese and air transported fruit and vegetables is restricted. Additional contribution to global warming derives from the cutting down of forest, particularly tropical rainforest in Brazil, to make cattle pastures.

Apart from their role in global atmospheric pollution, more localised pollution is commonly associated with cattle production enterprises. Nitrates in drinking water commonly result from the keeping of cattle near watercourses (Naylor, Humphrey, Kelley, Easter, & Iverson, 2018), with up to 80% of nitrogen inputs ending up in leached water, and only 20% being retained on the farm. Nitrogen efficiency of cattle production systems is low, especially if large quantities of nitrogen fertiliser are applied to pasture, which also require considerable energy to produce (40 GJ/t in the case of ammonia). Phosphorus emissions are significant when fertilizer inputs exceed offtake, which may contribute to eutrophication of lakes. In addition, phosphorus reserves worldwide are finite and diminishing.

Cattle may also despoil water sources, particular ponds and streams where water flow is limited (Wilson, Chislock, Yang, Barros, & Roberts, 2018). Effluent from stores of silage is very acidic and even more polluting to groundwater than excreta. Cattle slurry, silage effluent and dirty water from washing the parlour and its surroundings, all have a high biological oxygen demand, and they deplete the oxygen content of streams and rivers if allowed to enter and thereby threaten the wildlife.

Human Health

Overconsumption of red meat and milk fats has been linked to human health problems, partly due to the association with obesity. The major concerns are that a high intake of saturated fats leads to increased risks of heart disease and stroke, diabetes and some cancers – of the breast, prostate and colon, caused by stimulation of steroid production and meat lacking the protective effects of fibre and antioxidants. Cholesterol in the body may also be increased. However, although some studies link the consumption of red meat to cardiovascular problems, others do not, which means that the effects are not universal (Lippi, Mattiuzzi, & Sanchis-Goma, 2015). The links to cancers may arise from heme catalysed oxidations, leading to free radical damage (Tappel, 2007).

Red meat is also high in calories and low in fibre, potentially contributing to obesity and the ensuing risk of diabetes and many other diseases. The nature of the link between red meat consumption and obesity continues to be unclear. Although there is evidence that fish provides a greater feeling of satiety than beef, there is no overwhelming evidence that beef and dairy products result in less satiety than an isoenergetic intake of, for example, soya products (Uhe, Collier, & Odea, 1992).

At the other end of the human nutrition scale, consumption of red meat in undernourished people can provide a good source of protein, iron and adipose tissue, to the extent that their lives may be saved. In poverty-stricken areas of the world, iron-deficient anaemia is common, especially in women, and the consumption of red meat will help to rectify the problem (Shankar, Agrawal, Beadreault, Avula, Martorell, Osendarp,… McLean, 2017). Since the 1990s, the intake of milk and milk products has been increasing in children of one of the world's most undernourished nations – India, at the expense of the intake of cereals and millet (Shankar, Agrawal, Beadreault, Avula, Martorell, Osendarp,… McLean, 2017). In adults, milk and milk product consumption has not declined markedly since the 1990s, but the consumption of staple cereals and millets has, on average from 444 to 360 g/day. This supports the idea that staple foods in the most underdeveloped countries will be in increasingly short supply, as a result of devoting resources to feeding animals (Phillips, 2015). The Indian government forbids the slaughter of cattle, which is strictly taboo in the Hindu faith, and as a result about 600,000 unwanted cattle – one seventh of the Indian cattle population, are kept in shelters until they die naturally (Kennedy, Sharma, & Phillips, 2018). In a country with diminishing food resources for its human population, cattle keeping policy must be carefully considered if malnutrition is to be contained in the future.

Cattle production systems are also capable of developing and harbouring zoonoses that threaten the human population, such as tuberculosis. The emergence of bovine encephalopathy (mad cow disease) in the 1980s in the UK appears to have caused the death of approximately 200 people from a new variant Creutzfeldt-Jacob disease (mad cow) at the end of the last century. This affected people at a young age, with an unknown proportion of the UK population harbouring the disease. Prior to this, the emergence of measles from rinderpest created a scourge on the human population that lasted many centuries.

Changing Global Tastes

Diets are changing globally. This is not just due to increased global tourism, which introduces an affluent sector of the population to different diets in exotic locations, few of which include cattle products. One of the most popular locations and diets is that of the Mediterranean region. Links between the Mediterranean diet and reduced cardiovascular disease are apparent but poorly understood (d'Almeida, Sanches, Spillere, Zuchinali, & Souza, 2018).

Other social change is caused by increased replacement of human labour with robotic labour, providing more leisure time to engage in tourism and preparation of exotic dishes. People are living longer and the diet of the elderly needs to be more functional, avoiding foods with high saturated fat content, for example, which may cause cardiovascular disease. Reduced mobility in the growing number of elderly requires greater storage life in foods.

Developing countries in Asia with a growing affluence are increasingly turning to a Western-style diet, including cattle products. In the least developed countries of the world, milk consumption increased from 22 to 32 kg/head per year between 1977 and 2013. In China, it increased from 4 to 33 kg/head between 1985 and 2013 (FAO, 2014). As well as milk consumption, milk products are increasingly versatile and are used in many bakery products and on pizza.

ALTERNATIVES TO THE USE OF CATTLE IN HUMAN SOCIETY

The process of replacing cattle in human society began in the industrial revolution, when the use of cattle for work began to be substituted by machines. Nowadays, they are rarely used for traction of agricultural implements but are still employed for pulling carts and moving timber in forests, where access for machines is difficult and horses are not strong enough.

Cattle are also used for sport and this is repeatedly questioned and supported, mainly by older members of society who are more conscious of its traditional value. It is likely that this will die a natural death in Europe because of lack of interest and a growing awareness of the associated cruelty. In developing countries, the same process will take much longer.

The main use of cattle worldwide is for meat and milk production. Cow milk alternatives, in the form of soya, almond, rice and other milks are growing in production volume in western countries. These are associated with a health-conscious consumer, usually in the western markets, although soya milk consumption is a tradition in south-east Asia where land is scarce for keeping cattle.

A major use of cattle worldwide is for meat production. Meat alternatives should give the consumer the experience of meat consumption, but do not require the raising and slaughtering of animals for production. In the following section, we will discuss the potential uptake of two meat alternatives: plant-based meat and clean meat.

Plant-Based Meat

Plant-based meat, also referred to as meat substitutes, meat alternatives or meat replacers, is a product created from plant-based ingredients such as pulses (soy), grain or fungi (Hoek, Luning, Weijzen, Engels, Kok, & de Graaf, 2011). The production of plant-based meat uses less land and energy than farmed meat (Nijdam, Rood, & Westhoek, 2012). Additionally, those produced from grain are more carbon efficient, require less water and less land than beef products (Hoek, Luning, Weijzen, Engels, Kok, & de Graaf, 2011; Nijdam, Rood, & Westhoek, 2012). Moreover, plant-based meat does not require the raising and slaughtering of animals for production. From this perspective, plant-based meat provides an avenue for consumers to access meat-like products without the environmental or ethical issues addressed earlier in this chapter.

Despite this, the consumption of plant-based meat generally trails far behind that of farmed meat. While there has been recent growth, data shows that the consumption of meat substitutes in the UK and Europe is around 3.6% of the market value of farmed meat (Mintel 2013a, 2013b). Moreover, those who consume meat may do so more frequently than those who consume meat substitutes. For example, a report by Aurelia (2002) found that while 80% of Dutch consumers ate meat 3 times per week or more, those who consume meat substitutes tended to consume them less than twice per week. However, since that study the meat substitutes market has experienced rapid growth, and, is predicted to grow from $4.3 billion in 2018 to approximately $5.4 billion by 2022 (Mintel, 2015). In line with this, the meat free company, Linda McCartney, reportedly saw a 50% increase in sales volumes between 2014 and 2015 (Mintel, 2015).

While growth is promising, the farmed meat market still significantly outperforms that of plant-based meat. In response, recent years have seen exploration into potential barriers, though experimental approaches are limited. It appears that general attitudes towards the products are relatively negative for people who are not already consumers. This is most true after tasting (Ottenfeld, Bernstein, & Witte, 2008; de Boer & Aiking, 2011; Elzerman, Hoek, van Boekel, & Luning, 2011) and in response to repeated exposure (Hoek, Elzerman, Hageman, Kok, Luning, & de Graaf, 2013). Indeed, one study found that in a sample of 553 participants in the UK and the Netherlands, the majority (57%) were not even considering using meat substitutes (Hoek, Luning, Weijzen, Engels, Kok, & de Graaf, 2011), while another found that only 9% of the participants reported preferring plant-based to farmed meat (de Boer & Aiking, 2011). Additionally, research into meat replacements generally suggests that plant-based meat is preferred less than animal-based meat replacement options, such as eggs and cheese (Schösler, de Boer, & Boersema, 2012).

Overall, negative perceptions are related to a number of factors. Taste and price appear to be the two key barriers. Currently, meat alternatives are generally more expensive than farmed meat. Research has consistently found that price plays a major role in food purchasing decisions (EUFIC, 2006), including meat and meat substitutes, with several studies identifying price as a major barrier (Apostolidis & McLeay, 2016; Elzerman, Hoek, van Boekel, & Luning, 2011; Hoek, Pearson, James, Lawrence, & Friel, 2017). Indeed, one study in the UK found that price is the most influential factor in people's (un) willingness to purchase meat substitutes (Apostolidis & McLeay, 2016).

Another barrier is taste and sensory appeal. Research repeatedly finds that plant-based meat products are not considered to be as appealing as farmed meat in terms of sensory properties (Apostolidis & McLeay, 2016; Hoek, Luning, Weijzen, Engels, Kok, & de Graaf, 2011; Hoek, Pearson, James, Lawrence, & Friel, 2017). This has been found across a range of countries, including Australia (Hoek, Pearson, James, Lawrence, & Friel, 2017), the UK (Apostolidis & McLeay, 2016) and the Netherlands (Hoek, Luning, Weijzen, Engels, Kok, & de Graaf, 2011).

However, not all people have equal perceptions of plant-based meat. Several studies have identified personal factors that may influence perceptions of plant-based meat. It is preferred by younger people (de Boer & Aiking, 2011; Ottenfeld, Bernstein, & Witte, 2008, but see Schösler, de Boer, & Boersema, 2012), women (de Boer & Aiking, 2011) and those who are better educated (de Boer & Aiking, 2011; Schösler, de Boer, & Boersema, 2012). Another factor that contributes to acceptability of plant-based meat is people's current farmed meat consumption and attitudes towards farmed meat. In this respect, plant-based meat is preferred by those who are already meat restrictors (Hoek, Luning, Weijzen, Engels,

Kok, & de Graaf, 2011). To explore this, Hoek and colleagues (2011) divided consumers into heavy users of plant-based meat (those who eat it regularly), light/medium users (those that eat it only sometimes) and non-users (those who have never eaten it). Those who ate little or no plant-based meat (non-or-light/medium users) reported that lower sensory attractiveness and unfamiliarity were barriers to their acceptance of plant-based meat. These participants also tended to have more positive perceptions of, and consume more, farmed meat than those who were heavy users of plant-based meat. They were also found to be more neophobic; that is, they were more likely to avoid unfamiliar new foods.

This suggests that people who frequently consume meat would prefer meat substitutes that are similar to farmed meat in taste and texture. In contrast, people who eat little to no meat (meat restrictors) tend to prefer substitutes that are less similar to meat. Therefore, to improve sensory appeal to a wide range of consumers, plant-based meat products must be more similar to meat. In line with this, recent years have seen the emergence of plant-based meat products that strive to be as similar as possible to farmed meat – e.g. the Impossible Burger, and Beyond Meat, which are marketed to 'smell, sear and bleed as though they were made entirely of beef' (Gajanan, 2018). All of these products aim to replicate the meat experience as closely as possible. This indicates that the uptake of such a product is focused on current meat consumers, rather than those who are already meat restrictors.

Currently, the plant-based meat market is relatively small, though evidence suggests that it is growing. Despite this, taste and price appear to be the primary factors that influence people's consumption decisions. Therefore, to achieve a greater market share, plant-based 'meats' need to become cheaper and better replicas of farmed meat products. As this shift occurs it is likely that we will see the consumer market move away from meat restrictors to a more general consumer base. In this way, plant-based meat can provide an alternate meat product for those who enjoy the taste of meat, but have concerns about the ethical, environmental and health implications of farmed meat and meat production.

Clean Meat

Clean meat is another meat alternative, and it is also known as cultured meat, lab-grown meat or in-vitro meat. It was famously predicted by Winston Churchill in 1932, when he said: "Fifty years hence we shall escape the absurdity of growing a whole chicken in order to eat the breast or wing by growing these parts separately under a suitable medium" (Quote Investigator, 2018). His timeframe was grossly ambitious, but the concept was revolutionary and showed great foresight. In fact, muscle cells were first grown in cell culture by Russell Ross in 1971. Culturing of animal stem cells began in earnest in the 1990s, with patents filed in the USA and the Netherland in 1998 and 2001. The first edible sample was produced in 2002 and cultured meat was grown from turkey cells by NASA in 2006. The first lab-grown beef burger produced by Dr Mark Post at Maastricht University in 2013 and currently there are about 30 labs working on cultured meat worldwide.

Clean meat is meat grown from cell culture, rather than cut from a whole slaughtered animal. The cells are removed through a biopsy, and myosatellite stem cells are extracted and grown on an edible scaffold in a nutrient-rich culture medium (Post, 2014a, 2014b). The growing muscle may be 'exercised' to stimulate growth and it is finally ground up into strips for reconstitution into a burger, with the addition of flavour, iron and vitamins if necessary. In this way, it is biologically meat. To date, only ground meat products have been produced (hamburger patties and chicken nuggets), but start-up companies are currently attempting to produce whole meat products in the near future. It is not yet commercially available and experts have widely varying estimates on when it will be. However, one producer, Hampton Creek, claims that it will have products on the shelves by the end of the year 2018 (Simon, 2018).

Like plant-based meat, clean meat is considered more environmentally friendly than farmed meat. One study found producing 1000 kg of cultured meat uses 99% less land, 82–96% less water and produces 78–96% fewer greenhouse gas emissions (Tuomisto & de Mattos, 2011). It also requires 7–45% lower energy, making it more efficient than beef – in fact, only poultry production has lower energy use (Tuomisto & de Mattos, 2011). In this way, it provides a potential avenue for consumers to continue to eat the product that they want, without the environmental costs associated with farmed meat. It is also considered more ethical – not requiring the raising and slaughtering of animals, nor other undesirable practices involved in mass farming. It is essentially meat without the animal.

While this product holds potential, a major barrier to consumption will be the attitudes of some meat consumers (Bryant & Barnett, 2018). Similar to plant-based meats, there is significant resistance to the product, albeit motivated from different concerns. Research into attitudes to clean meat shows varying levels of consumer acceptance and engagement. Reported rates of willingness to consume clean meat vary from around one quarter to two thirds of participants in different samples (Post, 2014a; Verbeke, Sans & Van Loo, 2015; Wilks & Phillips, 2017). This wide variability likely stems from different populations, as well as variability in question and product framing. Nonetheless, there is consistently less acceptance of clean meat, compared with the number of individuals who currently eat farmed meat.

Several barriers to clean meat have been identified. As for plant-based meat, price and sensory appeal appear to be two key factors. Price in particular is consistently identified; people report being unwilling to pay more for clean meat than farmed meat, and, in many cases, being only willing to pay much or somewhat less for clean meat (Verbeke, Marcu, Rutsaert, Gaspar, Seibt, Fletcher, & Barnett, 2015; Verbeke, Sans & Van Loo, 2015; Wilks & Phillips, 2017; Wilks & Phillips, 2017). In the initial production, clean meat cost US $325,000 for one hamburger patty. However, already the price as dropped to just $11 (Eater, 2015; Javelosa, 2017; Shoemaker, 2017). Nonetheless, the capacity for clean meat to price-match farm-produced meat is unclear.

Sensory barriers, such as taste and appeal are also commonly cited. That is, people are concerned that clean meat will not taste as good as farmed meat (Marcu, Gaspar, Rutsaert, Seibt, Fletcher, Verbeke, & Barnett, 2015; Verbeke, Marcu, Rutsaert, Gaspar, Seibt, Fletcher, & Barnett, 2015; Verbeke, Sans & Van Loo, 2015; Wilks & Phillips, 2017; Wilks & Phillips, 2017). Given that the product is not yet generally available, it is not possible to conduct taste research as is done with plant-based meat. As such, the outcome of these perceptions is somewhat dependent on the products that become available. However, appeal perceptions may also be related to concerns about naturalness, which appear as a third major barrier to clean meat (Bekker, Fischer, Tobi, & van Trijp, 2017; Laestadius & Caldwell, 2015; Marcu, Gaspar, Rutsaert, Seibt, Fletcher, Verbeke, & Barnett, 2015; Verbeke, Marcu, Rutsaert, Gaspar, Seibt, Fletcher, & Barnett, 2015; Wilks & Phillips, 2017). In line with this theme, qualitative research has raised questions about the definition of 'real' meat and the notion of 'playing God' (Marcu, Gaspar, Rutsaert, Seibt, Fletcher, Verbeke, & Barnett, 2015), as well as the genetically engineered nature of the product (Laestadius & Caldwell, 2015). These concerns have been found to reduce the perceived acceptability of clean meat. One study found that participants perceived health risks (colon cancer) as less acceptable when caused by clean meat compared to identical health risks caused by farmed meat (Siegrist & Sütterlin, 2017; Siegrist, Sütterlin, & Hartmann, 2018). This negative perception was mediated by the perception of clean meat as unnatural (Siegrist & Sütterlin, 2017). This indicates that concerns about naturalness may be a challenging barrier for the clean meat industry to overcome.

Despite these barriers, there are also a number of positive perceptions of clean meat. It is generally perceived as more ethical and environmentally friendly than farmed meat (Marcu, Gaspar, Rutsaert, Seibt, Fletcher, Verbeke, & Barnett, 2015; Tucker, 2014; Verbeke, Sans & Van Loo, 2015; Wilks & Phillips, 2017). Perceived ethical benefits tend to be related to animal welfare, particularly reducing the suffering of farm animals (Verbeke, Marcu, Rutsaert, Gaspar, Seibt, Fletcher, & Barnett, 2015). Environmental benefits relate to the potentially lower resource requirements for clean meat relative to farmed meat - including water and land – as well as reducing greenhouse gas emissions and the carbon footprint associated with farming (Verbeke, Marcu, Rutsaert, Gaspar, Seibt, Fletcher, & Barnett, 2015; Verbeke, Sans & Van Loo, 2015; Wilks & Phillips, 2017). Several studies have found that clean meat is perceived as a viable alternative to world food production from conventional meat and a potential solution for world famine problems via increased protein productivity (Tucker, 2014; Wilks & Phillips, 2017).

When considering these perceptions, it is worth noting that, as highlighted by Verbeke, Sans and colleagues (2015), the perceived benefits are distal – they relate to the world and animals on a broad, conceptual scale. Reducing animal suffering, improving sustainability and reducing world hunger are all factors that do not influence an individual person directly. In contrast, the concerns are proximal – they affect the individual indirectly. Price is salient at the time of purchasing. Eating meat that one finds unappealing is an affective experience, and one that will likely be pervasive. As such, the clean meat industry must identify mechanisms to overcome such barriers and bolster positive perceptions for the product to be successfully integrated into the market.

Research has begun to examine this, though it remains in its infancy. Initial research has focused on informational approaches to improving perceptions generally. Verbeke, Sans and colleagues (2015) found that after being provided with basic information, around two thirds of consumers in their sample reported favourable attitudes to cultured meat, with around one quarter reporting willingness to try it. Another two thirds were unsure, and around 9% of the sample rejected the idea. However, upon being provided with further information about the environmental benefits of clean meat, 43% of consumers reported being willing to try it, while 51% reported that they would 'maybe'. This demonstrates the powerful role that information can play when addressing general perceptions – positive environmental outcomes are a major benefit of clean meat relative to farmed meat, and highlights that this appears to hold power in shifting people's perceptions. However, it must be noted that this study did not employ an experimental design, lacking controls, so interpretation should be made with caution.

Another study did employ an experimental design in an informational approach. Participants were given either positive or negative information about clean meat, revealing that positive information improved attitudes while negative information reduced attitudes (Bekker, Fischer, Tobi, & van Trijp, 2017). They also identified that these effects were generalizable – that is, positive information about a similar product, solar panels, resulted in more positive attitudes to clean meat overall. The authors also conducted a follow-up study, clarifying that this effect was not merely due to affect shift, but due to the content of the information provided. The authors chose solar panels as their focal technology as they considered them in the same sustainable product category. Thus, the attitude bolstering here is again related to environmental benefits.

In line with the findings of the two aforementioned studies, which find that environmental approaches appear to improve attitudes to clean meat, we see increasing demand for sustainable products generally. With a focus on environmental benefits, clean meat may take its place as a sustainable meat source for our future. To date, no informational approaches have considered focusing on other topics, such as highlighting ethical benefits of cultured meat. This suggests that, despite the immense ethical and welfare benefits of clean meat, animal welfare is not yet considered a worthwhile mechanism by which to bolster perceptions of cultured meat.

Whilst the perceived sustainability does appear to improve perceptions of clean meat, this may not be sufficient to provide genuine attitude shift. Genetically modified (GM) food is highly sustainable in terms of crop yield. Moreover, certain strains are even engineered to improve human health, such as golden rice which biosynthesises beta-carotene – a precursor to vitamin A, essential in combating malnutrition. Despite these benefits, there is considerable opposition to GM foods, which has persisted across decades (Frewer, van der Lans, Fischer, Reinders, Menozzi, Zhang, … Zimmermann, 2013; Gaskell, 1999; Scott, Inbar, & Rozin, 2016). Often these concerns are around the unnatural nature of GM products (Scott, Inbar, & Rozin, 2016), which mirrors the aforementioned concerns around clean meat. It is, thus, necessary, to consider the potential for the natural concerns identified thus far to be a major barrier to clean meat acceptance.

Up to now, only one study has tried to address this concern. Vivalt and Macdonald (2017) tested three different methods of addressing naturalness concerns around clean meat. The three strategies were: debunking unnatural (natural things can be bad), embrace unnatural (unnatural things can be good), or a descriptive norm (lots of other people are excited by it), as well as a control. Perceptions were examined immediately and again at a 10 week follow up. The results revealed that all three approaches improved attitudes to clean meat relative to the placebo in the immediate examination. However, only the embracing unnatural approach improved perceptions at a 10-week follow up, suggesting lasting attitude change. The authors argue that this approach provides a low-dissonance path to clean meat acceptance – consumers could continue to see unnatural as a bad concept, but make exemptions for certain products, classifying them as 'unnatural but safe'. This is the first research to examine methods to overcome negative perceptions of clean meat. It is still in initial phases, and significantly more work is needed to understand the intricacies of clean meat perceptions, as well as natural bias more generally. Nonetheless, these findings are promising for the future acceptance of clean meat.

Finally, research has begun to identify demographic predictors of attitudes to clean meat. Notably, the samples in these experiments are geographically distinct and thus are not comparable. However, in general perceptions of clean meat are more positive in young than old people, and in those who are politically liberal (left-wing) as opposed to those who are conservative (right-wing) (Tucker, 2014; Wilks & Phillips, 2017). It is likely that these two factors are related, given that younger people tend to be more liberal and older people more conservative.

Additionally, some research has found that males tend to prefer clean meat more than females (Wilks & Phillips, 2017), and that those from urban areas tend to prefer clean meat more than those from rural areas (Tucker, 2014). People's eating habits are also predictive of responses. Vegetarians and vegans tend to be less likely to report willingness to eat clean meat than meat eaters but identify more benefits of the product, while meat eaters report perceiving fewer benefits but are more willing to try clean meat (Wilks & Phillips, 2017). This indicates that the market for clean meat as a product may not be vegetarians, as this has historically been the focus of plant-based meat alternatives. Instead, like plant-based 'meat', clean meat's consumer target is meat eaters – those who enjoy the sensory experience of eating meat already. In this way, clean meat may offer an alternative to those who like to eat meat but may wish to avoid the environmental and ethical issues associated with farming.

Considering the evidence, it is clear that there are some positive perceptions of clean meat, and reasonable willingness to engage with it as a product. Nonetheless, there are still significant barriers to consumer acceptance, and research to address such barriers is still in its infancy. Initial attempts are promising, and with further research and understanding it is plausible that clean meat could become accepted within society – though the extent and timing of this acceptance is not yet clear. However, through this acceptance, clean meat may offer a viable alternative to farmed meat that is as closely matched as possible to currently produced meat.

SOLUTIONS AND RECOMMENDATIONS

In the face of increasing competition from alternatives to cattle products, it is important to conduct research to assess the impact of adoption of these alternatives on cattle farmers and their workers, particularly in developing countries. The associated industries, such as transport and marketing, feed preparation and supply, veterinary and medical provisions and reproductive management should all be included in such an assessment. The social impact and effects on rural land use should all be considered in detail, particularly in developing countries. Contributions to rural depopulation should similarly be counted and recommendations on how to deal with the ensuing social change are important.

FUTURE RESEARCH DIRECTIONS

Research that includes the operator or cattle manager is vital to discover what can be done to mitigate the concerns about cattle farming. Further research is necessary to understand the attitudes towards alternatives, including clean meat and soya products, meat and milk. Research to scale production of clean meat at low cost is needed, as well as further work on the price sensitivity of the products.

REFLECTION AND CONCLUDING REMARKS

The long period of human dependence on cattle may be coming to an end. However, in order to smooth the transition as effectively as possible, it is important to: (a) recognise the trends and what is likely to ensure in the future, and; (b) take effective steps to minimise any adverse impact on those directly and indirectly involved in the industry.

We end this chapter by returning to our original question: is there a future for cattle farming? The arguments that we have developed indicate that there is probably not a good future for cattle in the long term, but few would risk predicting when their demise will occur. However, we speculate that it is likely that alternatives to beef mince will be dominant in the western market by 2050, and milk alternatives either have that status already or are rapidly achieving it in many developed regions. In developing countries, the reliance of many subsistence farmers and pastoralists on cattle will render any changes difficult to achieve in the short term, even if governments find them to be desirable. In the long term, change in reliance on cattle will only come about through substantial social upheaval, but in the continent of Africa, which has a rapidly growing proportion of the global population, that may be seen as inevitable.

In the longer term, people's inherent selfishness in perpetuating the human species will eventually cause cattle to become a relic from the past, an important phase in human development. Then we can say, with echoes of the reverence of those initially responsible for domesticating cattle: for the support that cattle have given us over the last 10,000 years, well done, thou good and faithful servants.

REFERENCES

Apostolidis, C., & McLeay, F. (2016). Should we stop meating like this? Reducing meat consumption through substitution. *Food Policy*, *65*, 74–89. doi:10.1016/j.foodpol.2016.11.002

Aurelia. (2002). *A marketing report on meat substitutes in the Netherlands*. Amersfoort, The Netherlands: Aurelia. (in Dutch)

Bekker, G. A., Fischer, A. R. H., Tobi, H., & van Trijp, H. C. M. (2017). Explicit and implicit attitude toward an emerging food technology: The case of cultured meat. *Appetite*, *108*, 245–254. doi:10.1016/j.appet.2016.10.002 PMID:27717657

Bryant, C., & Barnett, J. (2018). Consumer acceptance of cultured meat: A systematic review. *Meat Science*, *143*, 8–17. doi:10.1016/j.meatsci.2018.04.008 PMID:29684844

D'Almeida, M., Sanches, L., Spillere, S. R., Zuchinali, P., & Souza, G. C. (2018). Mediterranean Diet and other dietary patterns in primary prevention of heart failure and changes in cardiac function markers: A systematic review. *Nutrients*, *10*(1), E58. doi:10.3390/nu10010058 PMID:29320401

de Boer, J., & Aiking, H. (2011). On the merits of plant-based proteins for global food security: Marrying macro and micro perspectives. *Ecological Economics*, *70*(7), 1259–1265. doi:10.1016/j.ecolecon.2011.03.001

Drewnowski, A. (2010). The Nutrient Rich Foods Index helps to identify healthy, affordable foods. *The American Journal of Clinical Nutrition*, *91*(4), 1095S–1101S. doi:10.3945/ajcn.2010.28450D PMID:20181811

Eat. (2015). *The meat of the future: How lab-grown meat is made*. YouTube. Retrieved from https://www.youtube.com/watch?v=u468xY1T8fw

Elzerman, J. E., Hoek, A. C., van Boekel, M. A. J. S., & Luning, P. A. (2011). Consumer acceptance and appropriateness of meat substitutes in a meal context. *Food Quality and Preference*, *22*(3), 233–240. doi:10.1016/j.foodqual.2010.10.006

Food and Agriculture Organization of the United Nations (FAO). (2014). *FAOSTAT: Food and agriculture data*. Retrieved from http://www.fao.org/faostat/en/#home

Frewer, L. J., van der Lans, I. A., Fischer, A. R. H., Reinders, M. J., Menozzi, D., Zhang, X., ... Zimmermann, K. L. (2013). Public perceptions of agri-food applications of genetic modification - A systematic review and meta-analysis. *Trends in Food Science & Technology*, *30*(2), 142–152. doi:10.1016/j.tifs.2013.01.003

Gajanan, M. (2018). The meat industry has some serious beef with those "bleeding" plant-based burgers. *Time*. Retrieved from http://time.com/5181524/impossible-burger-plant-based/

Gaskell, G. (1999). Worlds apart? The reception of genetically modified foods in Europe and the U.S. *Science*, *285*(5426), 384–387. doi:10.1126cience.285.5426.384 PMID:10411496

Godfray, H. C. J., Beddington, J. R., Crute, I. R., Haddad, L., Lawrence, D., Muir, J. F., ... Toulmin, C. (2010). Food security: The challenge of feeding 9 billion people. *Science, 327*(5967), 812–818. doi:10.1126cience.1185383 PMID:20110467

Hallstrom, E., Carlsson-Kanyama, A., & Borjesson, P. (2015). Environmental impact of dietary change: A systematic review. *Journal of Cleaner Production, 91*, 1–11. doi:10.1016/j.jclepro.2014.12.008

Hoek, A. C., Elzerman, J. E., Hageman, R., Kok, F. J., Luning, P. A., & de Graaf, C. (2013). Are meat substitutes liked better over time? A repeated in-home use test with meat substitutes or meat in meals. *Food Quality and Preference, 28*(1), 253–263. doi:10.1016/j.foodqual.2012.07.002

Hoek, A. C., Luning, P. A., Weijzen, P., Engels, W., Kok, F. J., & de Graaf, C. (2011). Replacement of meat by meat substitutes. A survey on person- and product-related factors in consumer acceptance. *Appetite, 56*(3), 662–673. doi:10.1016/j.appet.2011.02.001 PMID:21315123

Hoek, A. C., Pearson, D., James, S. W., Lawrence, M. A., & Friel, S. (2017). Shrinking the food-print: A qualitative study into consumer perceptions, experiences and attitudes towards healthy and environmentally friendly food behaviours. *Appetite, 108*, 117–131. doi:10.1016/j.appet.2016.09.030 PMID:27686818

Javelosa, J. (2017, February 21). *Lab-grown meat is healthier. It's cheaper. It's the future*. Futurism. Retrieved from https://futurism.com/were-5-years-away-from-lab-grown-meat-hitting-store-shelves

Kennedy, U., Sharma, A., & Phillips, C. J. C. (2018). The sheltering of unwanted cattle, experiences in India and implications for cattle industries elsewhere. *Animals (Basel), 8*(5), 64. doi:10.3390/ani8050064 PMID:29701646

Laestadius, L. I., & Caldwell, M. A. (2015). Is the future of meat palatable? Perceptions of in vitro meat as evidenced by online news comments. *Public Health Nutrition, 18*(13), 2457–2467. doi:10.1017/S1368980015000622 PMID:25818555

Leach, G. (1975). Energy and food production. *Food Policy, 1*(1), 62–73. doi:10.1016/0306-9192(75)90009-3

Leach, G. (1976). *Energy and food production*. Guildford, UK: Science and Technology Press for the International Institute for the Environment and Development.

Legesse, G., Cordeiro, M. R. C., Ominski, K. H., Beauchemin, K., Kroebel, K., McGeough, A. J., ... McAllister, T. A. (2018). Water use intensity of Canadian beef production in 1981 as compared to 2011. *The Science of the Total Environment, 619-620*, 1030–1039. doi:10.1016/j.scitotenv.2017.11.194 PMID:29734581

Lippi, G., Mattiuzzi, C., & Sanchis-Gomar, F. (2015). Red meat consumption and ischemic heart disease: A systematic literature review. *Meat Science, 108*, 32–36. doi:10.1016/j.meatsci.2015.05.019 PMID:26017245

Marcu, A., Gaspar, R., Rutsaert, P., Seibt, B., Fletcher, D., Verbeke, W., & Barnett, J. (2015). Analogies, metaphors, and wondering about the future: Lay sense-making around synthetic meat. *Public Understanding of Science (Bristol, England), 24*(5), 547–562. doi:10.1177/0963662514521106 PMID:24553438

Mekonnen, M. M., & Hoekstra, A. Y. (2012). A global assessment of the water footprint of farm animal products. *Ecosystems (New York, N.Y.), 15*(3), 401–415. doi:10.100710021-011-9517-8

Mintel International Group Limited. (2013a). M*eat-free and free-from foods – UK – September 2013.* Retrieved from https://store.mintel.com/meat-free-and-free-from-foods-uk-september-2013

Mintel International Group Limited. (2013b). *Red Meat – UK 2013.* Mintel Official Market Reports Store. Retrieved from https://store.mintel.com

Mintel International Group Limited. (2015). *Red Meat – UK. Mintel Official Market Reports Store.* Retrieved from https://store.mintel.com

Naylor, E., Humphrey, C., Kelley, T., Easter, L., & Iverson, G. (2018). Evaluation of nitrate concentrations and potential sources of nitrate in private water supply wells in North Carolina. *Journal of Environmental Health, 80*, 16–23.

Nijdam, D., Rood, T., & Westhoek, H. (2012). The price of protein: Review of land use and carbon footprints from life cycle assessments of animal food products and their substitutes. *Food Policy*, *37*(6), 760–770. doi:10.1016/j.foodpol.2012.08.002

Oliver, S. P., Murinda, S. E., & Jayarao, B. M. (2011). Impact of antibiotic use in adult dairy cows on antimicrobial resistance of veterinary and human pathogens: A comprehensive review. *Foodborne Pathogens and Disease*, *3*(3), 337–355. doi:10.1089/fpd.2010.0730 PMID:21133795

Ottenfeld, M., Bernstein, D., & Witte, C. L. (2008). An exploration of U.S. consumer perceptions and affect: Two forms of a soy-based food product. *Journal of Food Products Marketing*, *14*(3), 49–76. doi:10.1080/10454440801985936

Phillips, C. (2010). Animals, Quo Vadis? Welcome to a new, multidisciplinary, integrated, open access journal: Animals. *Animals (Basel)*, *1*(1), 1–3. doi:10.3390/ani1010001 PMID:26486211

Phillips, C. J. C. (2015). *The animal trade.* Wallingford, UK: CAB International. doi:10.1079/9781780643137.0000

Phillips, C. J. C. (2018). *Principles of cattle production* (3rd ed.). Wallingford, UK: CAB International. doi:10.1079/9781786392701.0000

Post, M. J. (2014a). An alternative animal protein source: Cultured beef. *Annals of the New York Academy of Sciences*, *1328*(1), 29–33. doi:10.1111/nyas.12569 PMID:25376889

Post, M. J. (2014b). Cultured beef: Medical technology to produce food. *Journal of the Science of Food and Agriculture*, *94*(6), 1039–1041. doi:10.1002/jsfa.6474 PMID:24214798

Qian, X., Sun, W., Gu, J., Wang, X., Sun, J., Yin, Y., & Duan, M. (2016). Variable effects of oxytetracycline on antibiotic resistance gene abundance and the bacterial community during aerobic composting of cow manure. *Journal of Hazardous Materials*, *315*, 61–69. doi:10.1016/j.jhazmat.2016.05.002 PMID:27179201

Quote Investigator. (2018). *We shall escape the absurdity of growing a whole chicken in order to eat the breast or wing*. Retrieved from https://quoteinvestigator.com/2017/01/22/meat/

Sadiq, M. B., Ramanoon, S. Z., Mansor, R., Syed-Hussain, S. S., & Mossadeq, W. M. S. (2017). Prevalence of lameness, claw lesions, and associated risk factors in dairy farms in Selangor, Malaysia. *Tropical Animal Health and Production, 49*(8), 1741–1748. doi:10.100711250-017-1387-4 PMID:28856534

Schösler, H., de Boer, J., & Boersema, J. J. (2012). Can we cut out the meat of the dish? Constructing consumer-oriented pathways towards meat substitution. *Appetite, 58*(1), 39–47. doi:10.1016/j.appet.2011.09.009 PMID:21983048

Scott, S. E., Inbar, Y., & Rozin, P. (2016). Evidence for absolute moral opposition to genetically modified food in the United States. *Perspectives on Psychological Science, 11*(3), 315–324. doi:10.1177/1745691615621275 PMID:27217243

Shankar, B., Agrawal, S., Beadreault, A. R., Avula, L., Martorell, R., Osendarp, S., ... McLean, M. S. (2017). Dietary and nutritional change in India: Implications for strategies, policies, and interventions. *Annals of the New York Academy of Sciences, 1395*(1), 49–59. doi:10.1111/nyas.13324 PMID:28306140

Shoemaker, N. (2017). Price of lab-grown burger falls from $325K to $11.36. *Big Think*. Retrieved from https://bigthink.com/ideafeed/answering-how-a-sausage-gets-made-will-be-more-complicated-in-2020

Siegrist, M., & Sütterlin, B. (2017). Importance of perceived naturalness for acceptance of food additives and cultured meat. *Appetite, 113*, 320–326. doi:10.1016/j.appet.2017.03.019 PMID:28315418

Siegrist, M., Sütterlin, B., & Hartmann, C. (2018). Perceived naturalness and evoked disgust influence acceptance of cultured meat. *Meat Science, 139*, 213–219. doi:10.1016/j.meatsci.2018.02.007 PMID:29459297

Simon, M. (2018, February 16). Lab-grown meat is coming, whether you like it or not. *Wired*. Retrieved from https://www.wired.com/story/lab-grown-meat/

Sutherland, I. A., & Leathwick, D. M. (2011). Anthelmintic resistance in nematode parasites of cattle: A global issue? *Trends in Parasitology, 4*(4), 176–181. doi:10.1016/j.pt.2010.11.008 PMID:21168366

Tappel, A. (2007). Heme of consumed red meat can act as a catalyst of oxidative damage and could initiate colon, breast and prostate cancers, heart disease and other diseases. *Medical Hypotheses, 68*(3), 562–564. doi:10.1016/j.mehy.2006.08.025 PMID:17045417

The European Food Information Council (EUFIC). (2006). *The determinants of food choice*. Retrieved from http://www.eufic.org/en/healthy-living/article/the-determinants-of-food-choice

Tilman, D., & Clark, M. (2014). Global diets link environmental sustainability and human health. *Nature, 515*(7528), 518–522. doi:10.1038/nature13959 PMID:25383533

Tiplady, C., Walsh, D. B., & Phillips, C. J. C. (2012). Cruelty to Australian cattle in Indonesian abattoirs – how the public responded to media coverage. *Journal of Agricultural & Environmental Ethics, 26*, 869–885. doi:10.100710806-012-9412-0

Tucker, C. (Ed.). (2018). *Advances in cattle welfare*. Duxford, UK: Woodhead Publishing.

Tucker, C. A. (2014). The significance of sensory appeal for reduced meat consumption. *Appetite, 81*, 168–179. doi:10.1016/j.appet.2014.06.022 PMID:24953197

Tuomisto, H. L., & de Mattos, M. J. T. (2011). Environmental impacts of cultured meat production. *Environmental Science & Technology, 45*(14), 6117–6123. doi:10.1021/es200130u PMID:21682287

Uhe, A. M., Collier, G. R., & Odea, K. (1992). A comparison of the effects of beef, chicken and fish-protein on satiety and amino-acid profiles in lean male subjects. *The Journal of Nutrition, 122*(3), 467–472. doi:10.1093/jn/122.3.467 PMID:1542005

Verbeke, W., Marcu, A., Rutsaert, P., Gaspar, R., Seibt, B., Fletcher, D., & Barnett, J. (2015). "Would you eat cultured meat?": Consumers' reactions and attitude formation in Belgium, Portugal and the United Kingdom. *Meat Science, 102*, 49–58. doi:10.1016/j.meatsci.2014.11.013 PMID:25541372

Verbeke, W., Sans, P., & Van Loo, E. J. (2015). Challenges and prospects for consumer acceptance of cultured meat. *Journal of Integrative Agriculture, 14*(2), 285–294. doi:10.1016/S2095-3119(14)60884-4

Vivalt, E., & Macdonald, B. (2017). *The impact of new products on ethical beliefs*. Retrieved from https://osf.io/4dczh

Weary, D. M., & von Keyserlingk, M. A. G. (2017). Public concerns about dairy-cow welfare: How should the industry respond? *Animal Production Science, 57*(7), 1201–1209. doi:10.1071/AN16680

Wilks, M., & Phillips, C. J. C. (2017). Attitudes to in vitro meat: A survey of potential consumers in the United States. *PLoS One, 12*(2), e0171904. doi:10.1371/journal.pone.0171904 PMID:28207878

Wilson, A. E., Chislock, M. F., Yang, Z., Barros, M. U. G., & Roberts, J. F. (2018). Pond bank access as an approach for managing toxic cyanobacteria in beef cattle pasture drinking water ponds. *Environmental Monitoring and Assessment, 190*(4), 247. doi:10.100710661-018-6595-5 PMID:29574498

Wolka, K., Sterk, G., Biazin, B., & Negash, M. (2012). Benefits, limitations and sustainability of soil and water conservation structures in Omo-Gibe basin, Southwest Ethiopia. *Land Use Policy, 73*, 1–10. doi:10.1016/j.landusepol.2018.01.025

Zebeli, Q., Ghareeb, K., Humer, E., Metzler-Zebeli, B. U., & Besenfelder, U. (2015). Nutrition, rumen health and inflammation in the transition period and their role on overall health and fertility in dairy cows. *Research in Veterinary Science, 103*, 126–136. doi:10.1016/j.rvsc.2015.09.020 PMID:26679807

ADDITIONAL READING

de Boer, J., & Aiking, H. (2011). On the merits of plant-based proteins for global food security: Marrying macro and micro perspectives. *Ecological Economics, 70*(7), 1259–1265. doi:10.1016/j.ecolecon.2011.03.001

Hoek, A. C., Luning, P. A., Weijzen, P., Engels, W., Kok, F. J., & de Graaf, C. (2011). Replacement of meat by meat substitutes. A survey on person- and product-related factors in consumer acceptance. *Appetite, 56*(3), 662–673. doi:10.1016/j.appet.2011.02.001 PMID:21315123

Nijdam, D., Rood, T., & Westhoek, H. (2012). The price of protein: Review of land use and carbon footprints from life cycle assessments of animal food products and their substitutes. *Food Policy, 37*(6), 760–770. doi:10.1016/j.foodpol.2012.08.002

Phillips, C. J. C. (2015). *The animal trade*. Oxford, United Kingdom: CAB International. doi:10.1079/9781780643137.0000

Phillips, C. J. C. (2018). *Principles of cattle production* (3rd ed.). Wallingford, United Kingdom: CAB International. doi:10.1079/9781786392701.0000

Post, M. J. (2014a). An alternative animal protein source: Cultured beef. *Annals of the New York Academy of Sciences, 1328*(1), 29–33. doi:10.1111/nyas.12569 PMID:25376889

Scott, S. E., Inbar, Y., & Rozin, P. (2016). Evidence for absolute moral opposition to genetically modified food in the United States. *Perspectives on Psychological Science, 11*(3), 315–324. doi:10.1177/1745691615621275 PMID:27217243

Shapiro, P. (2018). *Clean meat: How growing meat without animals will revolutionize dinner and the world*. New York, NY: Gallery Books.

Siegrist, M., & Sütterlin, B. (2017). Importance of perceived naturalness for acceptance of food additives and cultured meat. *Appetite, 113*, 320–326. doi:10.1016/j.appet.2017.03.019 PMID:28315418

Tuomisto, H. L., & de Mattos, M. J. T. (2011). Environmental impacts of cultured meat production. *Environmental Science & Technology, 45*(14), 6117–6123. doi:10.1021/es200130u PMID:21682287

Zimmermann, K. L. (2013). Public perceptions of agri-food applications of genetic modification - A systematic review and meta-analysis. *Trends in Food Science & Technology, 30*(2), 142–152. doi:10.1016/j.tifs.2013.01.003

KEY TERMS AND DEFINITIONS

Animal Welfare: The wellbeing of animals, particularly farm and domestic animals as well as animals kept in zoos; countries have different understanding and consequently different standards about what is considered good living conditions for animals; the concept is based on the understanding that animals are sentient beings.

Cattle: Large ruminant animals domesticated mainly for meat and milk; the female animals are called cows and the male animals are called bulls; they are held as property or raised for use, including slaughter for the production of meat.

Cattle Farming: A type of agriculture which is focused on the commercial production of cow milk and beef and veal meat.

Clean Meat/Cultured Meat/Lab-Grown Meat: Meat grown in-vitro from animal cells; it requires less resources and produces less pollution compared to livestock-based meat.

Meat: The flesh of an animal or plant which can be consumed as solid food.

Meat Alternatives (or Alternatives to Meat): Food products based on plants and lab-grown in-vitro meat (the latter is also described as clean or cultured meat); these products can also be described as meat analogues or meat substitutes.

Plant-Based Meat: A food alternative to animal meat which resembles it but is produced from plants.

Resource Use: Exploitation of the natural and social capital of the planet; some resources are finite and can be exhausted through continued use while others are renewable, and their reserves can be replenished.

Ruminant: A herbivore mammal which has a specialized stomach (called rumen) prior to digestion where the plant-based food is fermented through microbial actions; the origin of the word is from Latin and means chewing again.

Sustainable: Able of enduring and lasting for a long time by causing little or no damage to the natural environment whilst providing good livelihood and economic opportunities for people.

Chapter 14
Reconciling Not Eating Meat and Masculinity in the Marketing Discourse for New Food Alternatives

Diana Bogueva
Curtin University, Australia

Dora Marinova
Curtin University, Australia

ABSTRACT

Traditional hegemonic masculinity can be traced on the typical man's plate where meat represents the centerpiece. Meat consumption dominates the current marketing discourse which builds on masculinity to reinforce the stereotyped gender-based diets. In light of scientific evidence about the detrimental impacts of meat consumption on human wellbeing and environmental health, this chapter argues that men are at the crossroads where the concept of masculinity is being redefined. Their social role is similarly changing with new expectations for more sustainable diets which call for plant-based food choices and possibly lab-grown meat. Some men are endorsing these imperatives while others continue to succumb to social inertia. A new marketing discourse is needed which reconciles masculinity with not eating meat and encourages a transition to alternative dietary choices that are better for personal health, allow improved use of the planet's resources, and have less impact on climate change.

INTRODUCTION

Meat's association with manliness in Western cultures is based on a long-established socially constructed gender identity, a norm and a way for society to exert pressure on men's food selection to communicate their masculinity. Being a complex food choice, the consumption of meat, and in particular red meat, endorses a pleiad of meanings and traits contributing to creating perceptions about men's masculine identity (Adams, 1990; Fiddles, 1991; Rogers, 2008; Ruby & Heine, 2011; Rozin, Hormes, Faith, &

DOI: 10.4018/978-1-5225-7350-0.ch014

Wansink, 2012; Rothgerber, 2013; Meah, 2014; Schösler, de Boer, Boersema, & Aiking, 2015; Sumpter, 2015; Bogueva & Phau, 2016; Bogueva & Marinova, 2018). This is in contrast to the precarious feminine identity which gravitates around avoiding meat and preference for plant-based options (Prättälä, Paalanen, Grinberga, Helasoja, Kasmel, & Petkeviciene, 2007; Zhu, Brescoll, Newman, & Uhlmann, 2015; Bogueva & Marinova, 2018). In the West, the hegemonic meat culture is supported by the ultra-powerful livestock industry and associated lobby groups which constantly bombard the public through advertising, especially on the social media, influencing them to remain meat-hooked (Brester & Schroeder, 1995; McDermott, 2012; Shanker, 2015; Zaraska, 2016; Hanrahan, Elvery, McGhee, & Liddy, 2017; Hunt, 2017). The reinforcement of manliness through meat consumption messages deprives consumers of the ability to make informed and independent decisions about their eating preferences, including quantities and frequency, and forces them to separate food by gender appropriation. Splitting food into male and female manifests the acceptance of gendered dietary norms and also subconscious individual approval of such socially-established perceptions. Recent studies show that men tend to choose significantly more gender-normative masculine meal options, usually containing large portions with ample meat, while women do not object to lighter, vegetarian, plant-based foods (Sobal, 2005; Bartlett, 2010; Gal & Wilkie, 2010; Potts & Parry, 2010; McPhail, Beagan, & Chapman, 2012; Rothgerber, 2013; Cavazza, Guidetti, & Butera, 2015; Zhu, Brescoll, Newman, & Uhlmann, 2015; Bogueva, Raphaely, Marinova, & Marinova, 2017). In addition, a male choice to not consume meat is often perceived as an assault on the gender stereotype compromising masculinity and the male gender identity (Sobal, 2005; Gal & Wilkie, 2010; Bogueva & Marinova, 2018).

These studies illustrate a strong socio-psychological dependence of the male sex on the commonly established norm that eating meat is a sign of masculinity, muscle, testosterone, strength, power, dominance and aggression. Such qualities allow humans to kill mammals and make them part of the food chain. The inability or reluctance to engage in these hegemonic practices of masculinity is punishable by mockery, ridicule, hostility, discrimination and even labelling men's sexuality as female or homosexual – perceived by the masculine carnivorous male as having a lower place in the social hierarchy (Nathanson & Young, 2001, p. 203; Connell & Messerschmidt, 2005; Schrock & Schwalbe, 2009; Faludi, 2000, p. 41; Pierce, 2010; Pompper, 2010). According to Nath (2010, p. 266), "criticisms, and comments expressing fault, bewilderment or severe disapproval, are the principal tools that nonvegetarian men use to ensure observance of, or obedience to the established standard of consuming meat and animal products on a regular basis". Vegetarian food choices by a man are rarely tolerated by males who support the meat-eating hegemonic masculinity (Pierce, 2010) and see such diets as too extreme and not normal.

The marketing world builds and reinforces such stereotypes. However, we are now at a point in time when people's food choices are irreversibly affecting climate change, the use of the common resources of this planet and also negatively impacting the public health system (Raphaely & Marinova, 2016). Continuing marketing meat as a male option will be doing everybody injustice. A new portrayal of masculinity is required which should reconcile avoiding meat with the perceptions of power, strength and goodness.

This chapter examines the emerging alternative understanding of masculinity in response to the current crisis situation requiring non-animal based food choices to be seen as superior. Before we do this, we examine the role of marketing discourses in society and how they shape people's behavior. We then expand on the currently prevailing marketing masculinity discourse in relation to consuming meat. Finally, we discuss the need for alternative social shaping of masculinity which is better aligned with scientific evidence about human and environmental health. The emerging market of new meat alternatives is still unchartered waters for marketing. There is good opportunity to seize the momentum and build a discourse which reconciles masculinity with the current planetary health priorities.

MARKETING DISCOURSE

A discourse is a particular way of talking about an issue which is embedded in the current structures of power and dominant culture and institutions within society (Foucault, 1980; Weedon, 1987). Discourses essentially represent sets of statements (Ardley & Quinn, 2014, p. 99) which shape the social reality from a particular subjective point of view. These statements can be expressed through words but also visually with images. Contrary to physical coercion, discourses put psychological and social pressure on members of society by taking the status of representing the "truth" and with this influence people's thinking and practices. They establish a hegemonic world in which alternative thinking or practices are subjugated, suppressed and dismissed.

According to Foucault (1980), there are some discourses that do not allow for new knowledge to emerge while others enable alternative thinking to appear. Science discourses, for example, encourage the production of new knowledge, reflexivity and flexible adjustment to new scientific evidence. The legal system is an example of a discourse which suppresses alternative views. Overall discourses are linked to issues considered contentious from a political, cultural, social and even scientific point of view, and aim to intervene in some way (Gee & Handford, 2012). Power significantly influences the construction of meaning within society and the dominant discourses (Peeples, 2015).

Marketing is one of the tools with which those who hold power in society convey their influence and authority. Hence for marketing, discourses are not just a combination of content-free words, but they define social responses or the social responses that are targeted. In fact, "discourse is the story of reality as it is presented to us through media or other cultural texts" (Van den Bulck, 2010, p. 85). Discourse is not what is real, it actually produces what consumers understand to be real, so they can easily be hooked to it (Wood, 2015). It helps widespread perceptions and common understandings to rise (Lull, 2000, p. 173).

In our modern societies, mass media and advertising are considered to play a key role as a propagator of the marketing discourse. They promulgate lifestyles, signs and symbols made to convey certain meanings, social trends and forms of self-presentation. These ideas shape our understanding of truth and inevitably form our consumer behavior, actions and acceptance of particular consumption models. The media representations are believed to be constructive for the environment and the society we live in, but also influence the relationships with our physical bodies, sexuality and our own emotional needs (Gauntlett, 2002, p. 113). Food-related discourses are not an exception. Marketing constrains people, preventing them from doing or choosing things that are not in line with the socially accepted norms or dogma. For example, abstaining from meat consumption is seen as being outside the socially prevailing Western food practices. The discourse promoted through marketing does not unswervingly exert a direct influence on members of society, but rather has a hegemonic power that forces people to conform to certain modes of thinking and behavior.

Intertextual by nature, advertising usually draws its ideas from existing social knowledge and cultural trends. Gender ideology is an easy way to convey images and build meaning based on the natural difference between men and women. Advertising deliberately relies on widely assumed or narrowly held masculine codes, ideas, values and knowledge reproducing dominant social assumptions about masculinity to create "truth". The outcome is that marketing often maintains myths and reproduces dominant social assumptions about masculinity or femininity despite evidence to the contrary. Furthermore, when there is dominance of a certain commonly shared tradition in place, "individual actions do not have to be analyzed and thought about so much because choices are already prescribed by traditions" (Gauntlett, 2002, p. 96). Masculinity and femininity are already socially constructed and constituted in their discourse (Connell, 1995, p. 5).

Creating meat myths is a distinctive area of marketing impact (Bogueva & Phau, 2016) which benefits those with power, including economic power, as "high personal consumption rates… keep the system moving" (Heighton & Cunningham, 1984, p. 119). Marketing has long relied on a discourse emphasizing the link between meat and masculinity as a powerful myth to capture consumer behavior and support this dominant culture within society.

MASCULINITY MARKETING DISCOURSE

Culturally defined and further shaped by social interactions, masculinity is not a biological characteristic (Kimmel & Tissier-Desbordes, 1999), not a "fixed or unitary", but "invented category" (Nixon, 1997, p. 301) and is a traditional attribute of men. Media advertising and marketing thrive on masculinity and promote it heavily as reflecting the reality of society. The marketing discourses related to masculinity function as tools to shape the way male consumers perceive their masculine world and themselves as part of it, based on popular myths and traditions from previous generations of what is normal, typical, expectable and acceptable of being a man. For example, the concept of hegemonic masculinity is used in the mass media for representing diverse and selected images of men (Connell & Messerschmidt, 2005).

In relation to food, the culturally defined and socially institutionalized dominant masculinity discourse builds on men's love for meat. This marketing discourse is served very carefully by the media and advertising funded by the powerful livestock industry and is becoming a social clinch promoting manliness and excessive meat consumption. It narrowly portraits what it means to be a man (Nakayama, 2004) showing muscular role models and male stereotypes with undelaying suggestions of heavy meat consumption lifestyles as part of men's self-representation.

Marketing is pumped up not only with men roaring for meat like in the Burger King's "I am man" double whopper TV commercial (Burger King, 2006). It is also rife with stereotypes about both genders, defining them by their diet and the food they consume – men eat meat, women prefer fruits and vegetables (Rozin, Hormes, Faith, & Wansink, 2012). In media messages, norms about what constitutes a "proper" diet choice are often based on socially constructed concepts of gender. One of the main implications of this advertising and the overwhelming overarching stereotypes is the division it creates between the sexes and the expectations of certain behavior that often limit the actions and desires of an individual in deciding how to behave differently. As Rogers (2008, p. 282) argues, advertisements like the Burger King commercial that link directly meat and masculinity, are frequently shaped as opposed to feminized "environmental and animal rights movements", which tend to focus on less meat-based diets. In the same line of gender understanding, eating meat "allows one to be seen as masculine, and the avoidance of meat permits one to be viewed as feminine" (Sumpter, 2015, p. 104).

Masculinity is an integral part of the strategy of the food industry that also directs its products specifically to men and the satisfaction of machoism. The "Hungry-Man" frozen dinner line with a slogan "Eat Like a Man" offers ample quantities comfort food incorporating fried chicken, pulled pork, grilled bourbon steak strips and other animal-based proteins which come at a higher cost justifying the male expectations (Watrous, 2014). Stouffer's Fit Kitchen line similarly claims to have created meat-centered meal products for the health-conscious male lifestyle (Krasny, 2015). These are prime examples of male-oriented marketing and efforts for designing specific male products to capture the market share of the present-day pressed for time professional males.

The marketing industry cleverly plays with the man's ego, manipulating it with meat masculinity symbolism. In the advertisement for another product, the relationship between masculinity and meat is taken a step further by equating beef with man. Instead of serving man's palate, the "Fresh Meat Masculine Wipes" (Scotch Porter, 2017) are aimed at his phallic power and assisting his personal hygiene and care for his physical manly attributes – "balls" and "extra-large bratwurst". In this ad meat is taken literally as symbolizing men's potency and testosterone.

As some researchers argue, consumer culture can offer a platform for discursive resources and men can draw on the symbols of masculinity represented by media imagery to be treated as being "of a particular type" (Norman, 2011, p. 433). As such, the media can be regarded as a special structural resource in which the relationship between meat flesh and masculinity is created and recreated. Participation in consumer practices is one form in which men have focused to reshape their bodies to maintain the cultural standards and expectations of their masculinity status quo (Norman, 2011, p. 431), using diet and exercise.

The relationship between masculinity and meat in advertisements is often presented in unremitting oversimplified opposition to "feminine" food, such as vegetables (Burger King, 2006) and tofu (Hummer, 2006). Nowadays men are targeted and pursued with specific media messages about what constitutes "real" masculinity that is frequently defined in food advertisements "by the rejection of small portions, bourgeois aesthetics, quiche, and tofu, as well as by eating meat and performing acts of physical strength" (Rogers, 2008, p. 295).

Furthermore lately, the meat industry also uses the stereotyped characteristics of strength, power and high performance to influence women on consuming more animal proteins. For example, Meat & Livestock Australia (MLA) developed a series of advertisements "Beef: The Rival", "Beef: The Contender" and "Beef: The Challenger" which link women's poor performance and daily failures to insufficient red meat diets. The recommendation from all these three advertisements in the series is the feminine gender to shift to a beef-rich diet which should make women stronger, powerful and reliable like men with the message: "Missis nice guy, fight back, get the right fuel, and taste victory" (MLA, 2016). It is obvious that any financial benefits from such a shift will flow to the meat industry which will allow it to maintain its power within society.

Marketing discourses by the livestock industry penetrate the public realm not only through direct advertising but also through the softer suggestive approach of placing meat and other animal-based foods inside the imaginary world of movies and television programs. In recent years, specific cooking programs and food movies have also become extremely popular. Such eating and cooking scenes are designed to capture the viewers' attention and arouse their appetite (Hamadeh & Estepan, 2018). They reinforce the masculine stereotypes, their messages are persuasive, remain in people's memories (often accompanied with an attractive story line) and influence their eating preferences.

The masculinity discourse can be supplemented with other themes channeling different feel-good media messages, which build cultural expectations and create particular emotions. For example, in 2014 MLA launched its "You're Better on Beef" advertising campaign to convey to consumers that they will perform better if they eat red meat. The campaign was also aimed at reassuring consumers that there are no barriers to high beef consumption, such as health concerns (MLA, 2017). However, against increasing evidence to the contrary, in 2017 MLA switched to a new message built around greatness, namely

"Australian Beef. The Greatest", encouraging consumers to feel proud with their daily beef choices (MLA, 2017). This reinforces the images of men (and women) doing great things, puts pressure on them to support the best beef on earth and aims at counteracting the Australian population's changing lifestyles and dietary preferences.

The study by Norman (2011, p. 430) talks about prevailing discourses creating double-bind of masculinity among young men "where they are simultaneously incited to work on and transform their bodies into culturally recognizable ideals, while at the same time remaining distant and aloof to the size, shape, and appearance of their bodies". Such a double-bind is even more dangerous in relation to meat consumption as the scientific evidence points to numerous health problems generated by excessive intake of animal-based foods. The list of studies is indisputably long and includes:

- Heart disease (Burke & de Francisco, 2005; Micha, Wallace, & Mozaffarian, 2010; Bronzato & Durante, 2017);
- Diabetes (Micha, Wallace, & Mozaffarian, 2010; Shaw, 2012; Live science, 2012; Barnard, Levin, & Trapp, 2014; Talaei, Wang, Yuan, Pan, & Koh, 2017);
- Obesity (Wang & Beydoun, 2009; You & Henneberg, 2016);
- Cancers (Greger, 2012; IARC & WHO, 2015; WHO, 2018a, 2018b);
- Erectile dysfunction (Esposito, Giugliano, Maiorino, & Giugliano, 2010; Maiorino, Bellastella, & Esposito, 2015; Castleman, 2018);
- Reduced longevity and life expectancy (Pan, Sun, Bernstein, Schulze, Manson, Stampfer, Willett, & Hu, 2012; Kmietowicz, 2017; WHO, 2018b).

Whilst vested industry interests continue to argue for more evidence (NCBA, 2015) and bombard the consumer with images of meat and masculinity, the meat-based macho food is essentially killing men (Sax, 2016).

In fact, the situation is even worse when the environmental impacts of meat-rich diets are brought into the picture (Marinova & Raphaely, 2018; Sabaté & Jehi, 2018). Irrespective of the type of environmental impact considered – greenhouse gas emissions and climate change, land use changes and degradation, water use, pollution, biodiversity loss, administering of antibiotics, application of fossil fuels as a source of energy and as fertilizers, the consumption of meat "is associated with a heavier ecological footprint that exceeds the capacity of the planet and its ability to feed the global population" (Marinova & Raphaely, 2018, p. 13). Hence, the masculinity's relationship of social and cultural dominance (Cook, 2000) extends to people but also to other species and the planet itself.

The changing health and environmental awareness however impacts on the modern man's consumer choices which the dominant masculinity marketing discourse wants to avoid. Men are starting to question the nature of male power (Creed, 2003). A new alternative masculinity is emerging linked to creating positive images of this changing man who is concerned about his own health, reacts to other people's feelings and expresses his own emotions as well as endorses the notion of environmental care and climate change action as part of today's cultural expectations. These new alternatives are gradually disconnecting meat consumption and masculinity, establishing a new role for men. They require the reconciling of the marketing discourse to take away the power from the sections of society which are responsible for generating environmental harm, triggering climate change, causing animal suffering and deteriorating human wellbeing.

CHANGING MASCULINITY

The concept of masculinity is not static; it is not a fixed entity but a configuration of practices which change according to the historic and social settings. Masculinity models "in various ways, express widespread ideals, fantasies, and desires" (Connell & Messerschmidt, 2005, p. 838). They are linked to social expectations and respond to societal priorities.

Masculinity has recently undergone some gradual changes, especially in western societies including Australia, toward more health conscientious and sustainability oriented behavior (Bogueva & Marinova, 2018). It is slowly overcoming the traditional tough and silent man comportment (Courtenay, 2003; De Visser, Smith, & McDonnell, 2009; O'Neil, 2013; Reiner, 2016), endorsing a more emotional form (Forrest, 2010; Roberts, 2013; Holmes, 2015). In addition, men have become increasingly comfortable in displaying some typical feminine qualities, such as expressing feelings, being sensitive, emotionally honest, soft, nurturing, involved fathers in raising children (McMahon, 1999; Allen, 2007; Forrest 2010). These qualities are becoming more accepted in modern public culture (Reiner, 2016; Wood, 2009) and are not perceived as a weakness in the new masculinity values. Multiple studies show men's willingness to deviate from the old-fashioned gender emotional norms, stereotypes, cultural and societal assumptions about how they should behave and think about their identity. The media serving the meat industry are still acting blind for this newly emerging trend. Through advertising and marketing messages they tend to reinforce the dominant ideology of hegemonic masculinity failing to portray the changing cultural norms. Male machismo is visibly open for a change as men are slowly and continuously re-defining masculinity (Anderson, 2008; Forrest, 2010; Montes, 2013; Roberts, 2013, 2015; White & Peretz 2009, Bogueva & Marinova, 2018).

Old-fashioned masculinity and manhood are maybe in "crisis" (Clare, 2001; Rogers, 2008) due to men's diminished role as a family provider and bread earner, which nowadays is successfully taken or supported by the less passive, stronger, independent and confident woman (Popa & Gavriliu, 2015, p. 1203). Men's virility power is also weakened in relation to human reproduction, because of the changing societal norms with lower numbers of children and one-parent families as well as with the advancements in medicine. In the new culture of the 21st century, the modern man is ready to solve the identity crisis and renegotiate his place betraying the old masculinity architypes and reorienting himself towards his own health and planetary sustainability (Bogueva & Marinova, 2018). He is not afraid to embrace the change and also become a loving, caring, considerate, vulnerable, emotional, affectionate, happier, consultative, gentle, family-oriented and engaging man in contrast to the old-fashioned, socially embedded traditional male identity described with manliness, power, strengths, virility, possession, violence, aggression, authority, hegemony, supremacy, testosterone, independence, self-confidence, assertiveness and other manly traits (Kimmel & Tissier-Desbordes, 1999; Bogueva & Marinova, 2018).

In the new millennium, the conceptions of gender are slowly, but steadily being transformed. There is a clear transition in social beliefs about male traits with society considered to be in the middle of a "tectonic shift" in masculinity (Schiller, 2015) which makes it hard to untangle gender roles and norms, with many different manifestations of masculinity and femininity (Timke & O'Barr, 2017; Farrelly, 2018). Some media have also begun to abandon inculcated gender roles in favor of a more fluid (Monllos, 2016) and unstereotyped (Beaudoux, 2017) representation although gender differentiations exist in the way boys and girls represent themselves, including in the social media (Herring & Kapidzic, 2015). Within these changing trends, food marketing continues to "do gender" (Sobal, 2005) by favoring meat, especially red meat, as an archetypical masculine food, and fruits, vegetables and sweets as typical feminine choices (Wardle, Haase, Steptoe, Nillapun, Jonwutiwes, & Bellisie, 2004; Sobal, 2005, Nath, 2010; Arganini, Saba, Comitato, Virgili, & Turrini, 2012; Rozin, Hormes, Faith, & Wansink, 2012; Rothgerber, 2013; Vartanian 2015; Wong, 2017; Bogueva & Marinova, 2018).

The masculinity marketing discourse in relation to food does not reflect advances in dietary guidelines or health warnings. For example, the latest American dietary guidelines include a vegetarian diet as a legitimate choice for everybody and one of the three recommended healthy eating patterns (Mangels, 2018) while the World Health Organization warns about the carcinogenic risk associated with the consumption of red meat (WHO, 2015). Social construction of masculinity in western society is often linked to images of high performing athletes and sport itself has long been seen as a masculine domain (Drummond, 2002). After conducting a thorough analysis of the specific requirements of athletes at all levels of performance – from recreational to elite, Larson-Meyer (2018) concludes that their needs for energy, micronutrients, vitamins and minerals can be adequately met on a plant-based diet.

The discourse about whether meat, and the western type levels of high meat consumption, is beneficial or detrimental to human health and performance, continues to be very vivid not only in academic research but also in all forms of media. A study of the meat's representation in relation to human health in the mass media in the period from 2001 to 2015 shows that the discourse has intensified with a larger number of news articles and increasing sensationalism (Leroy, Brengman, Ryckbosch, & Scholliers, 2018). Although since 2003 the share of news items which report meat's association with health as negative has been consistently higher than those which see it as positive, Leroy et al. (2018, p. 345) describe the media debate as "rowdy", "dissonant", "histrionic", "with serious contradictions" and overall heterogeneous. Across all media platforms, the link with the deep-seated symbolism of masculinity, virility and strength persists in the news which portray meat as beneficial to human health (Leroy & Praet, 2015; Leroy, Brengman, Ryckbosch, & Scholliers, 2018).

Furthermore, the marketing discourse does not signal the grave concerns and existential threats to human life on this planet as a result of meat's enormous negative environmental impacts and potential to single-handedly push global warming beyond safe levels (Pelletier & Tyedmers, 2010). Some are even of the opinion that environmental factors alone cannot influence reduction in meat consumption, particularly in countries such as New Zealand where it is perceived to be essential part of the cultural identity and agricultural economy (Tucker, 2018). Against the strong opposition to reducing meat consumption because of vested interests and cultural inertia, government-funded social marketing campaigns educating and exposing the facts about animal-based foods, seem to be the best way to influence the individual consumers to "vote with their forks" (Guthman & Brown, 2016) or chopsticks.

According to Gelfer (2017., n.p.), the present-day connection between masculinity and the natural environment is a pharmakon – simultaneously beneficent and maleficent: "masculinity is one of the great drivers of environmental destruction, but it also has the potential to be one of its great saviors". The protection of the environment requires strength and commences with people's food choices, including the dietary preferences of men. Hultman (in MacGregor & Nicole Seymour, 2017, p. 12) summarizes the nexus between men and nature as follows: "Men are a big part of the environmental problem… especially white, wealthy, middle-aged men who travel too much, eat too much meat, and live in energy-intensive buildings. We need… a possible exit politics for men who want to change". An alternative masculinity marketing discourse about food should facilitate such an exit and allow men to become the great saviors we all need.

ALTERNATIVE MASCULINITY MARKETING DISCOURSE

Men who have chosen the truly sensible option to eat alternatives to meat and other animal-based products, face immense pressure and are usually subject to significant social censorship, resistance and ridicule from other people who question their masculinity (Nath, 2010). Their choice to avoid eating meat is perceived as a major lifestyle change, challenging and subverting conventional masculinity (Nath, 2010; McDonald, 2000). To avoid being ridiculed, many western men who have tried the vegetarian or vegan options, give up and conform with the dominant cultural and gastronomic norms. The fear of being ostracized, disliked, viewed in a negative light, excluded from taking up leadership positions in society (Browarnik, 2012) are some possible explanations about their inability to persist with actions which transgress the conventional male diet.

To the contrary, a man abandoning the consumption of meat should be seen as the boldest, masculine bravest act, given the strong pressure from the society to do otherwise. Despite the well-justified reasons for making the choice which could be based in environmental values, climate change-related concerns about future generations, health considerations, compassion and rejection of the exploitation of nonhuman animals, this option is not respected by other men and society at broad. It is considered as deviant practice and a non-masculine dietary restriction. Only the strongest and most courageous men – physically and mentally, can do this and many examples exist.

Iconic for their manliness men, as the boxing champion Mike Tyson (Neporent, 2013), the first vegan in combat sports martial art fighter Mac Danzig (Curreri, 2012), the weightlifter Kendrick Farris (Rodio, 2016), the nine-time Olympic gold medallist sprinter Karl Lewis (YouTube VegSource, 2006; Merchant, 2013), the actor, activist and politician Arnold Schwarzenegger (Chiorando, 2018a), other famous actors such as Joaquin Phoenix (Chiorando, 2018b), the "Hunger Games" co-stars the Australian actor Liam Hemsworth (Flores, 2017) and the American Woody Harrelson (Webber, 2018), the teen heartthrob of the 1990s and Hollywood royalty Leonardo DiCaprio (Chiorando, 2018c), are permanently opting out from meat. With their dietary shift, vegetarianism and veganism gain support as a means of masculine physical strength, courage and prowess.

Some of the most powerful men in the world – Bill Gates and Richard Branson, are now investing in alternatives to meat and animal products (Morgan, 2018). The lab-grown clean meat does not require crops to be fed to animals, nor land or water and could become a viable alternative to plant-based options – a disruptive innovation to the livestock sector. It seems that men are already more receptive than women to try such clean meat (Wilks & Philips, 2017). The marketing discourse needs to build on these alternative views and manifestations of masculinity and reconcile not eating meat with the images of innovation, power, goodness and responsibility. Mycek's (2015) analysis of vegetarian and vegan men shows that they are able to subvert the dominant eating style by positioning their choices based on scientific evidence and rational decision making. In other words, they are masculinizing what are perceived to be feminine practices (Mycek, 2015).

Environmental, climate change, health and ethical reasons are taking center stage in the new portrayal of masculinity. The muscle masculinity icon Arnold Schwarzenegger who years ago used to deliver lines, such as "you hit like a vegetarian" (BBC Newsbeat, 2015) shifted to a vegan diet for health reasons (Radke, 2018). Patrik Baboumian – Germany's strongest man (Tashjian, 2015) who holds several Guinness world records for completing the yoke walk with unprecedented weights (the latest being 560kg), is vegan because he wants to adhere to a cruelty-free diet. Environmental factors are challenging the wellbeing of all people on Earth and Arnold Schwarzenegger is now asking people to reduce meat intake as an essential part of the strategies for climate change mitigation through an advertisement that portrays the alternative masculinity marketing discourse. "Less meat, less heat, one life" is a marketing campaign directed by the filmmaker James Cameron which endorses China's initiative to reduce meat consumption by 50% and urges the Chinese and American audiences to curb their intake of animal proteins (Neff, 2016; Shoard, 2016).

The rise of interest in vegan and plant-based alternatives to meat in recent years can be considered as a movement towards understanding the need of preserving human and planetary health. According to the global online food delivery Just Eat, "going vegan" is the biggest food trend of 2018 with 33% of its 93,700 restaurant partners offering vegan options (Just Eat, 2018). With veganism slowly and surely becoming a mainstream dietary choice, other allied movements, such as flexitarianism (de Bakker & Dagevos, 2012; Raphaely & Marinova, 2014; Dagevos & Reinders; 2018), are also working on reducing the global carbon footprint by convincing people to decrease the amount of meat they consume. Given the health and environmental benefits from such a change, strong calls are being made for the need for social marketing (Bogueva & Marinova, 2018) which can be the mechanism to transform the dominant eating behavior. Messages reflecting the transformed masculinity are most suitable to be used in such social marketing of predominantly plant-based diets and food products.

It is easy for an alternative masculinity marketing discourse to build on the existing scientific evidence about bolstered performance and the traditional masculine qualities of logic and rationality. There is no manlier attribute valued by males more than testosterone. However, when it comes to testosterone the love for hegemonic meat consumption proves to be unfavorable to masculinity, as studies show that vegetarians have higher testosterone levels than carnivorous men (Allen, Appleby, Davey, & Key, 2000) and because of this manly hormone they have less cancer (Greger, 2013). Aside from the testosterone, vegan males have more pleasant, more attractive and less intense scents (Stryker, 2007) due to their diet. Red meat consumption in particularly despite its perceived hedonistic attractiveness is found to have a negative impact on body odor (Havlicek & Lenochova, 2006). Also, vegetarianism is scientifically shown to play a beneficial role in promoting health, weight loss and preventing obesity (Campbell & Campbell, 2006; Tonstad, 2009; Sabaté & Wien, 2010; Turner-McGrievy, 2017). The positives of plant-based alternatives to meat are more than clear.

New meat alternatives are still quite complex and incongruous in the minds of consumers who are yet to endorse them. They require greater acceptance but also better health awareness and widespread understanding of the environmental impacts of human diets. The alternative masculinity marketing discourse can emerge from a range of convincing factors which counteract the current social climate and have numerous implications for friends and family, dining out, control over the male body, symbolic connotations of food as power, cost and fad fashion, cultural associations of image, religious beliefs, nutritional links with health and health promotion, fear of the unknown and breaking up with tradition. If marketing professionals and health advocates want to counteract the traditionally strong meat associations with masculinity and replace them with alternatives that resemble red meat with blood and grill marks for a visual authenticity, they can use existing western metaphors influencing meat cognition and consumption (Rozin, Hormes, Faith, & Wansink, 2012).

Presenting meat alternatives in a way that is attractive to the increasingly health and environmentally conscious men is important but a general transition towards meat alternatives across the entire global population is required without delay. Many of the companies which produce meat alternatives deliberately design products targeting "the cravings of meat-lovers without compromising the sensory experience they enjoy… The vast majority of global consumers desire the taste and experience of meat" (Brown, 2018, n.p.). This trend however is changing fast and some believe that within one generation people will stop seeing meat as food (Brown, 2018).

Masculinity urgently needs a new alternative and exit point from the prevailing meat trajectory. Such an alternative should allow men to oppose and overcome stereotypes and myths built by the media and industrial marketing (Bogueva & Phau, 2016) and transition towards healthier and greener dietary choices with regard to human and planetary health. New meat alternatives are one of those challenges that are out of sight for the old-school masculinity. The consumption of meat alternatives should be considered as a request for men to take control over their own nutritional choice, despite the challenges of the dominant cultural and social rules and norms.

Although they are not entirely bloody and with the taste and structure of animal flesh, meat alternatives should be perceived as a better, healthier and more sustainable alternative to the manly meat. Certainly, success in this direction will vary from one man to another, precisely because of the guaranteed clash with traditional masculinity and love for meat. However, over time with the changing modern perceptions of the 21st century masculinity and with the improved possibilities meat alternatives present, different norms combined with new social and institutional power will emerge in support of better and more sustainable diets.

CONCLUSION

With the evolving representation of masculinity in response to the greatest challenges of the 21st century, it appears there is room for alternative dietary choices based on positive changes men can bring to this world. The modern man's consumption choices should be mandated by an individual desire in agreement with his own health priorities and environmental responsibilities. Any marketing discourse, including social marketing, needs to communicate these new messages and inspire men to embrace the alternatives to meat as part of manifesting their true masculinity. Marketing needs to acknowledge and accept its social role beyond selling products and beyond satisfying the requirements of those who pay for the service. It needs to broaden its responsibility and be prepared to support only messages which are scientifically underpinned, particularly as new knowledge emerges.

In marketing, "understanding the metaphor a consumer might have for a brand could move the art of positioning toward more of a science" (Rozin, Hormes, Faith, & Wansink, 2012, p. 641). Understanding the scientific evidence and the perception of masculinity could help find ways to promote meat alternatives in a more acceptable way. Reconciling not eating meat with masculinity is not a simple task and needs further research to clarify the marketing implications of meat alternatives in the male consumers' minds. It also requires a change in the dominant marketing discourse, which currently supports the existing industries and institutions within society and the power they exert. According to Foucault (Gutting, 1994, p. 12), discourses emerge from "an anonymous and polymorphous will to knowledge, capable of transformations and caught up in an identifiable play of dependence". They are trapped in historical inertia which can only be broken through the power of new scientific knowledge and courageous acts by those who are prepared to reject the dependence of the past and embrace change for a better future.

Accepting and adopting meat alternatives, vegetarian or vegan dietary options based on ethical, health and environmental concerns against social discontent is an act of bravery as well as goodness, strength and responsibility – the male qualities defining timeless masculinity. Eating livestock-based products does not need to be part of the essence describing masculinity and the marketing discourse can develop better messages that reflect the new meaning of being a man in this 21st century.

REFERENCES

Adams, C. J. (1990). *The sexual politics of meat: A feminist-vegetarian critical theory*. New York, NY: Continuum Print.

Allen, L. (2007). Sensitive and real macho all at the same time: Young heterosexual men and romance. *Men and Masculinities*, *10*(2), 137–152. doi:10.1177/1097184X05284221

Allen, N. E., Appleby, P. N., Davey, G. K., & Key, T. J. (2000). Hormones and diet: Low insulin-like growth factor-I, but normal bioavailable androgens in vegan man. *British Journal of Cancer*, *83*(1), 95–97. doi:10.1054/bjoc.2000.1152 PMID:10883675

Anderson, E. (2008). Inclusive masculinity in a fraternal setting. *Men and Masculinities*, *10*(5), 604–620. doi:10.1177/1097184X06291907

Ardley, B. C., & Quinn, L. (2014). Practitioner accounts and knowledge production: An analysis of three marketing discourses. *Marketing Theory*, *14*(1), 97–118. doi:10.1177/1470593113512322

Arganini, C., Saba, A., Comitato, R., Virgili, F., & Turrini, A. (2012). Gender differences in food choice and dietary intake in modern western societies. In J. Maddock (Ed.), *Public health social and behavioural health* (pp. 85–102). Rijeka, Croatia: InTech Open Access. doi:10.5772/37886

Barnard, N., Levin, S., & Trapp, C. (2014). Meat consumption as a risk factor for type 2 diabetes. *Nutrients*, *6*(2), 897–910. doi:10.3390/nu6020897 PMID:24566443

Bartlett, T. (2010). Manly meals and feminine foods. *The Chronicle of Higher Education*. Retrieved from https://www.chronicle.com/blogs/percolator/manly-mealsfeminine-foods/25576

BBC Newsbeat. (2015). Arnold Schwarzenegger: Stop eating meat and save the planet. *BBC*. Retrieved from http://www.bbc.co.uk/newsbeat/article/35038053/arnold-schwarzenegger-stop-eating-meat-and-save-the-planet

Beaudoux, V. G. (2017). How media sexism demeans women and fuels abuse by men like Weinstein. *The Conversation*. Retrieved from https://theconversation.com/how-media-sexism-demeans-women-and-fuels-abuse-by-men-like-weinstein-85789

Bogueva, D., & Marinova, D. (2018). What is more important: Perception of masculinity or personal health and the environment? In D. Bogueva, D. Marinova, & T. Raphaely (Eds.), *Handbook of research on social marketing and its influence on animal origin food product consumption* (pp. 148–162). Hershey, PA: IGI Global. doi:10.4018/978-1-5225-4757-0.ch010

Bogueva, D., Marinova, D., & Raphaely, T. (Eds.). (2018). *Handbook of research on social marketing and its influence on animal origin food product consumption*. Hershey, PA: IGI Global. doi:10.4018/978-1-5225-4757-0

Bogueva, D., & Phau, I. (2016). Meat myths and marketing. In T. Raphaely & D. Marinova (Eds.), *Impact of meat consumption on health and environmental sustainability* (pp. 264–276). Hershey, PA: IGI Global. doi:10.4018/978-1-4666-9553-5.ch015

Bogueva, D., Raphaely, T., Marinova, D., & Marinova, M. (2017) Sustainability social marketing. In J. Hartz-Karp & D. Marinova (Eds), Methods for sustainability research (pp. 280-291). Cheltenham, UK: Edward Elgar. doi:10.4337/9781786432735.00030

Brester, G. W., & Schroeder, T. C. (1995). The impacts of brand and generic advertising on meat demand. *American Journal of Agricultural Economics*, *77*(4), 969–979. doi:10.2307/1243820

Bronzato, S., & Durante, A. (2017). A contemporary review of the relationship between red meat consumption and cardiovascular risk. *International Journal of Preventive Medicine*, *8*(1), 40. doi:10.4103/ijpvm.IJPVM_206_16 PMID:28656096

Browarnik, B. (2012). *Attitudes toward male vegetarians: Challenging gender norms through food choices.* Psychology Honors Papers, 25. Retrieved from https://digitalcommons.conncoll.edu/cgi/viewcontent.cgi?referer=https://www.google.com.au/&httpsredir=1&article=1021&context=psychhp

Brown, J. (2018). The rise of the bloody mock meat movement: Fad, or the key to sustainability? *Independent.* Retrieved from https://www.independent.co.uk/life-style/food-and-drink/vegans-vegetarians-meat-substitutes-bloody-mock-meats-food-lifestyle-a8427281.html

Burger King. (2006). *I am man.* Retrieved from https://www.youtube.com/watch?v=vGLHlvb8skQ

Burke, M., & de Francisco, A. (2005). *Monitoring flows for health research 2005: Behind the global numbers.* Geneva, Switzerland: Global Forum for Health Research.

Campbell, T. C., & Campbell, T. M. II. (2006). *The China Study*. Dallas, TX: BenBella Books.

Castleman, M. (2018). *Stop poisoning your sex life*. Men's Journal. Retrieved from https://www.mensjournal.com/food-drink/stop-poisoning-your-sex-life/

Cavazza, N., Guidetti, M., & Butera, F. (2015). Ingredients of gender-based stereotypes about food: Indirect influence of food type, portion size and presentation on gendered intentions to eat. *Appetite*, *91*, 266–272. doi:10.1016/j.appet.2015.04.068 PMID:25934089

Chiorando, M. (2018a). *Arnold Schwarzenegger accused of 'Drinking the vegan kool-aid'*. Beef Magazine, Plant Based News. Retrieved from https://www.plantbasednews.org/post/arnold-schwarzenegger-drinking-vegan-kool-aid)

Chiorando, M. (2018b). *Actor Joaquin Phoenix shares 'profound' experience that turned him vegan.* Plant based news. Retrieved from https://www.plantbasednews.org/post/actor-joaquin-pheonix-shares-profound-experience-that-turned-him-vegan

Chiorando, M. (2018c). *Leonardo DiCaprio's plans to make vegan burgers mainstream.* Plant Based News. Retrieved from https://www.plantbasednews.org/post/revealed-leonardo-dicaprios-plans-vegan-burgers-mainstream

Clare, A. W. (2001). *On men: Masculinity in crisis.* London, UK: Arrow Books.

Connell, R. (1995). *Masculinities.* St Leonards, Australia: Allen & Unwin.

Connell, R. W., & Messerschmidt, J. W. (2005). Hegemonic masculinity: Rethinking the concept. *Gender & Society, 19*(6), 829–859. doi:10.1177/0891243205278639

Cook J. (2000). Men's magazines at the millennium: New spaces, new selves. *Continuum, 14*(2), 171-186. doi:10.1080/713657705

Courtenay, W. H. (2003). Key determinants of the health and well-being of men and boys. *International Journal of Men's Health, 2*(1), 1–30. doi:10.3149/jmh.0201.1

Creed, B. (2003). *Media matrix: Sexing the new reality.* Sydney, Australia: Allen and Unwin.

Curreri, F. (2012). *Mac Danzig's diet - the truth about vegan.* UFC. Retrieved from http://www.ufc.com/news/Mac-Danzig-Diet-The-Truth-About-Vegan?id=

Dagevos, H., & Reinders, M. (2018). Flexitarianism and social marketing: Reflections on eating meat in moderation. In D. Bogueva, D. Marinova, & T. Raphaely (Eds.), *Handbook of research on social marketing and its influence on animal origin food product consumption* (pp. 105–120). Hershey, PA: IGI Global. doi:10.4018/978-1-5225-4757-0.ch007

de Bakker, E., & Dagevos, H. (2012). Reducing meat consumption in today's consumer society: Questioning the citizen-consumer gap. *Journal of Agricultural & Environmental Ethics, 25*(6), 877–894. doi:10.100710806-011-9345-z

De Visser, R. O., Smith, J. A., & McDonnell, E. J. (2009). 'That's not masculine': Masculine capital and health-related behaviour. *Journal of Health Psychology, 14*(7), 1047–1058. doi:10.1177/1359105309342299 PMID:19786531

Drummond, M. J. N. (2002). Sport and images of masculinity: The meaning of relationships in the life course of "elite" male athletes. *Journal of Men's Studies, 10*(2), 129–141. doi:10.3149/jms.1002.129

Esposito, K., Giugliano, F., Maiorino, M. I., & Giugliano, D. (2010). Dietary factors, Mediterranean diet and erectile dysfunction. *Journal of Sexual Medicine, 7*(7), 2338–2345. doi:10.1111/j.1743-6109.2010.01842.x PMID:20487239

Faludi, S. (2000). *Stiffed: The betrayal of modern man.* London, UK: Vintage.

Farrelly, E. (2018). Gender equality: It takes a troubled woman to change a troubled world. *Sydney Morning Herald.* Retrieved from https://www.smh.com.au/opinion/gender-equality-it-takes-a-troubled-woman-to-change-a-troubled-world-20180125-h0oip9.html

Fiddles, N. (1991). *Meat: A natural symbol.* London, UK: Routledge.

Flores, B. (2017). Liam Hemsworth officially crowned Australia's sexiest vegan. *Livekindly*. Retrieved from https://www.livekindly.co/australias-sexiest-vegan-is-officially-liam-hemsworth/

Forrest, S. (2010). Young men in love: The (re)making of heterosexual masculinities through "serious" relationships. *Sexual and Relationship Therapy*, *25*(2), 206–218. doi:10.1080/14681991003660260

Foucault, M. (1980). *Power/knowledge*. Brighton, UK: Harvester.

Gal, D., & Wilkie, J. (2010). Real men don't eat quiche: Regulation of gender-expressive choices by men. *Social Psychological & Personality Science*, *1*(3). doi:10.1177/1948550610365003

Gauntlett, D. (2002). *Media, gender and identity: An introduction*. London, UK: Routledge.

Gee, J. P., & Handford, M. (n.d.). Introduction. In J. P. Gee & M. Handford (Eds.), The Routledge handbook of discourse analysis (pp. 1–6). London, UK: Routledge.

Gelfer, J. (2017). *Masculinity and the environment: A double-edged sword*. Retrieved from https://goodmenproject.com/featured-content/masculinity-and-the-environment-a-double-edged-sword-wcz/

Greger, M. (2012). *Poultry and penis cancer*. Retrieved from https://nutritionfacts.org/video/poultry-and-penis-cancer/

Greger, M. (2013). *Vegan men: More testosterone but less cancer*. Retrieved from https://nutritionfacts.org/2013/02/12/less-cancer-in-vegan-men-despite-more-testosterone/

Guthman, J., & Brown, S. (2016). Midas' not-so-golden touch: On the demise of methyl iodide as a soil fumigant in California. *Journal of Environmental Policy and Planning*, *18*(3), 324–341. doi:10.1080/1523908X.2015.1077441

Gutting, G. (Ed.). (1994). *Cambridge companion to Foucault*. Cambridge, UK: Cambridge University Press. doi:10.1017/CCOL9780521403320

Hamadeh, S., & Estepan, S. (2018). The visual discourse of food and its impact on health: Research and practical implications. *Acta Scientific Nutritional Health*, *2*(6), 12–20. Retrieved from https://actascientific.com/ASNH/pdf/ASNH-02-0084.pdf

Hanrahan, C., Elvery, S., McGhee, A., & Liddy, M. (2017) *Political donations: New data maps industries' web of influence*. ABC news. Retrieved from http://www.abc.net.au/news/2017-02-09/political-donations-industry-dataset/8229192

Havlicek, J., & Lenochova, P. (2006). The effect of meat consumption on body odor attractiveness. *Chemical Senses*, *31*(8), 747–752. doi:10.1093/chemse/bjl017 PMID:16891352

Heighton, E., & Cunningham, D. (1984). *Advertising in the broadcast and cable media*. Belmont, CA: Wadsworth Publishing.

Herring, S. C., & Kapidzic, S. (2015). Teens, gender, and self-presentation in social media. In J. D. Wright (Ed.), *International encyclopedia of social and behavioural sciences* (2nd ed.). Oxford, UK: Elsevier. doi:10.1016/B978-0-08-097086-8.64108-9

Holmes, M. (2015). Men's emotions: Heteromasculinity, emotional reflexivity, and intimate relationships. *Men and Masculinities*, *18*(2), 176–192. doi:10.1177/1097184X14557494

Hummer. (2006). *Tofu*. Retrieved from https://www.youtube.com/watch?v=lL4ZkYPLN38

Hunt, E. (2017). Meatonomics author says government working with meat and dairy industry to boost consumption. *The Guardian*. Retrieved from https://www.theguardian.com/science/2017/may/06/meatonomics-author-says-government-working-with-meat-and-dairy-industry-to-boost-consumption

International Agency for Research on Cancer (IARC) and World Health Organisation (WHO). (2015). *IARC Monographs evaluate consumption of red meat and processed meat*. Retrieved from http://www.iarc.fr/en/media-centre/pr/2015/pdfs/pr240_E.pdf

Just Eat. (2018). *'Going vegan' is predicted to be the biggest food trend of 2018*. Retrieved from https://www.riseofthevegan.com/blog/going-vegan-biggest-trend-for-2018

Kimmel, A. J., & Tissier-Desbordes, E. (1999). Males, masculinity, and consumption: An exploratory investigation. In B. Dubois, T. M. Lowrey, & L. J. Shrum (Eds.), *European advances in consumer research* (Vol. 4, pp. 243–251). Provo, UT: Association for Consumer Research. Retrieved from http://acrwebsite.org/volumes/11390/volumes/e04/E-04

Kmietowicz, Z. (2017). Red meat consumption is linked to higher risk of death from most major causes. *BMJ (Clinical Research Ed.)*, *357*. doi:10.1136/bmj.j2241

Krasny, J. (2015). Wait, are frozen dinner actually good now? *Esquire*. Retrieved from https://www.esquire.com/food-drink/food/a35481/stouffers-new-frozen-dinners/

Larson-Meyer, D. E. (2018). Nutritionally adequate vegetarian diets and athletic performance. In W. J. Craig (Ed.), *Vegetarian nutrition and wellness* (pp. 279–302). Boca Raton, FL: CRC Press.

Leroy, F., Brengman, M., Ryckbosch, W., & Scholliers, P. (2018). Meat in the post-truth era: Mass media discourses on health and disease in the attention economy. *Appetite*, *125*, 345–355. doi:10.1016/j.appet.2018.02.028 PMID:29486208

Leroy, F., & Praet, I. (2015). Meat traditions: The co-evolution of humans and meat. *Appetite*, *90*, 200-211. doi:10.1016/j.appet.2015.03.014

Live Science. (2012). *Want to live longer? Hold the red meat*. The Live Science. Retrieved from https://www.livescience.com/36195-red-meat-consumption-death-risk.html

Lull, J. (2000). *Media, communication, culture*. Cambridge, UK: Polity Press.

MacGregor, S., & Seymour, N. (Eds.). (2017). Men and nature: Hegemonic masculinities and environmental change. *RCC Perspectives: Transformations in Environment and Society*. Retrieved from http://www.environmentandsociety.org/sites/default/files/2017_i4_layout_final_1.pdf

Maiorino, M. I., Bellastella, G., & Esposito, K. (2015). Lifestyle modifications and erectile dysfunction: What can be expected? *Asian Journal of Andrology*, *17*(1), 5–10. doi:10.4103/1008-682X.137687 PMID:25248655

Mangels, R. (2018). Vegetarian diets: Trends in acceptance and perception. What do the dietary guidelines suggest? In W. J. Craig (Ed.), *Vegetarian nutrition and wellness* (pp. 1–12). Boca Raton, FL: CRC Press. doi:10.1201/b22003-1

Marinova, D., & Raphaely, T. (2018). Impact of vegetarian diets on the environment. In W. J. Craig (Ed.), *Vegetarian nutrition and wellness* (pp. 13–24). Boca Raton, FL: CRC Press.

McDermott, M. (2012). *Meat industry grossly outspends animal rights*. Veg Groups, Treehugger. Retrieved from https://www.treehugger.com/green-food/meat-industry-grossly-outspends-animal-rights-veg-groups.html

McDonald, B. (2000). Once you know something, you can't not know it: An empirical look at becoming vegan. *Society & Animals*, *8*(1), 1–23. doi:10.1163/156853000510961

McMahon, A. (1999). *Taking care of men: Sexual politics in the public mind*. Cambridge, UK: Cambridge University Press. doi:10.1017/CBO9780511597107

McPhail, D., Beagan, B., & Chapman, G. E. (2012). I don't want to be sexist but...: Denying and re-inscribing gender through food. *Food, Culture, & Society*, *15*, 473–489. doi:10.2752/175174412X13276629246046

Meah, A. (2014). Reconceptualising "masculinity" through men's contribution to domestic woodwork. In A. Gorman-Murray & P. Hopkins (Eds.), Masculinities and place (pp. 191-208). London, UK: Ashgate.

Meat and Livestock Australia (MLA). (2016). *New MLA ads show Australian women "You're Better on Beef"*. Meat and Livestock Australia. Retrieved from https://www.mla.com.au/news-and-events/industry-news/new-mla-ads-show-australian-women-youre-better-on-beef-/

Meat and Livestock Australia (MLA). (2017). *MLA launches Australian Beef the Greatest*. Meat and Livestock Australia. Retrieved from https://www.mla.com.au/news-and-events/industry-news/mla-launches-australian-beef-the-greatest/

Merchant, B. (2013). *9 superstar athletes who don't eat meat*. Mother nature network. Retrieved from https://www.mnn.com/food/healthy-eating/photos/9-superstar-athletes-who-dont-eat-meat/carl-lewis

Micha, R., Wallace, S. K., & Mozaffarian, D. (2010). Red and processed meat consumption and risk of incident coronary heart disease, stroke, and diabetes: A systematic review and meta-analysis. *Circulation*, *121*(21), 2271–2283. doi:10.1161/CIRCULATIONAHA.109.924977 PMID:20479151

Monllos, K. (2015). *Brands are throwing out gender norms to reflect a more fluid world*. Adweek. Retrieved from https://www.adweek.com/brand-marketing/brands-are-throwing-out-gender-norms-reflect-more-fluid-world-174070/

Montes, V. (2013). The role of emotions in the construction of masculinity: Guatemalan migrant men, transnational migration, and family relations. *Gender & Society*, *27*(4), 469–490. doi:10.1177/0891243212470491

Morgan, R. (2018). *Bill Gates and Richard Branson are betting lab-grown meat might be the food of the future*. CNBC Disruptor 50. Retrieved from https://www.cnbc.com/2018/03/23/bill-gates-and-richard-branson-bet-on-lab-grown-meat-startup.html

Mycek, M. K. (2015). *Meatless meals and masculinity: An examination of men's use of rationality and scientific research to explain their plant-based diets* (Master's dissertation). North Carolina State University. NCSU Libraries, 1-44. Retrieved from https://repository.lib.ncsu.edu/bitstream/handle/1840.16/10723/etd.pdf?sequence=1&isAllowed=y

Nakayama, T. (1989). Images of men in advertising. Center for Media Literacy. *Media & Values, 48*. Retrieved from http://www.medialit.org/reading-room/images-men-advertising#bio

Nath, J. (2010). Gendered fare? A qualitative investigation of alternative food and masculinities. *Journal of Sociology (Melbourne, Vic.)*, *47*(3), 261–278. doi:10.1177/1440783310386828

Nathanson, P., & Young, K. (2001). *Spreading misandry: The teaching of contempt for men in popular culture*. Montreal, Canada: McGill-Queen's University Press.

National Cattlemen's Beef Association (NCBA). (2015). *Science does not support international agency opinion on red meat and cancer*. Retrieved from http://www.beefusa.org/newsreleases1.aspx?newsid=5418

Neff, M. (2016). *Arnold Schwarzenegger wants you to eat less meat – for the sake of the planet's future*. One green planet. Retrieved from http://www.onegreenplanet.org/news/arnold-schwarzenegger-wants-you-to-eat-less-meat/

Neporent, L. (2013). *Mike Tyson on losing 140 pounds and Evander Holyfield's Ear*. ABC News. Retrieved from https://abcnews.go.com/Health/mike-tyson-veganism-evander-holyfields-ear/story?id=18958770

Nixon, S. (1997). Exhibiting masculinity. In S. Hall (Ed.), *Representation: Cultural representation and signifying practices* (pp. 291–336). London, UK: Sage.

Norman, M. E. (2011). Embodying the double-bind of masculinity: Young men and discourses of normalcy, health, heterosexuality, and individualism. *Men and Masculinities*, *14*(4), 430–449. doi:10.1177/1097184X11409360

O'Neil, J. M. (2013). Gender role conflict research 30 years later: An evidence-based diagnostic schema to assess boys and men in counselling. *Journal of Counseling and Development*, *91*(4), 490–498. doi:10.1002/j.1556-6676.2013.00122.x

Pan, A., Sun, Q., Bernstein, A. M., Schulze, M. B., Manson, J. A. E., Stampfer, M. J., ... Hu, F. B. (2012). Red meat consumption and mortality: Results from two prospective cohort studies. *Archives of Internal Medicine*, *172*(7), 555–563. doi:10.1001/archinternmed.2011.2287 PMID:22412075

Peeples, J. (2015). Discourse/rhetorical analysis approaches to environment, media, and communication. In A. Hansen & R. Cox (Eds.), *The Routledge handbook of environment and communication* (pp. 39–48). London, UK: Routledge.

Pelletier, N., & Tyedmers, P. (2010). Forecasting potential global environmental costs of livestock production 2000–2050. *Proceedings of the National Academy of Sciences of the United States of America*, *107*(43), 18371–18374. doi:10.1073/pnas.1004659107 PMID:20921375

Pierce, K. (2010). Men leave their own mark on veganism. *Boston Globe*. Retrieved from http://archive.boston.com/lifestyle/food/articles/2010/03/24/men_leave_their_own_mark_on_veganism/

Pompper, D. (2010). Masculinities, the metrosexual, and media images: Across dimensions of age and ethnicity. *Sex Roles*, *63*(9-10), 682–696. doi:10.100711199-010-9870-7

Popa, D., & Gavriliu, D. (2015). Gender representations and digital media. *Procedia: Social and Behavioral Sciences*, *180*, 1199–1206. doi:10.1016/j.sbspro.2015.02.244

Potts, A., & Parry, J. (2010). Vegan sexuality: Challenging heteronormative masculinity through meat-free sex. *Feminism & Psychology*, *20*(1), 53–72. doi:10.1177/0959353509351181

Prättälä, R., Paalanen, L., Grinberga, D., Helasoja, V., Kasmel, A., & Petkeviciene, J. (2007). Gender differences in the consumption of meat, fruit and vegetables are similar in Finland and the Baltic countries. *European Journal of Public Health*, *17*(5), 520–525. doi:10.1093/eurpub/ckl265 PMID:17194710

Radke, A. (2018). Arnold Schwarzenegger drinks the vegan Kool-Aid. *Beef Magazine*. Retrieved from http://www.beefmagazine.com/ranching/arnold-schwarzenegger-drinks-vegan-kool-aid

Raphaely, T., & Marinova, D. (2014). Flexitarianism: Decarbonising through flexible vegetarianism. *Renewable Energy*, *67*, 90–96. doi:10.1016/j.renene.2013.11.030

Raphaely, T., & Marinova, D. (Eds.). (2017). *Impact of meat consumption on health and environmental sustainability*. Hershey, PA: IGI Global.

Reiner, A. (2016). Teaching men to be emotionally honest. *The New York Times*. Retrieved from https://www.nytimes.com/2016/04/10/education/edlife/teaching-men-to-be-emotionally-honest.html

Roberts, S. (2013). Boys will be boys … Won't they? Change and continuities in contemporary young working-class masculinities. *Sociology*, *47*(4), 671–686. doi:10.1177/0038038512453791

Roberts, S. (Ed.). (2015). *Debating modern masculinities: change, continuity, crisis?* Basingstoke, UK: Palgrave Macmillan.

Rodio, M. (2016). The Vegan diet of American Olympic weightlifter Kendrick Farris. *Men's Journal*. Retrieved from: https://www.mensjournal.com/food-drink/vegan-diet-american-olympic-weightlifter-kendrick-farris/

Rogers, R. (2008). Beasts, burgers and hummers: Meat and the crisis of masculinity in contemporary television advertising. *Environmental Communication*, *2*(3), 281–301. doi:10.1080/17524030802390250

Rothgerber, H. (2013). Real men don't eat (vegetable) quiche: Masculinity and the justification of meat consumption. *Psychology of Men & Masculinity*, *14*(4), 363–375. doi:10.1037/a0030379

Rozin, P., Hormes, J. M., Faith, M. S., & Wansink, B. (2012). Is meat male? A quantitative multimethod framework to establish metaphoric relationships. *The Journal of Consumer Research*, *39*(3), 629–643. doi:10.1086/664970

Ruby, M. B., & Heine, S. J. (2011). Meat, morals, and masculinity. *Appetite*, *56*(2), 447–450. doi:10.1016/j.appet.2011.01.018 PMID:21256169

Sabaté, J., & Jehi, T. (2018). The sustainability of vegetarian diets. In W. J. Craig (Ed.), *Vegetarian nutrition and wellness* (pp. 25–44). Boca Raton, FL: CRC Press.

Sabaté, J., & Wien, M. (2010). Vegetarian diets and childhood obesity prevention. *The American Journal of Clinical Nutrition, 91*(5), 1525S–1529S. doi:10.3945/ajcn.2010.28701F PMID:20237136

Sax, D. (2016). How years of macho food marketing is killing men. *NY Magazine*. Retrieved from http://nymag.com/betamale/2016/06/macho-food-marketing-is-killing-men.htm

Schiller, L. (2015). *Inflection point: The tectonic shift of gender roles*. Kalw Radio, San Francisco. Retrieved from http://kalw.org/post/inflection-point-tectonic-shift-gender-roles#stream/0

Schösler, H., de Boer, J., Boersema, J. J., & Aiking, H. (2015). Meat and masculinity among young Chinese, Turkish and Dutch adults in the Netherlands. *Appetite, 89*, 152–159. doi:10.1016/j.appet.2015.02.013 PMID:25681291

Schrock, D., & Schwalbe, M. (2009). Men, masculinity, and manhood acts. *Annual Review of Sociology, 35*(1), 277–295. doi:10.1146/annurev-soc-070308-115933

Scotch Porter. (2017). Introducing fresh meat masculine wipes. *The Scotch Porter Journal*. Retrieved from http://journal.scotchporter.com/2017/11/introducing-fresh-meat-masculine-wipes/

Shanker, D. (2015). *The US meat industry's wildly successful, 40-year crusade to keep its hold on the American diet*. Quartz Media. Retrieved from https://qz.com/523255/the-us-meat-industrys-wildly-successful-40-year-crusade-to-keep-its-hold-on-the-american-diet/

Shaw, J. (2012). A diabetes link to meat. *Harvard Magazine*. Retrieved from https://harvardmagazine.com/2012/01/a-diabetes-link-to-meat

Shoard, C. (2016). Arnold Schwarzenegger and James Cameron urge people to eat less meat. *The Guardian*. Retrieved from https://www.theguardian.com/film/2016/jun/23/arnold-schwarzenegger-james-cameron-eat-less-meat-china

Sobal, J. (2005). Men, meat and marriage: Models of masculinity. *Food & Foodways, 13*(1-2), 135–158. doi:10.1080/07409710590915409

Stryker, J. (2007). Vegansexuality. *New York Times*. Retrieved from https://www.nytimes.com/2007/12/09/magazine/09vegansexuality.html?_r=0

Sumpter, K. C. (2015). Masculinity and meat consumption: An analysis through the theoretical lens of hegemonic masculinity and alternative masculinity theories. *Sociology Compass, 9*(2), 104–114. doi:10.1111oc4.12241

Talaei, M., Wang, Y. L., Yuan, J. M., Pan, A., & Koh, W. P. (2017). Meat, dietary heme iron, and risk of type 2 diabetes mellitus. *American Journal of Epidemiology, 186*(7), 824–833. doi:10.1093/aje/kwx156 PMID:28535164

Tashjian, L. (2015). *Vegan Strongman Patrik Baboumian breaks his own world record!* Clearly Veg. Retrieved from https://www.clearlyveg.org/blog/2015/09/23/vegan-strongman-patrik-baboumian-breaks-his-own-world-record/

Timke, E., & O'Barr, W. M. (2017). Representation of masculinity and femininity in advertising. *Advertising & Society Review, 17*(3-4). doi:10.1353/asr.2017.0004

Tonstad, S., Butler, T., Yan, R., & Fraser, G. E. (2009). Type of vegetarian diet, body weight, and prevalence of type 2 diabetes. *Diabetes Care*, *32*(5), 791–796. doi:10.2337/dc08-1886 PMID:19351712

Tucker, C. (2018). Using environmental imperatives to reduce meat consumption: Perspectives from New Zealand. *Kōtuitui: New Zealand Journal of Social Sciences*, *13*(1), 99–110. doi:10.1080/1177083X.2018.1452763

Turner-McGrievy, G., Mandes, T., & Crimarco, A. (2017). A plant-based diet for overweight and obesity prevention and treatment. *Journal of Geriatric Cardiology: JGC*, *14*(5), 369–374. doi:10.11909/j.issn.1671-5411.2017.05.002 PMID:28630616

Van Den Bulck, H. (2010). Qualitative content analysis. In C. Newbold, O. Boyd-Barrett, & H. Van Den Bulck (Eds.), *The media book* (pp. 83–88). London, UK: Bloomsbury.

Vartanian, L. R. (2015). Impression management and food intake: Current directions in research. *Appetite*, *86*, 74–80. doi:10.1016/j.appet.2014.08.021 PMID:25149198

Wang, Y., & Beydoun, M. A. (2009). Meat consumption is associated with obesity and central obesity among US adults. *International Journal of Obesity*, *33*(6), 621–628. doi:10.1038/ijo.2009.45 PMID:19308071

Wardle, J., Haase, A. M., Steptoe, A., Nillapun, M., Jonwutiwes, K., & Bellisie, F. (2004). Gender differences in food choice: The contribution of health beliefs and dieting. *Annals of Behavioral Medicine*, *27*(2), 107–116. doi:10.120715324796abm2702_5 PMID:15053018

Watrous, M. (2014). *The curious case of Hungry Man*. Baking Business. Retrieved from https://www.bakingbusiness.com/articles/26125-the-curious-case-of-hungry-man

Webber, J. (2018). *Woody Harrelson credits vegan diet for his high energy levels*. Live kindly. Retrieved from https://www.livekindly.co/woody-harrelson-credits-vegan-diet-for-his-high-energy-levels/

Weedon, C. (1987). *Feminist practice and poststructuralist theory*. Oxford, UK: Blackwell.

White, A., & Peretz, T. (2009). Emotions and redefining black masculinity: Movement narratives of two profeminist organizers. *Men and Masculinities*, *12*(4), 403–424. doi:10.1177/1097184X08326007

Wilks, M., & Philips, C. J. C. (2017). Attitudes to in vitro meat: A survey of potential consumers in the United States. *PLoS One*, *12*(2), e0171904. doi:10.1371/journal.pone.0171904 PMID:28207878

Wong, J. (2017). *Man made: A gendered food culture*. The Isthmus. Retrieved from http://www.theisthmus.com.au/2017/03/man-made-a-gendered-food-culture/

Wood, J. (2015). *Gendered lives: Communication, gender and culture*. Stanford, CT: Cengage Learning.

World Health Organisation (WHO). (2015). *Q&A on the carcinogenicity of the consumption of red meat and processed meat*. Retrieved from http://www.who.int/features/qa/cancer-red-meat/en/

World Health Organization (WHO). (2018a). *Cancer*. Retrieved from: http://www.who.int/en/news-room/fact-sheets/detail/cancer

World Health Organization (WHO). (2018b). *World Health Statistics 2018. Monitoring Health for the SDGs, sustainable development goals*. Retrieved from http://apps.who.int/iris/bitstream/handle/10665/272596/9789241565585-eng.pdf?ua=1

You, W., & Henneberg, M. (2016). Meat consumption providing a surplus energy in modern diet contributes to obesity prevalence: An ecological analysis. *BMC Nutrition*, *2*(1), 22. doi:10.118640795-016-0063-9

YouTube VegSource. (2006). *Carl Lewis: Olympic medals through the vegan diet.* Retrieved from https://www.youtube.com/watch?v=bOTETXwfIaY

Zaraska, M. (2016). *This is why you crave beef: Inside secrets of Big Meat's billion-dollar ad and lobbying campaigns.* Salon.com. Retrieved from https://www.salon.com/2016/04/03/this_is_why_you_crave_beef_inside_secrets_of_big_meats_billion_dollar_ad_and_lobbying_campaigns/

Zhu, L. L., Brescoll, V. L., Newman, G. E., & Uhlmann, E. L. (2015). Macho nachos. The implicit effects of gendered food packaging on preferences for healthy and unhealthy foods. *Social Psychology*, *46*(4), 182–196. doi:10.1027/1864-9335/a000226

ADDITIONAL READING

Adams, C. J. (1990). *The sexual politics of meat: A feminist-vegetarian critical theory.* New York, NY: Continuum Print.

Bogueva, D., & Marinova, D. (2018). What is more important: Perception of masculinity or personal health and the environment? In D. Bogueva, D. Marinova, & T. Raphaely (Eds.), *Handbook of research on social marketing and its influence on animal origin food product consumption* (pp. 148–162). Hershey, PA: IGI Global. doi:10.4018/978-1-5225-4757-0.ch010

Bogueva, D., Marinova, D., & Raphaely, T. (Eds.). (2018). *Handbook of research on social marketing and its influence on animal origin food product consumption.* Hershey, PA: IGI Global. doi:10.4018/978-1-5225-4757-0

Bogueva, D., Raphaely, T., Marinova, D., & Marinova, M. (2017) Sustainability social marketing. In J. Hartz-Karp, & D. Marinova, D. (Eds), Methods for sustainability research (pp. 280-291). Cheltenham, UK: Edward Elgar. doi:10.4337/9781786432735.00030

Connell, R. W., & Messerschmidt, J. W. (2005). Hegemonic masculinity: Rethinking the concept. *Gender & Society*, *19*(6), 829–859. doi:10.1177/0891243205278639

Newsbeat, B. B. C. (2015). Arnold Schwarzenegger: Stop eating meat and save the planet. BBC. Retrieved from http://www.bbc.co.uk/newsbeat/article/35038053/arnold-schwarzenegger-stop-eating-meat-and-save-the-planet

Schösler, H., de Boer, J., Boersema, J. J., & Aiking, H. (2015). Meat and masculinity among young Chinese, Turkish and Dutch adults in the Netherlands. *Appetite*, *89*, 152–159. doi:10.1016/j.appet.2015.02.013 PMID:25681291

Sumpter, K. C. (2015). Masculinity and meat consumption: An analysis through the theoretical lens of hegemonic masculinity and alternative masculinity theories. *Sociology Compass*, *9*(2), 104–114. doi:10.1111oc4.12241

Timke, E., & O'Barr, W. M. (2017). Representation of masculinity and femininity in advertising. *Advertising & Society Review*, *17*(3-4). doi:10.1353/asr.2017.0004

Wilks, M., & Philips, C. J. C. (2017). Attitudes to in vitro meat: A survey of potential consumers in the United States. *PLoS One*, *12*(2), e0171904. doi:10.1371/journal.pone.0171904 PMID:28207878

KEY TERMS AND DEFINITIONS

Advertising: A marketing communication paid by an interested body which is represented through text, sound, images, or combination of these and openly conveys particular ideas or messages.

Clean Meat/Cultured Meat/Lab-Grown Meat: Meat grown in-vitro from animal cells; it requires less resources and produces less pollution compared to livestock-based meat.

Discourse: A way of talking about an issue which represents the current structures of power, dominant culture, and institutions within society.

Flexitarian: A human diet which aims at reducing the consumption of all kinds of meat; the diet aims in particular to keep red meat consumption within the limit recommended by reputable health organizations – a maximum of 500 g of lean fresh meat; the term can also be used in reference to a person who follows such a diet.

Hegemonic: Ruling, dominant, or highly influential within a social context.

Macho: A man who is explicitly proud of his masculinity; the term is also associated with an assumption that such a man will be assertively and even aggressively displaying his masculinity.

Manliness: A set of traits, abilities, attributes, and qualities which characterize the male human species.

Masculinity: Manifestation of personal traits, behaviors, and social roles associated with men and boys; manifestation of manliness.

Meat Alternatives (or Alternatives to Meat): Food products based on plants and lab-grown in-vitro meat (the latter is also described as clean or cultured meat); these products can also be described as meat analogues or meat substitutes.

Myth: A widely held belief or set of ideas which are wrong and not based on convincing scientific evidence.

Social Marketing: Marketing which aims at inducing a behavioral change and maintaining such behavior for the greater social good, including benefits for the individual and society as a whole.

Stereotype: A generalized and simplified belief or image about a particular category of people.

Vegan: A human diet which excludes all animal-based products; the term can also be used in reference to a person who follows such a diet.

Vegetarian: A human diet which excludes all kinds of meat and fish; the term can also be used in reference to a person who follows such a diet.

Vested Interest: A personal reason for involvement in a particular activity or in supporting a particular idea because of the expectations for financial gain or other advantages.

Chapter 15
Collective Awareness Raising Towards a Plant-Based Diet Through Social Networking Sites

Weronika Kalamus
Global Deep Network, Poland

ABSTRACT

This chapter proposes that social media networking, social media platforms, and the internet facilitate a dietary shift towards a plant-based diet. The rise in human consciousness in the past few decades finds an expression in the plant-based revolution and social media platforms render a suitable space to manifest the collective rising in consciousness and empower communities to implement change. By examining the "power vs. force" theory, the chapter discusses its possible impact and analyses how the growth of awareness disrupts the acceptance of meat production and consumption. The second part of the chapter investigates the natural implementation of the "see–feel–change" concept through which pro-vegan online communities instigate changes across nations and modify public demand altering the whole international market of consumers. An empirical reflection on the growth of veganism based on an exploratory survey concludes the chapter.

INTRODUCTION

Every hour more than 6 million animals are slaughtered for food; annually around 70 billion animals lose their lives to become human food (Compassion in World Farming, 2013, p. 15). Approximately every 12 seconds a child dies from hunger or hunger- related causes in some of the poorest parts in the world (Poverty.com, 2018) while every 12 seconds a cow is killed for human consumption in a wealthy country (Pachirat, 2013, p. 9). According to Ritchie and Roser (2017, n.p.), only in China: "Increases in Chinese pigmeat production have been rapid, growing around 35-fold from 1.5 million tonnes in 1961 to 54 million tonnes in 2014"; poultry production has risen 28-fold and beef meat – 87-fold. In

DOI: 10.4018/978-1-5225-7350-0.ch015

the meantime, "[r]educing the meat production by merely 10% could release enough grain and another natural foods to feed 60 million people"(Moritz, 2009, p. 176).

An increasing number of scientists expose more robust evidence that the rapid growth of the industrialised meat production and its consumption is the major drive for global warming and species extinction (FAO, 2006; Bailey, Froggatt, & Wellesley, 2014). The Food and Agriculture Organization of the United Nations predicts that the Western style meat manufacture model which is rapidly spreading around the world will become a primary reason for conflicts over water resources over the next decades (FAO, 2006, p. 126).

Yet consumption of animal produce is still socially accepted, embedded as a natural, necessary part of our lives. Meat production is not only recommended but also enforced on people by most national and international government bodies. According to the report produced by UK's Chatham House, "[d] espite the clear case for action to tackle demand for meat and dairy products, there is remarkable lack of policies, initiatives or campaigns to do so" (Bailey, Froggatt, & Wellesley, 2014, p. 22).

Paradoxically in the last few decades, in parallel with the increasing demand for animal produce, the popularity of plant-based diet has also been growing. While the European Union remains one of the major meat consumers, vegetarianism and veganism are undoubtedly on the rise in Europe. Vegan products are becoming new profitable avenues yet to be fully explored. In some cases, profits from selling plant-based protein substitutes and vegan alternatives to dairy have tripled in the last few years. With the rise of veganism, the demand for plant-based substitutes to dairy and meat is increasing. In the USA, the sales of plant-based foods went up by 8.1% in 2017 to reach US$ 3.1 billion and according to Forbes there are significant business advantages to embrace this trend which is gradually becoming mainstream (Fox, 2017). According to the 2018 market research by the firm Euromonitor International, by 2020 Australia's packaged vegan food market will be worth A$215 million. It also found Australia was the third fastest growing vegan market in the world (Sanda, 2018). In France, in the last 20 years meat sales gradually declined as population is increasingly more concerned about the devastating environmental consequences associated with livestock, animal welfare and negative impact on health (Bréville, 2018).

The "new" consumers have many more options as restaurants specialising in vegetarian and fully plant-based foods are becoming part of the hospitality and leisure world. Data from Happy Cow, the social platform which shows all vegan and vegetarian restaurants and stores around the world, shows that only in Spain the number of fully veggie restaurants had tripled between the years 2011 and 2016. Studies show that nowadays 7.8% of the Spanish population is either vegan, vegetarian or flexitarian in response to global warming, animal right concerns and awareness regarding the link between meat and cancer (Lantern Consulting, 2017). The vegan trend is growing globally not only in terms of diet but also as a whole new lifestyle which will further influence the global economics.

This chapter analyses the paradox of rising meat consumption due to the strong propaganda from governmental and international organizations and companies often with vested interests, and the parallel growth of interest in the plant-based diet empowered by social media networks and platforms. The main aim is to explore the growth in popularity of the plant-based diet and its potential correlation with the rise in human awareness in the last decades. The "Power versus Force" theory (Hawkins, 2012) is analysed in regard to the current situation where meat consumption is still predominant on Earth while simultaneously veganism is rapidly gaining popularity. A study of the "Map of the scale of Consciousness" is used to detect how low levels of collective consciousness are established and embedded in our current carnivorous diet. The chapter's hypothesis is that higher levels of consciousness could be defined with the expansion of the current "green revolution".

The chapter also investigates how social media platforms and the broad access to the internet have acted as a major force in changing dietary customs on a global scale through education and emotional influence within the concept of "see–feel–change" (Heath & Heath, 2010, p. 106). It explores how people's spiritual inclination is reflected on acting upon witnessing the truth and experiencing feelings. In the last part of the chapter an empirical evidence is provided about the rise in veganism and vegetarianism based on an exploratory survey conducted in 2018.

SPIRITUAL CONTEXT OF MEAT CONSUMPTION

The omnipotent capitalism which humans have imposed on themselves is a system based on an external power understood in a five-dimensional reality where power is equal with the control and oppression of others (Zukav, 2004, p. 24; Hawkins, 2012). Human domination over nature through extraction, devastation and modification as well as the way one nation dominates over another is similar to the way humanity controls animals and treats them as non-sentient beings. The system that humanity has created lacks any spiritual inclination of love and compassion. Capitalism, at its basis, has commodified animals: "[In the Western world] The first form of capital and monetary exchange were sheep, goats and cattle for only they were consumable property with tangible worth. In fact, the word Capital derives from the Capita, Latin for 'head', as in head of cattle and sheep" (Tuttle, 2005, p. 18)

In this day and age, we are faced with rising inequality, gender issues, racial discrimination, poverty, and planetary degradation which all represent the low state of awareness characteristic for the majority of people. Thus, it is easier to understand that the entire global community which promotes animal rights and supports plant-based diets does this in opposition to the discriminatory system deeply rooted in human history. The spiritual approach seems quite opposite to the demanding systems we have created for ourselves and which praise material accumulation, control of others and greediness. By contrast, spiritual development points humans towards personal growth of wisdom, compassion, love, empathy and collaboration.

Hawkins (2012) in his book "Power versus Force" through the method of kinesiology developed a tool to measure human awareness and explore the possibility of human collective consciousness. In his studies, he designed a map of consciousness which "is a numerical scale whereby one can measure positive from negative, power from force and truth from falsehood" (Thompson, 2012, n.p.). The Hawkins' "Map of the scale of Consciousness" (see Figure 1) describes in a clear and simple way each person's different state of awareness based on their personal thoughts and feelings. Depending on the level of consciousness of a person, "force" or "power" will be used in everyday life creating the reality according to one's motives and intentions.

Hawkins (2012) claims that to understand the history of humanity, the pivotal matter is to realize the difference between "power" and "force". "Force" characterises the whole planetary system, where the stronger has control over the weaker and is based on fear and pride. By contrast, "power" characterises itself as a subtler element which is based on love, compassion and enlightenment, and therefore can have a great impact. According to Hawkins (2012), the power versus force concept can offer explanation about how in the past some single individuals have been capable of significantly influencing history with their own power against the force of omnipotent governments.

To understand the matter, the example of Mahatma Gandhi is described, whose levels of consciousness according to the Map oscillated on a very high level of 700 which equals states of enlightenment,

Figure 1. Map of scale of consciousness
Source: Hawkins (2012)

SCALE OF CONSCIOUSNESS						
View related to God	View related to Life	Level of Conciousness		Log	Dominant Emotion	Dominant Process
Self	Is	Enlightenment	↑	700-1000	Ineffable	Pure consiousness
All-being	Perfect	Peace	↑	600	Bliss	Illumination
One	Complete	Joy	↑	540	Serenity	Transfiguration
Loving	Benign	Love	↑	500	Reverence	Revelation
Wise	Meaningful	Reason	↑	400	Understanding	Abstraction
Merciful	Harmonious	Acceptance	↑	350	Foregiveness	Transcendence
Inspiring	Hopeful	Willingness	↑	310	Optimism	Intention
Enabling	Satisfactory	Neutrality	↑	250	Trust	Release
Petmitting	Feasible	Courage	↑	200	Affirmation	Empowerment
Indifferent	Demanding	Pride	↓	175	Scorn	Inflation
Vengeful	Antagonistic	Anger	↓	150	Hate	Aggression
Denying	Dissapointing	Desire	↓	125	Craving	Enslavement
Punitive	Frightening	Fear	↓	100	Anxiety	Withdrawal
Disdainful	Tragic	Grief	↓	75	Regret	Despondency
Condemning	Hopeless	Apathy	↓	50	Despair	Abdication
Vindictive	Evil	Guilt	↓	30	Blame	Desctruction
Despising	Miserable	Shame	↓	20	Humiliation	Elimination

peace and love. In comparison, the motivation of the British Empire was based on force characterised by greed and desire of domination. Observing the historical achievement of independence in India which was instigated by Mahatma Gandhi and his pacifist movement, the concept of power is insightful in explaining the level of influence one person had.

I argue that in the light of Hawkins' theory of power and force, any organization that supports animal rights and wellness of the planet Earth at its basis comes from power and its motivation is of compassion and love. Examined from this perspective, the whole industrialised meat production is fuelled by force and associated with lower levels of awareness. It might be of major interest to notice that India, one the most spiritually inclined places in the world, has one of the lowest meat consumption per capita.

According to the analysis by Oxford University, “meat consumption has been most marked in countries who have underwent a strong economic transition– per capita consumption in China has grown approximately 15-fold since 1961; rates in Brazil have nearly quadrupled. The major exception to this pattern has been India: dominant lactovegetarian preferences mean per capita meat consumption in 2013 was almost exactly the same as in 1961 at less than 4 kilograms per person” (Ritchie & Roser, 2017, n.p.).

To keep entire nations unconscious about the cruelty behind the animal industrialised farming and slaughtering the “populace must be shielded from direct exposure to the victims of the system. Otherwise they begin questioning the system or their participation in it” (Joy, 2011, p. 40). This is a reason why social networking sites and internet platforms are of major importance for a dietary shift. They allow direct exposure to some of the facts associated with industrial animal farming systems. I argue that internet-based communities naturally implement the SEE–FEEL–CHANGE (Kotter & Cohen, 2002) concept which is pivotal for their efficiency to help rise collective awareness and modify the socially ingrained dietary habits of the consumers.

HOW SOCIAL MEDIA NETWORKS AND PLATFORMS RISE HUMAN AWARENESS: SEE, FEEL AND CHANGE

Social networking sites and the internet-based platforms in conjunction represent excellent instruments to enable society to introduce new behavioural patterns in regard to eating habits. They can expose people to the range of urgent matters the carnivore diet poses on humanity: the devastation of the planet, cruelty induced on animals, deterioration of people’s health, obesity, rise in poverty, conflicts over water sources and many more.

Following Kotter’s theory, the pivotal basic grounds for change are: 1.) SEE – to show the truth, especially when it is problematic and uncomfortable, so that the public can feel the urgency of action (Kotter & Cohen, 2002); animal activist videos presenting cruelty against animals in the process of meat production, environmental damage caused by livestock, such as water bodies pollution, are just a few examples; 2.) FEEL – affect people’s feelings (Kotter & Cohen, 2002); videos, photos, publications and any other truthful materials that provoke feelings of sadness and compassion towards animals are essential in exposing the hidden or unnoticed reality, and; 3.) CHANGE – propose clear and easy solution/s to the problem that will lead to empowerment and action (Kotter & Cohen, 2002); stop eating meat and dairy products in exchange of plant-based meals is a practical and simple solution for most places around the world.

Below are two examples based on the use of social networking sites to create the Israeli phenomenon and the work of the Australian organisation Animals Australia. Many more similar examples can create a swell towards a transition to a plant-based diet.

Social Networking Sites: The Israeli Phenomenon

The case of Israel proves the remarkable impact of social networking sites (SNSs) for societal change. The idea that one video shared on someone SNSs’ profile can go viral by being “shared”, “liked” or “reacted to” and therefore provoke meat consumption reduction (Campbell, 2018, p. 140) has been found to work in reality. Gary Yourofsky – an animal rights activist from the United States of America is claimed to be a person who initiated the “green revolution in Israel”. Gary Yourofsky, 45 at the time and a Jewish himself "controversially calls slaughterhouses 'concentration camps' and compares the treatment of animals to the Holocaust" (Kirokva, 2015, n.p.). His video with Hebrew subtitles is one of the most shared videos in Israel and is claimed to be a catalyst for Israel becoming the fastest growing vegan community in the world. It is estimated that up to 5% of Israelis turned to a plant-based diet in recent years, many of whom because of watching Yourofsky's video (Kirokva, 2015). Commenced through the social media, the Israeli plant-based diet revolution has started to penetrate the country's society also at the governmental level. The Israeli army supports veganism not only by offering vegan food but also non-leather shoes; the idea of "meatless" Mondays gained the support of the prime minister Benjamin Netanyahu. "Domino's pizza" in Israel is the first and only franchise in the world that offers vegan pizza topping (Leenaert, 2017).

Although Israel still is one of the major meat consumers in the world and the large part of the change in dietary habits is happening around Tel Aviv, rather than the more conservative Jerusalem or other regions of the country, the recent rapid societal growth in the national community which supports a plant based diet and animal rights is still remarkable. Certainly, the fact that Israel has a small population of 8 million people and its spiritually-based tradition of kosher requirements already facilitate the changes in food consumption. Seemingly the elite of the Jewish community in many parts of the world also supports a plant-based diet and shows its concerns toward industrialised animal factories: "more than 70 rabbis from around the world signed a declaration urging Jews to choose veganism, saying it was a contradiction to claim that products made 'through a process that involves inordinate cruelty and barbarity toward animal life can truly be considered kosher in our world'" (Holms, 2018, n.p.).

Social Media: Animals Australia

Animals Australia is another interesting example about the use of social media in raising collective awareness about the plight of livestock animals. The organisation was established to work for eliminating cruelty against animals by exposing callous actions of the livestock industry within Australia as well as internationally. Its founder Lyn Whyte was a former police woman who uses her investigative skills to expose actions of cruelty against sentient beings. Starting from humble beginnings, the organisation now has 2 million individual supporters with whom it stays in touch through social networking platforms. It actively engages in organising campaigns against cruelty during animal slaughter, industrial animal raising practices, live export of sheep and cattle from Australia to other destinations, such as the Middle East, as well as to disassociate the cultural perception about being Australian with the consumption of meat. Crowd-funding is commonly used to finance the development of advertisements aired on the national television and displayed on internet sites.

Its mission also includes to represent the animals and their interests in front of government and other decision-making bodies. It works for strengthening the animal protection movement by educating, inspiring and empowering (Animals Australia, 2018). For example, one of the photos on Lyn Whyte's facebook page carries the title: "What's done in the dark, will be brought to the light" (Whyte, 2018).

Although Animals Australia does not explicitly campaign for plant-based diets, Lyn Whyte herself is a vegan. The organisation uses its power to influence a consciousness transition within the Australian society. It is not surprising then that Australia is the third fastest growing vegan market in the world (after United Arab Emirates and China) (AAP, 2018).

PLANT-BASED DIET: SHIFTING THE CONSCIOUSNESS, CHANGING THE SYSTEMS

Humanity is passing through a unique transformation in terms of changes in individual and collective consciousness (Zukav, 2002, p.12–13). The rise in veganism, not only in Israel and Australia, but in many other parts of the world, reflects this process.

Authors writing about spirituality stress the correlation between increase in awareness and taking the responsibility for one's own choices, decisions and actions (Dhiman & Marques, 2016). According to the Map of Consciousness Scale (see Figure 1), the rise in awareness leads us to the stage of courage where an individual starts acting (Hawkins, 2012, p. 78–79; 2013, p. 112). As such the SNSs and internet platforms act as help for people to become more aware by exposing cruelty induced to animals, environmental damage, or health implications from meat consumption. Compassion and empathy emerge from the awareness that everything is connected, that our joy is the joy of others and our sorrows are the sorrows of others (Zukav, 2002, p. 21).

Thus far the exuberant scale of the violence towards animals and the exploitation of the Earth seem to have reached its culmination point in the whole human history. Humanity affected the planet to a degree that now its degradation, climate change and rising pollution pose a serious threat to our own survival. However, O'Brien observes "climate change presents a unique opportunity for the introductions of the behaviours, systems, and technologies able to improve all aspects of life on Earth" (O'Brien, 2009, p. 2). The new dietary behaviour, enacted as a result of higher awareness, which embraces the plant-based diet has the potential to encourage innovative ways of living based on sharing, collaboration and compassion.

While in the last fifty years planet exploitation and cruelty towards animals have reached shocking stages, at the same time human awareness which oscillated on dangerously low levels for many centuries, in the 1980s transitioned to a higher scale of consciousness (Hawkins, 2012, p. 255). The interest in spirituality in recent decades (Ulluwishewa, 2015) possibly stimulated the higher awareness levels in humanity as a whole.

SLOWLY GROWING CHANGES IN THE DIETARY HABITS

To verify empirically the popularity of a plant-based diet, a short exploratory survey was conducted through the social media and in person during the "Couchsurfing" vegan and vegetarian dinner organised by the author through the online network www.couchsurfing.com. There were 36 participants in total of whom 9 identified themselves as vegan, 3 as vegetarians with the remaining 24 being meat-eaters, including 2 flexitarians. This was an exploratory survey which did not seek to demonstrate any statistical validity but wanted to canvass the issues surrounding the role of social networking sites in a transition to a plant-based diet. The majority of the participants, namely 20 people were interviewed personally or through social media networks while the other 16 people completed a paper-based questionnaire. All names of the participants are changed to protect their identity. Their ages ranged from 19 to 62 years old. The data was gathered over 6 days in September 2018 and the participation was entirely voluntary.

Description of the Interviewees

There was quite a large international spread in the represented countries – namely 21 (see Table 1). Their answers are discussed below using also excerpts from the interviews.

The first question the participants were asked was whether they believe or have witnessed veganism and vegetarianism becoming more popular in the last few years. The overwhelming majority of them (namely 35) answered positively and readily provided examples of friends or family members who were embracing such a new diet. Only one participant – Michal, 44, from Cracow in Poland, have not noticed veganism or vegetarianism becoming more popular and socially accepted. Table 2 presents arbitrarily chosen opinions from the survey participants to highlight possible current tendencies in regard to the popularity of veganism and vegetarianism.

Two Case Studies

The background and interview summary of two of the participants are explored in more depth in the following section. This allows some insights into the phenomenon of how the younger generation is turning around the plates of their parents.

Table 1. Overview of the survey participants

Country	Number of Participants
Poland	4
Australia	3
Brazil	3
Spain	3
UK	3
Belgium	2
Germany	2
Israel	2
Lebanon	2
USA	2
Austria	1
Belarus	1
Colombia	1
Georgia	1
Iraq	1
Italy	1
Portugal	1
Romania	1
Saudi Arabia	1
Turkey	1
Total	*36*

Table 2. Excerpts from the interviews

Name	Age	Country	Interview Excerpt
Hugo	24	Brazil	I eat meat. My mother became vegan. But it was too difficult for her to stick to this diet due to the lack of vegan restaurants in Porto Alegre – the city I am from. Now she is vegetarian and I see the vegetarian community is growing fast in my home town but unfortunately the market doesn't fulfil yet the public demand for plant-based alternatives
Diana	24	Brazil	I have been vegetarian for 2 years and 6 months and I currently live in Tweed Heads, Australia. I have a 2-month old baby and I'm looking forward to raise him as a vegetarian. Raising a vegetarian child will definitely be tough as people still judge the way you feed your baby and you feel you need to defend and explain yourself to everyone. Some people still believe that you miss out on protein and other nutrients by following a vegetarian diet even though that's a completely old-fashioned way of thinking and very judgmental. We are simply taking away dead bodies from our food, which should be everyone else's concern. It's a clean and passive decision. I'm not forcing him to be a vegetarian, it's quite the opposite. I will tell him exactly what he is eating and it will be his decision. I'm sure he won't like the idea of eating his friends. I can't say it was a hard transition for me as veganism in Australia is on the rise due to health and environmental concerns. You can find vegan and vegetarian options in every restaurant and market you go to (...) But returning to Brazil will definitely be a challenge as the vegan lifestyle is not a very known thing after all and the country is still one of the world's biggest meat producers.
Nir	41	Israel USA	I am the only vegan in between my friends but I don't feel any discrimination. (Well, very rarely I have fish, trying to cut it out from my diet completely.) In Tel Aviv every week there is a new vegan restaurant and definitely interest in veganism is rising, same as in New York. My friends and family try often vegan options.
Binar	28	Iraq	I eat meat but I see a change happening in Iraq. I think it's the influence of international visitors, because of them more people convert to vegetarianism. That is now socially accepted. This started happening maybe around 7 years ago, I think.
Josef	33	Lebanon	I lived in Montreal for 15 years. For some time, I had a vegan girlfriend. It was so easy for us to get vegan food. We could have different vegan options from various restaurants every day. I don't want to know about dying animals. I think in the future I will be vegetarian.
Marc	25	Lebanon	My cousin always tries to be vegetarian but she lives with her family in Beirut which always cooks meat. So, she is vegetarian few weeks, then meat eater, then vegetarian again (…) I think vegetarians and vegans…they are people with very strong personalities. They are going against all the stream, against all the society…Yes, veganism is growing in my country but slowly.
Sere	32	Spain	My father educated me to have respect for the animals but we were all meat eaters. I became vegetarian because of what happens to the animals. I read a lot about the industry. Then my father also has seen the videos and I was shocked one day he says "I'm vegetarian too". But my mum is still meat eater. She complains a lot that my dad stopped eating meat. They are living together.

Grazyna

The story of one of the participants – Grazyna (62), is particularly striking. While growing up in socialist Poland, Grazyna consumed very little animal products. The "pork", "beef" or "chicken" was served perhaps "once every two weeks for a special occasion. Grains, potatoes, and pulses were more common and vegetable soups were cooked on a bare bone". In Poland, with the post-communism capitalistic boom and all pro meat propaganda, animal products became a staple everyday food within an astonishing 10-year period of time! On Grazyna's table astounding amounts of sausages, hams, and cheeses were served up to 3 times a day. Her children were raised fully carnivorous, growing up with pets and not once questioning the consumption of animal products.

Grazyna's youngest daughter disliked eating meat from early childhood, yet its consumption was imposed on her by her mother who strongly believed that high amount of animal protein is essential for the proper development of her child. Six years ago, Grazyna's youngest daughter stopped eating meat. Being very sceptical at the beginning, after a year Grazyna also decided to become a vegetarian. Her daughter was a great influence and a source of information about how unhealthy their previous diet was and about the meat industry's negative impact on planet Earth.

Today Grazyna says: "I wish there were information before, when I was younger. I would never have fed my babies pork or beef! I am glad there is so much information nowadays easily accessible on the internet and also many interesting publications in the book stores. I feel that my daughter enlightened me!"

Paul

Paul, from Scotland had suffered major intestine operation before becoming vegetarian after his 32-year old son (already vegetarian for a few years) convinced him to try "this new diet only for a few weeks" to protect his digestive system from future troubles. Today Paul still consumes cheese but avoids meat at all cost.

Looking back, as his health improved dramatically – dropping meat taught him how to cook from scratch. It required him to "discover new tastes" and consume vegetables that he wouldn't have tried before.

Power Against Force

The purpose of this short survey was merely to draw attention to the paradox that is happening around the world and how indeed the plant-based revolution is rising against all the powerful propaganda and marketing of the consumption of animal products. Although more than two-thirds of the participants were meat eaters, all of them (with only one exception) confirmed that the plant-based diet is becoming more popular. Many expressed the possibility of becoming vegetarian in the future but have concerns about how to cook new meat-free meals and whether this type of diet would be sufficient for providing protein and other nutrients.

Most of the targeted participants had been travelling to at least a few countries in their lifetime, some of them being extensive travelers and also "digital nomads". Due to the small number of participants, it is impossible to draw general conclusions about how fast the vegan community is growing. Yet the certainty about the growth in popularity of the plant-based diet is clear.

The language used by the participants confirmed the two theories espoused in this chapter, namely "Power vs Force" and "see–feel–change". They all communicated stories of becoming aware about the negative impacts of meat consumption and how exposure to the facts is making them change their perceptions and in some cases, also their actions. It was similarly interesting to observe the language the participants used, including words, such as "enlightened", "love", "acceptance" and "willingness". The movement towards the top of the Map of Scale of Consciousness was evident, although there were no attempts made to actually measure the level where each participant is. In fact, this is not needed as what matters in the revolution of power against force is the direction towards making the planet a better place for all species.

CONCLUSION

The SNSs and internet platforms serve as an interconnectedness tool for humanity; they raise awareness at an individual, but also community, national and international levels and consequently change previously ingrained habits characterised with low awareness as is the case with dietary preferences. However, once the reality is exposed the barriers behind which the meat industry is currently hiding, are easy to demolish. It is easy to replicate and share experiences using SNSs and internet platforms and people are no longer willing to ignore their own dietary impact on the planet and other sentient beings. What is similarly important is that this interconnected world allows for new shared experiences to appear which build the transformative power of the new awareness.

There is a lot of scepticism, including from environmentalists and climate change scientists, about the willingness of people to change their dietary habits. In many ways, the older generation is the one that has grown in times of different values and global priorities. Armed with easy access to information, communication and support from the global community of digital nomads, many people are improving their awareness and embracing the new dietary habits. The rise of veganism and vegetarianism is a new fashion trend, but a fashion to stay. Irrespective as to how passionate those who have made the transition to a higher level of consciousness are, they are a minority compared to the alarming rapidly growing industrialised meat production and consumption which is spreading its tentacles across the globe. Yet, undoubtedly the pro plant-based diet global community is rising. Pushed by public demand, the vegan market is growing too and the governments slowly acknowledge the request to introduce vegetarian and vegan options in public schools and organizations. The hospitality industry is much quicker in responding to these public demands.

"We change the world not by what we say or do, but as a consequence of what we have become" – Hawkins (2012, p. 121). Analysing the theory of "power versus force", it could be argued that the vegan movement is characterised by a higher state of consciousness and its authentic power is based on empathy, compassion and love. It holds great potential to impact and change the history of humanity in the upcoming years.

REFERENCES

Animals Australia. (2017). *We are Australia's leading animal protection organization*. Retrieved September 14, 2018 from https://www.facebook.com/AnimalsAustralia/

Australian Associated Press. (2018). Vegan trend takes hold in Australia. *SBS News*. Retrieved September 14, 2018 from https://www.sbs.com.au/news/vegan-trend-takes-hold-in-australia

Bailey, R., Froggatt, A., & Wellesley, L. (2014). *Livestock – climate change's forgotten sector: Global public opinion on meat and dairy consumption*. London, UK: Chatman House.

Bréville, B. (2018). Meat goes out of style: A meat-free diet used to indicate poverty, but now the affluent adopt it to better their health, or to save animals and the planet. Instead of meat, what does the food industry have on the menu? *Le Monde Diplomatique*. Retrieved September 14, 2018 from https://mondediplo.com/2018/08/11animals

Campbell, J. (2018). Flipping the script: Creating mass change through social networking sites. In D. Bogueva, D. Marinova, & T. Raphaely (Eds.), *Handbook of research on social marketing and its influence on animal origin food product consumption* (pp. 135–147). Hershey, PA: IGI Global. doi:10.4018/978-1-5225-4757-0.ch009

Compassion in World Farming. (2013). *Strategic Plan 2013-2017: For kinder, fairer farming worldwide.* Retrieved September 14, 2018 from https://www.ciwf.org.uk/media/3640540/ciwf_strategic_plan_20132017.pdf

Dhiman, S., & Marques, J. (Eds.). (2016). *Sustainability and spirituality: New horizons and exemplary approaches.* Berlin, Germany: Springer International.

Food and Agriculture Organization of the United Nations (FAO). (2006). *Livestock's long shadow: Environmental issues and option.* Rome, Italy: FAO. Retrieved September 14, 2018 from http://www.fao.org/docrep/010/a0701e/a0701e00.HTM

Fox, K. (2017). Here's why you should turn your business vegan in 2018. *Forbes*. Retrieved September 14, 2018 from https://www.forbes.com/sites/katrinafox/2017/12/27/heres-why-you-should-turn-your-business-vegan-in-2018/#6d97f4182144

Hawkins, D. R. (2012). *Power vs force: The hidden determinates of human behavior.* Warsaw, Poland: Virgo. (in Polish)

Hawkins, D. R. (2013). *Letting go: The pathway of surrender.* London, UK: Hay House.

Heath, C., & Heath, D. (2010). *Switch: How to change things when change is hard.* New York, NY: Random House.

Holms, O. (2018). 'There is no kosher meat': The Israelis full of zeal for going vegan. *The Guardian International Edition.* Retrieved September 14, 2018 from https://www.theguardian.com/world/2018/mar/17/there-is-no-kosher-meat-the-israelis-full-of-zeal-for-going-vegan

Joy, M. (2011). *Why we love dogs, eat pigs and wear cows: An introduction to carnism.* San Francisco, CA: Conari Press.

Kirokva, D. (2015). How animal activist Gary Yourofsky turned '8% of Israel vegan' after comparing slaughterhouses to the Holocaust. *Metro*. Retrieved September 14, 2018 from https://metro.co.uk/2015/09/28/how-animal-rights-activist-gary-yourofsky-turned-8-of-israels-population-vegan-after-comparing-slaughterhouses-to-the-holocaust-5361205/

Kotter, J., & Cohen, D. (2002). *The heart of change: Real-life stories of how people change their organizations.* Boston, MA: Harvard Business Review Press.

Lantern Consulting. (2017). *The Green Revolution: Understanding the 'veggie' boom.* Retrieved September 14, 2018 from http://www.lantern.es/papers/the-green-revolution-entendiendo-el-auge-del-mundo-veggie

Leenaert, T. (2017). A vegan revolution in Israel: Fact or fiction? *The Vegan Strategist.* Retrieved September 14, 2018 from http://veganstrategist.org/2017/05/17/vegan-revolution-israel-fact-fiction/?doing_wp_cron=1536745141.3476879596710205078125

Moritz, A. (2009). *Timeless secrets of health & rejuvenation.* Brevard, NC: Ener-Chi Wellness Press.

O'Brien, J. (2009). *Opportunities beyond carbon: Looking forward to a sustainable world.* Melbourne, Australia: Melbourne University Press.

Pachirat, T. (2013). *Every twelve seconds: Industrialized slaughter and politics of sight.* New Haven, CT: Yale University Press.

Poverty.com. (2018). *Hunger and world poverty.* Retrieved September 14, 2018 from https://www.poverty.com

Ritchie, H., & Roser, M. (2018). *Meat and seafood production & consumption.* Our World in Data. Retrieved September 12, 2018 from https://ourworldindata.org/meat-and-seafood-production-consumption

Sanda, D. (2018). Vegan trend takes hold in Australia. *The Guardian.* Retrieved September 12, 2018 from http://www.theguardian.com.au/story/5316808/vegan-trend-takes-hold-in-australia/

Thompson, C. M. (2012). *The Map of Consciousness – Hawkin's scale.* Stankov's Universal Law Press. Retrieved September 14, 2018 fromhttp://www.stankovuniversallaw.com/2012/01/the-map-of-consciousness-hawkins-scale/

Tuttle, W. (2005). *The world peace diet: Eating for spiritual health and social harmony.* New York: Lantern Books.

Ulluwishewa, R. (2015). *Spirituality demystified: Understanding spirituality in rational terms.* Palmerston, New Zealand: Sairon Books.

Whyte, L. (2018). *Lyn Whyte (Animals Australia).* Retrieved September 14, 2018 from https://www.facebook.com/Lyn-White-Animals-Australia--162121637186671/

Zukav, G. (2004). *The seat of the soul.* Gdynia, Poland: Centrum.

KEY TERMS AND DEFINITIONS

Awareness: Knowledge, understanding, or perception about something.

Consciousness: The quality or state of being aware.

Diet: The food consumed by a person or an organism; the word also has an implied meaning that the food consumption is with a health-related aims although ethical considerations can also affect the choice of foods to be consumed.

Internet Platform: A computer connected to the internet through which the user can access and participate in many networked sites, including social media sites, such as Facebook, Twitter, Instagram, blogs, and various webpages.

Kinesiology: A practical philosophy and healing technology based on the understanding that the body has inner energy to heal itself but can be assisted to do so.

Map of the Scale of Consciousness: A stratified classification of the levels of consciousness from weak to strong, developed by David R. Hawkins.

Plant-Based Diet: A diet which is based on the consumption of different plants and plant-parts, such as fruit, vegetables, seeds, nuts, pulses, roots and tubers, and foods prepared from plant components, and avoids the consumption of foods of animal origin.

Social Networking Site: A website or another online platform created on the internet which allows the user to create public profile and connect with other users.

Spirituality: A broad concept which describes the connection a person has with the essence of being human.

Chapter 16
Leave No One Behind, Not Even the Animals: Implications for the New Meat Alternatives

Alexis J. Nagy
Curtin University, Australia

Dora Marinova
Curtin University, Australia

ABSTRACT

The sustainability agenda is a modern-day exercise in global ethics. Why then is animal welfare an absent policy within the ethical framework? Why do we continue to see farm animals only as food-related commodities? In this chapter, these issues are explored using case studies to support the emotional complexities of animals as well as the recent legal developments in animal personhood rights. The purpose of this chapter is to establish a logical and ethical argument to push the animal welfare agenda forward within the sustainable development conversation and provide a useful tool for future policy frameworks. This chapter is comprised of a comparative research methodology with the objectives to analyze, compare and contrast secondary research, and use case studies to establish an argument for the inclusion of animal welfare as an independent thread of human rights and provide implications for new meat alternatives together with recommendations for government and policymakers.

INTRODUCTION

This chapter explores the interconnected journey of ethics and moral philosophy throughout the human condition and its relationship to social justice as an ever-evolving global legal concept. Most notably the evolution of human rights and what it means to the current sustainable development goals are discussed and why it should be imperative that the current sustainability agenda initiates an 18th goal specifically focusing on animal welfare.

DOI: 10.4018/978-1-5225-7350-0.ch016

Why do we have human rights? This is not an entirely easy question to answer, and the concept of universal human rights is still only a very recent development in human history (Shestack, 1998). Homo sapiens has proven a long history as a war waging species. The only other animal on earth to come close to this type of inherent behavior are chimpanzees (pan troglodyte), who share 99% of their genetic make-up with homo sapiens, making them our closest living relative (Prüfer & Paabo, 2012). A 2005 genome sequencing study shows that humans also share 99% of their genetic make-up with the bonobo (pan paniscus) (Gibbons, 2012). However, the bonobo does not tend to exhibit the same war-like social qualities as the human and the chimpanzee (Prüfer & Paabo 2012). Remarkably there is only a 3% variation between these three species that can account for our similarities and differences, where humans share more common traits with both the bonobo and the chimpanzee, than the chimpanzee shares with the bonobo (Prüfer & Paabo, 2012).

Perhaps by acknowledging that there are only slight genetic variations between humans and their closest living non-human relatives, we might begin to consider bridging the gap between human and animal rights and take tentative steps towards understanding why the human can act with both peace and war. Theoretically speaking, universal rights to live free from fear and torture could arguably be a social philosophy the bonobo has practiced alongside their historical origins of evolution with human beings since their evolutionary entrance. However, the human animal has the necessary neural complexities that have allowed us as a species to develop and implement the Universal Declaration of Human Rights (United Nations,1948).

The concept of 'life, liberty and the pursuit of happiness' with unalienable human rights, was written into the United States Declaration of Independence of 1776 (National Archives Museum, 2018). Interestingly, both – the United States Declaration of Independence and the Universal Declaration of Human Rights, were drafted as a result of war. What is it about war that compels humans back to a moral philosophy with the intent of establishing peace? Perhaps it is the brutality of war that begets the resolution for peace. Still, it does not explain why human beings crave peace and as a result have turned to moral philosophy to establish a set of universal principles pertaining to human rights for all. Developments in the field of neuropsychology are expanding on a theory that the seat of empathy and learning for primates and some other animals may be found in the mechanism of mirror neurons (Corballis, 2015). The 14 clusters of mirror neurons give us the capacity to not only empathize with others, but to feel what they feel. Our mirror neurons mimic the action and emotion of what we see others do and feel leading towards empathy for their position (Coballis, 2015). From a cognitive and biological perspective, this could help explain why humans and other primate species have a tendency to socially navigate towards peaceful social settings, with some primates, including humans, experiencing intervals of war or social unrest.

This chapter focuses on the argument of animal sentience being the driving force to incorporate improved welfare policies on par with human rights. The argument for shared empathy and the non-human animal's ability to express deeper emotions and bonds could be a result of our shared mammalian heritage and neocortex. It is through our similarities and not our differences that the case for legal personhood rights for all sentient beings is discussed. This chapter concludes with recommendations addressing the steps needed to legitimize legal personhood for animals as a beneficial initiative of both the sustainability agenda and as a stronger arm of the United Nations animal welfare initiatives. The implications from this analysis is on the one hand that animals should be perceived as having rights beyond serving as human food – be it as meat or dairy products; and on the other, that people's food choices will be better if animal-based options are left out of the menu.

BACKGROUND

As it has evolved in today's interpretation, the sustainability agenda could arguably be described as a modern-day exercise in global ethics from almost every conceivable angle – all, except for the gaping hole that represents animal welfare. The initiative of the Sustainable Development Goals is highlighted in a simple, yet effective slogan: 'Leave no one behind' (United Nations, 2015). However, many are being left behind, many whose rights are not yet acknowledged, many who suffer under extreme conditions and whose voices, pleas and cries for help are all but ignored. Their voice is extinguished behind an impenetrable wall of legal statutes and corporate interests. These voices are further silenced and repackaged in our supermarkets portraying an image of the happy cow or pleasant-pig reaching out through clever marketing to invite the consumer to feel confident and pleased with their purchase (Morris, 2011). It implies that no suffering exists behind this packaging. In reality, their voice is not so different to ours, and in many ways, we still share an ongoing history of oppression, rape, murder, torture and slavery (Singer, 1975).

According to Jeffery Sachs (2015, p.1), "sustainable development is a central concept for our age. It is both a way to understand the world and a method for solving problems". However, the sustainable development agenda "has systematically neglected animal welfare" (Rawles, 2006, p.208). The problem of animal welfare within the sustainability agenda is not just that it is lacking in a formal framework. In fact, because it is lacking, the role of ethics towards animals, especially those within the animal agricultural industry, risks perpetuating existing frameworks that do not include their welfare and interests (Buller & Morris, 2008; Miele & Lever, 2014).

Many key concepts in moral philosophy, ethics, psychology, the human/animal distinction, biology, and legal frameworks need to be explored to try and piece together a suitable framework for animal welfare that can be adopted into the Sustainable Development Goals 2030. To do this the notion of legal personhood needs to be accepted. Historically there exist a few countries of common law which have already enacted some form of legal personhood rights to specific species of animals. For instance, since 2002 under the German basic law, paragraph 20A all animals have the right to be "respected by the state and to have their dignity protected" (Connolly, 2002, n.p.). Under the Animal Welfare Act 1999 New Zealand granted great apes legal rights (Rook, 2009). In 2013, India granted non-human person rights to cetaceans (Coelho, 2013). For this reason, if countries entering into partnerships for the progression of the Sustainable Development Goals can of their own accord afford legal or close to legal personhood rights for non-human animals, then there already exists a space for this framework to benefit the sustainability agenda.

To do this the framework would need to consider the rights of non-human animals as a stand-alone goal in an effort to keep in line with the progressive and inclusive policies so fundamental to the Sustainable Development Goals. According to Mary Anne Warren – an American philosopher, for a being to have personhood status, it should meet five criteria (Warren, 1973). Firstly, it should have consciousness; secondly, they need to be able to reason; thirdly, they should be self-motivated; fourthly, they must have the capacity to communicate; and fifthly, they should possess the presence of self-awareness (Warren, 1973). Non-human animals have shown evidence for some, if not all, of the key elements pertaining to consciousness, especially with regards to their moral positions and actions that could be argued worthy of consideration for inclusion into some form of a personhood legal platform. Such a platform could very much be established with the sustainability agenda leading the way by introducing a stand-alone goal with the sole focus of expanding animal welfare rights throughout our globalised communities, and their actors.

However, there are many powerful obstacles in place that can only be rectified with a change to the legal status of non-human animals from a product to legal person. In a 2017 article for the Guardian, the journalist Elle Hunt discusses the Australian government's involvement to boost the public consumption of meat and dairy to support the industry. Hunt (2017, n.p.) states: "the legal counsel for Voiceless, Sarah Margo, said every federal animal welfare initiative had been dismantled by the Abbott government. It had withdrawn funding for the Australian animal welfare strategy, disbanded its advisory committee and dissolved an animal welfare subdivision within the Department of Agriculture". This is an attempt to completely silence animal welfare debates within the Australian government, thereby providing a disguised platform for horrific animal abuse to benefit profit margins.

With the sale of beef considerably declining since the 1970s in Australia (Taylor & Butt, 2017) as well as in other wealthy countries (Marinova & Raphaely, 2018b), it would appear that the livestock industries are colluding with their governments to drive up the sale of meat, even though research has shown increased consumption could lead to cancer in the consumer (Simon, 2013, 2017). Further to this, Greenpeace International's 2018 report (Greenpeace, 2018a) states that "global meat and dairy production and consumption must be cut in half by 2050 to avoid dangerous climate change and keep the Paris Agreement on track… If left unchecked, agriculture is projected to produce 52% of global greenhouse gas emissions in the coming decades, 70% of which will come from meat and dairy" (Greenpeace International, 2018, n.p.). Greenpeace (2018a, p.31) also argues for ecological livestock farming where the land is shared between humans, livestock and other species, warns that intensive systems require larger external inputs related to the supply of feed and most importantly represent "an ethical and animal welfare question". The argument for animal welfare rests heavily on pursuing the idea of establishing legal personhood rights for non-human animals, and why this movement is so important to be considered alongside the sustainability agenda.

LEGAL PERSONHOOD

Animals raised for food within the livestock agricultural industry, who have been cared for and loved by their human owners, exhibit similar emotions of love and empathy. Studies into pig intelligence show that these domestic animals have a similar cognitive capacity to children (Dooling, 2014; Bekoff, 2015). Many farmers discuss the bond they form with their livestock and express a certain degree of love for their animals who appear to reciprocate this affection (Bock, van Huik, Prutzer, Kling Eveillard, & Dockes, 2007; Agriculture Victoria, 2015).

In order for this bond to happen, the non-human animal would need to share many personality traits akin to humans. Ethically, for this reason sentient non-human animals should have proper representation from a legal perspective and this should be part of the sustainability agenda – a framework that seeks to "strengthen universal peace in larger freedom" and represents "a plan of action for people, planet and prosperity" unifying moral concepts of dignity and healing the Earth (Sustainable Development Knowledge Platform, n.d., n.p.). Such a framework should also eliminate fear and suffering for non-human animals through a legal pathway towards personhood.

The concept of personhood is a widely philosophized and political notion stemming from legal terms. It allows for not only rights, but also the capacity to engage with legal entities, such as governments and industry, to enter into contracts with each other (O'Brian, 2013). A legal person however, is not necessarily a biological human. It is an entity that holds the capacity for legal rights, which can include corporations, religious institutions and now natural habitats and in some cases non-human animals, such as great apes and cetaceans (Wise, 2010). For the most part, being a biological human is typically sufficient to claim personhood rights – that is, rights to freedom, liberty, citizenship and equality. These are the fundamental principles guiding the UN Sustainable Development Goals. The right to education, gender equality, food and healthcare all ascribe to the rights of the person to legally enact their freedom to live in a dignified manner (United Nations, 2015).

However, legal personhood does not only engage the rights of the human species. It incorporates legal entities that operate as groups. For example, any corporation is considered a "legal person" separate and distinctive from the group of people who form it (Investopedia, 2018). Where the body politic is the collective legal right of citizens to a state, jurisprudence lends to the notions of the natural person as a political entity with rights as it does to the juridical person that is a nonhuman legal entity (Dyschkant, 2015). Several, if not all 17, Sustainable Development Goals would fall into one or the other category of legal personhood status. An example is legal personhood as a political and legal governing platform which allows entities from Industry, Innovation and Infrastructure (Goal 9) to operate and engage with entities from Partnerships for the Goals (Goal 17). Without such legal rights, interconnected partnerships through enterprise could and would fall victim to corruption with no avenue for legal recourse.

The legal system which is fundamentally designed to protect human interests does not by default protect all humans, all of the time. It protects the powerful human interests first, those that are often associated with government, corporate, religious and economic interests. The concept of "juristic person" evolved with human development in conjunction to the human individual as holding unique power and being an acting agent of their own legal and moral obligations and responsibilities (Koessler, 1949). This is perhaps why the legal notion of person does not by its own statement protect just the biological person, but also the person as a collective of human interests (Koessler, 1949). This is how corporations, governing bodies, and other entities can claim the rights to personhood under the legal framework of the legal person. In fact, the legal person can often act in a manner distinctively different from the human being[1].

It appears that the first step in moving towards improving animal welfare is by closing the gap between human and non-human animals in the eyes of the law. Initiatives like the Nonhuman Rights Project, which concentrates its attention on great apes and cetaceans (Wise, 2010), have important campaigns and litigations in process. However, their focus is not on nonhuman animals held captive within the livestock agricultural industry (Cupp, 2015). The push to have nonhuman animals with higher cognition recognized and granted a legal vessel with which to effect personhood rights status, under the responsibility of a legal person, will in time provide a pathway for animals within the livestock agricultural industry to follow (Choplin, 2016).

At the moment, the pursuit of personhood rights for nonhuman animals is being litigated on a singular case-by-case basis, in the hopes that when one case is granted personhood rights, animal rights lawyers will find it easier to litigate on the behalf of other nonhuman animals and expect to win a similar result, as set by the judicial precedence. Once in one legal case a nonhuman animal is being granted personhood rights, this will allow animal rights lawyers to argue for all animals of that species, and not just the individual animal, to be granted similar personhood rights (Animal People Inc., 2018). From there this will expand to other species of nonhuman animals and eventually pave the way for the most abused nonhuman animals from the livestock industry where they are being treated only as a source of food.

On the other hand, people already know that livestock and domesticated animals exhibit attachment and love towards their owners. The issue that requires recognition is that any animal, including wild and ferocious species, can experience feelings and should not be subjected to suffering and torture.

Setting that initial legal precedence is most challenging and complex. Legal institutions are aware that once this precedence is set, it will force the entire system into a domain it has never been before – that of acknowledging the welfare and suffering of nonhuman animals as akin to our own as a human species.

Although in the western world the legal system is held in high respect, it often fails to protect the weakest members even of the human society. For example, institutional racism is an implicit practice that has long been identified as serving the social, political and economic interests of only one group (or race) within society (Carmichael & Hamilton, 1967). The legal system has in the past exerted and continues to exert institutional racism to Aboriginal and Indigenous peoples in almost all countries across the globe[2]. Similarly, the legal system is a bastion of institutional speciesism – a moral philosophy or a worldview which subjugates the nonhuman animal species (Brügger, 2018). This is why it is fundamental that the sustainability agenda – the most ambitious transformative global and universal agenda, also includes the welfare of the nonhuman animals.

SUSTAINABLE DEVELOPMENT AGENDA

The Sustainable Development Goals represent the development of a form of universal law or approach that can be used globally to govern the implementations of economic, human and environmental rights. They have huge hurdles to overcome before the gap of injustice is effectively closed. Nevertheless, there have already been small pockets of success in poverty alleviation, healthcare improvements and gender equality. The Sustainable Development Goals Report for 2018 shows that extreme poverty has fallen by a third of what it was in 1990, maternal mortality is down 37% since 2000, and child marriages in Southern Asia have dropped by over 40% since the year 2000 (UN, 2018). However, the report also highlights slow or negative changes in many categories and by its own admission, progress has not been rapid enough to achieve the goals by 2030 (UN, 2018). The rise in hunger is of particular interest in this regard with the proportion of undernourished increasing to 11% in 2016 (from 10.6% in 2015), representing 815 million people, and high to moderately high food prices present in 26 countries in 2016 (UN, 2018).

On the surface, one would think that using nonhuman animals as food for humans is contributing towards the reduction of world hunger. In reality, denigrating and defaming sentient animals to being food sources are contributing to global injustice, power imbalances and misappropriation of resources. The livestock industry is a prime reason for global hunger. The existing global "economic and political system props up and accelerates industrial livestock production" (Greenpeace, 2018b, p.33) which on the one hand, allows the wealthier countries and richer sections of society to overconsume animal proteins and fats resulting in obesity and prevalence of non-communicable diseases, such as diabetes type 2, and on the other, a large amount of the produced grains are given to food animals instead of directly feeding people. The scientific evidence suggests that eliminating animal products from human diet will result in feeding billions more people on this planet without requiring additional use of natural resources, with reduced environmental costs, improved human health and liberated nonhuman species. On a vegan diet, today the Earth can feed 12.8 billion people – an additional 5 billion (Marinova & Raphaely, 2018a). With eliminating the long inefficient and wasteful food chain based on feeding the animal first and then eating the animal product, the competition for grains will also decrease and food prices will drop making the achievement of Goal 2 more realistic.

The 17 Sustainable Development Goals provide a very important platform to humanity with which to attempt to address the many issues plaguing humankind. It is this very platform that is most important, as it provides a vessel with which to begin unifying people from across the world in an attempt to find solutions for the betterment and peaceful progression of the human race. It is this vessel that makes it imperative for the sustainability agenda to consider and adopt an 18th goal focusing entirely on animal welfare or animal liberation. This will not automatically change the current status of animal agriculture, but will at least provide a global public platform in line with our wider humanitarian initiatives with which to address the real concerns regarding sentient animal suffering.

This 18th goal can provide an important framework where global voices can unite and advocate together as a unified force, regardless of their colour, gender or religion. It will give people from around the world the opportunity to become unified in their discourse on modern day moral philosophy as it is currently applied to industry, trade and legal expectations of states and governments. Further to this, it can provide policymakers from around the globe the opportunity to learn from each other in how to move forward and implement legal strategies where empathy is the main focus. The 18th goal in animal welfare humanizes humankind and sets the stage for higher global empathy towards each other and the other species (Kant, 1963).

Environmental ethics drives the Sustainable Development Goal 15 Life on Earth, part of which includes the protection of endangered species and animals under threat from poaching. However, Goal 15 does not frame this protection as the rights to life of the animal, but rather as a natural resource within the non-human environment. It is still very anthropocentric, it places the nonhuman animal in the realm of commodity, carefully steering away and intentionally ignoring the sentience of the living animal. This is problematic and indicative of why an 18th goal in the pursuit of animal welfare is so important.

The thinking behind Goal 15 casts a shadow over society where turning a blind eye is legitimated – the nonhuman animal is property, commodity and akin to a plant that is not sentient in nature. The non-human animal exists in our social consciousness and practices through the sustainable development goals as a resource to be exploited for human benefit. Although the views of the 18th century's ethical philosopher Immanuel Kant on the animal align with the human–nonhuman typology, he made a very convincing argument for the ethical treatment of sentient life. Kant argued that from a moral philosophic point of view, the human being would benefit from treating animals with respect and care as to do otherwise might lead the human to also behave cruelly towards other human beings (Kant, 1963, p.240). Kant's views are echoed in modern psychology, where a strong link has been shown between those who engage in acts of animal cruelty and violence towards other humans and other forms of antisocial behavior (Gullone & Arkow, 2012). Environmental ethics in its own right has fundamental benefits to offer sustainable development; however, the ethic complexities in relation to sentient life need a deeper understanding and higher moral grounds described by Singer (1973) as animal liberation. Below are six case studies which help build such a higher ground understanding through demonstrating the personhood and the existence of mirror neurons in some unexpected animals.

ANIMAL PERSONHOOD CASE STUDIES

These case studies, which can also be seen as case stories, were selected among many other recounts about the animal world and its magic. As in the case of the legal system, if the reader finds one of them convincing in relation to the nonhuman animal's right to personhood, all the others will simply follow creating a convincing and logical argument that humans and nonhumans share the same genetic material and sentient behavior but also the same rights to be free from suffering.

Case Study 1: Pocho the Crocodile and His Human Father

Roger Horrocks hosted the National Geographic Documentary entitled *Touching the Dragon* released in 2013. In this documentary, he tells the story of Pocho the crocodile.

A young wild crocodile Pocho had been shot in the head and left to die by a farmer in Costa Rica in 1989. Pocho was found by a local man nicknamed Chito who then spent the next three years nursing the young reptile back to health. Upon Pocho's full recovery, Chito released him back into the wild only to awake the next morning to find the crocodile sleeping on his veranda. Chito would go on to make further attempts to release Pocho back into the wild only to find the crocodile back on his front porch the next morning.

From this sparked a unique bond between Chito and Pocho that would span decades before the crocodile's death due to natural causes in 2011 (Levin, 2011). What makes this story unique is the bond that was formed between Pocho and Chito during the course of his life. Pocho, a five meter fully grown male crocodile, had bonded with Chito and come to seek out his attention and love. Chito was able to form an intimate friendship with Pocho. The crocodile would allow Chito to enter the water with him, play with him and feed him. Chito would make sound vibrations with his hand in the water to signal his arrival to Pocho.

Over 20 years Chito had developed a well-established form of communication with Pocho that included using vibrations in the water, eye contact, hand signals as well as verbal cues. What is really remarkable is the tone of voice Chito used to communicate his love and warmth to Pocho and the response from the crocodile was gentle and loving. The argument for Chito and Pocho's unique and unusual relationship between a mammal and a reptile is said to have been able to happen based on the early neural developmental pathways the crocodile developed as a result of his close proximity to his human father (Horrocks, 2013; DadoTheGoodVillain, 2013). Without this early neural pathway, it is argued that it might have been impossible for Chito and Pocho to develop a strong and bonded relationship, given that reptile brains are thought to not involve empathy or attachment. However, this is clearly not the case between Pocho and Chito.

This is one very intriguing and relevant example of the human-animal relationship and the capacity for love, as it would seem, to transcend the species' barrier (much as it were in the unusual instances of feral children being raised by animals). There is something unique to these paternalistic relationships that has yet to be explored within its full capacity, as human condition is to think animals are a little more than biological robots working simply off instinct.

A crocodile's instinct would be to kill anything that it might deem as food. As this relationship spanned over twenty years, it would seem that Pocho had every opportunity to feast on his human father, to leave his human father and pave out his own path in life. However, Pocho's clear choice was to remain with his human father throughout his life even though he was given every opportunity to play out his role of crocodile within the South American jungle.

It would be very easy to downplay the bonded relationship between Pocho and Chito as a unique and unusual one-off human-animal relationship between a wild predator and his prey, given that Pocho was raised by Chito from infancy. However, could a similar form of empathy arise from an adult animal typically known as a predator in the wild with little human contact?

Case Study 2: The Kidnapped Girl Rescued by Adult Lions From Her Captors

In 2005, a group of three lions reportedly rescued a 12-year-old girl who had been kidnapped by a group of men in rural south-west Ethiopia (BBC News, 2005; NBC News, 2005; The Guardian, 2005). The girl was taken by the group of men in Addis Ababa, Ethiopia on her way home from school and had been beaten repeatedly by them.

Tilahun Kassa, a local government official, corroborated official police reports saying that kidnapping for the purpose of child marriage is common in the area (NBC News, 2005). Sgt. Wondimu Wedajo reported that the girl was found accompanied by three lions who stood guard over her for half a day until police, who were in pursuit of the kidnappers had arrive (BBC News, 2005; NBC News, 2005; The Guardian, 2005). When police found the girl, Sgt. Wondimu Wedajo said: "They stood guard until we found her and then they just left her like a gift and went back into the forest" (BBC News, 2005; NBC News, 2005). The girl told police that the men had beaten her repeatedly, but that the group of three lions had not harmed her (BBC News, 2005; NBC News, 2005; The Guardian, 2005). Police found the girl 350 miles southwest of Addis Ababa on the outskirts of the provincial capital Bita Genet (NBC News, 2005).

The news outlet NBC News spoke to Stuart Williams from the rural development ministry who said: "A young girl whimpering could be mistaken for the mewing sound from a lion cub, which could explain why they didn't eat her" (NBC News, 2005). Colonel Lemma Legesse, a game hunter, believed that "they were probably preparing to eat her but were intercepted by the police and the others" (BBC News, 2005). Who the 'others' are that Colonel Legesse is referring to is unclear, perhaps the kidnappers themselves? This might seem unlikely given that the girl had spent half a day with the three lions and the only harm that had befallen her was at the hands of her human captors.

It would be impossible to really know what the three lions' intentions were that day, all that is clear is that had it not been for them, this girl might never have been reunited safely with her family. This is an incredible account and one that could easily be anthropomorphized into a heroic tale of the animal predator fending off the human predator in an altruistic effort to save the damsel in distress – damsel to the human predators, child in all other respects. The account does not specify whether the three lions were male or female, and although this might not appear relevant on surface value, it does become important when considering abnormal lioness behavior towards other species that would normally be considered food for this incredible female hunter.

Case Study 3: The Lioness and the Oryx Calf

In 2002, Kenyan wildlife experts were astonished to find a lioness who had adopted a newborn oryx calf in the Samburu National Park (Astill, 2002). Witnesses were amazed to see the lioness grooming and caring for the calf whilst also allowing it to feed from its biological mother before reclaiming the calf again under her protection (Astill, 2002; Wildlife African Foundation, 2002). The lioness and the calf roamed freely together with the lioness protecting it from other male lions (Astill, 2002; Wildlife African Foundation, 2002). Unfortunately, the calf did not survive as a male lion caught and killed it whilst the lioness slept.

Shortly after the lioness lost her calf, she adopted another oryx calf (Astill, 2002; Wildlife African Foundation, 2002). This is a highly unusual behavior for a wild lioness to exhibit, and no scientific behavioral explanation has been offered to account for the motivation of the lioness (Astill, 2002; Wildlife African Foundation, 2002). The only assumption is that the lioness' maternal instincts had somehow overridden her natural predatory behavioral patterns (Astill, 2002; Wildlife African Foundation, 2002).

From the past two examples, it would seem that the lion has the ability to feel empathy and awareness and project this towards other species. In more recent times, the conservationist and wildlife reserve ranger Gerry van Der Walt was taking a group out on safari when they came across a lioness who had just made a fresh kill of a red hartebeest (van Der Walt, 2008). The group watched on taking video and photos of the scene until they noticed that the lioness was behaving very strangely. It soon became apparent that the lioness had realized that her kill was pregnant. Instead of feasting on the unborn fetus, the lioness removed it from the mother's stomach and gently placed in on the ground nudging it in a caring manner. This left van Der Walt and his group astonished and dumbfounded to explain what they were witnessing (van Der Walt, 2008; Daily Mail, 2009).

The question remains, what underlying psychological process was prompting these lionesses to behave so uncharacteristically? Also, can other species of predator in the wild exhibit a similar response to young animals they would typically hunt and kill for food?

Case Study 4: The Leopard and the Baby Baboon

In 2006, the conservationists and filmmakers Dereck and Beverly Joubert released their award-winning documentary *Eye of the Leopard*. The documentary filmmakers follow a young leopard named Legadema through her journey in life. In one particular episode Legadema hunts and kills a baboon. At the time of the kill Legadema is unaware that clinging to the baboon was a young baby. Unexpectedly, instead of killing the young baboon, Legadema appears to react surprised before her maternal instincts kicked in and instead she abandons her kill to protect the baby baboon (Wildlife Films, 2014).

The young baboon was groomed and protected from hyenas gathering under the tree where Legadema sought refuge. Young and weak without its mother, the baby baboon had fallen several times from the tree whereupon Legadema would race down to scoop it up and bring it back up to the safety of the tree. Unfortunately, by the morning the baby baboon had passed away and Legadema was forced to abandon its body, untouched and uneaten (Joubert, Joubert, Apostal, & Bowman, 2006; Huffington Post, 2013).

These case studies are compelling evidence for something we would call empathy in humans. The next case study takes a closer look into some of the ways our closest living primate relatives exhibite foresight and planning.

Case Study 5: Koko the Western Lowland Gorilla Uses Sign-Language to Discuss Different Family Planning Strategy With a Developmental Psychologist

There are many arguments for the superiority of the human species above all other nonhuman animals. It is important to draw attention to those cases that could suggest that the human–animal difference can in some ways be broken down. Koko the Gorilla is a famous western lowland gorilla brought to the San

Francisco Zoo in 1971 as a one year old. She is famous for learning how to sign more than one thousand words taught to her by her trainer and behavioral psychologist Penny Patterson.

Koko would go on to become the face of The Gorilla Foundation and koko.org as well as a YouTube sensation for videos showing how highly intelligent and aware she is by her ability to converse with humans through sign language. One of the most famous clips of Koko is where she meets the late acting legend Robin Williams, who also happens to be her favorite actor as she is known for enjoying his comedy movies. Koko can be seen teasing Robin Williams and asking him to tickle her, where in response to the actor's tickles, she smiles and acts coy (Kokoflix, 2014).

The gorilla was also well-known for her maternal nature where she adopted a kitten and expressed great anguish and grief when it was unexpectedly killed by a car (Kluger, 2018). Koko would repeatedly ask Patterson for her own baby and used dramatic play to pretend her toy dolls were alive (Patterson, 2002). She would even go as far as to manipulate the hands of her baby doll to make it sign, much the same way a human child might speak for their toy doll.

In one compelling 2012 video released by the official Gorilla Foundation's YouTube channel, Koko is presented with four alternate plans for having a baby. Koko can be seen listening to Patterson carefully before deciding on the option that suits her best, displaying a clear ability to plan and project into the future. Sadly, Koko never realized her dreams of becoming a mother and died in her sleep at the age of 46 in 2018 (Kluger, 2018).

Case Study 6: Steven Wise and the Nonhuman Animal Rights Project

Steven Wise is an American Animal Rights Lawyer, originally from Michigan and co-founder of the Nonhuman Animal Rights Project. In 2016, he released his documentary called *Unlocking the Cage*, which follows his legal staff as they attempt to argue habeas corpus – a legal right for a person under arrest to be brought to court, in the state of New York.

Wise often laments how nonhuman animals should be entitled to equality but because they cannot enter into contracts, essentially you can make them slaves for their whole lives (Wise, Pennebaker, & Hegedus, 2016). Together with his legal team, Wise took on the State for New York on behalf of two chimpanzees Leo and Hercules and argued for their freedom by invoking the Great Writ: Habeas Corpus recourse in court to extend the law to consider Leo and Hercules as legal persons, thereby giving them personhood rights to freedom, life and liberty (Wise, Pennebaker, & Hegedus, 2016). The habeas corpus writ is a legal recourse that was most famously used to free individual slaves in the United States. The historical significance this writ has to freeing predominantly black slaves, makes this legal recourse particularly problematic, especially given that it is now being used to try and ascribe limited personhood rights for two chimpanzees.

Steven Wise and his team faced this problematic conundrum when invoking the habeas corpus writ in court where the judge could not overlook the potential sensitivity issues pertaining to racial discrimination (Wise, Pennebaker, & Hegedus, 2016). As a result, Leo and Hercules would not, like many enacting the writ before them, find their freedom this way (Wise, Pennebaker, & Hegedus, 2016). Leo and Hercules are but two chimpanzees of many nominated as potential candidates to be legally represented by Steven Wise through the habeas corpus recourse. They were at the time detained in the Stoney Brook Medical Facility in New York and used as research subjects (Wise, Pennebaker, & Hegedus, 2016). Steven Wise and his team would argue habeas corpus – Latin for "you have the body", in an attempt to get a court of common law to make a ground-breaking decision to grant legal personhood rights for the first time to a nonhuman animal (Wise, Pennebaker, & Hegedus, 2016).

The primary necessity for enacting the habeas corpus writ and not an animal welfare case is that Wise and his team specifically sought to bring about a precedence for nonhuman animals to be considered in a court of law as a legal person with limited abilities. This provides the nonhuman animal with the application of some access to legal human rights as a legal person (Marceau & Wise, 2018).

Of course, primates are not human by fact of DNA, however, they do meet all other criteria for sentient consciousness and should therefore be at the very least considered to be granted access to limited legal human rights applications under the law. This includes the right not to be imprisoned in a medical facility against their will and exposed to testing that may not meet the ethical standards for a human participant, or the right to live in peace as they would be sent to a sanctuary instead of being confined in cages all day.

Habeas corpus is a powerful legal strategy used by Steven Wise and his team who have since founded the Nonhuman Rights Project and taken on other sentient animal legal campaigns (Marceau & Wise, 2018). As Steven Wise says in his 2016 documentary *Unlocking the Cage* (Wise, Pennebaker, & Hegedus, 2016), legal personhood is given to nonhumans only when it is in the human's interest.

What should also be considered is how personhood rights for non-sentient or conscious entities came into existence as it extends from the human interest. Throughout history it has been a common human practice to conquer other regions through force for the economic development of power by land, subjects and wealth. This goes as far back as Pope Innocent IV (1243-1254), who had incorporated the idea of the persona ficta, or fictional person without soul or body, to imbue monasteries and institutions as a collective with personhood rights to protect these institutional bodies from individual actions (Dewby, 1926; Robinson, 2010) and in this way protect the overall human interests of power, growth and economy of the Roman Church (Dewy, 1926; Robinson, 2010). Here we see that collective interests are just as powerful, if not more, than the individual interest of those who hold claim to the legal personhood status. This is where the great writ: habeas corpus challenges the power and/or interest of the state to hold the body of an individual, albeit, an individual of the natural person.

Kristen Epps' 2017 article entitled *Habeas Corpus, The Fugitive Slave Law, and Executive Authority*, states that the implementation of habeas corpus to rescue a detainee from state or federal custody harkens back to the enslaved people detained under the Fugitive Slave Law of 1850. This law was indented to protect slaveholders' property interests and reinforce a pro-slavery interpretation of the US Constitution. The argument made here in relation to the use of the habeas corpus writ with regards to slavery specifically in America, is to draw a comparison between the economic interests of slave owners, or stakeholders, and the use of the law with which to maintain ownership of the body of another for the collective interest of maintaining power in either trade, institution or for economic benefit. These stakeholder interests are discussed where one is human and the other is a nonhuman animal.

It is also imperative to break free from the speciesist thought process that binds our societies to a form of cognitive dissonance where we collectively ignore the suffering of one species over another for fear of offense. The habeas corpus recourse effectively is a legal strategy to free the body of an innocent that is enslaved for profit.

This is an important point for the global livestock agricultural industry which is frighteningly lucrative and will be almost impossible to challenge just on the basis or merit of moral philosophy alone. The only effective way humanity can collectively move forward and successfully take on big corporate interests within the livestock agricultural industry is by pursuing the notion of granting legal personhood rights to all sentient animals. Whether it is through habeas corpus or any other legal strategy, is not the most important issue – it is in creating a legal precedence from which to build the argument. However, as is always the case, profit margins trump environmental and humanitarian agency. By granting legal personhood rights to all sentient nonhuman animals there will exist a solid and legal framework with which to effect a real change for the benefit of our world, its environment and the humans who interpret it.

THE 18TH SUSTAINABLE DEVELOPMENT GOAL

Sustainable development as the new paradigm which is replacing the now outdated 20th century modernist narrative of progress (Myerson & Rydin, 1996), has a duty of care to incorporate animal welfare and animal rights initiatives across the globe. This could be done as an 18th goal in the Sustainable Development Goals agenda with the sole purpose of shifting people's perceptions about food and establishing specific targets. It will not be a fast solution, but a slow and progressive step in the right direction for addressing the problems we as species inhabiting this planet face collectively. Through the work of champions, such as Steven Wise, the US is setting the stage to become a leader of a fundamental acknowledgment of the necessity to shift our moral and legal awareness in line with our philosophical values of freedom and liberty for all, leaving no one behind – even the animals.

The World Animal Protection Organization has been campaigning the United Nations for years to adopt a Universal Declaration on Animal Welfare (World Society for the Protection of Animals, 2007). This initiative seeks to provide a universal declaration that promotes protecting animal welfare, reducing suffering in sentient animals, as well as leading towards better human health, social development, proper disaster management, reduction in poverty and hunger, responding to climate change and achieving a transition to sustainability (World Society for the Protection of Animals, 2007). To date the United Nations have yet to adopt the particular model. However, in 2018 the European Commission released a report on the impact of animal welfare international activities on the globalized world (European Commission, 2018). The report acknowledges the need for "a solid and science-based legislative model on animal welfare" which "should also be disseminated internationally" (European Commission, 2018, p.1).

Whilst this is a positive step forward, both the World Animal Protection Organization and the European Commission still only address the welfare of animals solely as food producing units within the livestock agricultural industry in order to establish a level playing field across all countries. Hence, the welfare of animals is addressed only from the point of view of their economic value as a commodity destined for human consumption. Again, the nonhuman animal is significantly reduced in its intrinsic value as it relates to human interests.

Through the sustainability agenda, the United Nations have the opportunity as well as the moral imperative to recognize the sentience of animals and acknowledge their intrinsic value outside the human interests. There needs to be real recognition of the nonhuman animal and its ability to feel pain as well as the considerable suffering imposed on that animal when treated as a commodity. The sustainability agenda needs to counteract the institutional speciesism of the legal system which is likely to continue to delay the recognition of the status of body ownership or habeas corpus to nonhuman animals. Without an overarching global agenda to improve the treatment of nonhuman animals, livestock will continue to remain exactly this – live stock, that is goods and merchandise for distribution and sale which are alive and satisfy human interest based on economic transactions through corporate enterprise. The task of the sustainability agenda is to take on the role of innovative governance with regards to animal welfare – not as a commodity traded on domestic and global markets, but as sentient beings with a purely intrinsic value.

Such an 18th goal would steer away from the current position of subjugation, exploitation and suffering experienced by animals as sentient beings without choice. It will deliver numerous benefits for the human animals, such as better nutrition and health, less environmental harm and higher moral ethics. Such an approach would differ from the legal system where the animal will be considered case by case within the limits of its individual circumstances, as is the process of litigations on behalf of individual animals initiated by the Nonhuman Rights Project. It will rather view and recognize the individual needs of the animal as a collective enterprise worthy of legal personhood rights and representation. Collectivizing the interests of nonhuman animals in a similar fashion to those of a corporation, which has already been given legal personhood rights, will open the way to a new worldview of the place, role and responsibilities of human animals of this planet.

It is not to say that animals for food will cease to exist under the proposed Goal 18 within the sustainability agenda. However, their numbers would be significantly reduced and will be given better and more dignifying living conditions. Industrial meat production in intensive factory farms will have to be phased out. Interest and business opportunities in plant-based food alternatives will spur. This proposal is not perfect and contains many legal and philosophical implications that our current worldview and legal systems are yet to find ways to deal with. However, such a goal will instigate the necessary small incremental steps forward to allow time for legislation and industries to evolve, shake off their speciesism and develop better and more humanitarian approaches to feeding the global population.

IMPLICATIONS FOR NEW MEAT ALTERNATIVES

Reduction in the current levels of meat and dairy consumption is a top priority for countries and individuals where personal intake exceeds the scientifically established healthy levels. In the case of red meat, this limit is set at 26 kg per person per year or the equivalent to no more than 70 g per day (WCR/AICR, 2007)[3]. There will be a natural switch to more plant-based foods, such as vegetables, fruits, grains, nuts, legumes, beans, seeds, tubers and roots as well as coagulated and fermented products based on plant milks, such as tofu and bean curds – the list is too long to be exhaustive. This will ignite the interest in the new plant-based meat alternatives especially developed to replace products traditionally based on meat, eggs and dairy. There will be abundance of business opportunities to replace the current agricultural practices, jobs and products.

Another avenue for change is to increase government support for lab-grown meat research and industries. Clean meat holds a lot of promise "to revolutionize the food system" (The Good Food Institute, n.d., n.p.) by drastically reducing the requirements for land, water and the use of antibiotics. As the current technologies still rely on the use of small samples of animal cells, it is paramount that these animals are given the chance to live a normal and dignified life.

The 18th goal would help put a stop on intensive industrial systems of meat and dairy production which exploit the animals together with the commons and pollute the air, soil and waters on this planet. Policies, subsidies and public spending currently channeled to animal-based products, including reliefs for drought, flood and fires, will have to be diverted to plant-rich foods and innovative options sourced from ecologically aware producers. Withdrawing subsidies to farmers raising livestock for slaughter is the first step for governments to make. The message has never been more pertinent than it is now, the public purse should not be subsidizing a declining and destructive industry, nor should the government be pushing a meat-eating agenda, especially since science shows that processed meat is a class 1 carcinogen with red meat coming in behind as a 2A carcinogen (Hunt, 2017; Simon, 2017).

Further ethical consideration should be given to livestock related legislation. For example, legislation that prohibits exposure of negative farming practices should be repelled and reporting abuse of food animals should be decriminalized. Such practices will make sure producers are being scrutinized and liable (Hunt, 2017). A shocking Australian example is when US-style ag-gag laws came into effect in the state of New South Wales under the radar of public opinion in 2017 (Sydney Criminal Lawyers, 2017). Such laws are supposed to protect the agricultural industry, but their practical effect is to inhibit the public coverage of animal cruelty and thereby enable big farming operations to treat animals as they please (Sydney Criminal Lawyers, 2017).

The introduction of an 18th goal will further progress the sustainability agenda related to other SDGs, particularly No Hunger (Goal 2) under which some countries are dealing with undernourishment while others are having to combat growing obesity rates. Recognizing the rights of animals and their ability to feel pain, suffer and show empathy is not just a matter of biological mirror neurons – it is part of a much broader sustainability global agenda with beneficial implications for all.

CONCLUSION

The stage is set for success and it is only a matter of time before the issue of personhood for nonhuman animals becomes a moral dilemma of the past. There will come a time when future generations will look back on the atrocities inflicted on nonhuman animals and be ashamed of how long it took for our moral species to bring about social justice to be in line with our legal representations. Regret will not take away the fact that we have blatantly abused sentient beings denigrating them to food commodities when healthier and better alternatives exist.

In order to make today's practices a thing of the past however, there is an important role for global moral and policy initiatives like the UN Sustainable Development Goals 2030 to get onboard and provide a worldwide outreach program for all countries to measure their progress in this respect. No place should be left behind, no one should be left behind, not even the animals. Developing an 18th goal within the sustainability agenda will provide a much-needed platform for countries across the globe to find unity, support and partnership through innovation and industry to further the animal welfare agenda and help align our legal systems with our moral values. This proposal will not immediately offer all nonhuman animals the right to personhood status, but it will at least provide an important ethical framework from which all countries can begin civil negotiations on behalf of the sentient beings used to drive economies and feed our nations. In time, this important platform can measure and document the historical movement towards greater animal welfare on par with the global initiative to seek dignity and freedom for human beings within the original ethical framework from where sustainability originated.

There is no bigger hope for the world today than for the people to activate their mirror neurons for all sentient beings.

REFERENCES

Animal People Inc. (2018). *Animal rights law and the pursuit of nonhuman personhood.* Animal People Forum. Retrieved from http://animalpeopleforum.org/2018/08/01/animal-rights-law-pursuit-nonhuman-personhood/

Astill, J. (2002). Lioness adopts another antelope. *The Guardian.* Retrieved from https://www.theguardian.com/world/2002/feb/17/jamesastill.theobserver

BBC News. (2005). *Kidnapped girl 'rescued' by lions.* Retrieved from http://news.bbc.co.uk/2/hi/africa/4116778.stm

Bekoff, M. (2015). Pigs are intelligent, emotional, and cognitively complex. *Psychology Today.* Retrieved from https://www.psychologytoday.com/us/blog/animal-emotions/201506/pigs-are-intelligent-emotional-and-cognitively-complex

Bock, B. B., van Huik, M. M., Prutzer, M., Kling Eveillard, F., & Dockes, A. (2007). Farmer' relationship with different animals: The importance of getting close to the animals – case studies of French, Swedish and Dutch cattle, pig and poultry farmers. *International Journal of Sociology of Agriculture and Food, 15*(3), 108–125.

Brügger, P. (2018). It's the speciesism, stupid! Animal abolition, environmentalism, and the mass media. In D. Bogueva, D. Marinova, & T. Raphaely (Eds.), *Handbook of research on social marketing and its influence on animal origin food product consumption* (pp. 92–104). Hershey, PA: IGI Global. doi:10.4018/978-1-5225-4757-0.ch006

Buller, H. J., & Morris, C. (2009). Beasts of a different burden: Agricultural sustainability and farm animals. In S. Seymour, R. Fish, & S. Watkins (Eds.), *Transdisciplinary perspectives on sustainable farmland management* (pp. 135–148). Wallingford, UK: Centre for Agriculture and Biosciences International.

Carmichael, S., & Hamilton, C. V. (1967). *Black power: Politics of liberation.* New York, NY: Vintage.

Choplin, L. (2016). *Animal-exploiting industries, take note.* Nonhuman Rights. Retrieved from https://www.nonhumanrights.org/blog/animal-exploiting-industries/

Coelho, S. (2013). *Dolphins gain unprecedented protection in India.* DW. Retrieved from https://www.dw.com/en/dolphins-gain-unprecedented-protection-in-india/a-16834519

Connolly, K. (2002). German animals given legal rights. *The Guardian.* Retrieved from https://www.theguardian.com/world/2002/jun/22/germany.animalwelfare

Corballis, M. (2015). Mirror neurons, theory of. In J. D. Wright (Ed.), *International encyclopedia of the social & behavioral sciences* (pp. 582–588)., doi:10.1016/B978-0-08-097086-8.23204-2

Cupp, R. L. (2015). Human responsibility, not legal personhood, for nonhuman animals. *The Federalist Society, 16*(2), 29–35.

DadoTheGoodVillain. (2013). *The man who swims with crocodiles.* Retrieved from https://www.youtube.com/watch?v=I7fZZUfvx0s

Daily Mail. (2009). *Lioness showed remorse realising killed pregnant antelope.* Retrieved from http://www.dailymail.co.uk/news/article-1165832/The-lioness-showed-remorse-realising-killed-pregnant-antelope.html

Dewby, J. (1926). The historic background of corporate legal personality. *The Yale Law Journal, 35*(6), 655–673. doi:10.2307/788782

Dooling, B. (2014). *Who's smarter – pigs or kids?* One Green Planet. Retrieved from https://www.onegreenplanet.org/news/whos-smarter-pigs-or-kids-video/

Dyschkant, A. (2015). Legal personhood: How we are getting it wrong. *University of Illinois Law Review, 5*, 2075–2110.

Epps, K. (2017). Habeas corpus, the fugitive slave law, and executive authority. *The Journal of Civil War Era.* Retrieved from https://journalofthecivilwarera.org/2017/02/habeas-corpus-fugitive-slave-law-executive-authority/

European Commission. (2018). *Report from the Commission to the European Parliament and the Council: On the impact of animal welfare international activities on the competitiveness of European livestock producers in a globalized world.* Retrieved from https://ec.europa.eu/food/sites/food/files/animals/docs/aw_international_ publication-report_en.pdf

Gibbons, A. (2012). Bonobos join chimps as closest human relatives. *Science AAAS.* Retrieved from http://www. sciencemag.org/news/2012/06/bonobos-join-chimps-closest-human-relatives

Greenpeace. (2018a). *Less is more: Reducing meat and dairy for a healthier life and planet.* Scientific background of the Greenpeace vision of the meat and dairy system towards 2050. Retrieved from https://storage.googleapis.com/p4-production-content/international/wp-content/uploads/2018/03/6942c0e6-longer-scientific-background.pdf

Greenpeace. (2018b). *Less is more: Reducing meat and dairy for a healthier life and planet.* The Greenpeace vision of the meat and dairy system towards 2050. Retrieved from https://storage.googleapis.com/p4-production-content/international/wp-content/uploads/2018/03/698c4c4a-summary_greenpeace-livestock-vision-towards-2050.pdf

Greenpeace International. (2018). *Greenpeace calls for decrease in meat and dairy production and consumption for a healthier planet.* Retrieved from https://www.greenpeace.org/international/press-release/15111/greenpeace-calls-for-decrease-in-meat-and-dairy-production-and-consumption-for-a-healthier-planet/

Gullone, E. (2012). Historical and current conceptualizations of animal cruelty. In E. Gullone & P. Arkow (Eds.), *Animal cruelty, antisocial behaviour, and aggression: More than a link* (pp. 5–15). Basingstoke, UK: Palgrave Macmillan. doi:10.1057/9781137284549_2

Horrocks, R., Foster, C., & Foster, D. (2013). *Touching the Dragon* [DVD]. Natural History Unit Africa: National Geographic.

Huffington Post. (2013). *Leopard cub adopts her prey's baby, shows what we can learn from the animal world.* Retrieved from https://www.huffingtonpost.com/2013/12/04/leopard-adoptsbabybaboo_n_4386147.html

Hunt, E. (2017). Meatonomics author says government working with meat and dairy industry to boost consumption. *The Guardian*. Retrieved from https://www.theguardian.com/science/2017/may/06/meatonomics-author-says-government-working-with-meat-and-dairy-industry-to-boost-consumption

Investopedia. (2018). *Corporation*. Retrieved from https://www.investopedia.com/terms/c/corporation.asp

Joubert, B., Joubert, D., Apostal, J., & Bowman, J. (2006). *Eye of the leopard*. Gaborone, Botswana: Wildlife Films Botswana, National Geographic.

Kant, I. (1963). *Lectures on ethics*. New York, NY: Harper & Row Publishers. Retrieved from https://www.uta.edu/philosophy/faculty/burgess-jackson/Duties%20Towards%20Animals%20and%20Spirits%20(1930,%201963).pdf

Kluger, J. (2018). Koko the Gorilla wasn't human, but she taught us so much about ourselves. *Time*. Retrieved from http://time.com/5318710/koko-gorilla-life/

Koessler, M. (1949). The person in imagination or persona ficta of the corporation. *Louisiana Law Review*, *9*(4), 435–449.

Kokoflix. (2012, September 11). *Koko Selects Adoption Options*. Retrieved from https://www.youtube.com/watch?v=xSBI_JMemqY

Kokoflix. (2014). *Koko's tribute to Robin Williams*. Retrieved fromhttps://www.youtube.com/watch?v=I9I_QvEXDv0

Levin, M. (2011). Croc Pacho. *The Tico Times*. Retrieved from http://www.ticotimes.net/2011/10/17/costa-rica-says-goodbye-to-famous-croc-pocho

Marceau, J. F., & Wise, S. (2018). *Exonerating the innocent: Habeas for nonhuman animals*. University of Denver Sturm College of Law Working Paper No. 18-16. Retrieved from https://papers.ssrn.com/sol3/papers.cfm?abstract_id=3176104

Marinova, D., & Raphaely, T. (2018a). Impact of vegetarian diets on the environment. In W. J. Craig (Ed.), *Vegetarian nutrition and wellness* (pp. 13–24). Boca Raton, FL: CRC Press.

Marinova, D., & Raphaely, T. (2018b). Taxing meat and animal food products. In D. Bogueva, D. Marinova, & T. Raphaely (Eds.), *Handbook of research on social marketing and its influence on animal origin food product consumption* (pp. 121–134). Hershey, PA: IGI Global. doi:10.4018/978-1-5225-4757-0.ch008

Miele, M., & Lever, J. (2014). Improving animal welfare in Europe: Cases of comparative bio-sustainabilities. In T. Marsden & A. Morely (Eds.), *Sustainable food systems: Building a new paradigm* (pp. 143–165). Abingdon, UK: Routledge.

Morris, M. (2011). *Factory farming and animal liberation in New Zealand*. Wellington, New Zealand: Earl of Seacliff Art Workshop.

Myerson, G., & Rydin, Y. (1996). Sustainable development: the implications of the global debate for landuse planning. In S. Buckingham-Hatfield & B. Evans (Eds.), *Environmental planning and sustainability* (pp. 19–34). Chichester, UK: Wiley.

National Archives Museum. (2018). *Declaration of Independence: A Transcription.* Retrieved from https://www.archives.gov/founding-docs/declaration-transcript

National Health and Medical Research Council (NHMRC). (2013). *Australian dietary guidelines.* Canberra, Australia: NHMRC and Department of Health and Ageing, Australian Government. Retrieved from https:// www.nhmrc.gov.au/guidelines-publications/n55 and https://www.eatforhealth.gov.au/sites/default/files/files/the_guidelines/n55_australian _dietary_guidelines.pdf

NBC News. (2005). *Ethiopian girl reportedly guarded by lions.* Retrieved from http://www.nbcnews.com/id/8305836/ns/world_news-africa/t/ethiopian-girl-reportedly-guarded-lions/#.W24ohdgzY6g

O'Brien, C. (2013). I trade, therefore I am: Legal personhood in the European Union. *Common Market Law Review, 50*(6), 1643–1684.

Patterson, P. (2002). *Koko wants to have a baby.* The Gorilla Foundation. Retrieved from https://www.koko.org/node/1889

Prüfer, K., Munch, K., Hellmann, I., Akagi, K., Miller, J. R., Walenz, B., ... Pääbo, S. (2012). The bonobo genome compared with the chimpanzee and human genomes. *Nature, 486*(7404), 527–531. doi:10.1038/nature11128 PMID:22722832

Rawles, K. (2006). Sustainable development and animal welfare: The neglected dimension. In J. D'Silva & J. Turner (Eds.), *Animals, ethics, and trade: The challenge of animal sentience* (pp. 208–216). London, UK: Earthscan.

Robinson, J. W. (2010). *William of Ockham's early theory of property rights: Sources, texts, and contexts.* Heritage Branch, Centre for Medieval Studies University of Toronto, Ottawa. Retrieved from http://individual.utoronto.ca/jwrobinson/articles/j Robinson_phd-thesis_ockham.pdf

Rook, D. (2009). Should great apes have 'human rights'? *Web Journal of Current Legal Issues, 1*, 1-10. Retrieved from http://www.bailii.org/uk/other/journals/WebJCLi/2009/issue1/ rook1.html

Sachs, J. D. (2015). *The age of sustainable development.* New York, NY: Columbia University Press. doi:10.7312ach17314

Shestack, J. (1998). The philosophic foundations of human rights. *Human Rights Quarterly, 20*(2), 201–234. doi:10.1353/hrq.1998.0020

Simon, D. R. (2013). *Meatonomics: How the rigged economics of meat and dairy make you consume too much–and how to eat better, live longer, and spend smarter.* San Francisco, CA: Conari Press.

Simon, D. R. (2017). Why is government sponsoring meat products that can cause cancer? *The Sydney Morning Herald.* Retrieved from https://www.smh.com.au/ opinion/why-is-government-sponsoring-meat-products-that-can-cause-cancer-20170503-gvy5bd.html

Singer, P. (1975). *Animal liberation: A new ethics for our treatment of animals.* London, UK: Harper Collins.

Sustainable Development Knowledge Platform. (n.d.). *Transforming our world: The 2030 Agenda for Sustainable Development.* Retrieved from https://sustainabledevelopment.un.org/post2015/transformingourworld

Sydney Criminal Lawyers. (2017). *NSW government silences animal rights activists.* Retrieved from https://www.sydneycriminallawyers.com.au/blog/nsw-government-silences-animal-rights-activists/

Taylor, E., & Butt, A. (2017). Three charts on Australia's declining taste for beef and growing appetite for chicken. *The Conversation.* Retrieved from https://the conversation.com/three-charts-on-australias-declining-taste-for-beef-and-growing-appetite-for-chicken-78100

The Good Food Institute. (n.d.). *The future of meat.* Retrieved from https://cleanmeat.org

The Guardian. (2005). *Lions save girl from kidnappers.* Retrieved from https://www.theguardian.com/world/2005/jun/22/3

United Nations. (1948). *The universal declaration of human rights.* General Assembly Resolution 217A (1948, December 10). Retrieved from http://www.un.org/en/universal-declaration-human-rights/

United Nations. (2015). *Transforming our world: The 2030 agenda for sustainable development.* General Assembly (2015, September 25). Retrieved from http://www.un.org/en/ development/desa/population/migration/generalassembly/docs/globalcompact/A_RES_70_1_E.pdf

United Nations. (2018). *The Sustainable Development Goals Report 2018.* Statistical Division. Retrieved from https://unstats.un.org/sdgs/report/2018/overview/

Van Der Walt, G. (2008). *Brutal or amazing. You decide…* Retrieved from http://photo-africa.blogspot.com/2008/09/brutal-or-amazing-you-decide.html

Victoria, A. (2015). *How farmers view their animals.* Retrieved from http://agriculture.vic.gov.au/agriculture/livestock/sheep/sheep-notes-newsletters/sheep-notes-autumn-2015/how-farmers-view-their-animals#

Warren, M. A. (1973). On the moral and legal status of abortion. *The Monist, 57*(4), 1–9. PMID:11661013

Wildlife African Foundation. (2002). *Kenyan lioness adopts another oryx.* Retrieved from https://www.awf.org/news/kenyan-lioness-adopts-another-oryx

Wildlife Films. (2014). *Incredible leopard and baby baboon interaction.* Retrieved from https://www.youtube.com/watch?v=ugi4x8kZJzk

Wise, S. (2010). Legal personhood and the Nonhuman Rights Project. *Animal Law, 17*(1), 1–12.

Wise, S., Pennebaker, D. A., & Hegedus, C. (2016). *Unlocking the cage. Documentary Film.* New York, NY: HBO Documentary Films.

World Cancer Research Fund and American Institute for Cancer Research (WCRF/AICR). (2007). *Food, nutrition, physical activity and the prevention of cancer: A global perspective.* Washington, DC: AICR. Retrieved from http://www.aicr.org/assets/docs/pdf/reports/Second_Expert_Report.pdf

World Society for the Protection of Animals. (2007). *Universal declaration on animal welfare: An agreement among people and nations to recognize the animals are sentient and can suffer, to respect their welfare needs and to end animal cruelty – for good.* Retrieved from https://www.worldanimalprotection.ca/sites/default/files/ca_-_en_files/case_for_a_udaw_tcm22-8305.pdf

ADDITIONAL READING

Bock, B. B., van Huik, M. M., Prutzer, M., Kling Eveillard, F., & Dockes, A. (2007). Farmer' relationship with different animals: The importance of getting close to the animals – case studies of French, Swedish and Dutch cattle, pig and poultry farmers. *International Journal of Sociology of Agriculture and Food*, *15*(3), 108–125.

Brügger, P. (2018). It's the speciesism, stupid! Animal abolition, environmentalism, and the mass media. In D. Bogueva, D. Marinova, & T. Raphaely (Eds.), *Handbook of research on social marketing and its influence on animal origin food product consumption* (pp. 92–104). Hershey, PA: IGI Global. doi:10.4018/978-1-5225-4757-0.ch006

Brügger, P., Marinova, D., & Raphaely, T. (2016). Animal production and consumption: An ethical educational approach. In T. Raphaely & D. Marinova (Eds.), *Impact of meat consumption on health and environmental sustainability* (pp. 295–312). Hershey, PA: IGI Global. doi:10.4018/978-1-4666-9553-5.ch017

DadoTheGoodVillain. (2013). *The man who swims with crocodiles*. Retrieved from https://www.youtube.com/watch?v=I7fZZUfvx0s

Dyschkant, A. (2015). Legal personhood: How we are getting it wrong. *University of Illinois Law Review*, *5*, 2075–2110.

Greenpeace (2018). *Less is more: Reducing meat and dairy for a healthier life and planet.* Scientific background of the Greenpeace vision of the meat and dairy system towards 2050. Retrieved from https://storage.googleapis.com/p4-production-content/international/wp-content/uploads/2018/03/6942c0e6-longer-scientific-background.pdf

Simon, D. R. (2013). *Meatonomics: How the rigged economics of meat and dairy make you consume too much–and how to eat better, live longer, and spend smarter*. San Francisco, CA: Conari Press.

United Nations. (2018). *The Sustainable Development Goals Report 2018*. Statistical Division. Retrieved from https://unstats.un.org/sdgs/report/2018/overview/

Wise, S. (2010). Legal personhood and the Nonhuman Rights Project. *Animal Law*, *17*(1), 1–12.

Wise, S., Pennebaker, D. A., & Hegedus, C. (2016). *Unlocking the cage. Documentary Film.* New York, NY: HBO Documentary Films.

KEY TERMS AND DEFINITIONS

Case Study: A research method which relies on detailed description of a particular subject, object, or event within its context; it does not allow for statistical generalization and its validity is judged based on shared common characteristics and value systems.

Damsel: A young women who has not been married.

Habeas Corpus: Translated from Latin means "produce the body"; this legal writ is used to challenge incarceration without a valid reason. If granted, this legal writ can result in (1) release from custody, (2) reduction in the term of imprisonment, (3) an order declaring the conditions of confinement being illegal, and (4) a declaration of rights.

Institutional Racism: A form of racism (or belief in superiority of one race above the others) expressed through the existing institutions within society and supported by its political system.

Human Rights: Norms and behavior protected as normal or through the legal system on the basis of certain moral principles.

Speciesism: A belief that humans are superior to nonhuman animals; a form of prejudice, analogous to racism and sexism, based on morally irrelevant differences.

Sustainable Development Goals (SDGs): A set of 17 goals (with accompanying targets and indicators) set by the United Nations as a global agenda for 2016 to 2030; they address current global challenges and although each goal aims at a particular issue they are interconnected and applicable to any place on Earth; areas covered by the SDGs include poverty, hunger, health, education, gender equality, water and sanitation, affordable and clean energy, jobs and economic growth, innovation and infrastructure, inequalities, sustainable cities and communities, responsible production and consumption, climate action, life below water and on land, peace and justice, and partnerships for sustainability.

ENDNOTES

1 For example, the 2003 documentary *The Corporation* describes this legal person acting in a manner that would be regarded and penalized as a dangerously psychopathic behavior but is considered the norm for the business world.

2 Between 1860 and 1960, the Aboriginal Protection Board (and its different incarnations, such as the Aborigines Welfare Board) in Australia was established and operated as a statutory legal authority under the Aboriginal Protection Act with almost full control over the lives of Aboriginal people depriving them from basic human rights.

3 In Australia, this limit is set at 65 g per day or 23.7 kg per person per year (NHMRC, 2013).

Conclusion

There is a lot of innovation and new technologies coming up in relation to energy, transportation, buildings, industry and all sectors of the economy, to combat climate change, improve environmental performance and reverse some of the negative impacts people's activities have had on the land, water and air of this planet. We are hopeful that once the importance of such issues is recognised, human creativity has no boundaries. The field of medicine and health care is constantly presenting us with examples of how we can do things better, improve living conditions and expand life expectancies across the globe.

It is time to apply such creativity, imagination and resourcefulness to food. Our everyday decisions are impacting not only our own bodies but also the health of the global commons and all other species on this planet. However, every single person can make a difference to the quality of their own life and that of others by showing kindness and consideration, by educating and exposing themselves to the latest scientific evidence, by being aware of all options and looking for those which are least harmful when decisions about food are being made. By comparison with the other economic sectors, when it comes to food we are less trapped in the inertia of old technological trajectories and are free to execute our choice every single day with every meal we take. However, we need to break free from the currently prevailing marketing and social discourses which make behaviour changes difficult. Instead of blaming others, we can take full responsibility and move to a different level of planetary consciousness. We no longer have to accept animal products as an inevitability but instead see food "as invitation to build, innovate and create change, a pathway that awakens creativity, compassion and genius" (Hawken, 2017, p. xi).

The work of all contributors to this book is confirming this message. We have choices between proven, traditionally used plants with amazing nutritional values and new product developments which are even more astounding. The important thing is to recognise that it is time for alternatives to livestock foods – alternatives that bring excitement and hope for a better life for all on this planet. By understanding the barriers that keep us hooked to the old meat-rich diets, we can find the drivers for change in the nutritional benefits of plant-based and non-animal food sources, we can reconsider social norms, including the meaning of masculinity, the pleasure of eating and the place of animals in the world we live in. We can make informed decisions whether insects or clean meat are a better option than plant-derived proteins as well as see the role of microbial proteins and soybeans in relation to food security. The business world is already responding to the changes in consumer preferences and the acknowledgement that it is time for a dietary shift with new food products which have the co-benefits of being environmentally better and more beneficial for human health. This new food trajectory has just begun but is fast gaining momentum as a global phenomenon supported by information, investments, research and development, new knowledge and skills, connections and social networks. It is driven by the rise in collective awareness and the realisation that at the heart of a sustainability transition lies the desire for goodness and a new ethics model.

Although a lot of the knowledge about food comes from the past, the book definitely looks into the future – a better future that we can imagine which redefines the place and power of the human species.

Thanks for reading the book and sharing its agenda for change.

Dora Marinova
Curtin University, Australia

Diana Bogueva
Curtin University, Australia

Talia Raphaely
Curtin University, Australia

Kurt Schmidinger
University of Vienna, Austria

REFERENCES

Hawken, P. (Ed.). (2017). *Drawdown: The most comprehensive plan ever proposed to reverse global warming*. New York, NY: Penguin Books.

Compilation of References

A soya point. (1975, February 14). *The Guardian*, p. 20. Retrieved from the ProQuest database: https://www.proquest.com/libraries/academic/databases/

Aaronson, S. (2000). Fungi. In K. F. Kiple & K. C. Ornelas (Eds.), *The Cambridge world history of food* (pp. 313–336). Cambridge, UK: Cambridge University Press.

Aasen, A. J., Eimhjellen, K. E., Liaaen-Jensen, S., Heinegård, D., Balaban, A. T., & Craig, J. C. (1969). An Extreme Source of beta-Carotene. *Acta Chemica Scandinavica*, *23*, 2544–2545. doi:10.3891/acta.chem.scand.23-2544 PMID:5368546

Abbas, C. A. (2006). Production of antioxidants, aromas, colours, flavours, and vitamins by yeasts. In A. Querol & G. H. Fleet (Eds.), *Yeasts in food and beverages* (pp. 285–334). Berlin, Germany: Springer. doi:10.1007/978-3-540-28398-0_10

Abugoch James, L. E. (2009). Quinoa (Chenopodium quinoa Willd.): Composition, chemistry, nutritional, and functional properties. *Advances in Food and Nutrition Research*, *58*, 1–31. doi:10.1016/S1043-4526(09)58001-1 PMID:19878856

Aburto, N. J., Hanson, S., Gutierrez, H., Hooper, L., Elliott, P., & Cappuccio, F. P. (2013a). Effect of increased potassium intake on cardiovascular risk factors and disease: Systematic review and meta-analyses. *British Medical Journal*, *346*, f1326. doi:10.1136/bmj.f1326 PMID:23558164

Acapsil. (n.d.). *What is the difference between microbiome, microbiota and microflora in relation to wounds?* Retrieved from http://acapsil.com/en-gb/2016/12/28/what-is-the-difference-between-microbiome-microbiota-and-microflora-in-relation-to-wounds/

Acharjee, S., & Sarmah, B. K. (2013). Biotechnologically generating 'super chickpea' for food and nutritional security. *Plant Science*, *207*, 108–116. doi:10.1016/j.plantsci.2013.02.003 PMID:23602105

Adams, C. J. (1990). *The sexual politics of meat: A feminist-vegetarian critical theory*. New York, NY: Continuum Print.

Adamson, M. W. (2004). *Food in medieval times*. Westport, CT: Greenwood Press.

Adedayo, M. R., Ajiboye, E. A., Akintunde, J. K., & Odaibo, A. (2011). Single cell proteins: As nutritional enhancer. *Advances in Applied Science Research*, *2*(5), 396–409.

Ademolu, K. O., Idowu, A. B., & Olatunde, G. O. (2010). Nutritional value assessment of variegated grasshopper, Zonocerus variegatus (L.) (Acridoidea: Pygomorphidae) during post-embryonic development. *African Entomology*, *18*(2), 360–364. doi:10.4001/003.018.0201

Ahenkora, K., Arthur, K., & Banahene, S. (2012). Marketing strategy for soybean products: Familiarity breeds content. *International Journal of Marketing Studies*, *4*(3), 186–189. doi:10.5539/ijms.v4n3p186

Aidoo, K. E., Rob Nout, M. J., & Sarkar, P. K. (2005). Occurrence and function of yeasts in Asian indigenous fermented foods. *FEMS Yeast Research*, *6*(1), 30–39. doi:10.1111/j.1567-1364.2005.00015.x PMID:16423068

Al Loman, A., & Ju, L. K. (2017). Enzyme–based processing of soybean carbohydrate: Recent developments and future prospects. *Enzyme and Microbial Technology, 106*, 35–47. doi:10.1016/j.enzmictec.2017.06.013 PMID:28859808

Alajaji, S. A., & El–Adawy, T. A. (2006). Nutritional composition of chickpea (Cicer arietinum L.) as affected by microwave cooking and other traditional cooking methods. *Journal of Food Composition and Analysis, 19*(8), 806–812. doi:10.1016/j.jfca.2006.03.015

Aleksandrowicz, L., Green, R., Joy, E. J. M., Smith, P., & Haines, A. (2016). The impacts of dietary change on greenhouse gas emissions, land use, water use, and health: A systematic review. *PLoS One, 11*(11), e0165797. doi:10.1371/journal.pone.0165797 PMID:27812156

Alexander, P., Brown, C., Arneth, A., Dias, C., Finnigan, J., Moran, D., & Rounsevell, M. D. A. (2017). Could consumption of insects, cultured meat or imitation meat reduce global agricultural land use? *Global Food Security, 15*, 22–32. doi:10.1016/j.gfs.2017.04.001

Alican, A. (2017). *Soğuk pres tekniği ile elde edilen chia tohumu atıklarının salata sosu üretiminde kullanılması*. Yıldız Teknik Üniversitesi.

Allen, L. (2007). Sensitive and real macho all at the same time: Young heterosexual men and romance. *Men and Masculinities, 10*(2), 137–152. doi:10.1177/1097184X05284221

Allen, N. E., Appleby, P. N., Davey, G. K., & Key, T. J. (2000). Hormones and diet: Low insulin-like growth factor-I, but normal bioavailable androgens in vegan man. *British Journal of Cancer, 83*(1), 95–97. doi:10.1054/bjoc.2000.1152 PMID:10883675

Altschul, A. (Ed.). (1974). *New protein foods (Vol. 1A Technology)*. New York, NY: Academic Press.

American Heart Association. (2017). *Top 25 foods that add the most sodium to your diet*. Retrieved from https://sodiumbreakup.heart.org/top_25_foods_that_add_the_most_sodium_to_your_diet

Anderson, E. (2008). Inclusive masculinity in a fraternal setting. *Men and Masculinities, 10*(5), 604–620. doi:10.1177/1097184X06291907

Angold, R., Beech, G., & Taggart, J. (1989). *Food biotechnology*. Cambridge, UK: Cambridge University Press.

Animal People Inc. (2018). *Animal rights law and the pursuit of nonhuman personhood*. Animal People Forum. Retrieved from http://animalpeopleforum.org/2018/08/01/animal-rights-law-pursuit-nonhuman-personhood/

Animals Australia. (2017). *We are Australia's leading animal protection organization*. Retrieved September 14, 2018 from https://www.facebook.com/AnimalsAustralia/

Anupama, & Ravindra, P. (2000). Value-added food: Single cell protein. *Biotechnology Advances, 18*(6), 459–479. doi:10.1016/S0734-9750(00)00045-8

Apostolidis, C., & McLeay, F. (2016a). It's not vegetarian, it's meat-free! Meat eaters, meat reducers and vegetarians and the case of Quorn in the UK. *Social Business, 6*(3), 267–290. doi:10.1362/204440816X14811339560938

Apostolidis, C., & McLeay, F. (2016b). Should we stop meating like this? Reducing meat consumption through substitution. *Food Policy, 65*, 74–89. doi:10.1016/j.foodpol.2016.11.002

Ardhana, M. M., & Fleet, G. H. (2003). The microbial ecology of cocoa bean fermentations in Indonesia. *International Journal of Food Microbiology, 86*(1), 87–99. doi:10.1016/S0168-1605(03)00081-3 PMID:12892924

Ardley, B. C., & Quinn, L. (2014). Practitioner accounts and knowledge production: An analysis of three marketing discourses. *Marketing Theory, 14*(1), 97–118. doi:10.1177/1470593113512322

Arganini, C., Saba, A., Comitato, R., Virgili, F., & Turrini, A. (2012). Gender differences in food choice and dietary intake in modern western societies. In J. Maddock (Ed.), *Public health social and behavioural health* (pp. 85–102). Rijeka, Croatia: InTech Open Access. doi:10.5772/37886

Astill, J. (2002). Lioness adopts another antelope. *The Guardian*. Retrieved from https://www.theguardian.com/world/2002/feb/17/jamesastill.theobserver

Aurelia. (2002). *A marketing report on meat substitutes in the Netherlands*. Amersfoort, The Netherlands: Aurelia. (in Dutch)

Australian Associated Press. (2018). Vegan trend takes hold in Australia. *SBS News*. Retrieved September 14, 2018 from https://www.sbs.com.au/news/vegan-trend-takes-hold-in-australia

Azzam, A. M. (1992). Production of metabolites, industrial enzymes, amino acids. *Journal of Environmental Science & Engineering*, *56*, 67–99.

Bacon, L., & Krpan, D. (2018). (Not) Eating for the environment: The impact of restaurant menu design on vegetarian food choice. *Appetite*, *125*, 190–200. doi:10.1016/j.appet.2018.02.006 PMID:29428546

Bailey, R., Froggatt, A., & Wellesley, L. (2014). *Livestock – climate change's forgotten sector: Global public opinion on meat and dairy consumption*. London, UK: Chatman House.

Banjo, A. D., Lawal, O. A., & Sangonuga, E. A. (2006). The nutritional value of fourteen species of edible insects in southern western Nigeria. *African Journal of Biotechnology*, *5*(3), 201–298. doi:10.5897/AJB05.250

Bao, X. F., Wang, X. S., Dong, Q., Fang, J. N., & Li, X. Y. (2002). Structural features of immunologically active polysaccharides from Ganoderma lucidum. *Phytochemistry*, *59*(2), 175–181. doi:10.1016/S0031-9422(01)00450-2 PMID:11809453

Barclay, A. W., Petocz, P., McMillan-Price, J., Flood, V. M., Prvan, T., Mitchell, P., & Brand-Miller, J. C. (2008). Glycemic index, glycemic load, and chronic disease risk: A meta-analysis of observational studies. *The American Journal of Clinical Nutrition*, *87*(3), 627–637. doi:10.1093/ajcn/87.3.627 PMID:18326601

Bar–El Dadon, S., Abbo, S., & Reifen, R. (2017). Leveraging traditional crops for better nutrition and health – The case of chickpea. *Trends in Food Science & Technology*, *64*, 39–47. doi:10.1016/j.tifs.2017.04.002

Barennes, H., Phimmasane, M., & Rajaonarivo, C. (2015). Insect consumption to address undernutrition, a national survey on the prevalence of insect consumption among adults and vendors in Laos. *PLoS One*, *10*(8), 1–16. doi:10.1371/journal.pone.0136458 PMID:26317772

Barker, D., Fitzpatrick, M. P., & Dierenfeld, E. S. (1998). Nutrient composition of selected whole invertebrates. *Zoo Biology*, *17*(2), 123–134. doi:10.1002/(SICI)1098-2361(1998)17:2<123::AID-ZOO7>3.0.CO;2-B

Barret, J. R. (2006). The Science of soy: What do we really know? *Environmental Health Perspectives*, *114*(6), A352–A358. Retrieved from https://ehp.niehs.nih.gov/doi/pdf/10.1289/ehp.114-a352 PMID:16759972

Bartlett, T. (2010). Manly meals and feminine foods. *The Chronicle of Higher Education*. Retrieved from https://www.chronicle.com/blogs/percolator/manly-mealsfeminine-foods/25576

Bawa, A. S., & Anilakumar, K. R. (2013). Genetically modified foods: Safety, risks and public concerns – a review. *Journal of Food Science and Technology*, *50*(6), 1035–1046. doi:10.100713197-012-0899-1 PMID:24426015

Baye, K. (2014). *Teff: Nutrient composition and health benefits*. Ethiopia Strategy Support Program. Workig Paper 67. Retrieved from https://www.researchgate.net/profile/Kaleab_Baye2/publication/266316373_Teff_Nutrient_Composition_and_Health_Benefits/links/542c24320cf277d58e8b0b42.pdf

BBC News. (2005). *Kidnapped girl 'rescued' by lions*. Retrieved from http://news.bbc.co.uk/2/hi/africa/4116778.stm

BBC Newsbeat. (2015). Arnold Schwarzenegger: Stop eating meat and save the planet. *BBC*. Retrieved from http://www.bbc.co.uk/newsbeat/article/35038053/arnold-schwarzenegger-stop-eating-meat-and-save-the-planet

Beardsworth, A., & Bryman, A. (1999). Meat consumption and vegetarianism among young adults in the UK: An empirical study. *British Food Journal*, *101*(4), 289–300. doi:10.1108/00070709910272169

Beardsworth, A., & Bryman, A. (2004). Meat consumption and meat avoidance among young people: An 11-year longitudinal study. *British Food Journal*, *106*(4), 313–327. doi:10.1108/00070700410529573

Beaudoux, V. G. (2017). How media sexism demeans women and fuels abuse by men like Weinstein. *The Conversation*. Retrieved from https://theconversation.com/how-media-sexism-demeans-women-and-fuels-abuse-by-men-like-weinstein-85789

Becker, E. W. (2007). Micro-algae as a source of protein. *Biotechnology Advances*, *25*(2), 207–210. doi:10.1016/j.biotechadv.2006.11.002 PMID:17196357

Becker, E. W., & Venktaraman, L. V. (1982). *Biotechnology and exploitation of algae: The Indian approach. A comprehensive report on the cultivation and utilization of microalgae performed at the Central Food Technological Research Institute, Mysore, India*. Eschborn, Germany: German Agency for Technical Cooperation.

Beelman, R. B., Royse, D. J., & Chikthimmah, N. (2003). Bioactive components in button mushroom Agaricus bisporus (J. Lge) Imbach (Agaricomycetideae) of nutritional, medicinal, and biological importance. *International Journal of Medicinal Mushrooms*, *5*(4), 321–338. doi:10.1615/InterJMedicMush.v5.i4.10

Beer, E. (2015). *"No success model" for new protein ingredients*. Retrieved from http://www.foodnavigator.com/Market-Trends/No-success-model-for-new-protein-ingredients

Bekatorou, A., Psarianos, C., & Koutinas, A. A. (2006). Production of food grade yeasts. *Food Technology and Biotechnology*, *44*(3), 407–415.

Bekker, G. A., Fischer, A. R., Tobi, H., & van Trijp, H. C. (2017). Explicit and implicit attitude toward an emerging food technology: The case of cultured meat. *Appetite*, *108*, 245–254. doi:10.1016/j.appet.2016.10.002 PMID:27717657

Bekker, G. A., Tobi, H., & Fischer, A. R. H. (2017). Meet meat: An explorative study on meat and cultured meat as seen by Chinese, Ethiopians and Dutch. *Appetite*, *114*, 82–92. doi:10.1016/j.appet.2017.03.009 PMID:28323057

Bekoff, M. (2015). Pigs are intelligent, emotional, and cognitively complex. *Psychology Today*. Retrieved from https://www.psychologytoday.com/us/blog/animal-emotions/201506/pigs-are-intelligent-emotional-and-cognitively-complex

Belay, A., Ota, Y., Miyakawa, K., & Shimamatsu, H. (1993). Current knowledge on potential health benefits of Spirulina. *Journal of Applied Phycology*, *5*(2), 235–241. doi:10.1007/BF00004024

Belluco, S., Losasso, C., Maggioletti, M., Alonzi, C. C., Paoletti, M. G., & Ricci, A. (2013). Edible insects in a food safety and nutritional perspective: A critical review. *Comprehensive Reviews in Food Science and Food Safety*, *12*(3), 296–313. doi:10.1111/1541-4337.12014

Ben-Amotz, A., & Avron, M. (1980). Glycerul β-carotene and dry al gal meal production by commercial cultivation of Dunaltella. In G. Shelef & C. J. Soeder (Eds.), *Algal biomass: Production and use* (pp. 603–610). Amsterdam: Elsevier.

Bennington-Castro, J. (2017). How crickets could help save the planet. *NBC News*. Retrieved from https://www.nbcnews.com/mach/environment/how-eating-crickets-could-help-save-planet-n721416

Bernard, N., Levin, S., & Trapp, C. (2014). Meat consumption as a risk factor for type 2 diabetes. *Nutrients, 6*(2), 897–891. doi:10.3390/nu6020897 PMID:24566443

Besler, T., Rakıcıoglu, N., Ayaz, A., Demirel Büyüktuncer, Z., Özel, G. H., Samur Eroğlu, G., . . . Yürük, A. (2015). *Türkiye'ye özgü besin ve beslenme rehberi.* Ankara. Retrieved from http://www.bdb.hacettepe.edu.tr/TOBR_kitap.pdf

Bhupathiraju, S. N., Tobias, D. K., Malik, V. S., Pan, A., Hruby, A., Manson, J. E., ... Hu, F. B. (2014). Glycemic index, glycemic load, and risk of type 2 diabetes: Results from 3 large US cohorts and an updated meta-analysis. *The American Journal of Clinical Nutrition, 100*(1), 218–232. doi:10.3945/ajcn.113.079533 PMID:24787496

Birt, D. F., Boylston, T., Hendrich, S., Jane, J. L., Hollis, J., Li, L., ... Whitley, E. M. (2013). Resistant starch: Promise for improving human health. *Advances in Nutrition, 4*(6), 587–601. doi:10.3945/an.113.004325 PMID:24228189

Blanchflower, D. G., Oswald, A. J., & Stewart–Brown, S. (2012). Is psychological well-being linked to the consumption of fruit and vegetables? *Social Indicators Research, 114*(3), 785–801. doi:10.100711205-012-0173-y

Blanchflower, D. G., Oswald, A. J., & Stewart–Brown, S. (2012). *Is psychological well-being linked to the consumption of fruit and vegetables? Working Paper 18469.* Cambridge, MA: National Bureau of Economic Research Working Paper Series. Retrieved from http://www.nber.org/papers/w18469.pdf doi:10.3386/w18469

Bloomberg. (2018). *Company overview of five seasons ventures sarl.* Company Report. Retrieved April 03, 2018 from https://www.bloomberg.com/profiles/companies/1578659D:FP-five-seasons-ventures-sarl

Boada, L. D., Henríquez-Hernández, L. A., & Luzardo, O. P. (2016). The impact of red and processed meat consumption on cancer and other health outcomes: Epidemiological evidences. *Food and Chemical Toxicology, 92*, 236–244. doi:10.1016/j.fct.2016.04.008 PMID:27106137

Boa, E. (2004). *Wild edible fungi: A global overview of their use and importance to people.* Rome, Italy: Food and Agriculture Organization of the United Nations. Retrieved from http://www.fao.org/docrep/007/y5489e/y5489e00.htm#TopOfPage

Bock, B. B., van Huik, M. M., Prutzer, M., Kling Eveillard, F., & Dockes, A. (2007). Farmer' relationship with different animals: The importance of getting close to the animals – case studies of French, Swedish and Dutch cattle, pig and poultry farmers. *International Journal of Sociology of Agriculture and Food, 15*(3), 108–125.

Bodenheimers, F. S. (1951). Insets as human food. The Huage, The Netherlands: Springer-Science+Business Media. doi:10.1007/978-94-017-6159-8

Bodnar, A. (2018). The scary truth behind fear of GMOs. *Biology Fortified.* Retrieved September 12, 2018 from https://www.biofortified.org/2018/02/scary-truth-gmo-fear/

Boehm, J. K., Williams, D. R., Rimm, E. B., Ryff, C., & Kubzansky, L. D. (2013). Association between optimism and serum antioxidants in the midlife in the United States study. *Psychosomatic Medicine, 75*(1), 2–10. doi:10.1097/PSY.0b013e31827c08a9 PMID:23257932

Boeing, H., Bechthold, A., Bub, A., Ellinger, S., Haller, D., Kroke, A., ... Watzl, B. (2012). Critical review: Vegetables and fruit in the prevention of chronic diseases. *European Journal of Nutrition, 51*(6), 637–663. doi:10.100700394-012-0380-y PMID:22684631

Bogueva, D., Marinova, D., & Raphaely, T. (Eds.). (2018). Handbook of research on social marketing and its influence on animal origin food product consumption. Hershey, PA: IGI Global.

Bogueva, D., Raphaely, T., Marinova, D., & Marinova, M. (2017) Sustainability social marketing. In J. Hartz-Karp & D. Marinova (Eds), Methods for sustainability research (pp. 280-291). Cheltenham, UK: Edward Elgar. doi:10.4337/9781786432735.00030

Bogueva, D., & Marinova, D. (2018). What is more important perception of masculinity or personal health and the environment? In D. Bogueva, D. Marinova, & T. Raphaely (Eds.), *Handbook of research on social marketing and its influence on animal origin food product consumption* (pp. 148–162). Hershey, PA: IGI Global. doi:10.4018/978-1-5225-4757-0.ch010

Bogueva, D., Marinova, D., & Raphaely, T. (2017). Reducing meat consumption: The case for social marketing. *Asia Pacific Journal of Marketing and Logistics*, *29*(3), 477–500. doi:10.1108/APJML-08-2016-0139

Bogueva, D., Marinova, D., & Raphaely, T. (Eds.). (2018). *Handbook of research on social marketing and its influence on animal origin food product consumption*. Hershey, PA: IGI Global. doi:10.4018/978-1-5225-4757-0

Bogueva, D., & Phau, I. (2016). Meat myths and marketing. In T. Raphaely & D. Marinova (Eds.), *Impact of meat consumption on health and environmental sustainability* (pp. 264–276). Hershey, PA: IGI Global. doi:10.4018/978-1-4666-9553-5.ch015

Böhm, I., Ferrari, A., & Woll, S. (2017). *In-vitro-Fleisch: Eine technische Vision zur Lösung der Probleme der heutigen Fleischproduktion und des Fleischkonsums? Institut für Technikfolgenabschätzung und Systemanalyse*. Karlruhe, Germany: ITAS.

Bojanic, A. (2011). *La quinua : Cultivo milenario para contribuir a la seguridad alimentaria mundial*. Proinpa, 58. Food and Agriculture Organization of the United Nations.

Bonny, S. P., Gardner, G. E., Pethick, D. W., & Hocquette, J. F. (2017). Artificial meat and the future of the meat industry. *Animal Production Science*, *57*(11), 2216–2223. doi:10.1071/AN17307

Borkhataria, C. (2017). Would you eat a burger made from insects? Mealworm-based food line set to hit grocery stores in Switzerland next week. *Daily Mail Australia*. Retrieved from https://www.dailymail.co.uk/sciencetech/article-4790414/Food-insects-sold-Swiss-supermarket.html

Borowitzka, M. A. (1999). Commercial production of microalgae: Ponds, tanks, tubes and fermenters. *Journal of Biotechnology*, *70*(1–3), 313–321. doi:10.1016/S0168-1656(99)00083-8

Bouvard, V., Loomis, D., Guyton, K. Z., Grosse, Y., El Ghissassi, F., & Benbrahim-Tallaa, L. (1599–1600). … Straif, K. on behalf of the International Agency for Research on Cancer Monograph Working Group (2015) Carcinogenicity of consumption of red and processed meat. *Lancet*, *16*(16). doi:10.1016/S1470-2045(15)00444-1

Bozakouk, A. H. (2002). *Acid hydrolysis of Phragmites austral is powder for production of single cell protein by* Candida utilis (Master thesis). Garyounis University, Benghazi, Libya.

Boze, H., Moulin, G., & Galzy, P. (1992). Production of food and fodder yeasts. *Critical Reviews in Biotechnology*, *12*(1–2), 65–86. doi:10.3109/07388559209069188 PMID:1733522

Branson, R. (2018). Clean meat is the future of meat. *Virgin*. Retrieved from https://www.virgin.com/richard-branson/clean-meat-future-meat

Braude, R., Hosking, Z. D., Mitchell, K. G., Plonka, S., & Sambrook, I. E. (1977). Pruteen, a new source of protein for growing pigs. I. Metabolic experiment: Utilization of nitrogen. *Livestock Production Science*, *4*(1), 79–89. doi:10.1016/0301-6226(77)90022-7

Breene, K. (2016). *Food security and why it matters*. World Economic Forum. Retrieved from https://www.weforum.org/agenda/2016/01/food-security-and-why-it-matters/

Breene, W. M. (1990). Nutritional and medicinal value of specialty mushrooms. *Journal of Food Protection, 53*(10), 883–894. doi:10.4315/0362-028X-53.10.883

Bresman, H., & Rao, V. D. (2017). A survey of 19 countries shows how generations X, Y, and Z are — and aren't — different. *Harvard Business Review*. Retrieved from https://hbr.org/2017/08/a-survey-of-19-countries-shows-how-generations-x-y-and-z-are-and-arent-different

Brester, G. W., & Schroeder, T. C. (1995). The impacts of brand and generic advertising on meat demand. *American Journal of Agricultural Economics, 77*(4), 969–979. doi:10.2307/1243820

Bréville, B. (2018). Meat goes out of style: A meat-free diet used to indicate poverty, but now the affluent adopt it to better their health, or to save animals and the planet. Instead of meat, what does the food industry have on the menu? *Le Monde Diplomatique*. Retrieved September 14, 2018 from https://mondediplo.com/2018/08/11animals

Brockes, E. (2000, December 1). Cheat meat. *The Guardian*, pp. A6–A7. Retrieved from the ProQuest database: https://www.proquest.com/libraries/academic/databases/

Bronzato, S., & Durante, A. (2017). A contemporary review of the relationship between red meat consumption and cardiovascular risk. *International Journal of Preventive Medicine, 8*(1), 40. doi:10.4103/ijpvm.IJPVM_206_16 PMID:28656096

Brookes, G., & Barfoot, P. (2017). Farm income and production impacts of using GM crop technology 1996–2015. *Biotechnology in Agriculture and the Food Chain, 8*(3), 156–193. doi:10.1080/21645698.2017.1317919 PMID:28481684

Browarnik, B. (2012). *Attitudes toward male vegetarians: Challenging gender norms through food choices*. Psychology Honors Papers, 25. Retrieved from https://digitalcommons.conncoll.edu/cgi/viewcontent.cgi?referer=https://www.google.com.au/&httpsredir=1&article=1021&context=psychhp

Brown, J. (2018). The rise of the bloody mock meat movement: Fad, or the key to sustainability? *Independent*. Retrieved from https://www.independent.co.uk/life-style/food-and-drink/vegans-vegetarians-meat-substitutes-bloody-mock-meats-food-lifestyle-a8427281.html

Brown, M. J. (2016). How short-chain fatty acids affect health and weight. *Healthline*. Retrieved from https://www.healthline.com/nutrition/short-chain-fatty-acids-101

Brown, L. R. (2011). *Full planet, empty plates: The new geopolitics of food scarcity. Earth Policy Institute*. New York, NY: W. W. Norton & Company.

Brugård Konde, Å., Bjerselius, R., Haglund, L., Jansson, A., Pearson, M., Färnstrand, J. S., & Johansson, A. (2015). *Swedish dietary guidelines - risk and benefit management report*. Retrieved March 16, 2018 from http://www.fao.org/3/a-az907e.pdf

Brügger, P. (2018). It's the speciesism, stupid! Animal abolition, environmentalism, and the mass media. In D. Bogueva, D. Marinova, & T. Raphaely (Eds.), *Handbook of research on social marketing and its influence on animal origin food product consumption* (pp. 92–104). Hershey, PA: IGI Global. doi:10.4018/978-1-5225-4757-0.ch006

Brügger, P., Marinova, D., & Raphaely, T. (2016). Animal production and consumption: an ethical educational approach. In T. Raphaely & D. Marinova (Eds.), *Impact of meat consumption on health and environmental sustainability* (pp. 295–312). Hershey, PA: IGI Global. doi:10.4018/978-1-4666-9553-5.ch017

Bryant, C., & Barnett, J. (2018). Consumer acceptance of cultured meat: A systematic review. *Meat Science, 143*, 8–17. doi:10.1016/j.meatsci.2018.04.008 PMID:29684844

Buckingham, L. (1999, September 10). United Biscuits hit by GM fears. *The Guardian*, p. 27. Retrieved from the ProQuest database: https://www.proquest.com/libraries/academic/databases/

Bukkens, S. G. F. (1997). The nutritional value of edible insects. *Ecology of Food and Nutrition*, *36*(2-4), 287–319. doi:10.1080/03670244.1997.9991521

Bukkens, S. G. F. (2005). Insects in the human diet: Nutritional aspects. In M. G. Paoletti (Ed.), *Ecological implications of minilivestock: potential of insects, rodents, frogs and snails* (pp. 545–577). Enfield, NH: Science Publishers.

Buller, H. J., & Morris, C. (2009). Beasts of a different burden: Agricultural sustainability and farm animals. In S. Seymour, R. Fish, & S. Watkins (Eds.), *Transdisciplinary perspectives on sustainable farmland management* (pp. 135–148). Wallingford, UK: Centre for Agriculture and Biosciences International.

Burger King. (2006). *I am man*. Retrieved from https://www.youtube.com/watch?v=vGLHlvb8skQ

Burke, M., & de Francisco, A. (2005). *Monitoring flows for health research 2005: Behind the global numbers*. Geneva, Switzerland: Global Forum for Health Research.

Burley, V. J., Paul, A. W., & Blundell, J. E. (1993). Influence of a high-fibre food (myco-protein) on appetite: Effects on satiation (within meals) and satiety (following meals). *European Journal of Clinical Nutrition*, *47*, 409–409. PMID:8396025

Buscemi, F. (2014). From killing cows to culturing meat. *British Food Journal*, *116*(6), 952–964. doi:10.1108/BFJ-11-2012-0288

Busch, S. (2018). USDA protein requirements in grams. *Healthy Eating, SF Gate*. Retrieved from http://healthyeating.sfgate.com/usda-protein-requirements-grams-8619.html

Byrne, C. S., Chambers, E. S., Morrison, D. J., & Frost, G. (2015). The role of short chain fatty acids in appetite regulation and energy homeostasis. *International Journal of Obesity*, *39*(9), 1331–1338. doi:10.1038/ijo.2015.84 PMID:25971927

Çaglarirmak, N., Ünal, K., & Ötles, S. (2002). Nutritional value of wild edible mushrooms collected from the Black Sea region of Turkey. *Micologa Aplicada International*, *14*(1), 1–5.

Campbell, J. (2018). Flipping the script: Creating mass change through social networking sites. In D. Bogueva, D. Marinova, & T. Raphaely (Eds.), *Handbook of research on social marketing and its influence on animal origin food product consumption* (pp. 135–147). Hershey, PA: IGI Global. doi:10.4018/978-1-5225-4757-0.ch009

Campbell, T. C., & Campbell, T. M. (2006). *The China Study: The most comprehensive study of nutrition ever conducted and the startling implications for diet, weight loss and long-term health*. Dallas, TX: Benbella Books.

Campbell, T. C., & Campbell, T. M. II. (2006). *The China Study*. Dallas, TX: BenBella Books.

Campo, E., del Arco, L., Urtasun, L., Oria, R., & Ferrer–Mairal, A. (2016). Impact of sourdough on sensory properties and consumers' preference of gluten–free breads enriched with teff flour. *Journal of Cereal Science*, *67*, 75–82. doi:10.1016/j.jcs.2015.09.010

Caplice, E., & Fitzgerald, G. F. (1999). Food fermentations: Role of microorganisms in food production and preservation. *International Journal of Food Microbiology*, *50*(1-2), 131–149. doi:10.1016/S0168-1605(99)00082-3 PMID:10488849

Capua, I., & Alexander, D. J. (2006). The challenge of avian influenza to the veterinary community. *Avian Pathology*, *35*(3), 189–205. doi:10.1080/03079450600717174 PMID:16753610

Carbonero, E. R., Gorin, P. A. J., Iacomini, M., Sassaki, G. L., & Smiderle, F. R. (2008). Characterization of a heterogalactan: Some nutritional values of the edible mushroom Flammulina velutipes. *Food Chemistry*, *108*(1), 329–333. doi:10.1016/j.foodchem.2007.10.029

Carlozzi, P. (2003). Dilution of solar radiation through "culture" lamination in photobioreactor rows facing south–north: A way to improve the efficiency of light utilization by cyanobacteria (*Arthrospira platensis*). *Biotechnology and Bioengineering*, *81*(3), 305–315. doi:10.1002/bit.10478 PMID:12474253

Carmichael, S., & Hamilton, C. V. (1967). *Black power: Politics of liberation*. New York, NY: Vintage.

Carolan, M. (2018). Big data and food retail: Nudging out citizens by creating dependent consumers. *Geoforum*, *90*, 142–150. doi:10.1016/j.geoforum.2018.02.006

Carrasco–Castilla, J., Hernández–Álvarez, A. J., Jiménez–Martínez, C., Jacinto–Hernández, C., Alaiz, M., Girón–Calle, J., ... Dávila–Ortiz, G. (2012). Antioxidant and metal chelating activities of peptide fractions from phaseolin and bean protein hydrolysates. *Food Chemistry*, *135*(3), 1789–1795. doi:10.1016/j.foodchem.2012.06.016 PMID:22953924

Carrington, D. (2010). Insects could be the key to meeting food needs of growing global population. *The Guardian*. Retrieved from https://www.theguardian.com/environment/2010/aug/01/insects-food-emissions

Carroll, K. K., & Braden, L. M. (1985). Dietary fat and mammary carcinogenesis. *Nutrition and Cancer*, *6*(4), 254–259. doi:10.1080/01635588509513831 PMID:6443636

Caruso, M. C., Favati, F., Di Cairano, M., Galgano, F., Labella, R., Scarpa, T., & Condelli, N. (2018). Shelf–life evaluation and nutraceutical properties of chia seeds from a recent long–day flowering genotype cultivated in Mediterranean area. *Lebensmittel-Wissenschaft + Technologie*, *87*, 400–405. doi:10.1016/j.lwt.2017.09.015

Castleman, M. (2018). *Stop poisoning your sex life*. Men's Journal. Retrieved from https://www.mensjournal.com/food-drink/stop-poisoning-your-sex-life/

Cavallo, C., & Materia, V. C. (2018). Insects or not insects? Dilemmas or attraction for young generations: A case in Italy. *International Journal on Food System Dynamics*, *9*(3), 226–239. doi:10.18461/ijfsd.v9i3.932

Cavazza, N., Guidetti, M., & Butera, F. (2015). Ingredients of gender-based stereotypes about food: Indirect influence of food type, portion size and presentation on gendered intentions to eat. *Appetite*, *91*, 266–272. doi:10.1016/j.appet.2015.04.068 PMID:25934089

Central People's Government of the People's Republic of China. (2018b). *Domestic soybeans should give full play to food consumption advantages*. Retrieved September 13, 2018 from http://www.gov.cn/zhengce/2018-05/21/content_5292362.htm

Central People's Government of the People's Republic of China. (2018b). *Figure: The report about China's agricultural forecast shows that China's soybeans planting areas will be restoratively increasing in 2018*. Retrieved September 13, 2018 from http://www.gov.cn/xinwen/2018-04/20/content_5284607.htm

Cerritos, R., & Cano-Santana, Z. (2008). Harvesting grasshoppers Sphenarium purpurascens in Mexico for human consumption: A comparison with insecticidal control for managing pest outbreaks. *Crop Protection Journal*, *27*(3–5), 473–480. doi:10.1016/j.cropro.2007.08.001

Chai, J. Y., Shin, E. H., Lee, S. H., & Rim, H. J. (2009). Foodborne intestinal flukes in Southeast Asia. *Korean Journal of Parasitology*, *47*(Suppl), S69–S102. doi:10.3347/kjp.2009.47.S.S69 PMID:19885337

Chang, R. (2017). Survey: Generation Z's best students seek careers in STEM and healthcare. *The Journal*. Retrieved from https://thejournal.com/articles/2017/07/12/survey-generation-zs-best-students-seek-careers-in-stem-and-healthcare.aspx

Charlebois, S., McCormick, M., & Juhasz, M. (2016). Meat consumption and higher prices: Discrete determinants affecting meat reduction or avoidance amidst retail price volatility. *British Food Journal*, *118*(9), 2251–2270. doi:10.1108/BFJ-03-2016-0121

Chaumont, D. (1993). Biotechnology of algal biomass production: A review of systems for outdoor mass culture. *Journal of Applied Phycology*, *5*(6), 593–604. doi:10.1007/BF02184638

Chen, X., Feng, Y., Zhang, H., & Chen, Z. (2010). Review of the nutritive value of edible insects. In P. B. Durst, D. V. Johnson, R. N. Leslie & K. Shono (Eds), *Forest insects as food: Humans bite back* (pp. 85–92). Bangkok, Thailand: Food and Agriculture Organization of the United Nations. Retrieved from http://www.fao.org/docrep/012/i1380e/i1380e00.pdf

Chen, J.-H., Song, J., Chen, Y., Ding, Q., Peng, A., & Mao, L. (2016). The effect of vegan protein–based diets on metabolic parameters, expressions of adiponectin and its receptors in wistar rats. *Nutrients*, *8*(10), 643. doi:10.3390/nu8100643 PMID:27763537

Chen, P. P., Wongsiri, S., Jamyanya, T., Rinderer, T. E., Vongsamanode, S., Matsuka, M., ... Oldroyd, B. P. (1998). Honey bees and other edible insects used as human food in Thailand. *American Entomologist*, *44*(1), 24–29. doi:10.1093/ae/44.1.24

Chen, X., Feng, Y., & Chen, Z. (2009). Common edible insects and their utilization in China. *Entomological Research*, *39*(5), 299–303. doi:10.1111/j.1748-5967.2009.00237.x

Cheredar, T. (2013). Google's Sergey Brin buys a $325,000 synthetic hamburger. *Venture Beat*. Retrieved from https://venturebeat.com/2013/08/05/googles-sergey-brin-backs-synthetic-hamburger-meat-project/

Chiles, R. M. (2013). If they come, we will build it: In vitro meat and the discursive struggle over future agrofood expectations. *Agriculture and Human Values*, *30*(4), 511–523. doi:10.100710460-013-9427-9

China Economy Net. (2018) *Domestic soybeans should play a leading role in food consumption. This year's output may hit a new high in 10 years.* Retrieved September 13, 2018 from http://www.ce.cn/cysc/sp/info/201805/21/t20180521_29192313.shtml

China Industrial Information Network. (2018). *Analysis on China's soybeans' production and consumption in 2018 including export and import.* Retrieved September 13, 2018 from http://www.chyxx.com/industry/201805/638760.html

Chiorando, M. (2018a). *Arnold Schwarzenegger accused of 'Drinking the vegan kool-aid'*. Beef Magazine, Plant Based News. Retrieved from https://www.plantbasednews.org/post/arnold-schwarzenegger-drinking-vegan-kool-aid)

Chiorando, M. (2018b). *Actor Joaquin Phoenix shares 'profound' experience that turned him vegan.* Plant based news. Retrieved from https://www.plantbasednews.org/post/actor-joaquin-pheonix-shares-profound-experience-that-turned-him-vegan

Chiorando, M. (2018c). *Leonardo DiCaprio's plans to make vegan burgers mainstream.* Plant Based News. Retrieved from https://www.plantbasednews.org/post/revealed-leonardo-dicaprios-plans-vegan-burgers-mainstream

Choplin, L. (2016). *Animal-exploiting industries, take note.* Nonhuman Rights. Retrieved from https://www.nonhumanrights.org/blog/animal-exploiting-industries/

Christensen, D. L., Orech, F. O., Mungai, M. N., Larsen, T., Friis, H., & Aagaard-Hansen, J. (2006). Entomophagy among the Luo of Kenya: A potential mineral source? *International Journal of Food Sciences and Nutrition*, *57*(3–4), 198–203. doi:10.1080/09637480600738252 PMID:17127470

Chung, Y. C. A. (2010). Edible insects and entomophagy in Borneo. In P. B. Durst, D. V. Johnson, R. N. Leslie & K. Shono (Eds), *Forest insects as food: Humans bite back* (pp. 141–150). Bangkok, Thailand: Food and Agriculture Organization of the United Nations. Retrieved September 12, 2018 from http://www.fao.org/docrep/012/i1380e/i1380e00.pdf

Chung, H.-J., Liu, Q., Hoover, R., Warkentin, T. D., & Vandenberg, B. (2008). In vitro starch digestibility, expected glycemic index, and thermal and pasting properties of flours from pea, lentil and chickpea cultivars. *Food Chemistry*, *111*(2), 316–321. doi:10.1016/j.foodchem.2008.03.062 PMID:26047429

Churchill, W. (1931). Fifty years hence. *Strand Magazine, UK*. Retrieved Sepember 10, 2018 from http://rolandanderson.se/Winston_Churchill/Fifty_Years_Hence.php

Cicatiello, C., De Rosa, B., Franco, S., & Lacetera, N. (2016). Consumer approach to insects as food: Barriers and potential for consumption in Italy. *British Food Journal, 118*(9), 2271–2286. doi:10.1108/BFJ-01-2016-0015

Clare, A. W. (2001). *On men: Masculinity in crisis*. London, UK: Arrow Books.

Clark, M., & Tilman, D. (2017). Comparative analysis of environmental impacts of agricultural production systems, agricultural input efficiency, and food choice. *Environmental Research Letters, 12*(6), 064016. doi:10.1088/1748-9326/aa6cd5

Clune, S., Crossin, E., & Verghese, K. (2017). Systematic review of greenhouse gas emissions for different fresh food categories. *Journal of Cleaner Production, 140*(Part 2), 766–783. doi:10.1016/j.jclepro.2016.04.082

Coelho, S. (2013). *Dolphins gain unprecedented protection in India*. DW. Retrieved from https://www.dw.com/en/dolphins-gain-unprecedented-protection-in-india/a-16834519

Cohen, R., Persky, L., & Hadar, Y. (2002). Biotechnological applications and potential of wood-degrading mushrooms of the genus Pleurotus. *Applied Microbiology and Biotechnology, 58*(5), 582–594. doi:10.100700253-002-0930-y PMID:11956739

Cokkizgin, A., & Shtaya, M. J. Y. (2013). Lentil: Origin, cultivation techniques, utilization and advances in transformation. *Agricultural Science, 1*(1), 55–62. doi:10.12735/as.v1i1p55

Collavo, A., Glew, R. H., Huang, Y.-S., Chuang, L. T., Bosse, R., & Paoletti, M. G. (2005). House cricets small-scale farming. In M. G. Paoletti (Ed.), *Ecological implications of minilivestock: potential of insects, rodents, frogs and snails* (pp. 519–544). Enfield, NH: Science Publishers.

Collingham, L. (2012). *The taste of war: World War Two and the battle for food*. London, UK: Penguin.

Collins, R. (1975, February 6). Describe a spun protein chop. *The Guardian*, p. 1. Retrieved from the ProQuest database: https://www.proquest.com/libraries/academic/databases/

Comerford, K. B., & Pasin, G. (2016). Emerging evidence for the importance of dietary protein source on glucoregulatory markers and type 2 diabetes: Different effects of dairy, meat, fish, egg, and plant protein foods. *Nutrients, 8*(8), 446. doi:10.3390/nu8080446 PMID:27455320

Committee on the Scientific Evaluation of Dietary Reference Intakes. (1998). *Dietary reference intakes: A risk assessment model for establishing upper intake levels for nutrients*. Washington, DC: National Academies Press (USA). Retrieved September 11, 2018 from https://www.ncbi.nlm.nih.gov/books/NBK45189/

Committee on World Food Security. (2012). *Global strategic framework for food security and nutrition*. Rome, Italy: Food and Agriculture Organization of the United Nations. Retrieved from http://www.fao.org/docrep/meeting/026/ME498E.pdf

Compassion in World Farming. (2013). *Strategic Plan 2013-2017: For kinder, fairer farming worldwide*. Retrieved September 14, 2018 from https://www.ciwf.org.uk/media/3640540/ciwf_strategic_plan_20132017.pdf

Connell, R. (1995). *Masculinities*. St Leonards, Australia: Allen & Unwin.

Connell, R. W., & Messerschmidt, J. W. (2005). Hegemonic masculinity: Rethinking the concept. *Gender & Society, 19*(6), 829–859. doi:10.1177/0891243205278639

Connolly, K. (2002). German animals given legal rights. *The Guardian*. Retrieved from https://www.theguardian.com/world/2002/jun/22/germany.animalwelfare

Cook it raw. (2015). *Menu*. Retrieved from http://www.cookitraw.org/home/

Cook J. (2000). Men's magazines at the millennium: New spaces, new selves. *Continuum, 14*(2), 171-186. doi:10.1080/713657705

Cook, J., Oreskes, N., Doran, P. T., Anderegg, W. R. L., Verheggen, B., Maibach, E. W., ... Green, S. A. (2016). Consensus on consensus: A synthesis of consensus estimates on human-caused global warming. *Environmental Research Letters, 11*(4), 048002. doi:10.1088/1748-9326/11/4/048002

Cooper, A. J., Forouhi, N. G., Ye, Z., Buijsse, B., Arriola, L., Balkau, B., ... Wareham, N. J. (2012). Fruit and vegetable intake and type 2 diabetes: EPIC-InterAct prospective study and meta-analysis. *European Journal of Clinical Nutrition, 66*(10), 1082–1092. doi:10.1038/ejcn.2012.85 PMID:22854878

Corballis, M. (2015). Mirror neurons, theory of. In J. D. Wright (Ed.), *International encyclopedia of the social & behavioral sciences* (pp. 582–588)., doi:10.1016/B978-0-08-097086-8.23204-2

Corpet, D. E. (2011). Red meat and colon cancer: Should we become vegetarians, or can we make meat safer? *Meat Science, 89*(3), 310–316. doi:10.1016/j.meatsci.2011.04.009 PMID:21558046

Council for Environment and Infrastructure (Rli). (2018). *Sustainable and Healthy: Working Together towards a Sustainable Food System*. The Hague, the Netherlands: Council for Environment and Infrastructure.

Courtenay, W. H. (2003). Key determinants of the health and well-being of men and boys. *International Journal of Men's Health, 2*(1), 1–30. doi:10.3149/jmh.0201.1

Cowspiracy. (2014). *The facts*. Retreived September 11, 2018 from http://www.cowspiracy.com/facts/

Craggs, R. J., Adey, W. H., Jessup, B. K., & Oswald, W. J. (1996). A controlled stream mesocosm for tertiary treatment of sewage. *Ecological Engineering, 6*(1–3), 149–169. doi:10.1016/0925-8574(95)00056-9

Craig. (Ed.). (2018). *Vegetarian nutrition and wellness*. Boca Raton, FL: CRC Press.

Craig, W. J. (2010). Nutrition concerns and health effects of vegetarian diets. *Nutrition in Clinical Practice, 25*(6), 613–620. doi:10.1177/0884533610385707 PMID:21139125

Craig, W. J., & Mangels, A. R. (2009). Position of the American Dietetic Association: Vegetarian diets. *Journal of the American Dietetic Association, 109*(7), 1266–1282. doi:10.1016/j.jada.2009.05.027 PMID:19562864

Craig, W. J., & Saunders, A. V. (2018). Critical nutrients in plant-based diet. In W. J. Craig (Ed.), *Vegetarian nutrition and wellness* (pp. 213–230). Boca Raton, FL: CRC Press. doi:10.1201/b22003

Creed, B. (2003). *Media matrix: Sexing the new reality*. Sydney, Australia: Allen and Unwin.

Cristiani, G., & Monnet, V. (2001). Food micro-organisms and aromatic ester synthesis. *Sciences des Aliments, 21*(3), 211–230. doi:10.3166da.21.211-230

Cupp, R. L. (2015). Human responsibility, not legal personhood, for nonhuman animals. *The Federalist Society, 16*(2), 29–35.

Curreri, F. (2012). *Mac Danzig's diet - the truth about vegan*. UFC. Retrieved from http://www.ufc.com/news/Mac-Danzig-Diet-The-Truth-About-Vegan?id=

D'Almeida, M., Sanches, L., Spillere, S. R., Zuchinali, P., & Souza, G. C. (2018). Mediterranean Diet and other dietary patterns in primary prevention of heart failure and changes in cardiac function markers: A systematic review. *Nutrients, 10*(1), E58. doi:10.3390/nu10010058 PMID:29320401

D'Onfro, J. (2016). *These are the future tech products that Alphabet's Eric Schmidt thinks will change the world.* Retrieved April 01, 2018 from http://uk.businessinsider.com/eric-schmidt-technologies-that-could-change-the-world-2016-6?r=US&IR=T

DadoTheGoodVillain. (2013). *The man who swims with crocodiles.* Retrieved from https://www.youtube.com/watch?v=I7fZZUfvx0s

Dagevos, H. (2016). Exploring flexitarianism: Meat reduction in a meat-centred food culture. In T. Raphaely & D. Marinova (Eds.), *Impact of meat consumption on health and environmental sustainability* (pp. 233–243). Hershey, PA: IGI Global. doi:10.4018/978-1-4666-9553-5.ch012

Dagevos, H., & Reinders, M. J. (2018). Flexitarianism and social marketing: Reflections on eating meat in moderation. In D. Bogueva, D. Marinova, & T. Raphaely (Eds.), *Handbook of research on social marketing and its influence on animal origin food product consumption* (pp. 105–120). Hershey, PA: IGI Global. doi:10.4018/978-1-5225-4757-0.ch007

Dagevos, H., & Voordouw, J. (2013). Sustainability and meat consumption: Is reduction realistic? Sustainability: Science. *Practice & Policy*, *9*, 60–69. doi:10.1080/15487733.2013.11908115

Dagnelie, P. C., Schuurman, A. G., Goldbohm, R. A., & Van den Brandt, P. A. (2004). Diet, anthropometric measures and prostate cancer risk: A review of prospective cohort and intervention studies. *BJU International*, *93*(8), 1139–1150. doi:10.1111/j.1464-410X.2004.04795.x PMID:15142129

Dahl, W. J., Agro, N. C., Eliasson, A. M., Mialki, K. L., Olivera, J. D., Rusch, C. T., & Young, C. N. (2017). Health benefits of fiber fermentation. *Journal of the American College of Nutrition*, *36*(2), 127–136. doi:10.1080/07315724.2016.1188737 PMID:28067588

Daily Mail. (2009). *Lioness showed remorse realising killed pregnant antelope.* Retrieved from http://www.dailymail.co.uk/news/article-1165832/The-lioness-showed-remorse-realising-killed-pregnant-antelope.html

Darling, A. L., Millward, D. J., Torgerson, D. J., Hewitt, C. E., & Lanham-New, S. A. (2009). Dietary protein and bone health: A systematic review and meta-analysis. *The American Journal of Clinical Nutrition*, *90*(6), 1674–1692. doi:10.3945/ajcn.2009.27799 PMID:19889822

Datar, I., & Betti, M. (2010). Possibilities for an in vitro meat production system. *Innovative Food Science & Emerging Technologies*, *11*(1), 13–22. doi:10.1016/j.ifset.2009.10.007

Dawson, K. A. (2002). Not just bread and beer: New applications for yeast and yeast products in human health and nutrition. In T. P. Lyons & F. A. Jaques (Eds.), *Nutritional biotechnology in the feed and food industry* (pp. 225–232). Nottingham, UK: Nottingham University Press.

de Bakker, E., & Dagevos, H. (2012). Reducing meat consumption in today's consumer society: Questioning the citizen-consumer gap. *Journal of Agricultural & Environmental Ethics*, *25*(6), 877–894. doi:10.100710806-011-9345-z

de Boer, J., & Aiking, H. (2011). On the merits of plant-based proteins for global food security: Marrying macro and micro perspectives. *Ecological Economics*, *70*(7), 1259–1265. doi:10.1016/j.ecolecon.2011.03.001

De Boer, J., Schösler, H., & Boersema, J. J. (2013). Motivational differences in food orientation and the choice of snacks made from lentils, locusts, seaweed or "hybrid" meat. *Food Quality and Preference*, *28*(1), 32–35. doi:10.1016/j.foodqual.2012.07.008

de Campo, C., dos Santos, P. P., Costa, T. M. H., Paese, K., Guterres, S. S., Rios, A. de O., & Flôres, S. H. (2017). Nanoencapsulation of chia seed oil with chia mucilage (Salvia hispanica L.) as wall material: Characterization and stability evaluation. *Food Chemistry*, *234*, 1–9. doi:10.1016/j.foodchem.2017.04.153 PMID:28551210

De Visser, R. O., Smith, J. A., & McDonnell, E. J. (2009). 'That's not masculine': Masculine capital and health-related behaviour. *Journal of Health Psychology, 14*(7), 1047–1058. doi:10.1177/1359105309342299 PMID:19786531

De Vuyst, L., & Neysens, P. (2005). The sourdough microflora: Biodiversity and metabolic interactions. *Trends in Food Science & Technology, 16*(1–3), 43–56. doi:10.1016/j.tifs.2004.02.012

Deamici, K. M., Santos, L. O., & Costa, J. A. V. (2018). Magnetic field action on outdoor and indoor cultures of Spirulina: Evaluation of growth, medium consumption and protein profile. *Bioresource Technology, 249*, 168–174. doi:10.1016/j.biortech.2017.09.185 PMID:29040851

Defoliart, G. R. (1995). Edible insects as minilivestock. *Biodiversity and Conservation, 4*(3), 306–321. doi:10.1007/BF00055976

DeFoliart, G. R. (1999). Insects as food: Why the Western attitude is important. *Annual Review of Entomology, 44*(1), 21–50. doi:10.1146/annurev.ento.44.1.21 PMID:9990715

Degreef, J. (1990). Inventaire, valeur alimentaire et culture de champignons da Shaba. Memoire fin d'études, Chaire d'Ecologie et Phytosociologie, Faculté des Sciences Agrononiques, de Gembloux, Belgium.

Department for Environment, Food & Rural Affairs (DEFRA). (2008). *A framework for proenvironmental behaviours.* Retrieved April 10, 2018 from https://assets.publishing.service.gov.uk/government/uploads/system/uploads/attachment_data/file/69277/pb13574-behaviours-report-080110.pdf

Department of Environment, Food, and Rural Affairs (DEFRA). (2017). *Family Food 2015.* London, UK: DEFRA. Retrieved February 24, 2018, from https://www.gov.uk/government/uploads/system/uploads/attachment_data/file/597667/Family_Food_2015-09mar17.pdf

Derbyshire, E. J. (2017). Flexitarian diets and health: A review of the evidence-based literature. *Frontiers in Nutrition, 3*, 55. doi:10.3389/fnut.2016.00055 PMID:28111625

Deroy, O., Reade, B., & Spence, C. (2015). The insectivore's dilemma, and how to take the West out of it. *Food Quality and Preference, 44*, 44–55. doi:10.1016/j.foodqual.2015.02.007

Deshpande, S. S., & Cheryan, M. (1984). Effects of phytic acid, divalent cations and their interactions on a-amylase activity. *Journal of Food Science, 49*(2), 516–519. doi:10.1111/j.1365-2621.1984.tb12456.x

Deutsche Gesellschaft für Ernährung. (2018). *10 guidelines of the German Nutrition Society (DGE) for a wholesome diet.* Deutsche Gesellschaft für Ernährung e. V. Retrieved September 11, 2018 from https://www.dge.de/ernaehrungspraxis/vollwertige-ernaehrung/10-regeln-der-dge/10-guidelines-of-the-german-nutrition-society/

Dewby, J. (1926). The historic background of corporate legal personality. *The Yale Law Journal, 35*(6), 655–673. doi:10.2307/788782

Dhiman, S., & Marques, J. (Eds.). (2016). *Sustainability and spirituality: New horizons and exemplary approaches.* Berlin, Germany: Springer International.

Di Ghionno, L., Marconi, O., Sileoni, V., De Francesco, G., & Perretti, G. (2017). Brewing with prolyl endopeptidase from Aspergillus niger: The impact of enzymatic treatment on gluten levels, quality attributes and sensory profile. *International Journal of Food Science & Technology, 52*(6), 1367–1374. doi:10.1111/ijfs.13375

Dibb, S., & Simkin, L. (2001). *The marketing casebook: Cases and concepts* (2nd ed.). London, UK: Thomson.

DiCello, M. C., Myc, A., Baker, J. R. Jr, & Baldwin, J. L. (1999). Anaphylaxis after ingestion of carmine colored foods: Two case reports and a review of the literature. *Allergy and Asthma Proceedings*, *20*(6), 377–382. doi:10.2500/108854199778251816 PMID:10624494

Dobermann, D., Swift, J. A., & Field, L. M. (2017). Opportunities and hurdles of edible insects for food and feed. *British Nutrition Foundation Nutrition Bulletin*, *42*(4), 293–308. doi:10.1111/nbu.12291

Donato, M., Vilela, M. H., & Bandarra, N. M. (2003). Fatty acids, sterols, α-tocopherol and total carotenoids composition of *Diacronema vlkianum*. *Journal of Food Lipids*, *10*(4), 267–276. doi:10.1111/j.1745-4522.2003.tb00020.x

Dooling, B. (2014). *Who's smarter – pigs or kids?* One Green Planet. Retrieved from https://www.onegreenplanet.org/news/whos-smarter-pigs-or-kids-video/

Draaiisma, R. B., Wijffels, R. H., Slegers, P. M., Brentner, L. B., Roy, A., & Barbosa, M. J. (2013). Food commodities from microalgae. *Current Opinion in Biotechnology*, *24*(2), 169–177. doi:10.1016/j.copbio.2012.09.012 PMID:23084075

Drewnowski, A. (2010). The Nutrient Rich Foods Index helps to identify healthy, affordable foods. *The American Journal of Clinical Nutrition*, *91*(4), 1095S–1101S. doi:10.3945/ajcn.2010.28450D PMID:20181811

Dreyer, J. J., & Wehmeyer, A. S. (1982). On the nutritive value of Mopanie worms. *South African Journal of Science*, *78*, 33–35.

Drummond, M. J. N. (2002). Sport and images of masculinity: The meaning of relationships in the life course of "elite" male athletes. *Journal of Men's Studies*, *10*(2), 129–141. doi:10.3149/jms.1002.129

Du Bois, C. M., Tan, C.-B., & Mintz, S. W. (2008). *The world of soy*. National University of Singapore: National University of Singapore Press.

Durst, P., & Johnson, D. (2010). Edible forest insects: Exploring new horizons and traditional practices. In P. B. Durst, D. V. Johnson, R. N. Leslie & K. Shono (Eds), *Forest insects as food: Humans bite back* (pp. 1–4). Bangkok, Thailand: Food and Agriculture Organization of the United Nations. Retrieved September 12, 2018 from http://www.fao.org/docrep/012/i1380e/i1380e00.pdf

Dyschkant, A. (2015). Legal personhood: How we are getting it wrong. *University of Illinois Law Review*, *5*, 2075–2110.

Eat. (2015). *The meat of the future: How lab-grown meat is made*. YouTube. Retrieved from https://www.youtube.com/watch?v=u468xY1T8fw

Edelman, J., Fewell, A., & Solomons, G. L. (1983). Myco-protein – a new food. *Nutrition Abstracts and Reviews*, *53*(6), 471–480.

Eighth China International Exhibition of Soyfood Processing Technology and Equipment. (2018). *Exhibition network*. Soy Products Professional Association of the National Food Industry Association of China. Retrieved September 13, 2018 from http://www.onezh.com/web/index_54336.html

Ekmekcioglu, C., Wallner, P., Kundi, M., Weisz, U., Haas, W., & Hutter, H.-P. (2018). Red meat, diseases, and healthy alternatives: A critical review. *Critical Reviews in Food Science and Nutrition*, *58*(2), 247–261. doi:10.1080/10408398.2016.1158148 PMID:27128451

Elander, R. P., & Lowe, D. A. (1994). Fungal biotechnology: An overview. In K. Arora, L. Ajello, & K. G. Mukerji (Eds.), *Handbook of applied mycology* (pp. 1–34). New York, NY: Marcel Dekker.

Elemo, B. O., Elemo, G. N., Makinde, M. A., & Erukainure, O. L. (2011). Chemical evaluation of African palm weevil, Rhychophorus phoenicis, larvae as a food source. *Journal of Insect Science*, *11*(146), 1–6. doi:10.1673/031.011.14601 PMID:22236060

Elif, Ö. (2017). *Erişte üretiminde farklı oran ve kombinasyonlarda karabuğday, amarant ve kinoa unlarının kullanım imkanları*. Necmettin Erbakan Üniversitesi.

Elzerman, J. E., Hoek, A. C., Van Boekel, M. A., & Luning, P. A. (2011). Consumer acceptance and appropriateness of meat substitutes in a meal context. *Food Quality and Preference*, *22*(3), 233–240. doi:10.1016/j.foodqual.2010.10.006

Elzerman, J. E., Hoek, A. C., van Boekel, M. J., & Luning, P. A. (2015). Appropriateness, acceptance and sensory preferences based on visual information: A web-based survey on meat substitutes in a meal context. *Food Quality and Preference*, *42*, 56–65. doi:10.1016/j.foodqual.2015.01.010

Elzerman, J. E., van Boekel, M. A., & Luning, P. A. (2013). Exploring meat substitutes consumer experiences and contextual factors. *British Food Journal*, *115*(5), 700–710. doi:10.1108/00070701311331490

English, N. (2017). 10 complete proteins vegans need to know about. *Greatist*. Retrieved from https://greatist.com/health/complete-vegetarian-proteins

Environmental Working Group. (2011). *Meat-eater's guide to climate change + health. Lifecycle assessments: Methodology and results.* Retrieved from http://static.ewg.org/reports/2011/meateaters/pdf/methodology_ewg_meat_eaters_guide_to_health_and_climate_2011.pdf

Enzing, C., Ploeg, M., Barbosa, M., & Sijtsma, L. (2014). *Microalgae-based products for the food and feed sector: An outlook for Europe*. JRC Scientific and Policy Reports. Institute for Prospective Technological Studies. Joint Research Centre. European Commission. Retrieved from ftp://ftp.jrc.es/pub/EURdoc/JRC85709.pdf

Epps, K. (2017). Habeas corpus, the fugitive slave law, and executive authority. *The Journal of Civil War Era*. Retrieved from https://journalofthecivilwarera.org/2017/02/habeas-corpus-fugitive-slave-law-executive-authority/

Eshel, G., Shepon, A., Makov, T., & Milo, R. (2014). Land, irrigation water, greenhouse gas, and reactive nitrogen burdens of meat, eggs, and dairy production in the United States. *Proceedings of the National Academy of Sciences of the United States of America*, *111*(33), 11996–12001. doi:10.1073/pnas.1402183111 PMID:25049416

Esposito, K., Giugliano, F., Maiorino, M. I., & Giugliano, D. (2010). Dietary factors, Mediterranean diet and erectile dysfunction. *Journal of Sexual Medicine*, *7*(7), 2338–2345. doi:10.1111/j.1743-6109.2010.01842.x PMID:20487239

Euromonitor International. (2011, August 16). *The war on meat: How low-meat and no-meat diets are impacting the consumer markets*. Retrieved September 11, 2018 from https://blog.euromonitor.com/2011/08/the-war-on-meat-how-low-meat-and-no-meat-diets-are-impacting-consumer-markets.html

Euromonitor International. (2017a). *Free from Meat: Euromonitor from trade sources/national statistics* (Database). Euromonitor International. Retrieved from https://www.euromonitor.com/

Euromonitor International. (2017b). *Plant-based Protein: Assessing Demand for Sustainable Alternatives* (Briefing). Retrieved from https://www.euromonitor.com/plant-based-protein-assessing-demand-for-sustainable-alternatives/report

European Commission. (2005). *Social values, science and technology*. Brussels: Directorate General Research, Directorate General Press and Communication. Retrieved September 10, 2018 from http://ec.europa.eu/commfrontoffice/publicopinion/archives/ebs/ebs_225_report_en.pdf

European Commission. (2018). *Report from the Commission to the European Parliament and the Council: On the impact of animal welfare international activities on the competitiveness of European livestock producers in a globalized world.* Retrieved from https://ec.europa.eu/food/sites/food/files/animals/docs/aw_international_ publication-report_en.pdf

European Food Safety Authority (EFSA). (2011). Scientific opinion on the public health risks of bacterial strains producing extended-spectrum β-lactamases and/or AmpC β-lactamases in food and food-producing animals. *EFSA Journal*, *9*(8), 95. doi:10.2903/j.efsa.2011.2322

European Food Safety Authority (EFSA). (2015). Risk profile related to production and consumption of insects as food and feed. *EFSA Journal*, *13*(10), 4257. doi:10.2903/j.efsa.2015.4257

European Parliament and Council. (2015). Regulation (EU) 2015/2283 of the European Parliament and of the Council of 25 November 2015 on novel foods, amending Regulation (EU) No 1169/2011 of the European Parliament and of the Council and repealing Regulation (EC) No 258/97 of the European Parliament and of the Council and Commission Regulation (EC) No 1852/2001.

Evans, J., Alemu, M. H., Flore, R., Frøst, M. B., Halloran, A., Jensen, A. B., ... Eilenberg, J. (2015). "Entomophagy": An evolving terminology in need of review. *Journal of Insects as Food and Feed*, *1*(4), 293–305. doi:10.3920/JIFF2015.0074

Ewbank, R., Ray, P., Kim-Madslien, F., & Hart, B. (Eds.). (1999). *Management and welfare of farm animals.* Wheathampstead, UK: Universities Federation for Animal Welfare.

Ezeh, A. C., Bongaarts, J., & Mberu, B. (2012). Global population trends and policy options. *Lancet*, *380*(9837), 142–148. doi:10.1016/S0140-6736(12)60696-5 PMID:22784532

Faludi, S. (2000). *Stiffed: The betrayal of modern man.* London, UK: Vintage.

Farrelly, E. (2018). Gender equality: It takes a troubled woman to change a troubled world. *Sydney Morning Herald.* Retrieved from https://www.smh.com.au/opinion/gender-equality-it-takes-a-troubled-woman-to-change-a-troubled-world-20180125-h0oip9.html

Fenger, M. H., Aschemann-Witzel, J., Hansen, F., & Grunert, K. G. (2015). Delicious words – Assessing the impact of short storytelling messages on consumer preferences for variations of a new processed meat product. *Food Quality and Preference*, *41*, 237–244. doi:10.1016/j.foodqual.2014.11.016

Feng, Y., Chen, X., Zhao, M., He, Z., Sun, L., Wang, C., & Ding, W. (2017). Edible insects in China: Utilization and prospects. *Insect Science*, *25*(2), 184–198. doi:10.1111/1744-7917.12449 PMID:28225201

Fernqvist, F., & Ekelund, L. (2014). Credence and the effect on consumer liking of food–A review. *Food Quality and Preference*, *32*, 340–353. doi:10.1016/j.foodqual.2013.10.005

Fiddles, N. (1991). *Meat: A natural symbol.* London, UK: Routledge.

Filonow, A. B., Vishniac, H. S., Anderson, J. A., & Janisiewicz, W. J. (1996). Biological control of *Botrytis cinerea* in apple by yeasts from various habitats and their putative mechanisms of antagonism. *Biological Control*, *7*(2), 212–220. doi:10.1006/bcon.1996.0086

Finke, M. D. (2002). Complete nutrient composition of selected invertebrates commonly fed to insectivores. *Zoo Biology*, *21*, 269–285. doi:10.1002/zoo.10031

Finke, M. D. (2007). Estimate of chitin in raw whole insects. *Zoo Biology*, *26*(2), 105–115. doi:10.1002/zoo.20123 PMID:19360565

Finnigan, T. J. (2011). Mycoprotein: Origins, production and properties. In G. O. Phillips & P. A. Williams (Eds.), *Handbook of Food Proteins* (pp. 335–352). Cambridge, UK: Woodhead Publishing. doi:10.1533/9780857093639.335

Firestone, R., Rowe, C. J., Modi, S. N., & Sievers, D. (2017). The effectiveness of social marketing in global health: A systematic review. *Health Policy and Planning*, *32*(1), 110–124. doi:10.1093/heapol/czw088 PMID:27476502

Fischer, S., Wilckens, R., Jara, J., Aranda, M., Valdivia, W., Bustamante, L., ... Obal, I. (2017). Protein and antioxidant composition of quinoa (Chenopodium quinoa Willd.) sprout from seeds submitted to water stress, salinity and light conditions. *Industrial Crops and Products*, *107*, 558–564. doi:10.1016/j.indcrop.2017.04.035

Flint, H. J., Scott, K. P., Louis, P., & Duncan, S. H. (2012). The role of the gut microbiota in nutrition and health. *Nature Reviews. Gastroenterology & Hepatology*, *9*(10), 577–589. doi:10.1038/nrgastro.2012.156 PMID:22945443

Flores, B. (2017). Liam Hemsworth officially crowned Australia's sexiest vegan. *Livekindly*. Retrieved from https://www.livekindly.co/australias-sexiest-vegan-is-officially-liam-hemsworth/

Fontenot, J. P. (1999). Nutrient recycling: The North American experience — review. *Asian-Australasian Journal of Animal Sciences*, *12*(4), 642–650. doi:10.5713/ajas.1999.642

Food and Agriculture Organization of the United Nations (FAO), International Fund for Agricultural Development (IFAD), United Nations' Children's FUND (UNICEF), World Food Programme (WFP), & World Health Organization (WHO). (2017). *The state of food security and nutrition in the world 2017: Building resilience for peace and food security*. Rome, Italy: FAO. Retrieved from www.fao.org/3/a-I7695e.pdf

Food and Agriculture Organization of the United Nations (FAO). (2006). *Livestock's long shadow: Environmental issues and option*. Rome, Italy: FAO. Retrieved September 14, 2018 from http://www.fao.org/docrep/010/a0701e/a0701e00.HTM

Food and Agriculture Organization of the United Nations (FAO). (2006). *Livestock's long shadow: Environmental issues and options*. Rome, Italy: Food and Agriculture Organization of the United Nations (FAO). Retrieved September 12, 2018 from http://www.fao.org/docrep/010/a0701e/a0701e.pdf

Food and Agriculture Organization of the United Nations (FAO). (2009). *How to feed the world in 2015*. Retrieved from http://www.fao.org/fileadmin/templates/wsfs/docs/expert_paper/How_to_Feed_the_World_in_2050.pdf

Food and Agriculture Organization of the United Nations (FAO). (2009). *How to feed the world in 2050?* Rome, Italy: FAO. Retrieved from http://www.fao.org/fileadmin/templates/wsfs/docs/expert_paper/How_to_Feed_the_World_in_2050.pdf

Food and Agriculture Organization of the United Nations (FAO). (2010). *The state of addressing food insecurity in protracted crises*. Retrieved from http://www.fao.org/catalog/inter–e.htm

Food and Agriculture Organization of the United Nations (FAO). (2013). *Edible insects: Future prospects for food and fed security*. Rome, Italy: FAO. Retrieved September 12, 2018 from http://www.fao.org/docrep/018/i3253e/i3253e.pdf

Food and Agriculture Organization of the United Nations (FAO). (2014). *FAOSTAT: Food and agriculture data*. Retrieved from http://www.fao.org/faostat/en/#home

Food and Agriculture Organization of the United Nations (FAO). (2016). *The contribution of insects to food security, livelihoods and the enviroment*. Rome, Italy: FAO. Retrieved from http://www.fao.org/docrep/018/i3264e/i3264e00.pdf

Food and Agriculture Organization of the United Nations (FAO). (2016). *The state of food and agriculture: Climate change, agriculture and food security*. Rome, Italy: FAO. Retrieved September 11, 2018 from http://www.fao.org/3/a-i6030e.pdf

Food and Agriculture Organization of the United Nations (FAO). (2016). *The state of food and agriculture: Climate change, agriculture and food security*. Rome, Italy: FAO. Retrieved September 13, 2018 from http://www.fao.org/3/a-i6030e.pdf

Food and Agriculture Organization of the United Nations (FAO). (2017). *The state of food and agriculture. Leveraging food systems for inclusive rural transition.* Rome, Italy: FAO. Retrieved September 12, 2018 from http://www.fao.org/3/a-i7658e.pdf

Food and Agriculture Organization of the United Nations (FAO). (2018). *FAOSTATS: Food Supply - Livestock and Fish Primary Equivalent.* Retrieved from http://www.fao.org/faostat/en/#data/CL

Food and Agriculture Organization of the United Nations (FAO). (2018). *Food and agriculture data.* Retrieved from http://www.fao.org/faostat/en/#home

Food and Agriculture Organization of the United Nations (FAO). (2018). *Food-based dietary guidelines.* Retrieved September 11, 2018 from http://www.fao.org/nutrition/education/food-dietary-guidelines/regions/en/

Food and Agriculture Organization of the United Nations/World Health Organization (FAO/WHO). (2001). *Human vitamin and mineral requirements.* Rome, Italy: Food and Nutrition Division, FAO. Retrieved from http://www.fao.org/3/a-y2809e.pdf

Food Business Net. (2016). *China's soybeans consumption reached to about 100 million tones, with 80% imported, what are they use for?* Retrieved September 13, 2018 from http://news.21food.cn/33/2772440.htmlhttp://news.21food.cn/33/2772440.html

Food Think Tank. (2014). *Between Meadow and Forest.* Retrieved from http://foodthinktank.pl/miedzylakaalasem (in Polish)

Forrest, S. (2010). Young men in love: The (re)making of heterosexual masculinities through "serious" relationships. *Sexual and Relationship Therapy, 25*(2), 206–218. doi:10.1080/14681991003660260

Forss, K. G., Gadd, G. O., Lundell, R. O., & Williamson, H. W. (1974). *Process for the manufacture of protein-containing substances for fodder, foodstuffs and technical applications.* US Patent Office, Patent No. 3809614.

Foucault, M. (1980). *Power/knowledge.* Brighton, UK: Harvester.

Fox, K. (2017). Here's why you should turn your business vegan in 2018. *Forbes.* Retrieved September 14, 2018 from https://www.forbes.com/sites/katrinafox/2017/12/27/heres-why-you-should-turn-your-business-vegan-in-2018/#6d97f4182144

Frazzetto, G. (2003). White biotechnology. *European Molecular Biology Organization (EMBO). Reports, 4*(9), 835–837. doi:10.1038j.embor.embor928

Frewer, L. J., van der Lans, I. A., Fischer, A. R. H., Reinders, M. J., Menozzi, D., Zhang, X., ... Zimmermann, K. L. (2013). Public perceptions of agri-food applications of genetic modification - A systematic review and meta-analysis. *Trends in Food Science & Technology, 30*(2), 142–152. doi:10.1016/j.tifs.2013.01.003

Frey, C. N. (1930). History and development of the modern yeast industry. *Industrial & Engineering Chemistry, 22*(11), 1154–1162. doi:10.1021/ie50251a012

Froggatt, A., & Wellesley, L. (2016). *China shows way with new diet guidelines on meat.* Chatham House. Retrieved September 13, 2018 from https://www.chathamhouse.org/expert/comment/china-shows-way-new-diet-guidelines-meat

Funk, C., & Rainie, L. (2016). *Public and scientists' views on science and society.* Pew Research Centre Internet & Technology. Retrieved September 13, 2018 from http://www.pewinternet.org/2015/01/29/public-and-scientists-views-on-science-and-society/

Gabriel, A., Victor, N., & du Preez James, C. (2014). Cactus pear biomass, a potential lignocellulose raw material for single cell protein production (SCP): A review. *International Journal of Current Microbiology and Applied Sciences*, *3*(7), 171–197.

Gajanan, M. (2018). The meat industry has some serious beef with those "bleeding" plant-based burgers. *Time*. Retrieved from http://time.com/5181524/impossible-burger-plant-based/

Gal, D., & Wilkie, J. (2010). Real men don't eat quiche: Regulation of gender-expressive choices by men. *Social Psychological & Personality Science*, *1*(3). doi:10.1177/1948550610365003

Ganesan, K., & Xu, B. (2017). Polyphenol–rich dry common beans (Phaseolus vulgaris L.) and their health benefits. *International Journal of Molecular Sciences*, *18*(11), 2331. doi:10.3390/ijms18112331 PMID:29113066

Garcia-Garibay, M., Gómez-Ruiz, L., Cruz-Guerrero, A. E., & Bárzana, E. (2014). Single-cell protein/The algae. In C. Batt & C. A. Batt (Eds.), *Encyclopedia of food microbiology* (2nd ed.; pp. 425–430). Cambridge, MA: Academic Press; doi:10.1016/B978-0-12-384730-0.00309-8

Garden Gourmet. (2018). *Philosophy*. Retrieved March 25, 2018 from https://www.gardengourmet.com/about

Garnett, T. (2014). *What is a sustainable healthy diet?* Food Climate Research Network. Retrieved from https://www.fcrn.org.uk/sites/default/files/fcrn_what_is_a_sustainable_healthy_diet_final.pdf

Garnett, T., Mathewson, S., Angelides, P., & Borthwick, F. (2015). *Policies and actions to shift eating patterns: What works? A review of the evidence of the effectiveness of interventions aimed at shifting diets in more sustainable and healthy directions*. Oxford, UK: Food Climate Research Network and Chatham House.

Garofalo, C., Osimani, A., Milanovic, V., Taccari, M., Cardinali, F., Aquilanti, L., ... Clementi, F. (2017). The microbiota of marketed processed edible insects as revealed by high-throughput sequencing. *Food Microbiology*, *62*, 15–22. doi:10.1016/j.fm.2016.09.012 PMID:27889142

Gasco, L., Finke, M., & van Huis, A. (2018). Can diets containing insects promote animal health? *Journal of Insects as Food and Feed*, *4*(1), 1–4. doi:10.3920/JIFF2018.x001

Gaskell, G. (1999). Worlds apart? The reception of genetically modified foods in Europe and the U.S. *Science*, *285*(5426), 384–387. doi:10.1126cience.285.5426.384 PMID:10411496

Gauntlett, D. (2002). *Media, gender and identity: An introduction*. London, UK: Routledge.

Gaylor, M. O., Harvey, E., & Hale, R. C. (2012). House crickets can accumulate polybrominated diphenyl ethers (PBDEs) directly from polyurethane foam common in consumer products. *Chemosphere*, *86*(5), 500–505. doi:10.1016/j.chemosphere.2011.10.014 PMID:22071374

Gebremariam, M. M., Zarnkow, M., & Becker, T. (2012). Teff (Eragrostis tef) as a raw material for malting, brewing and manufacturing of gluten–free foods and beverages: A review. *Journal of Food Science and Technology*, *51*(11), 2881–2895. doi:10.100713197-012-0745-5 PMID:26396284

Gee, J. P., & Handford, M. (n.d.). Introduction. In J. P. Gee & M. Handford (Eds.), The Routledge handbook of discourse analysis (pp. 1–6). London, UK: Routledge.

Gelfer, J. (2017). *Masculinity and the environment: A double-edged sword*. Retrieved from https://goodmenproject.com/featured-content/masculinity-and-the-environment-a-double-edged-sword-wcz/

Gere, A., Székely, G., Kovács, S., Kokai, Z., & Sipos, L. (2017). Readiness to adopt insects in Hungary: A case study. *Food Quality and Preference*, *59*, 81–86. doi:10.1016/j.foodqual.2017.02.005

Ghavidel, R. A., & Prakash, J. (2006). Effect of germination and dehulling on functional properties of legume flours. *Journal of the Science of Food and Agriculture*, *86*(8), 1189–1195. doi:10.1002/jsfa.2460

Giaretta, D., Lima, V. A., & Carpes, S. T. (2017). (in press). Improvement of fatty acid profile in breads supplemented with Kinako flour and chia seed. *Innovative Food Science & Emerging Technologies*. doi:10.1016/j.ifset.2017.11.010

Gibbons, A. (2012). Bonobos join chimps as closest human relatives. *Science AAAS*. Retrieved from http://www. sciencemag.org/news/2012/06/bonobos-join-chimps-closest-human-relatives

Giménez–Bastida, J. A., Piskuła, M., & Zieliński, H. (2015). Recent advances in development of gluten–free buckwheat products. *Trends in Food Science & Technology*, *44*(1), 58–65. doi:10.1016/j.tifs.2015.02.013

Giovannucci, E., Rimm, E. B., Colditz, G. A., Stampfer, M. J., Ascherio, A., Chut, C. C., & Willett, W. C. (1993). A prospective study of dietary fat and risk of prostate cancer. *Journal of the National Cancer Institute*, *85*(19), 1571–1579. doi:10.1093/jnci/85.19.1571 PMID:8105097

Giovannucci, E., Rimm, E. B., Stampfer, M. J., Colditz, G. A., Ascherio, A., & Willett, W. C. (1994). Intake of fat, meat, and fiber in relation to risk of colon cancer in men. *Cancer Research Journal*, *54*(9), 2390–2397. PMID:8162586

Giudici, K. V., & Weaver, C. M. (2018). Plant-based diets and risk of osteoporosis. In W. J. Craig (Ed.), *Vegetarian nutrition and wellness* (pp. 93–111). Boca Raton, FL: CRC Press.

Glick-Bauer, M., & Yeh, M. (2014). The health advantage of a vegan diet: Exploring the gut microbiota connection. *Nutrients*, *6*(11), 4822–4838. doi:10.3390/nu6114822 PMID:25365383

Global Market Insights Inc. (2016). *Edible insects market size by product (beetles, caterpillars, grasshoppers, bees, wasps, ants, scale insects & true bugs), by application (flour, protein bars, snacks), industry analysis report, regional outlook (U.S., Belgium, Netherlands, UK, France, China, Thailand, Vietnam, Brazil, Mexico): Application potential, price trends, competitive market share & forecast, 2016 – 2023*. Retrieved April, 12, 2018 from https:// www. gminsights. com/ industry-analysis/edible-insects-market

Glotz, J. (2018, March 3). Alternative proteins stir up animal passions over meat. *The Grocer*, p. 16. Retrieved from the Factiva database: https://www.dowjones.com/products/factiva/

Godfray, H. C. J., Beddington, J. R., Crute, I. R., Haddad, L., Lawrence, D., Muir, J. F., ... Toulmin, C. (2010). Food security: The challenge of feeding 9 billion people. *Science*, *327*(5967), 812–818. doi:10.1126cience.1185383 PMID:20110467

Goldberg, I. (1985). *Single cell protein*. Berlin, Germany: Springer. doi:10.1007/978-3-642-46540-6

Gong, L., Cao, W., Chi, H., Wang, J., Zhang, H., Liu, J., & Sun, B. (2018). Whole cereal grains and potential health effects: Involvement of the gut microbiota. *Food Research International*, *103*, 84–102. doi:10.1016/j.foodres.2017.10.025 PMID:29389647

Gour, S., Mathur, N., Singh, A., & Pradeep, B. (2015). Single cell protein production: A review. *International Journal of Current Microbiology and Applied Sciences*, *4*(9), 251–262.

Graczyk, T. K., Knight, R., & Tamang, L. (2005). Mechanical transmission of human protozoan parasites by insects. *Clinical Microbiology Reviews*, *18*(1), 128–132. doi:10.1128/CMR.18.1.128-132.2005 PMID:15653822

Graf, B. L., Rojas–Silva, P., Rojo, L. E., Delatorre–Herrera, J., Baldeón, M. E., & Raskin, I. (2015). Innovations in health value and functional food development of quinoa (Chenopodium quinoa Willd.). *Comprehensive Reviews in Food Science and Food Safety*, *14*(4), 431–445. doi:10.1111/1541-4337.12135 PMID:27453695

Grand View Research. (2015). *Global beef market size is expected to reach $2151 billion by 2020*. Retrieved from https://www.grandviewresearch.com/press-release/global-beef-market

Green Eatz. (n.d.). *Food's carbon footprint*. Retrieved from http://www.greeneatz.com/foods-carbon-footprint.html

Greenpeace International. (2018). *Greenpeace calls for decrease in meat and dairy production and consumption for a healthier planet*. Retrieved from https://www.greenpeace.org/international/press-release/15111/greenpeace-calls-for-decrease-in-meat-and-dairy-production-and-consumption-for-a-healthier-planet/

Greenpeace. (2018). *Less is more: Reducing meat and dairy for a healthier life and planet*. Scientific background of the Greenpeace vision of the meat and dairy system towards 2050. Retrieved from https://storage.googleapis.com/p4-production-content/international/wp-content/uploads/2018/03/6942c0e6-longer-scientific-background.pdf

Greenpeace. (2018a). *Less is more: Reducing meat and dairy for a healthier life and planet*. Scientific background of the Greenpeace vision of the meat and dairy system towards 2050. Retrieved from https://storage.googleapis.com/p4-production-content/international/wp-content/uploads/2018/03/6942c0e6-longer-scientific-background.pdf

Greenpeace. (2018b). *Less is more: Reducing meat and dairy for a healthier life and planet*. The Greenpeace vision of the meat and dairy system towards 2050. Retrieved from https://storage.googleapis.com/p4-production-content/international/wp-content/uploads/2018/03/698c4c4a-summary_greenpeace-livestock-vision-towards-2050.pdf

Greenwood, D. C., Threapleton, D. E., Evans, C. E., Cleghorn, C. L., Nykjaer, C., Woodhead, C., & Burley, V. J. (2013). Glycemic index, glycemic load, carbohydrates, and type 2 diabetes: Systematic review and dose-response meta-analysis of prospective studies. *Diabetes Care, 36*(12), 166–171. doi:10.2337/dc13-0325 PMID:24265366

Greger, M. (2012). *Poultry and penis cancer*. Retrieved from https://nutritionfacts.org/video/poultry-and-penis-cancer/

Greger, M. (2013). *Vegan men: More testosterone but less cancer*. Retrieved from https://nutritionfacts.org/2013/02/12/less-cancer-in-vegan-men-despite-more-testosterone/

Gregory, J. (2007). *Of Victorians and vegetarians: The Vegetarian movement in nineteenth-century Britain*. London, UK: Tauris Academic Studies.

Grewal, H., Kalra, K., & Kahlom, S. (1990). Improvement of lipase production at different stirring speeds and oxygen levels. *Journal of Research, 1*, 90–96.

Grunert, K. G., Hieke, S., & Wills, J. (2014). Sustainability labels on food products: Consumer motivation, understanding and use. *Food Policy, 44*, 177–189. doi:10.1016/j.foodpol.2013.12.001

Guarner, F., & Malagelada, J.-R. (2003). Gur flora in health and disease. *Lancet, 361*(9356), 512–519. doi:10.1016/S0140-6736(03)12489-0 PMID:12583961

Gullone, E. (2012). Historical and current conceptualizations of animal cruelty. In E. Gullone & P. Arkow (Eds.), *Animal cruelty, antisocial behaviour, and aggression: More than a link* (pp. 5–15). Basingstoke, UK: Palgrave Macmillan. doi:10.1057/9781137284549_2

Guo, X., Raphaely, T., & Marinova, D. (2016). China's growing meat demands, implications for sustainability. In T. Raphaely & D. Marinova (Eds.), *Impact of meat consumption on health and sustainability* (pp. 221–233). Hershey, PA: IGI Global. doi:10.4018/978-1-4666-9553-5.ch011

Guthman, J., & Brown, S. (2016). Midas' not-so-golden touch: On the demise of methyl iodide as a soil fumigant in California. *Journal of Environmental Policy and Planning, 18*(3), 324–341. doi:10.1080/1523908X.2015.1077441

Gutting, G. (Ed.). (1994). *Cambridge companion to Foucault*. Cambridge, UK: Cambridge University Press. doi:10.1017/CCOL9780521403320

Guynup, S., & Ruggia, N. (2004). *For most people eating bugs is only natural*. National Geographic Channel, National Geographic. Retrieved from https://news.nationalgeographic.com/news/2004/07/0715_040715_tvinsectfood.html

Haagsman, H. P., Hellingwerf, K. J., & Roelen, B. A. J. (2009). *Production of animal proteins by cell systems. Desk study on cultured meat ("Kweekvlees")*. Utrecht: University Utrecht, Faculteit DiergeneeskFaculty of Veterarinary Medicine. Retrieved September 10, 2018 from http://citeseerx.ist.psu.edu/viewdoc/download?doi=10.1.1.732.9877&rep=rep1&type=pdf

Haddad, E. H. (2018). Vegetarian diet and risk of cardiovascular disease. In W. J. Craig (Ed.), *Vegetarian nutrition and wellness* (pp. 45–70). Boca Raton, FL: CRC Press.

Haider, M., & El-Hassy, M. (2000). Biodegradation of orange and pineapple peels for production of single cell protein. *Garyounis Journal of Science, 1*, 7–23.

Hall, C., Hillen, C., & Garden Robinson, J. (2017). Composition, nutritional value, and health benefits of pulses. *Cereal Chemistry Journal, 94*(1), 11–31. doi:10.1094/CCHEM-03-16-0069-FI

Hall, D. O., Fernández, F. G., Guerrero, E. C., Rao, K. K., & Grima, E. M. (2003). Outdoor helical tubular photobioreactors for microalgal production: Modeling of fluid-dynamics and mass transfer and assessment of biomass productivity. *Biotechnology and Bioengineering, 82*(1), 62–73. doi:10.1002/bit.10543 PMID:12569625

Halliday, J. (2018). Flexitarians drive growth for meat substitutes in the EU. *Bites*. Retrieved September 11, 2018 from https://bites.figlobal.com/euromonitor-cultured-meat-flexitarian/

Halloran, A., Hanboonsong, Y., Roos, N., & Bruun, S. (2017). Life cycle assessment of cricket farming in north-eastern Thailand. *Journal of Cleaner Production, 156*, 83–94. doi:10.1016/j.jclepro.2017.04.017

Hallstrom, E., Carlsson-Kanyama, A., & Borjesson, P. (2015). Environmental impact of dietary change: A systematic review. *Journal of Cleaner Production, 91*, 1–11. doi:10.1016/j.jclepro.2014.12.008

Hamadeh, S., & Estepan, S. (2018). The visual discourse of food and its impact on health: Research and practical implications. *Acta Scientific Nutritional Health, 2*(6), 12–20. Retrieved from https://actascientific.com/ASNH/pdf/ASNH-02-0084.pdf

Hanboonsong, Y., & Durst, B. P. (2014). *Edible insects in Lao PDR: Building on tradition to enhance food security*. RAP Publication 2014/12, Food and Agriculture Organization of the United Nations (FAO), Regional Office for Asia and the Pacific, Bangkok, Thailand, 56. Retrieved September 12, 2018 from http://www.fao.org/3/a-i3749e.pdf

Hanboonsong, Y., Jamjanya, T., & Durst, P. B. (2013). *Six-legged livestock: Edible insect farming, collection and marketing in Thailand*. Bangkok, Thailand: Food and Agriculture Organization of the United Nations, Regional Office for Asia and the Pacific. Retrieved September 12, 2018 from http://www.fao.org/3/a-i3246e.pdf

Hande, A. (2015). *Hidrotermal işlemlerin karabuğday nişastasının fizikokimyasal özellikleri ve dirençli nişasta miktarı üzerindeki etkisi*. İstanbul Teknik Üniversitesi.

Hannon, M., Gimpel, J., Tran, M., Rasala, B., & Mayfield, S. (2010). Biofuels from algae: Challenges and potential. *Biofuels, 1*(5), 763–784. doi:10.4155/bfs.10.44 PMID:21833344

Han, R., Shin, J. T., Kim, J., Choi, S. Y., & Kim, W. Y. (2017). An overview of the South Korean edible insect food industry: Challenges and future pricing/promotion strategies. *Entomological Research, 47*(3), 141–151. doi:10.1111/1748-5967.12230

Hanrahan, C., Elvery, S., McGhee, A., & Liddy, M. (2017) *Political donations: New data maps industries' web of influence*. ABC news. Retrieved from http://www.abc.net.au/news/2017-02-09/political-donations-industry-dataset/8229192

Hartmann, C., Ruby, M. B., Schmidt, P., & Siegrist, M. (2018). Brave, health-conscious, and environmentally friendly: Positive impressions of insect food product consumers. *Food Quality and Preference*, *68*, 64–71. doi:10.1016/j.foodqual.2018.02.001

Hartmann, C., Shi, J., Giusto, A., & Siegrist, M. (2015). The psychology of eating insects: A cross-cultural comparison between Germany and China. *Food Quality and Preference*, *44*, 148–156. doi:10.1016/j.foodqual.2015.04.013

Hartmann, C., & Siegrist, M. (2017). Consumer perception and behaviour regarding sustainable protein consumption: A systematic review. *Trends in Food Science & Technology*, *61*, 11–25. doi:10.1016/j.tifs.2016.12.006

Hartmann, C., & Siegrist, M. (2017). Insects as food: Perception and acceptance. Findings from current research. *Ernährungs-Umschau*, *64*(3), 44–50. doi:10.4455/eu.2017.010

Harun, R., Singh, M., Forde, G. M., & Danquah, M. K. (2010). Bioprocess engineering of microalgae to produce a variety of consumer products. *Renewable & Sustainable Energy Reviews*, *14*(3), 1037–1047. doi:10.1016/j.rser.2009.11.004

Hase, R., Oikawa, H., Sasao, C., Morita, M., & Watanabe, Y. (2000). Photosynthetic production of microalgal biomass in a raceway system under greenhouse conditions in Sendai city. *Journal of Bioscience and Bioengineering*, *89*(2), 157–163. doi:10.1016/S1389-1723(00)88730-7 PMID:16232719

Havlicek, J., & Lenochova, P. (2006). The effect of meat consumption on body odor attractiveness. *Chemical Senses*, *31*(8), 747–752. doi:10.1093/chemse/bjl017 PMID:16891352

Hawken, P. (Ed.). (2017). *Drawdown: The most comprehensive plan ever proposed to reverse global warming*. New York, NY: Penguin Books.

Hawkins, D. R. (2012). *Power vs force: The hidden determinates of human behavior*. Warsaw, Poland: Virgo. (in Polish)

Hawkins, D. R. (2013). *Letting go: The pathway of surrender*. London, UK: Hay House.

Hayat, I., Ahmad, A., Masud, T., Ahmed, A., & Bashir, S. (2014). Nutritional and health perspectives of beans (Phaseolus vulgaris L.): An overview. *Critical Reviews in Food Science and Nutrition*, *54*(5), 580–592. doi:10.1080/10408398.2011.596639 PMID:24261533

Heath, C., & Heath, D. (2010). *Switch: How to change things when change is hard*. New York, NY: Random House.

Heffernan, O. (2017). Sustainability: A meaty issue. *Nature*, *544*(7651), S18–S20. doi:10.1038/544S18a PMID:28445446

Hefnawy, T. H. (2011). Effect of processing methods on nutritional composition and anti–nutritional factors in lentils (Lens culinaris). *Annals of Agricultural Science*, *56*(2), 57–61. doi:10.1016/j.aoas.2011.07.001

Heighton, E., & Cunningham, D. (1984). *Advertising in the broadcast and cable media*. Belmont, CA: Wadsworth Publishing.

Helble, S., & Wind, C. (2017). Food based on edible insects is a developing new market in Europe and also a challenge for the official food control authorities. *Tierärztliche Umschau*, *72*(4), 135–139.

Henchion, M., McCarthy, M., Resconi, V. C., & Troy, D. (2014). Meat consumption: Trends and quality matters. *Meat Science*, *98*(3), 561–568. doi:10.1016/j.meatsci.2014.06.007 PMID:25060586

Herring, S. C., & Kapidzic, S. (2015). Teens, gender, and self-presentation in social media. In J. D. Wright (Ed.), *International encyclopedia of social and behavioural sciences* (2nd ed.). Oxford, UK: Elsevier. doi:10.1016/B978-0-08-097086-8.64108-9

Heskey, C. E. (2018). The role of vegetarian diets in weight management. In W. J. Craig (Ed.), *Vegetarian nutrition and wellness* (pp. 113–169). Boca Raton, FL: CRC Press.

Hever, J., & Cronise, R. J. (2017). Plant-based nutrition for healthcare professionals: Implementing diet as a primary modality in the prevention and treatment of chronic disease. *Journal of Geriatric Cardiology: JGC*, *14*(5), 355–368. doi:10.11909/j.issn.1671-5411.2017.05.012 PMID:28630615

Hicks, T. M., Knowles, S. O., & Farouk, M. M. (2018). Global provision of red meat for flexitarian diets. *Frontiers in Nutrition*, *5*, 50. doi:10.3389/fnut.2018.00050 PMID:29963555

Hincks, J. (2018). The world is headed for a food security crisis. Here's how we can avert it. *Time Magazine*. Retrieved from http://time.com/5216532/global-food-security-richard-deverell/

Hocquette, J. F. (2016). Is in vitro meat the solution for the future? *Meat Science*, *120*, 167–176. doi:10.1016/j.meatsci.2016.04.036 PMID:27211873

Hoek, A. C., Elzerman, J. E., Hageman, R., Kok, F. J., Luning, P. A., & de Graaf, C. (2013). Are meat substitutes liked better over time? A repeated in-home use test with meat substitutes or meat in meals. *Food Quality and Preference*, *28*(1), 253–263. doi:10.1016/j.foodqual.2012.07.002

Hoek, A. C., Luning, P. A., Weijzen, P., Engels, W., Kok, F. J., & de Graaf, C. (2011). Replacement of meat by meat substitutes. A survey on person- and product-related factors in consumer acceptance. *Appetite*, *56*(3), 662–673. doi:10.1016/j.appet.2011.02.001 PMID:21315123

Hoek, A. C., Luning, P. A., Weijzen, P., Engels, W., Kok, F. J., & De Graaf, C. (2011). Replacement of meat by meat substitutes: A survey on person- and product-related factors in consumer acceptance. *Appetite*, *42*, 265–272. doi:10.1016/j.appet.2003.12.003 PMID:21315123

Hoek, A. C., Pearson, D., James, S. W., Lawrence, M. A., & Friel, S. (2017). Healthy and environmentally sustainable food choices: Consumer responses to point-of-purchase actions. *Food Quality and Preference*, *58*, 94–106. doi:10.1016/j.foodqual.2016.12.008

Hoek, A. C., Pearson, D., James, S. W., Lawrence, M. A., & Friel, S. (2017). Shrinking the food-print: A qualitative study into consumer perceptions, experiences and attitudes towards healthy and environmentally friendly food behaviours. *Appetite*, *108*, 117–131. doi:10.1016/j.appet.2016.09.030 PMID:27686818

Hoek, A. C., van Boekel, M. A., Voordouw, J., & Luning, P. A. (2011). Identification of new food alternatives: How do consumers categorize meat and meat substitutes? *Food Quality and Preference*, *22*(4), 371–383. doi:10.1016/j.foodqual.2011.01.008

Hoffman, J. R., & Falvo, M. J. (2004). Protein - Which is Best? *Journal of Sports Science & Medicine*, *3*(3), 118–130. PMID:24482589

Hoffmann, J. P. (1998). Wastewater treatment with suspended and nonsuspended algae. *Journal of Phycology*, *34*(5), 757–763. doi:10.1046/j.1529-8817.1998.340757.x

Holmes, M. (2015). Men's emotions: Heteromasculinity, emotional reflexivity, and intimate relationships. *Men and Masculinities*, *18*(2), 176–192. doi:10.1177/1097184X14557494

Holms, O. (2018). 'There is no kosher meat': The Israelis full of zeal for going vegan. *The Guardian International Edition.* Retrieved September 14, 2018 from https://www.theguardian.com/world/2018/mar/17/there-is-no-kosher-meat-the-israelis-full-of-zeal-for-going-vegan

Homburg, C., Koschate, N., & Hoyer, W. D. (2005). Do satisfied customers really pay more? A study of the relationship between customer satisfaction and willingness to pay. *Journal of Marketing, 69*(2), 84–96. doi:10.1509/jmkg.69.2.84.60760

Horan, F. (1974). Meat analogs. In A. M. Altschul (Ed.), New protein foods (Vol. 1A Technology, pp. 366–413). New York, NY: Academic Press. doi:10.1016/B978-0-12-054801-9.50014-4

Horrocks, R., Foster, C., & Foster, D. (2013). *Touching the Dragon* [DVD]. Natural History Unit Africa: National Geographic.

Hosie, R. (2017). Flexitarianism predicted as key food trend for 2017. *Independent.* Retrieved September 11, 2018 from https://www.independent.co.uk/life-style/food-and-drink/flexitarianism-predicted-as-key-food-trend-for-2017-vegetarian-less-meat-a7465156.html

Hosking, D. E., Eramudugolla, R., & Anstey, K. J. (2017). The MIND diet is associated with reduced incidence of 12-year cognitive impairment in an Australian setting. *Alzheimer's & Dementia, 13*(7), 1429–1430. doi:10.1016/j.jalz.2017.06.2234

House, J. (2016). Consumer acceptance of insect-based foods in the Netherlands: Academic and commercial implications. *Appetite, 107*, 47–58. doi:10.1016/j.appet.2016.07.023 PMID:27444958

Hu, D., Huang, J., Wang, Y., Zhang, D., & Qu, Y. (2014). Fruits and vegetables consumption and risk of stroke: A meta-analysis of prospective cohort studies. *Stroke, 45*(6), 1613–1619. doi:10.1161/STROKEAHA.114.004836 PMID:24811336

Huffington Post. (2013). *Leopard cub adopts her prey's baby, shows what we can learn from the animal world.* Retrieved from https://www.huffingtonpost.com/2013/12/ 04/leopard-adoptsbabybaboo_n_4386147.html

Hummer. (2006). *Tofu.* Retrieved from https://www.youtube.com/watch?v=lL4ZkYPLN38

Hunt, E. (2017). Meatonomics author says government working with meat and dairy industry to boost consumption. *The Guardian.* Retrieved from https://www.theguardian.com/science/2017/may/06/meatonomics-author-says-government-working-with-meat-and-dairy-industry-to-boost-consumption

Hwangbo, J., Hong, E. C., Jang, A., Kang, H. K., Oh, J. S., Kim, B. W., & Park, B. S. (2009). Utilization of house fly-maggots, a feed supplement in the production of broiler chickens. *Journal of Environmental Biology, 30*, 609–614. PMID:20120505

Hyner, C. (2015). A leading cause of everything: One industry that is destroying our planet and our ability to thrive on it. *Harvard Environmental Law Review.* Retrieved September 19, 2018 from http://harvardelr.com/2015/10/26/elrs-a-leading-cause-of-everything-one-industry-that-is-destroying-our-planet-and-our-ability-to-thrive-on-it/#_ftn1

Illgner, P., & Nel, E. (2000). The geography of edible insects in Sub-Saharan Africa: A study of the Mopane Caterpillar. *The Geographical Journal, 166*(4), 336–351. doi:10.1111/j.1475-4959.2000.tb00035.x

Ingenbleek, P., & Zhao, Y. (2018). *The Vegetarian Butcher: On its way to become the world's biggest "meat" producer?* [unpublished manuscript]. Wageningen: Marketing and Consumer Behaviour Group, Wageningen University.

Intergovernmental Panel on Climate Change (IPCC). (2014). *Climate Change 2014: Mitigation of Climate Change. Contribution of Working Group III to the Fifth Assessment Report of the Intergovernmental Panel on Climate Change.* Cambridge, UK: Cambridge University Press. Retrieved from http://www.ipcc.ch/pdf/assessment-report/ar5/wg3/ipcc_wg3_ar5_full.pdf

International Agency for Research on Cancer (IARC) and World Health Organisation (WHO). (2015). *IARC Monographs evaluate consumption of red meat and processed meat*. Retrieved from http://www.iarc.fr/en/media-centre/pr/2015/pdfs/pr240_E.pdf

International Agency for Research on Cancer (IARC). (2015). *IARC Monographs evaluate consumption of red meat and processed meat*. International Agency for Research on Cancer. Press release No. 240. World Health Organization. Retrieved from https://www.iarc.fr/en/media-centre/pr/2015/pdfs/pr240_E.pdf

Investopedia. (2018). *Corporation*. Retrieved from https://www.investopedia.com/terms/c/corporation.asp

Ippolito, A., El Ghaouth, A., Wilson, C. L., & Wisniewski, M. (2011). Control of postharvest decay of apple fruit by *Aureobasidium pullulans* and induction of defense responses. *Postharvest Biology and Technology*, *19*(3), 265–272. doi:10.1016/S0925-5214(00)00104-6

Jallinoja, P., Niva, M., & Latvala, T. (2016). Future of sustainable eating? Examining the potential for expanding bean eating in a meat-eating culture. *Futures*, *83*, 4–14. doi:10.1016/j.futures.2016.03.006

Jamet, J. P., & Chaumet, J. M. (2016). Soybean in China: Adapting to the liberalization. *OCL (Oilseeds & Fats, Crops and Lipids)*, *23*(6), D604. doi:10.1051/ocl/2016044

Janssen, R. H., Vincken, J.-P., van den Broek, L. A. M., Fogliano, V., & Lakemond, C. M. M. (2017). Nitrogen-to-protein conversion factors for three edible insects: Tenebrio molitor, Alphitobius diaperinus, and Hermetia illucens. *Journal of Agricultural and Food Chemistry*, *65*(11), 2275–2278. doi:10.1021/acs.jafc.7b00471 PMID:28252948

Javelosa, J. (2017, February 21). *Lab-grown meat is healthier. It's cheaper. It's the future*. Futurism. Retrieved from https://futurism.com/were-5-years-away-from-lab-grown-meat-hitting-store-shelves

Ji, K. M., Zhan, Z. K., Chen, J. J., & Liu, Z. G. (2008). Anaphylactic shock caused by silkworm pupa consumption in China. *Allergy*, *63*(10), 1405–1410. doi:10.1111/j.1398-9995.2008.01838.x PMID:18782121

Johns Hopkins Bloomberg School of Public Health. (n.d.). *Health & environmental implications of U.S. meat consumption & production*. Johns Hopkins University. Retrieved September 11, 2018 from https://www.jhsph.edu/research/centers-and-institutes/johns-hopkins-center-for-a-livable-future/projects/meatless_monday/resources/meat_consumption.html

Johnson, C. R., Thavarajah, D., Combs, G. F. Jr, & Thavarajah, P. (2013). Lentil (Lens culinaris L.): A prebiotic–rich whole food legume. *Food Research International*, *51*(1), 107–113. doi:10.1016/j.foodres.2012.11.025

Johnson, L., Myers, D., & Burden, D. (1992a). Early uses of soy protein in Far East, U.S. *INFORM – International News on Fats, Oils and Related Materials*, *3*(3), 282–290.

Johnson, L., Myers, D., & Burden, D. (1992b). Soy protein's history, prospects in food, feed. *INFORM – International News on Fats, Oils and Related Materials*, *3*(4), 429–444.

Johnston, J. (2017). Can consumers buy alternative foods at a big box supermarket? *Journal of Marketing Management*, *33*(7–8), 662–671. doi:10.1080/0267257X.2017.1297033

Jones, T. (2017). *14 healthy foods that are high in potassium*. Healthline. Retrieved from https://www.healthline.com/nutrition/high-potassium-foods#section3

Jongema, Y. (2017). *List of edible insects of the world (April 1, 2017)*. Wageningen University. Retrieved March 30, 2018 from http://www.wageningenur.nl/en/Expertise-Services/Chair-groups/Plant-Sciences/Laboratory-of-Entomology/Edible-insects/Worldwide-species-list.htm

Joshi, M., Timilsena, Y., & Adhikari, B. (2017). Global production, processing and utilization of lentil: A review. *Journal of Integrative Agriculture, 16*(12), 2898–2913. doi:10.1016/S2095-3119(17)61793-3

Joshi, V. K., & Kumar, S. (2015). Meat analogues: Plant based alternatives to meat products – a review. *International Journal of Food and Fermentation Technolology, 5*(2), 107–119. doi:10.5958/2277-9396.2016.00001.5

Joubert, B., Joubert, D., Apostal, J., & Bowman, J. (2006). *Eye of the leopard.* Gaborone, Botswana: Wildlife Films Botswana, National Geographic.

Joy, M. (2010). *Why we love dogs, eat pigs and wear cows: An introduction to carnism.* San Francisco, CA: Red Wheel/Weiser.

Just Eat. (2018). *'Going vegan' is predicted to be the biggest food trend of 2018.* Retrieved from https://www.riseofthevegan.com/blog/going-vegan-biggest-trend-for-2018

Just Food. (2000, December 11). *Vegetarian and meat-free foods.* Retrieved April 10, 2018, from https://www.just-food.com/analysis/vegetarian-and-meat-free-foods_id94010.aspx

Kagi, M. K., Wuthrich, B., & Johansson, S. G. (1994). Campari-orange anaphylaxis due to carmine allergy. *Lancet, 344*(8914), 60–61. doi:10.1016/S0140-6736(94)91083-9 PMID:7912327

Kantar Worldpanel. (2016). *The meat reducer.* Custom Market Research.

Kant, I. (1963). *Lectures on ethics.* New York, NY: Harper & Row Publishers. Retrieved from https://www.uta.edu/philosophy/faculty/burgess-jackson/Duties%20Towards%20Animals%20and%20Spirits%20(1930,%201963).pdf

Karim, M. M., Ashrafuzzaman, M., & Hossain, M. A. (2016). Effect of planting time on the growth and yield of chia (Salvia hispanica L.). *Asian Journal of Medical and Biological Research, 1*(3), 502. doi:10.3329/ajmbr.v1i3.26469

Kellershohn, J., & Russell, I. (2015). Yeast biotechnology. In R. Rai V (Ed.), Advances in food biotechnology. Chichester, UK: Wiley Blackwell. doi:10.1002/9781118864463.ch18

Kennedy, U., Sharma, A., & Phillips, C. J. C. (2018). The sheltering of unwanted cattle, experiences in India and implications for cattle industries elsewhere. *Animals (Basel), 8*(5), 64. doi:10.3390/ani8050064 PMID:29701646

Kenward, E. (2017). *Special report: The rise in "flexitarian" lifestyles.* Food Ingredients 1st. Retrieved September 11, 2018 from https://www.foodingredientsfirst.com/news/special-report-the-rise-in-flexitarian-lifestyles.html

Khachatourians, G. G., & Arora, D. K. (2002). *Applied mycology and biotechnology* (Vol. 2). Amsterdam: Elsevier.

Kiger, C. J. (1959). Etude de la composition chimique et de la valeur alimentaire de 57 espèces de champignons supérieurs. *Révue de Mycologie, 24,* 161–170.

Kimmel, A. J., & Tissier-Desbordes, E. (1999). Males, masculinity, and consumption: An exploratory investigation. In B. Dubois, T. M. Lowrey, & L. J. Shrum (Eds.), *European advances in consumer research* (Vol. 4, pp. 243–251). Provo, UT: Association for Consumer Research. Retrieved from http://acrwebsite.org/volumes/11390/volumes/e04/E-04

Kim, Y. M. (2014). *The 50 ways to cook edible insects.* Seoul, South Korea: Bumwoo.

Kirokva, D. (2015). How animal activist Gary Yourofsky turned '8% of Israel vegan' after comparing slaughterhouses to the Holocaust. *Metro.* Retrieved September 14, 2018 from https://metro.co.uk/2015/09/28/how-animal-rights-activist-gary-yourofsky-turned-8-of-israels-population-vegan-after-comparing-slaughterhouses-to-the-holocaust-5361205/

Kirsop, B. (n.d.). Food from micro-organisms. *Proceedings of the Nutrition Society, 44*(1), 13–17.

Kluger, J. (2018). Koko the Gorilla wasn't human, but she taught us so much about ourselves. *Time*. Retrieved from http://time.com/5318710/koko-gorilla-life/

Klunder, H. C., Wolkers-Rooijackers, J., Korpela, J. M., & Nout, M. J. R. (2012). Microbiological aspects of processing and storage of edible insects. *Food Control*, *26*(2), 628–631. doi:10.1016/j.foodcont.2012.02.013

Kmietowicz, Z. (2017). Red meat consumption is linked to higher risk of death from most major causes. *BMJ (Clinical Research Ed.)*, *357*. doi:10.1136/bmj.j2241

Koessler, M. (1949). The person in imagination or persona ficta of the corporation. *Louisiana Law Review*, *9*(4), 435–449.

Ko, J. L., Hsu, C. I., Lin, R. H., Kao, C. L., & Lin, J. Y. (1995). A new fungal immunomodulatory protein, FIP-fve isolated from the edible mushroom, Flammulina velutipes and its complete amino acid sequence. *European Journal of Biochemistry*, *228*(2), 244–249. doi:10.1111/j.1432-1033.1995.tb20256.x PMID:7705335

Kokoflix. (2012, September 11). *Koko Selects Adoption Options*. Retrieved from https://www.youtube.com/watch?v=xSBI_JMemqY

Kokoflix. (2014). *Koko's tribute to Robin Williams*. Retrieved fromhttps://www.youtube.com/watch?v=I9I_QvEXDv0

Kolonel, L. N. (1996). Nutrition and prostate cancer. *Cancer Causes & Control*, *7*(1), 83–44. doi:10.1007/BF00115640 PMID:8850437

Kotecki, P. (2018). The Sergey Brin-backed startup behind the first lab-grown burger just got a boost in the race to transform the meat industry. *Business Insider Australia*. Retrieved from https://www.businessinsider.com.au/sergey-brin-backed-startup-boost-race-transform-meat-industry-2018-7?r=US&IR=T

Kotter, J., & Cohen, D. (2002). *The heart of change: Real-life stories of how people change their organizations*. Boston, MA: Harvard Business Review Press.

Kouřimská, L., & Adámková, A. (2016). Nutritional and sensory quality of edible insects. *NFS Journal*, *4*, 22–26. doi:10.1016/j.nfs.2016.07.001

Kowitt, B. (2017a). Bill Gates and Richard Branson are investing in this clean meat startup. *Fortune*. Retrieved from: http://fortune.com/2017/08/23/bill-gates-richard-branson-invest-meat/

Kowitt, B. (2017b). Silicon Valley and the search for Meatless meat. *Fortune*. Retrieved from http://fortune.com/2017/12/19/silicon-valley-meatless-meat/

Krasny, J. (2015). Wait, are frozen dinner actually good now? *Esquire*. Retrieved from https://www.esquire.com/food-drink/food/a35481/stouffers-new-frozen-dinners/

Kuhad, R. C., Singh, A., Tripathi, K. K., Saxena, R. K., & Eriksson, K. E. L. (1997). Microorganisms as an alternative source of protein. *Nutrition Reviews*, *55*(3), 65–75. doi:10.1111/j.1753-4887.1997.tb01599.x PMID:9170892

Kumar, P., Chatli, M. K., Mehta, N., Singh, P., Malav, O. P., & Verma, A. K. (2017). Meat analogues: Health promising sustainable meat substitutes. *Critical Reviews in Food Science and Nutrition*, *57*(5), 923–932. doi:10.1080/10408398.2014.939739 PMID:25898027

Kung, S. J., Fenemore, B., & Potter, P. C. (2011). Anaphylaxis to mopane worms (Imbrasia belina). *Annals of Allergy, Asthma & Immunology*, *106*(6), 538–540. doi:10.1016/j.anai.2011.02.003 PMID:21624756

Kurpad, A. V., & Vaz, M. (2000). Protein and animo acid requirements in the elderly. *European Journal of Clinical Nutrition*, *54*(S3), S131–S142. doi:10.1038j.ejcn.1601035 PMID:11041085

Kuzmanova, S., Vandeska, E., Dimitrovski, A., & Doneva, D. (1989). Microbial procedures from utilization of food industry wastes. In *DECHEMA biotechnology conferences*. Retrieved from https://www.cabdirect.org/cabdirect/abstract/19911368140

Kye, S. Y., & Park, K. (2014). Health-related determinants of happiness in Korean adults. *International Journal of Public Health, 59*(5), 731–738. doi:10.100700038-014-0588-0 PMID:25033934

Laestadius, L. I., & Caldwell, M. A. (2015). Is the future of meat palatable? Perceptions of in vitro meat as evidenced by online news comments. *Public Health Nutrition, 18*(13), 2457–2467. doi:10.1017/S1368980015000622 PMID:25818555

Lähteenmaki-Uutela, A., Grmelová, N., Hénault-Ethier, L., Deschamps, M. H., Vandenberg, G. W., Zhao, A., ... Nemane, V. (2017). Insects as food and feed: Laws of the European Union, United States, Canada, Mexico, Australia, and China. *European Food and Feed Law Review, 12*(1), 22–37. Retrieved from https://www.researchgate.net/publication/316280459/download

Lang, T., & Heasman, M. (2004). *Food wars: The global battle for mouths, minds and markets.* London: Earthscan.

Lantern Consulting. (2017). *The Green Revolution: Understanding the 'veggie' boom.* Retrieved September 14, 2018 from http://www.lantern.es/papers/the-green-revolution-entendiendo-el-auge-del-mundo-veggie

Larson-Meyer, D. E. (2018). Nutritionally adequate vegetarian diets and athletic performance. In W. J. Craig (Ed.), *Vegetarian nutrition and wellness* (pp. 279–302). Boca Raton, FL: CRC Press.

Lau, P. S., Tam, N. F. Y., & Wong, Y. S. (1997). Wastewater nutrients (N and P) removal by carrageenan and alginate immobilized Chlorella vulgaris. *Environmental Technology, 18*(9), 945–951. doi:10.1080/09593331808616614

Leach, G. (1975). Energy and food production. *Food Policy, 1*(1), 62–73. doi:10.1016/0306-9192(75)90009-3

Leach, G. (1976). *Energy and food production.* Guildford, UK: Science and Technology Press for the International Institute for the Environment and Development.

Learmonth, E. (1963). The future of soya as human food in the United Kingdom. *Soybean Digest*, 14–16.

Lee, K. P., Simpson, S. J., & Wilson, K. (2008). Dietary protein-quality influences melanization and immune function in an insect. *Functional Ecology, 22*(6), 1052–1061. doi:10.1111/j.1365-2435.2008.01459.x

Leenaert, T. (2017). A vegan revolution in Israel: Fact or fiction? *The Vegan Strategist.* Retrieved September 14, 2018 from http://veganstrategist.org/2017/05/17/vegan-revolution-israel-fact-fiction/?doing_wp_cron=1536745141.3476879596710205078125

Lefevre, M., Kris-Etherton, P. M., Zhao, G., & Tracy, R. P. (2004). Dietary fatty acids, hemostasis, and cardiovascular disease risk. *Journal of the Academy of Nutrition and Dietetics, 104*(3), 410–419. doi:10.1016/j.jada.2003.12.022 PMID:14993864

Legesse, G., Cordeiro, M. R. C., Ominski, K. H., Beauchemin, K., Kroebel, K., McGeough, A. J., ... McAllister, T. A. (2018). Water use intensity of Canadian beef production in 1981 as compared to 2011. *The Science of the Total Environment, 619-620*, 1030–1039. doi:10.1016/j.scitotenv.2017.11.194 PMID:29734581

Leksawasdi, P. (2001). Bamboo caterpillar in Thailand. *Khon Kaen Agriculture Journal, 29*(1), 15–21. (in Thai)

Lensvelt, E. J. S., & Steenbekkers, L. P. (2014). Exploring consumer acceptance of entomophagy: A survey and experiment in Australia and the Netherlands. *Ecology of Food and Nutrition, 53*(5), 543–561. 865 doi:10.1080/03670244.2013.879

Leon-Guzman, M. F., Silva, I., & Lopez, M. G. (1997). Proximate chemical composition, free amino acid contents and free fatty acid contents of some edible mushrooms from Querétaro, Mexico. *Journal of Agricultural and Food Chemistry, 45*(11), 4329–4332. doi:10.1021/jf970640u

Leroy, F., Brengman, M., Ryckbosch, W., & Scholliers, P. (2018). Meat in the post-truth era: Mass media discourses on health and disease in the attention economy. *Appetite, 125*, 345–355. doi:10.1016/j.appet.2018.02.028 PMID:29486208

Leroy, F., & Praet, I. (2015). Meat traditions. The co-evolution of humans and meat. *Appetite, 90*, 200–211. doi:10.1016/j.appet.2015.03.014 PMID:25794684

Levin, M. (2011). Croc Pacho. *The Tico Times*. Retrieved from http://www.ticotimes.net/2011/10/17/costa-rica-says-goodbye-to-famous-croc-pocho

Li, C., Zhu, Y., Wang, Y., Zhu, J. S., Chang, J., & Kritchevsky, D. (1998). Monascus purpureus-fermented rice (red yeast rice): A natural food product that lowers blood cholesterol in animal models of hypercholesterolemia. *Nutrition Research (New York, N.Y.), 18*(1), 71–81. doi:10.1016/S0271-5317(97)00201-7

Li, G., & Zhu, F. (2018). Quinoa starch: Structure, properties, and applications. *Carbohydrate Polymers, 181*, 851–861. doi:10.1016/j.carbpol.2017.11.067 PMID:29254045

Li, M., Fan, Y., Zhang, X., Hou, W., & Tang, Z. (2014). Fruit and vegetable intake and risk of type 2 diabetes mellitus: Meta-analysis of prospective cohort studies. *BMJ Open, 4*(11), e005497. doi:10.1136/bmjopen-2014-005497 PMID:25377009

Lima, G., Ippolito, A., Nigro, F., & Salerno, M. (1997). Effectiveness of *Aureobasidium pullulans* and *Candida oleophila* against postharvest strawberry rots. *Postharvest Biology and Technology, 10*(2), 169–178. doi:10.1016/S0925-5214(96)01302-6

Lin, D., Lu, W., Kelly, A. L., Zhang, L., Zheng, B., & Miao, S. (2017). Interactions of vegetable proteins with other polymers: Structure–function relationships and applications in the food industry. *Trends in Food Science & Technology, 68*, 130–144. doi:10.1016/j.tifs.2017.08.006

Linda McCartney's Foods. (2016). *Linda McCartney's – 25 Years* (Promotional Video). Retrieved July 21, 2017, from http://www.lindamccartneyfoods.co.uk/our-story/our-journey/

Link, R. (2017). The 7 best plant sources of omega-3 fatty acids. *Healthline*. Retrieved from https://www.healthline.com/nutrition/7-plant-sources-of-omega-3s

Lippi, G., Mattiuzzi, C., & Cervellin, G. (2016). Meat consumption and cancer risk: A critical review of published meta-analysis. *Critical Reviews in Oncology/Hematology, 97*, 1–14. doi:10.1016/j.critrevonc.2015.11.008 PMID:26633248

Lippi, G., Mattiuzzi, C., & Sanchis-Gomar, F. (2015). Red meat consumption and ischemic heart disease: A systematic literature review. *Meat Science, 108*, 32–36. doi:10.1016/j.meatsci.2015.05.019 PMID:26017245

Litchfield, J. H. (1983). Single-cell proteins. *Science, 219*(4585), 740–746. doi:10.1126cience.219.4585.740 PMID:17814036

Liu, C. P., & Lin, L. P. (2001). Ultrastructural study and lipid formation of Isochrysis sp. CCMP1324. *Botanical Bulletin of Academia Sinica, 42*, 207–214.

Liu, Z., Ho, S., Hao, Y., Chen, Y., Woo, J., Wong, S. Y., ... Ling, W. (2016). Randomised controlled trial of effect of whole soy replacement diet on features of metabolic syndrome in postmenopausal women: Study protocol. *BMJ Open, 6*(9), e012741. doi:10.1136/bmjopen-2016-012741 PMID:27678545

Live Science. (2012). *Want to live longer? Hold the red meat*. The Live Science. Retrieved from https://www.livescience.com/36195-red-meat-consumption-death-risk.html

Li, Y., Horsman, M., Wang, B., Wu, N., & Lan, C. Q. (2008). Effects of nitrogen sources on cell growth and lipid accumulation of green alga Neochloris oleoabundans. *Applied Microbiology and Biotechnology*, *81*(4), 629–636. doi:10.100700253-008-1681-1 PMID:18795284

Llorente, P., Marquina, D., Santos, A., Peinado, J. M., & Spencer-Martins, I. (1997). Effect of salt on the killer phenotype of yeasts from olive brines. *Applied and Environmental Microbiology*, *63*, 1165–1167. PMID:9055432

Loef, M., & Walach, H. (2012). Fruit, vegetables and prevention of cognitive decline or dementia: A systematic review of cohort studies. *The Journal of Nutrition, Health & Aging*, *16*(7), 626–630. doi:10.100712603-012-0097-x PMID:22836704

Lokeshwari, R. K., & Shantibala, T. (2010). A review on the fascinating world of insect resources: Reason for thoughts. *Psyche*, *11*, 207570. doi:10.1155/2010/207570

Looy, H., Dunkel, F. V., & Wood, J. R. (2014). How then shall we eat? Insect-eating attitudes and sustainable foodways. *Agriculture and Human Values*, *31*(1), 1–11. doi:10.100710460-013-9450-x

López, D. N., Galante, M., Robson, M., Boeris, V., & Spelzini, D. (2018). Amaranth, quinoa and chia protein isolates: Physicochemical and structural properties. *International Journal of Biological Macromolecules*, *109*, 152–159. doi:10.1016/j.ijbiomac.2017.12.080 PMID:29247732

Los, F. G. B., Zielinski, A. A. F., Wojeicchowski, J. P., Nogueira, A., & Demiate, I. M. (2018). Beans (Phaseolus vulgaris L.): Whole seeds with complex chemical composition. *Current Opinion in Food Science*, *19*, 63–71. doi:10.1016/j.cofs.2018.01.010

Lowes, K. F., Shearman, C. A., Payne, J., MacKenzie, D., Archer, D. B., Merry, R. J., & Gasson, M. J. (2000). Prevention of yeast spoilage in feed and food by the yeast mycocin HMK. *Applied and Environmental Microbiology*, *66*(3), 1066–1076. doi:10.1128/AEM.66.3.1066-1076.2000 PMID:10698773

Lu, D., Pan, C., Ye, C., Duan, H., Xu, F., Yin, L., … Zhang, S. (2017). Meta-analysis of soy consumption and gastrointestinal cancer risk. *Nature Scientific Reports, 7*, 4048. Retrieved September 13, 2018 from https://www.nature.com/articles/s41598-017-03692-y

Lull, J. (2000). *Media, communication, culture*. Cambridge, UK: Polity Press.

Luna–Vital, D. A., Mojica, L., González de Mejía, E., Mendoza, S., & Loarca–Piña, G. (2015). Biological potential of protein hydrolysates and peptides from common bean (Phaseolus vulgaris L.): A review. *Food Research International*, *76*, 39–50. doi:10.1016/j.foodres.2014.11.024

Luo, Z. Y. (1997). Insects as food in China. *Ecology of Food and Nutrition*, *36*(2–4), 201–207. doi:10.1080/03670244.1997.9991515

MacClancy, J., Jeya Henry, C., & Macbeth, H. (2007). *Consuming the inedible: Neglected dimensions of food choice*. New York, NY: Berghahn Books.

MacEvilly, C. (2000). Bugs in the system. *Nutrition Bulletin*, *25*(4), 267–268. doi:10.1046/j.1467-3010.2000.00068.x

MacGregor, S., & Seymour, N. (Eds.). (2017). Men and nature: Hegemonic masculinities and environmental change. *RCC Perspectives: Transformations in Environment and Society*. Retrieved from http://www.environmentandsociety.org/sites/default/files/2017_i4_layout_final_1.pdf

Maghari, B. M., & Ardekani, A. M. (2011). Genetically modified foods and social concerns. *Avicenna Journal of Medical Biotechnology*, *3*(3), 109–117. PMID:23408723

Maiorino, M. I., Bellastella, G., & Esposito, K. (2015). Lifestyle modifications and erectile dysfunction: What can be expected? *Asian Journal of Andrology*, *17*(1), 5–10. doi:10.4103/1008-682X.137687 PMID:25248655

Malav, O. P., Talukder, S., Gokulakrishnan, P., & Chand, S. (2013). Meat analog: A review. *Critical Reviews in Food Science and Nutrition*, *55*(9), 1241–1245. doi:10.1080/10408398.2012.689381 PMID:24915320

Malik, V. S., Li, Y., Tobias, D. K., Pan, A., & Hu, F. B. (2016). Dietary protein intake and risk of type 2 diabetes in US men and women. *American Journal of Epidemiology*, *183*(8), 715–728. doi:10.1093/aje/kwv268 PMID:27022032

Malthus, R. (1798). *Essay on the principle of population*. Retrieved from http://www.esp.org/books/malthus/population/malthus.pdf

Mangels, R. (2018). Vegetarian diets: Trends in acceptance and perception. What do the dietary guidelines suggest? In W. J. Craig (Ed.), *Vegetarian nutrition and wellness* (pp. 1–12). Boca Raton, FL: CRC Press. doi:10.1201/b22003-1

Marceau, J. F., & Wise, S. (2018). *Exonerating the innocent: Habeas for nonhuman animals*. University of Denver Sturm College of Law Working Paper No. 18-16. Retrieved from https://papers.ssrn.com/sol3/papers.cfm?abstract_id=3176104

Marcu, A., Gaspar, R., Rutsaert, P., Seibt, B., Fletcher, D., Verbeke, W., & Barnett, J. (2015). Analogies, metaphors, and wondering about the future: Lay sense-making around synthetic meat. *Public Understanding of Science (Bristol, England)*, *24*(5), 547–562. doi:10.1177/0963662514521106 PMID:24553438

Marineli, R. da S., Lenquiste, S. A., Moraes, É. A., & Maróstica, M. R. Jr. (2015). Antioxidant potential of dietary chia seed and oil (Salvia hispanica L.) in diet–induced obese rats. *Food Research International*, *76*, 666–674. doi:10.1016/j.foodres.2015.07.039 PMID:28455051

Marinova, D., & Raphaely, T. (2018). Impact of vegetarian diets on the environment. In W. J. Craig (Ed.), *Vegetarian nutrition and wellness* (pp. 13–24). Boca Raton, FL: CRC Press.

Marinova, D., & Raphaely, T. (2018b). Taxing meat and animal food products. In D. Bogueva, D. Marinova, & T. Raphaely (Eds.), *Handbook of research on social marketing and its influence on animal origin food product consumption* (pp. 121–134). Hershey, PA: IGI Global. doi:10.4018/978-1-5225-4757-0.ch008

Mari-Sanchis, A., Gea, A., Basterra-Gortari, F. J., Martinez-Gonzalez, M. A., Beunza, J. J., & Bes-Rastrollo, M. (2016). Meat consumption and risk of developing type 2 diabetes in the SUN Project: A highly educated middle-class population. *PLoS One*, *11*(7), e0157990. doi:10.1371/journal.pone.0157990 PMID:27437689

Marsall, P., & Marinova, D. (2019). Health benefits of eating more plant foods and less meat. In D. Bogueva, D. Marinova, T. Raphaely, & K. Schmidinger (Eds.), Environmental, health and business opportunities in the new meat alternatives market. Hershey, PA: IGI Global.

Marsh, K., Zeuschner, C., & Saunders, A. (2012). Health implications of a vegetarian diet: A review. *American Journal of Lifestyle Medicine*, *6*(3), 250–267. doi:10.1177/1559827611425762

Marsh, K., Zeuschner, C., & Saunders, A. (2016). Red meat and health: Evidence regarding red meat, health, and chronic disease risk. In T. Raphaely & D. Marinova (Eds.), *Impact of meat consumption on health and environmental sustainability* (pp. 131–177). Hershey, PA: IGI Global. doi:10.4018/978-1-4666-9553-5.ch008

Marsh, K., Zeuschner, C., & Saunders, A. (2018). The health impact of eating foods of animal origin: Evidence regarding animal foods, health, and disease risk. In D. Bogueva, D. Marinova, & T. Raphaely (Eds.), *Handbook of research on social marketing and its influence on animal origin food product consumption* (pp. 283–297). Hershey, PA: IGI Global. doi:10.4018/978-1-5225-4757-0.ch002

Marti, A., Marengo, M., Bonomi, F., Casiraghi, M. C., Franzetti, L., Pagani, M. A., & Iametti, S. (2017). Molecular features of fermented teff flour relate to its suitability for the production of enriched gluten–free bread. *Lebensmittel-Wissenschaft + Technologie, 78*, 296–302. doi:10.1016/j.lwt.2016.12.042

Mason, P., & Lang, T. (2017). *Sustainable diets: How ecological nutrition can transform consumption and the food system.* Abingdon, UK: Routledge.

Matassa, S., Boon, N., Pikaar, I., & Verstraete, W. (2014). Microbial protein: Future sustainable food supply route with low environmental footprint. *Microbial Biotechnology, 9*(5), 568–575. doi:10.1111/1751-7915.12369 PMID:27389856

Materia, V. C., & Cavallo, C. (2015). Insetti per l'alimentazione umana: Barriere e drivers per l'accettazione da parte dei consumatori. *Rivista di Economia Agraria, LXX*(2), 139–161. doi:10.13128/REA-18004

Matondo, F. K., Takaisi, K., Nkuadiolandu, A. B., Kazadi Lukusa, A., & Aloni, M. N. (2016). Spirulina supplements improved the nutritional status of undernourished children quickly and significantly: Experience from Kisantu, the Democratic Republic of the Congo. *International Journal of Pediatrics, 1296414*, 1–5. doi:10.1155/2016/1296414 PMID:27777589

Mattick, C. S., Landis, A. E., Allenby, B. R., & Genovese, N. J. (2015). Anticipatory life cycle analysis of in vitro biomass cultivation for cultured meat production in the United States. *Environmental Science & Technology, 49*(19), 11941–11949. doi:10.1021/acs.est.5b01614 PMID:26383898

Mattisson, I., Wirfält, E., Johansson, U., Gullberg, B., Olsson, H., & Berglund, G. (2004). Intakes of plant foods, fibre and fat and risk of breast cancer – a prospective study in the Malmö Diet and Cancer cohort. *British Journal of Cancer, 90*(1), 122–127. doi:10.1038j.bjc.6601516 PMID:14710218

Ma, X. Q., Verkuil, J. M., Reinbach, H. C., & Meinert, L. (2017). Which product characteristics are preferred by Chinese consumers when choosing pork? A conjoint analysis on perceived quality of selected pork attributes. *Food Science & Nutrition, 5*(3), 770–775. doi:10.1002/fsn3.457 PMID:28572967

McCarney, L. J. (1975). Communication problems in the marketing of synthetic meats. *European Journal of Marketing, 9*(3), 188–197. doi:10.1108/EUM0000000005067

McDermott, M. (2012). *Meat industry grossly outspends animal rights.* Veg Groups, Treehugger. Retrieved from https://www.treehugger.com/green-food/meat-industry-grossly-outspends-animal-rights-veg-groups.html

McDonald, B. (2000). Once you know something, you can't not know it: An empirical look at becoming vegan. *Society & Animals, 8*(1), 1–23. doi:10.1163/156853000510961

McMahon, A. (1999). *Taking care of men: Sexual politics in the public mind.* Cambridge, UK: Cambridge University Press. doi:10.1017/CBO9780511597107

McPhail, D., Beagan, B., & Chapman, G. E. (2012). I don't want to be sexist but...: Denying and re-inscribing gender through food. *Food, Culture, & Society, 15*, 473–489. doi:10.2752/175174412X13276629246046

Meah, A. (2014). Reconceptualising "masculinity" through men's contribution to domestic woodwork. In A. Gorman-Murray & P. Hopkins (Eds.), Masculinities and place (pp. 191-208). London, UK: Ashgate.

Meat and Livestock Australia (MLA). (2016). *New MLA ads show Australian women "You're Better on Beef"*. Meat and Livestock Australia. Retrieved from https://www.mla.com.au/news-and-events/industry-news/new-mla-ads-show-australian-women-youre-better-on-beef-/

Meat and Livestock Australia (MLA). (2017). *MLA launches Australian Beef the Greatest*. Meat and Livestock Australia. Retrieved from https://www.mla.com.au/news-and-events/industry-news/mla-launches-australian-beef-the-greatest/

Megido, R. C., Gierts, C., Blecker, C., Brostaux, Y., Haubruge, E., Alabi, T., & Fransis, F. (2016). Consumer acceptance of insectbased alternative meat products in Western countries. *Food Quality and Preference*, *52*, 237–243. doi:10.1016/j.foodqual.2016.05.004

Megido, R. C., Sablon, L., Geuens, M., Brostaux, Y., Alabi, T., Blecker, C., ... Francis, F. (2014). Promising attitude for entomophagy development. *Journal of Sensory Studies*, *29*(1), 14–20. doi:10.1111/joss.12077

Mehlomakulu, N. N., Setati, M. E., & Divol, B. (2014). Characterization of novel killer toxins secreted by wine-related non-*Saccharomyces* yeasts and their action on *Brettanomyces* spp. *International Journal of Food Microbiology*, *188*, 83–91. doi:10.1016/j.ijfoodmicro.2014.07.015 PMID:25087208

Mekonnen, M. M., & Hoekstra, A. Y. (2012). A global assessment of the water footprint of farm animal products. *Ecosystems (New York, N.Y.)*, *15*(3), 401–415. doi:10.100710021-011-9517-8

Menga, V., Amato, M., Phillips, T. D., Angelino, D., Morreale, F., & Fares, C. (2017). Gluten–free pasta incorporating chia (Salvia hispanica L.) as thickening agent: An approach to naturally improve the nutritional profile and the in vitro carbohydrate digestibility. *Food Chemistry*, *221*, 1954–1961. doi:10.1016/j.foodchem.2016.11.151 PMID:27979185

Menozzi, D., Sogari, G., Veneziani, M., Simoni, E., & Mora, C. (2017a). Eating novel foods: An application of the theory of planned behaviour to predict the consumption of an insect-based product. *Food Quality and Preference*, *59*, 27–34. doi:10.1016/j.foodqual.2017.02.001

Menozzi, D., Sogari, G., Veneziani, M., Simoni, E., & Mora, C. (2017b). Explaining the intention to consume an insect-based product: A cross-cultural comparison. In A. H. Seyal & M. N. A. Rahman (Eds.), *Theory of planned behavior: New research* (pp. 201–215). New York, NY: Nova Publishers.

Merchant, B. (2013). *9 superstar athletes who don't eat meat*. Mother nature network. Retrieved from https://www.mnn.com/food/healthy-eating/photos/9-superstar-athletes-who-dont-eat-meat/carl-lewis

Mesías, M., Holgado, F., Márquez–Ruiz, G., & Morales, F. J. (2016). Risk/benefit considerations of a new formulation of wheat–based biscuit supplemented with different amounts of chia flour. *Lebensmittel-Wissenschaft + Technologie*, *73*, 528–535. doi:10.1016/j.lwt.2016.06.056

Messina, M. (2016). Soy and health update: Evaluation of the clinical and epidemiologic literature. *Nutrients*, *8*(12), 754. doi:10.3390/nu8120754 PMID:27886135

Michail, N. (2016). *Investors worth $1.25trn urge industry to switch to sustainable plant proteins*. Foodnavigator. Retrieved March 01, 2018 from https://www.foodnavigator.com/Article/2016/09/27/Investors-worth-1.25trn-urge-industry-to-switch-to-sustainable-plant-proteins

Micha, R., Michas, G., & Mozaffarian, D. (2012). Unprocessed red and processed meats and risk of coronary artery disease and type 2 diabetes–an updated review of the evidence. *Current Atherosclerosis Reports*, *14*(6), 515–524. doi:10.100711883-012-0282-8 PMID:23001745

Micha, R., Wallace, S. K., & Mozaffarian, D. (2010). Red and processed meat consumption and risk of incident coronary heart disease, stroke, and diabetes: A systematic review and meta-analysis. *Circulation, 121*(21), 2271–2283. doi:10.1161/CIRCULATIONAHA.109.924977 PMID:20479151

Middlemiss, L. (2018). *Sustainable consumption: Key issues*. Abingdon, UK: Routledge. doi:10.4324/9781315628035

Miele, M., & Lever, J. (2014). Improving animal welfare in Europe: Cases of comparative bio-sustainabilities. In T. Marsden & A. Morely (Eds.), *Sustainable food systems: Building a new paradigm* (pp. 143–165). Abingdon, UK: Routledge.

Migala, J. (2018). *What is the beyond burger and is it good for you?* Women's Health. Retrieved September 11, 2018 from https://www.womenshealthmag.com/food/a21566428/beyond-meat-burger-ingredients/

Milan–Carrillo, J., Valdez–Alarcon, C., Gutierrez–Dorado, R., Cárdenas–Valenzuela, O. G., Mora–Escobedo, R., Garzón–Tiznado, J. A., & Reyes–Moreno, C. (2007). Nutritional properties of quality protein maize and chickpea extruded based weaning food. *Plant Foods for Human Nutrition (Dordrecht, Netherlands), 62*(1), 31–37. doi:10.100711130-006-0039-z PMID:17243010

Milledge, J. J. (2012). Microalgae-commercial potential for fuel, food and feed. *Food Science and Technology (Campinas), 26*(1), 28–30.

Miller, G. T. Jr. (1985). *Living in the environment* (4th ed.). Belmont, CA: Wadswork Publishing.

Miller, S. A., & Dwyer, J. T. (2001). Evaluating the safety and nutritional value of mycoprotein. *Food Technology, 55*(7), 42–47.

Milman, O., & Leavenworth, S. (2016). China's plan to cut meat consumption by 50% cheered by climate campaigners: New dietary guidelines could reduce greenhouse gas emissions by 1bn tonnes by 2030, and could lessen country's problems with obesity and diabetes. *The Guardian*. Retrieved September 13, 2018 https://www.theguardian.com/world/2016/jun/20/chinas-meat-consumption-climate-change

Ministry of Agriculture and Rural Affairs of the People's Republic of China. (2016). *Ministry of Agriculture's guide on promotion of soybeans' production*. Retrieved September 13, 2018 from http://www.moa.gov.cn/govpublic/ZZYGLS/201604/t20160412_5091357.htm

Mintel International Group Limited. (2013a). *Meat-free and free-from foods – UK – September 2013*. Retrieved from https://store.mintel.com/meat-free-and-free-from-foods-uk-september-2013

Mintel International Group Limited. (2013b). *Red Meat – UK 2013*. Mintel Official Market Reports Store. Retrieved from https://store.mintel.com

Mintel International Group Limited. (2015). *Red Meat – UK. Mintel Official Market Reports Store*. Retrieved from https://store.mintel.com

Mintel. (2004). *Meat-free foods – UK – December 2004* (Market Research). London, UK: Mintel Group Ltd.

Mintel. (2006a). *Attitudes towards vegetarianism – UK – September 2006* (Market Research). London, UK: Mintel Group Ltd.

Mintel. (2006b). *Meat-free foods – UK - December 2006* (Market Research). London, UK: Mintel Group Ltd.

Mintel. (2010). *Meat-free foods – UK – December 2010*. Market Research Report. Mintel. Retrieved April 11, 2018, from http://academic.mintel.com/display/480957/#

Mintel. (2013). *Meat alternatives – US – June 2013*. Market Research Report. Mintel. Retrieved April 03, 2018, from https://store.mintel.com/meat-alternatives-us-june-2013

Mintel. (2017). *Meat-free foods – UK – May 2017* (Market Research). London, UK: Mintel Group Ltd.

Mintel. (2017). *Meat-free foods – UK – May 2017*. Market Research Report. Mintel. Retrieved April 04, 2018, from http://academic.mintel.com/display/796253/#

Mintel. (2018a). *France: Carrefour acquires majority stake in meal kit company*. Market Research Report. Mintel. Retrieved April 01, 2018 from http://academic.mintel.com/sinatra/oxygen_academic/display/id=887703?highlight#hit1

Mintel. (2018b). *Walmart pushes its meal kit business with plans to roll out to 2,000 stores this year*. Market Research Report. Mintel. Retrieved April 01, 2018 from http://academic.mintel.com/sinatra/oxygen_academic/display/id=885731?highlight#hit1

Mitsuhashi, J. (2010). Insects as traditional foods in Japan. *Ecology of Food and Nutrition*, *36*(2–4), 187–189. doi:10.1080/03670244.1997.9991514

Mlcek, J., Rop, O., Borkovcova, M., & Bednarova, M. (2014). A comprehensive look at the possibilities of edible insects as food in Europe – a review. *Polish Journal of Food and Nutrition Sciences*, *64*(3), 147–157. doi:10.2478/v10222-012-0099-8

Molinari, R., Costantini, L., Timperio, A. M., Lelli, V., Bonafaccia, F., Bonafaccia, G., & Merendino, N. (2017). Tartary buckwheat malt as ingredient of gluten–free cookies. *Journal of Cereal Science*, *80*, 37–43. doi:10.1016/j.jcs.2017.11.011

Møller, G., Sluik, D., Ritz, C., Mikkilä, V., Raitakari, O. T., Hutri–Kähönen, N., ... Raben, A. (2017). A protein diet score, including plant and animal protein, investigating the association with HbA1c and eGFR–The PREVIEW Project. *Nutrients*, *9*(7), 763. doi:10.3390/nu9070763 PMID:28714926

Monllos, K. (2015). *Brands are throwing out gender norms to reflect a more fluid world*. Adweek. Retrieved from https://www.adweek.com/brand-marketing/brands-are-throwing-out-gender-norms-reflect-more-fluid-world-174070/

Montalcini, T., De Bonis, D., Ferro, Y., Carè, I., Mazza, E., Accattato, F., ... Pujia, A. (2015). High vegetable fats intake is associated with high resting energy expenditure in vegetarians. *Nutrients*, *7*(7), 5933–5947. doi:10.3390/nu7075259 PMID:26193314

Montes, V. (2013). The role of emotions in the construction of masculinity: Guatemalan migrant men, transnational migration, and family relations. *Gender & Society*, *27*(4), 469–490. doi:10.1177/0891243212470491

Moorthi, R. N., Vorland, C. J., & Hill Gallant, K. M. (2017). Diet and diabetic kidney disease: Plant versus animal protein. *Current Diabetes Reports*, *17*(3), 15. doi:10.100711892-017-0843-x PMID:28271467

Morgan, R. (2018). *Bill Gates and Richard Branson are betting lab-grown meat might be the food of the future*. CNBC Disruptor 50. Retrieved from https://www.cnbc.com/2018/03/23/bill-gates-and-richard-branson-bet-on-lab-grown-meat-startup.html

Moritz, A. (2009). *Timeless secrets of health & rejuvenation*. Brevard, NC: Ener-Chi Wellness Press.

Morris, M. (2011). *Factory farming and animal liberation in New Zealand*. Wellington, New Zealand: Earl of Seacliff Art Workshop.

Morris, M. C., Tangney, C. C., Wang, Y., Sacks, F. M., Bennett, D. A., & Aggarwal, N. T. (2015). MIND diet associated with reduced incidence of Alzheimer's disease. *Alzheimer's & Dementia*, *11*(9), 1007–1014. doi:10.1016/j.jalz.2014.11.009 PMID:25681666

Mosaic Company. (2018). *Crop nutrition: Nitrogen*. Retrieved September 12, 2018 from https://www.cropnutrition.com/efu-nitrogen

Mouat, M. J., & Prince, R. (2018). Cultured meat and cowless milk: On making markets for animal-free food. *Journal of Cultural Economics, 11*(4), 315–329. doi:10.1080/17530350.2018.1452277

Murata, Y., Shimamura, T., Tagami, T., Takatsuki, F., & Hamuro, J. (2002). The skewing to Th1 induced by lentinan is directed through the distinctive cytokine production by macrophages with elevated intracellular glutathione content. *International Immunopharmacology, 2*(5), 673–689. doi:10.1016/S1567-5769(01)00212-0 PMID:12013506

Mycek, M. K. (2015). *Meatless meals and masculinity: An examination of men's use of rationality and scientific research to explain their plant-based diets* (Master's dissertation). North Carolina State University. NCSU Libraries, 1-44. Retrieved from https://repository.lib.ncsu.edu/bitstream/handle/1840.16/10723/etd.pdf?sequence=1&isAllowed=y

Myers, J. (2016). *China consumes more than a quarter of the world's meat. The government wants to change that.* World Economic Forum. Retrieved September 11, 2018 from https://www.weforum.org/agenda/2016/06/china-consumes-more-than-a-quarter-of-the-worlds-meat-the-government-wants-to-change-that/

Myers, S. S., Gaffikin, L., Golden, C. D., Ostfeld, R. S., Redford, K. H., & Ricketts, T. H. … Osofsky, S. A. (2013). Human health impact of ecosystem alteration. *Proceedings of the National Academy of Sciences of the United States of America, 110*(47), 18753–18760. 10.1073/pnas.1218656110

Myerson, G., & Rydin, Y. (1996). Sustainable development: the implications of the global debate for landuse planning. In S. Buckingham-Hatfield & B. Evans (Eds.), *Environmental planning and sustainability* (pp. 19–34). Chichester, UK: Wiley.

Mytton, O., Nnoaham, K., Eyles, H., Scarborough, P., & Ni Mhurchu, C. (2014). Systematic review and meta-analysis of the effect of increased vegetable and fruit consumption on body weight and energy intake. *BMC Public Health, 14*(1), 886–897. doi:10.1186/1471-2458-14-886 PMID:25168465

Nakagaki, B. J., & Defoliart, G. R. (1991). Comparison of diets for mass-rearing Acheta domesticus (Orthoptera: Gryllidae) as a novelty food, and comparison of food conversion efficiency with values reported for livestock. *Journal of Economic Entomology, 84*(3), 891–896. doi:10.1093/jee/84.3.891

Nakayama, T. (1989). Images of men in advertising. Center for Media Literacy. *Media & Values, 48*. Retrieved from http://www.medialit.org/reading-room/images-men-advertising#bio

Nam, K. C., Jo, C., & Lee, M. (2010). Meat products and consumption culture in the East. *Meat Science, 86*(1), 95–102. doi:10.1016/j.meatsci.2010.04.026 PMID:20510536

Naozuka, J., & Oliveira, P. V. (2012). Cooking effects on iron and proteins content of beans (Phaseolus Vulgaris L.) by GF AAS and MALDI–TOF MS. *Journal of the Brazilian Chemical Society, 23*(1), 156–162. doi:10.1590/S0103-50532012000100022

Nathanson, P., & Young, K. (2001). *Spreading misandry: The teaching of contempt for men in popular culture*. Montreal, Canada: McGill-Queen's University Press.

Nath, J. (2011). Gendered fare? A qualitative investigation of alternative food and masculinities. *Journal of Sociology (Melbourne, Vic.), 47*(3), 261–278. doi:10.1177/1440783310386828

National Agricultural Market Research Centre of the Agricultural University of China. (2018). *China's soybeans consumption grew over 4%*. Retrieved September 13, 2018 from http://finance.sina.com.cn/money/future/agri/2018-02-02/doc-ifyremfz3770325.shtml

National Archives Museum. (2018). *Declaration of Independence: A Transcription*. Retrieved from https://www.archives.gov/founding-docs/declaration-transcript

National Bureau of Statistics of China. (1997–2016). *China statistical yearbook*. Retrieved September 13, 2018 from http://www.stats.gov.cn/english/statisticaldata/annualdata/

National Bureau of Statistics. (2018). *Rural residents' per capita consumptions of soybeans, meat and poultry*. Retrieved September 13, 2018 from http://www.stats.gov.cn

National Cattlemen's Beef Association (NCBA). (2015). *Science does not support international agency opinion on red meat and cancer*. Retrieved from http://www.beefusa.org/newsreleases1.aspx?newsid=5418

National Health and Medical Research Council (NHMRC). (2006). *Nutrient reference values for Australia and New Zealand (2006)*. Canberra, Australia: National Health and Medical Research Council, Australian Government Department of Health and Ageing, New Zealand Ministry of Health.

National Health and Medical Research Council (NHMRC). (2013). *Australian dietary guidelines*. Canberra, Australia: NHMRC and Department of Health and Ageing, Australian Government. Retrieved from https:// www.nhmrc.gov.au/guidelines-publications/n55 and https://www.eatforhealth.gov.au/sites/default/files/files/the_guidelines/n55_australian_dietary_guidelines.pdf

Navruz–Varli, S., & Sanlier, N. (2016). Nutritional and health benefits of quinoa (Chenopodium quinoa Willd.). *Journal of Cereal Science*, *69*, 371–376. doi:10.1016/j.jcs.2016.05.004

Naylor, E., Humphrey, C., Kelley, T., Easter, L., & Iverson, G. (2018). Evaluation of nitrate concentrations and potential sources of nitrate in private water supply wells in North Carolina. *Journal of Environmental Health*, *80*, 16–23.

NBC News. (2005). *Ethiopian girl reportedly guarded by lions*. Retrieved from http://www.nbcnews.com/id/8305836/ns/world_news-africa/t/ethiopian-girl-reportedly-guarded-lions/#.W24ohdgzY6g

Neff, M. (2016). *Arnold Schwarzenegger wants you to eat less meat – for the sake of the planet's future*. One green planet. Retrieved from http://www.onegreenplanet.org/news/arnold-schwarzenegger-wants-you-to-eat-less-meat/

Nehete, J. Y., Bhambar, R. S., Narkhede, M. R., & Gawali, S. R. (2013). Natural proteins: Sources, isolation, characterization and applications. *Pharmacognosy Reviews*, *7*(14), 107–116. doi:10.4103/0973-7847.120508 PMID:24347918

Neporent, L. (2013). *Mike Tyson on losing 140 pounds and Evander Holyfield's Ear*. ABC News. Retrieved from https://abcnews.go.com/Health/mike-tyson-veganism-evander-holyfields-ear/story?id=18958770

Newby, P. K. (2009). Plant foods and plant-based diets: Protective against childhood obesity? *The American Journal of Clinical Nutrition*, *89*(5), 1572S–1587S. doi:10.3945/ajcn.2009.26736G

News Desk. (2017). *UK meat substitutes sales to grow by 25% in four years*. FoodBevMedia. Retrieved September 11, 2018 from https://www.foodbev.com/news/uk-meat-substitutes-sales-grow-25-four-years/

Nickelsburg, M. (2017). Bill Gates invests in veggie burger that 'bleeds' like beef to feed the masses and save the planet. *GeekWire*. Retrieved from https://www.geekwire.com/2017/bill-gates-invests-veggie-burger-bleeds-like-beef-feed-masses-save-planet/

Nielsen. (2015). *The sustainability imperative*. Nielsen Research. Retrieved from http://www.nielsen.com/au/en/insights/reports/2015/the-sustainability-imperative.html

Nielsen, D. S., Hønholt, S., Tano-Debrah, K., & Jespersen, L. (2005). Yeast populations associated with Ghanaian cocoa fermentations analysed using denaturing gradient gel electrophoresis (DGGE). *Yeast (Chichester, England)*, *22*(4), 271–284. doi:10.1002/yea.1207 PMID:15704234

Nielsen, L. V., Kristensen, M. D., Klingenberg, L., Ritz, C., Belza, A., Astrup, A., & Raben, A. (2018). Protein from meat or vegetable sources in meals matched for fibre content has similar effects on subjective appetite sensations and energy intake – a randomized acute cross–over meal test study. *Nutrients, 10*(1). doi:10.3390/nu10010096

Nie, Y., Luo, F., & Lin, Q. (2018). Dietary nutrition and gut microflora: A promising target for treating disease. *Trends in Food Science & Technology, 75*, 72–80. doi:10.1016/j.tifs.2018.03.002

Nijdam, D., Rood, T., & Westhoek, H. (2012). The price of protein: Review of land use and carbon footprints from life cycle assessments of animal food products and their substitutes. *Food Policy, 37*(6), 760–770. doi:10.1016/j.foodpol.2012.08.002

Nixon, S. (1997). Exhibiting masculinity. In S. Hall (Ed.), *Representation: Cultural representation and signifying practices* (pp. 291–336). London, UK: Sage.

Nonaka, K. (2010). Cultural and commercial roles of edible wasps in Japan. In P. B. Durst, D. V. Johnson, R. N. Leslie & K. Shono (Eds), *Forest insects as food: Humans bite back* (pp. 123–130). Bangkok, Thailand: Food and Agriculture Organization of the United Nations. Retrieved September 12, 2018 from http://www.fao.org/docrep/012/i1380e/i1380e00.pdf

Nordqvist, C. (2017). Everything you need to know about vitamin B-12. *Medical News Today*. Retrieved from https://www.medicalnewstoday.com/articles/219822.php?sr

Norman, M. E. (2011). Embodying the double-bind of masculinity: Young men and discourses of normalcy, health, heterosexuality, and individualism. *Men and Masculinities, 14*(4), 430–449. doi:10.1177/1097184X11409360

North Carolina (NC) Soybean Producers Association. (2014). *History of soybean*. Retrieved September 12, 2018 from http://ncsoy.org/media-resources/history-of-soybeans/

Nosworthy, M. G., Medina, G., Franczyk, A. J., Neufeld, J., Appah, P., Utioh, A., ... House, J. D. (2018). Effect of processing on the in vitro and in vivo protein quality of red and green lentils (Lens culinaris). *Food Chemistry, 240*, 588–593. doi:10.1016/j.foodchem.2017.07.129 PMID:28946315

Nout, M. R., & Boekhout, T. (2003). Traditional fermented products from Africa, Latin America and Asia. In T. Boekhout & V. Robert (Eds.), *Yeasts in food: Beneficial and detrimental aspects* (pp. 451–473). Hamburg, Germany: Behr's Verlag. doi:10.1533/9781845698485.451

NutritionFacts. (2018). *Heme iron*. Retrieved from https://nutritionfacts.org/topics/heme-iron/

O'Brien, C. (2013). I trade, therefore I am: Legal personhood in the European Union. *Common Market Law Review, 50*(6), 1643–1684.

O'Brien, J. (2009). *Opportunities beyond carbon: Looking forward to a sustainable world*. Melbourne, Australia: Melbourne University Press.

O'Doherty Jensen, K., & Holm, L. (1999). Preferences, quantities and concerns: Socio-cultural perspectives on the gendered consumption of foods. *European Journal of Clinical Nutrition, 53*(5), 351–359. doi:10.1038j.ejcn.1600767 PMID:10369489

O'Donnell, K., Cigelnik, E., & Casper, H. H. (1998). Molecular phylogenetic, morphological, and mycotoxin data support reidentification of the quorn mycoprotein fungus as *Fusarium venenatum*. *Fungal Genetics and Biology, 23*(1), 57–67. doi:10.1006/fgbi.1997.1018 PMID:9501477

O'Neil, J. M. (2013). Gender role conflict research 30 years later: An evidence-based diagnostic schema to assess boys and men in counselling. *Journal of Counseling and Development, 91*(4), 490–498. doi:10.1002/j.1556-6676.2013.00122.x

Obopile, M., & Seeletso, T. G. (2013). Eat or not eat: An analysis of the status of entomophagy in Botswana. *Food Security*, *5*(6), 817–824. doi:10.100712571-013-0310-8

Office for National Statistics. (2016). *Families and households in the UK: 2016*. Retrieved September 10, 2018 from https://www.ons.gov.uk/releases/familiesandhouseholdsintheuk2016

Okezie, O. A., Kgomotso, K. K., & Letswiti, M. M. (2010). Mopane worm allergy in a 36-year-old woman: A case report. *Journal of Medical Case Reports*, *4*(1), 42. doi:10.1186/1752-1947-4-42 PMID:20205892

Olaizola, M. (2003). Commercial development of microalgal biotechnology: From the test tube to the marketplace. *Biomolecular Engineering*, *20*(4–6), 459–466. doi:10.1016/S1389-0344(03)00076-5 PMID:12919832

Oliveira, A. P., de, & Mateó, B. (2018). Effect of cooking on the bioaccessibility of essential elements in different varieties of beans (Phaseolus vulgaris L.). *Journal of Food Composition and Analysis*, *67*, 135–140. doi:10.1016/j.jfca.2018.01.012

Oliveira–Alves, S. C., Vendramini–Costa, D. B., Betim Cazarin, C. B., Maróstica Júnior, M. R., Borges Ferreira, J. P., Silva, A. B., ... Bronze, M. R. (2017). Characterization of phenolic compounds in chia (Salvia hispanica L.) seeds, fibre flour and oil. *Food Chemistry*, *232*, 295–305. doi:10.1016/j.foodchem.2017.04.002 PMID:28490078

Oliver, R. L. (1980). A cognitive model of the antecedents and consequences of satisfaction decisions. *JMR, Journal of Marketing Research*, *17*(4), 460–469. doi:10.2307/3150499

Oliver, S. P., Murinda, S. E., & Jayarao, B. M. (2011). Impact of antibiotic use in adult dairy cows on antimicrobial resistance of veterinary and human pathogens: A comprehensive review. *Foodborne Pathogens and Disease*, *3*(3), 337–355. doi:10.1089/fpd.2010.0730 PMID:21133795

Oonincx, D. G. A. B., & de Boer, I. J. M. (2012). Environmental impact of the production of mealworms as a protein source for humans – a life cycle assessment. *PLoS One*, *7*(12), 51145. doi:10.1371/journal.pone.0051145 PMID:23284661

Oonincx, D. G. A. B., van Itterbeeck, J., Heetkamp, M. J. W., van den Brand, H., van Loon, J. J. A., & van Huis, A. (2010). An exploration on greenhouse gas and ammonia production by insect species suitable for animal or human comsumption. *PLoS One*, *5*(12), e14445. doi:10.1371/journal.pone.0014445 PMID:21206900

Oonincx, D. G., & Dierenfeld, E. S. (2012). An investigation into the chemical composition of alternative invertebrate prey. *Zoo Biology*, *31*(1), 40–54. doi:10.1002/zoo.20382 PMID:21442652

Oren, A. (2005). A hundred years of Dunaliella research: 1905–2005. *Saline Systems*, *1*(1), 2. doi:10.1186/1746-1448-1-2 PMID:16176593

Osho, A. (1995). Production of metabolites, industrial enzymes, amino acids, vitamins, single cell protein. *Journal of Research*, *6*, 521–529.

Otaegui-Arrazola, A., Amiano, P., Elbusto, A., Urdaneta, E., & Martínez-Lage, P. (2014). Diet, cognition, and Alzheimer's disease: Food for thought. *European Journal of Nutrition*, *53*(1), 1–23. doi:10.100700394-013-0561-3 PMID:23892520

Ottenfeld, M., Bernstein, D., & Witte, C. L. (2008). An exploration of U.S. consumer perceptions and affect: Two forms of a soy-based food product. *Journal of Food Products Marketing*, *14*(3), 49–76. doi:10.1080/10454440801985936

Ovando, C. A., de Carvalho, J. C., Vinícius de Melo Pereira, G., Jacques, P., Soccol, V. T., & Soccol, C. R. (2018). Functional properties and health benefits of bioactive peptides derived from Spirulina: A review. *Food Reviews International*, *34*(1), 34–51. doi:10.1080/87559129.2016.1210632

Owens, J. (2016). *Vegetarian propaganda and the Great War* (Project Report, BA History). Lancaster, UK: Lancaster University. Retrieved September 10, 2018 from https://www.academia.edu/31737175/Vegetarian_propaganda_and_the_Great_War

Pachirat, T. (2013). *Every twelve seconds: Industrialized slaughter and politics of sight*. New Haven, CT: Yale University Press.

Pan, A., Sun, Q., Bernstein, A. M., Schulze, M. B., Manson, J. A. E., Stampfer, M. J., ... Hu, F. B. (2012). Red meat consumption and mortality: Results from two prospective cohort studies. *Archives of Internal Medicine*, *172*(7), 555–563. doi:10.1001/archinternmed.2011.2287 PMID:22412075

Paoletti, M. G., Norberto, L., Damini, R., & Musumeci, S. (2007). Human gastric juice contains chitinase that can degrade chitin. *Annals of Nutrition & Metabolism*, *51*(3), 244–251. doi:10.1159/000104144 PMID:17587796

Parafati, L., Vitale, A., Restuccia, C., & Cirvilleri, G. (2015). Biocontrol ability and action mechanism of food-isolated yeast strains against *Botrytis cinerea* causing postharvest bunch rot of table grape. *Food Microbiology*, *47*, 85–92. doi:10.1016/j.fm.2014.11.013 PMID:25583341

Pascucci, S., & De-Magistris, T. (2013). Information bias condemning radical food innovators? The case of insect-based products in the Netherlands. *The International Food and Agribusiness Management Review*, *16*(3), 1–16. Retrieved from https://ageconsearch.umn.edu/bitstream/156421/2/20130022.pdf

Pasiakos, S. M., Agarwal, S., Lieberman, H. R., & Fulgoni, V. L. III. (2015). Sources and amounts of animal, dairy, and plant protein intake of US adults in 2007–2010. *Nutrients*, *7*(8), 7058–7069. doi:10.3390/nu7085322 PMID:26308049

Pasquet, V., Chérouvrier, J. R., Farhat, F., Thiéry, V., Piot, J. M., Bérard, J. B., ... Picot, L. (2011). Study on the microalgal pigments extraction process: Performance of microwave assisted extraction. *Process Biochemistry*, *46*(1), 59–67. doi:10.1016/j.procbio.2010.07.009

Passport. (2016). *Future meat alternatives beyond plant-based: In-vitro or 3D printed meat*. Market Research Report. Passport. Retrieved March 28, 2018, from https://www.portal.euromonitor.com/portal/analysis/tab

Patterson, P. (2002). *Koko wants to have a baby*. The Gorilla Foundation. Retrieved from https://www.koko.org/node/1889

Payne, C. (2018, January 18). Edible insects: Do insects actually taste any good? *BBC News*. Retrieved from: https://www.bbc.co.uk/news/world-42639877

Pearson, A. M. (1976). Meat extenders and substitutes. *Bioscience*, *26*(4), 249–256. doi:10.2307/1297347

Peeples, J. (2015). Discourse/rhetorical analysis approaches to environment, media, and communication. In A. Hansen & R. Cox (Eds.), *The Routledge handbook of environment and communication* (pp. 39–48). London, UK: Routledge.

Pelletier, N., & Tyedmers, P. (2010). Forecasting potential global environmental costs of livestock production 2000–2050. *Proceedings of the National Academy of Sciences of the United States of America*, *107*(43), 18371–18374. doi:10.1073/pnas.1004659107 PMID:20921375

Penarrendonda, J. L. (2017). *Could AI help to create a meat-free world?* Retrieved March 15, 2018 from http://www.bbc.com/future/story/20171214-could-ai-help-create-a-meat-free-world

Pennino, M., Dierenfeld, E. S., & Behler, J. L. (1991). Retinol, alpha-tocopherol and proximate nutrient composition of invertebrates used as feed. *International Zoo Yearbook*, *30*(1), 143–149. doi:10.1111/j.1748-1090.1991.tb03477.x

Pessoa, J. A., Mancilha, I. M., & Sato, S. (1996). Cultivation of Candida tropicalis in sugarcane hemicellulosic hydrolysate for microbial protein production. *Journal of Biotechnology*, *51*(1), 83–88. doi:10.1016/0168-1656(96)01572-6

Petersson, S., & Philippou, E. (2016). Mediterranean diet, cognitive function, and dementia: A systematic review of the evidence. *Advances in Nutrition, 7*(5), 889–904. doi:10.3945/an.116.012138 PMID:27633105

Pew Research Centre. (2011). *The generation gap and the 2012 election.* Retrieved from http://www.people-press.org/2011/11/03/section-8-domestic-and-foreign-policy-views/

Phillips, C. (2010). Animals, Quo Vadis? Welcome to a new, multidisciplinary, integrated, open access journal: Animals. *Animals (Basel), 1*(1), 1–3. doi:10.3390/ani1010001 PMID:26486211

Phillips, C. J. C. (2015). *The animal trade.* Wallingford, UK: CAB International. doi:10.1079/9781780643137.0000

Phillips, C. J. C. (2018). *Principles of cattle production* (3rd ed.). Wallingford, UK: CAB International. doi:10.1079/9781786392701.0000

Piazza, J., Ruby, M. B., Loughnan, S., Luong, M., Kulik, J., Watkins, H. M., & Seigerman, M. (2015). Rationalizing meat consumption. The 4Ns. *Appetite, 91*, 114–128. doi:10.1016/j.appet.2015.04.011 PMID:25865663

Pierce, K. (2010). Men leave their own mark on veganism. *Boston Globe.* Retrieved from http://archive.boston.com/lifestyle/food/articles/2010/03/24/men_leave_their_own_mark_on_veganism/

Pino-Moreno, J. M., & Ramos-Elorduy, J. J. (2006). Algunos ejemplos de aprovechamiento comercial de varios insectos comestibles y medicinales. In E. Estrada, J. N. Romero, A. M. Equihua, C. L. Luna L., & J. L. A. Rosas (Eds.), Entomología mexicana (pp. 524-533). Mexico City, Mexico: Sociedad Mexicana de Entomología.

Piqueras, J. A., Kuhne, W., Vera–Villarroel, P., van Straten, A., & Cuijpers, P. (2011). Happiness and health behaviours in Chilean college students: A cross-sectional survey. *BMC Public Health, 11*(1), 443. doi:10.1186/1471-2458-11-443 PMID:21649907

Pompper, D. (2010). Masculinities, the metrosexual, and media images: Across dimensions of age and ethnicity. *Sex Roles, 63*(9-10), 682–696. doi:10.100711199-010-9870-7

Ponis, E., Probert, I., Veron, B., Mathieu, M., & Robert, R. (2006). New microalgae for the Pacific oyster Crassostrea gigas larvae. *Aquaculture (Amsterdam, Netherlands), 253*(1–4), 618–627. doi:10.1016/j.aquaculture.2005.09.011

Poore, J., & Nemecek, T. (2018). Reducing food's environmental impacts through producers and consumers. *Science, 360*(6392), 987–992. doi:10.1126cience.aaq0216 PMID:29853680

Popa, D., & Gavriliu, D. (2015). Gender representations and digital media. *Procedia: Social and Behavioral Sciences, 180*, 1199–1206. doi:10.1016/j.sbspro.2015.02.244

Post, M. J. (2014a). An alternative animal protein source: Cultured beef. *Annals of the New York Academy of Sciences, 1328*(1), 29–33. doi:10.1111/nyas.12569 PMID:25376889

Post, M. J. (2014b). Cultured beef: Medical technology to produce food. *Journal of the Science of Food and Agriculture, 94*(6), 1039–1041. doi:10.1002/jsfa.6474 PMID:24214798

Post, M. J. (2018). Proteins in cultured beef. In R. Y. Yada (Ed.), *Proteins in food processing* (2nd ed.; pp. 289–298). Cambridge, UK: Woodhead Publishing. doi:10.1016/B978-0-08-100722-8.00012-7

Potts, A., & Parry, J. (2010). Vegan sexuality: Challenging heteronormative masculinity through meat-free sex. *Feminism & Psychology, 20*(1), 53–72. doi:10.1177/0959353509351181

Poverty.com. (2018). *Hunger and world poverty.* Retrieved September 14, 2018 from https://www.poverty.com

Prabhat, S. (2017). *Difference between yeast and fungus*. Retrieved from http://www.differencebetween.net/science/nature/difference-between-yeast-and-fungus/

Prasannan, A. (2018). *Meat substitute market by product type (tofu-based, tempeh-based, TVP-based, seitan-based, Quorn-based, and others), source (soy-based, wheat-based, mycoprotein, and other sources), category (frozen, refrigerated, and shelf stable) –global opportunity analysis and industry forecast, 2018-2025*. Allied Market Research. Retrieved September 11, 2018 from https://www.alliedmarketresearch.com/meat-substitute-market

Prättälä, R., Paalanen, L., Grinberga, D., Helasoja, V., Kasmel, A., & Petkeviciene, J. (2007). Gender differences in the consumption of meat, fruit and vegetables are similar in Finland and the Baltic countries. *European Journal of Public Health*, *17*(5), 520–525. doi:10.1093/eurpub/ckl265 PMID:17194710

Pratt, R., Daniels, T. C., Eiler, J. J., Gunnison, J. B., Kumler, W. D., Oneto, J. F., ... Smith, J. H. C. (1944). Chlorellin, an antibacterial substance from chlorella. *Science*, *99*(2574), 351–352. doi:10.1126cience.99.2574.351 PMID:17750208

Premalatha, M., Abbasi, T., Abbasi, T., & Abbasi, S. A. (2011). Energy-efficient food production to reduce global warming and ecodegradation: The use of edible insects. *Renewable & Sustainable Energy Reviews*, *15*(9), 4357–4360. doi:10.1016/j.rser.2011.07.115

Pribis, P. (2018). The role of legumes in maintaining health. In W. J. Craig (Ed.), *Vegetarian nutrition and wellness* (pp. 191–209). Boca Raton, FL: CRC Press.

Pritchett, J., Thilmany, D., & Johnson, K. (2005). Animal disease economic impacts: A survey of literature and typology of research approaches. *The International Food and Agribusiness Management Review*, *8*(1), 23–45. Retrieved from https://www.ifama.org/resources/Documents/v8i1/Pritchett-Thilmany-Johnson.pdf

Probst, L., Frideres, L., Pedersen, P., & Amato, F. (2015). Sustainable, safe and nutritious food. New nutrient sources. *Business Innovation Observatory*. Retrieved April 01, 2018, from https://ec.europa.eu/docsroom/documents/13426/attachments/1/translations/en/renditions/native

Prüfer, K., Munch, K., Hellmann, I., Akagi, K., Miller, J. R., Walenz, B., ... Pääbo, S. (2012). The bonobo genome compared with the chimpanzee and human genomes. *Nature*, *486*(7404), 527–531. doi:10.1038/nature11128 PMID:22722832

Qian, X., Sun, W., Gu, J., Wang, X., Sun, J., Yin, Y., & Duan, M. (2016). Variable effects of oxytetracycline on antibiotic resistance gene abundance and the bacterial community during aerobic composting of cow manure. *Journal of Hazardous Materials*, *315*, 61–69. doi:10.1016/j.jhazmat.2016.05.002 PMID:27179201

Quintana Pacheco, D. A., Sookthai, D., Wittenbecher, C., Graf, M. E., Stübel, R., Johnson, T., ... Kühn, T. (2018). Red meat consumption and risk of cardiovascular diseases—is increased iron load a possible link? *The American Journal of Clinical Nutrition*, *107*(1), 113–119. doi:10.1093/ajcn/nqx014 PMID:29381787

Quist, J. (2007). *Backcasting for a sustainable future: The impact after ten years*. Delft, The Netherlands: Eburon Publishers.

Quote Investigator. (2018). *We shall escape the absurdity of growing a whole chicken in order to eat the breast or wing*. Retrieved from https://quoteinvestigator.com/2017/01/22/meat/

Radke, A. (2018). Arnold Schwarzenegger drinks the vegan Kool-Aid. *Beef Magazine*. Retrieved from http://www.beefmagazine.com/ranching/arnold-schwarzenegger-drinks-vegan-kool-aid

Radmer, R. J., & Parker, B. C. (1994). Commercial applications of algae: Opportunities and constraints. *Journal of Applied Phycology*, *6*(2), 93–98. doi:10.1007/BF02186062

Raheem, D., Carrascosa, C., Oluwole, B. O., Nieuwland, M., Saraiva, A., Millán, R., & Raposo, A. (2018). Traditional consumption of and rearing edible insects in Africa, Asia and Europe. *Critical Reviews in Food Science and Nutrition*, 1–20. doi:10.1080/10408398.2018.1440191 PMID:29446643

Rahman, M. J., de Camargo, A. C., & Shahidi, F. (2017). Phenolic and polyphenolic profiles of chia seeds and their in vitro biological activities. *Journal of Functional Foods*, *35*, 622–634. doi:10.1016/j.jff.2017.06.044

Ramdath, D., Padhi, E., Sarfaraz, S., Renwick, S., & Duncan, A. (2017). Beyond the cholesterol–lowering effect of soy protein: A review of the effects of dietary soy and its constituents on risk factors for cardiovascular disease. *Nutrients*, *9*(4), 324. doi:10.3390/nu9040324 PMID:28338639

Ramos-Elorduy, J. (1998). *Creepy crawly cuisine: The gourmet guide to edible insects*. Paris, France: Park Street Press.

Ramos-Elorduy, J., Moreno, J. M. P., Prado, E. E., Perez, M. A., Otero, J. L., & de Guevara, O. L. (1997). Nutritional value of edible insects from the state of Oaxaca, Mexico. *Journal of Food Composition Analisys*, *10*(2), 142–157. doi:10.1006/jfca.1997.0530

Ranganathan, J. (2013). *The global food challenge explained in 18 Graphics*. World Resources Institute. Retrieved from https://www.wri.org/blog/2013/12/global-food-challenge-explained-18-graphics

Raphael, A. (1969, October 27). Meatless meat—synthetic way to the US heart. *The Guardian*, p. 3. Retrieved from the ProQuest database https://www.proquest.com/libraries/academic/databases/

Raphaely & Marinova. (Eds.). (2016a). *Impact of meat consumption on health and environmental sustainability*. Hershey, PA: IGI Global.

Raphaely, T., & Marinova, D. (2014). Flexitarianism: A more moral dietary option. *International Journal of Sustainable Society*, *6*(1/2), 189–211. doi:10.1504/IJSSOC.2014.057846

Raphaely, T., & Marinova, D. (2014). Flexitarianism: Decarbonising through flexible vegetarianism. *Renewable Energy*, *67*, 90–96. doi:10.1016/j.renene.2013.11.030

Raphaely, T., & Marinova, D. (Eds.). (2016). *Impact of meat consumption on health and environmental sustainability*. Hershey, PA: IGI Global. doi:10.4018/978-1-4666-9553-5

Raphaely, T., Marinova, D., Crisp, G., & Panayotov, J. (2013). Flexitarianism (flexible or part-time vegetarianism): A user-based dietary choice for improving personal, population and planetary wellbeing. *International Journal of User-Driven Healthcare*, *3*(3), 34–58. doi:10.4018/ijudh.2013070104

Raubenheimer, D., & Rothman, J. M. (2013). Nutritional ecology of entomophagy in humans and other primates. *Annual Review of Entomology*, *58*(1), 141–160. doi:10.1146/annurev-ento-120710-100713 PMID:23039342

Rawles, K. (2006). Sustainable development and animal welfare: The neglected dimension. In J. D'Silva & J. Turner (Eds.), *Animals, ethics, and trade: The challenge of animal sentience* (pp. 208–216). London, UK: Earthscan.

Reinehr, C. O., & Costa, J. A. V. (2006). Repeated batch cultivation of the microalga Spirulina platensis. *World Journal of Microbiology & Biotechnology*, *22*(9), 937–943. doi:10.100711274-006-9138-0

Reiner, A. (2016). Teaching men to be emotionally honest. *The New York Times*. Retrieved from https://www.nytimes.com/2016/04/10/education/edlife/teaching-men-to-be-emotionally-honest.html

Renault, D., Laparie, M., McCauley, S. J., & Bonte, D. (2018). Environmental adaptations, ecological filtering, and dispersal central to insect invasions. *Annual Review of Entomology*, *63*(1), 345–368. doi:10.1146/annurev-ento-020117-043315 PMID:29029589

Repo–Carrasco–Valencia, R. A.-M., & Serna, L. A. (2011). Quinoa (Chenopodium quinoa, Willd.) as a source of dietary fibre and other functional components. *Food Science and Technology (Campinas), 31*(1), 225–230. doi:10.1590/S0101-20612011000100035

Reuters. (2018a). *Brazil boasts world's second largest genetically modified crop area: ISAAA. Environment.* Retrieved September 13, 2018 from https://www.reuters.com/article/us-brazil-gmo/brazil-boasts-worlds-second-largest-genetically-modified-crop-area-isaaa-idUSKBN1JN1KW

Reuters. (2018b). *CORRECTED-China launches "emergency" campaign to boost soy output.* Retrieved September 13, 2018 from https://www.reuters.com/article/china-soybeans/china-launches-emergency-campaign-to-boost-soy-output-idUSL3N1SA32I

Reuters. (2018c). *China grants more subsidies to soy farmers as it cuts corn stocks.* Retrieved September 13, 2018 from https://www.reuters.com/article/us-china-agriculture-subsidies/china-grants-more-subsidies-to-soy-farmers-as-it-cuts-corn-stocks-idUSKCN1HA176

Reynolds, M. (2018). The clean meat industry is racing to ditch its reliance on foetal blood. *Wired.* Retrieved from http://www.wired.co.uk/article/scaling-clean-meat-serum-just-finless-foods-mosa-meat

Richter, C. K., Skulas–Ray, A. C., Champagne, C. M., & Kris–Etherton, P. M. (2015). Plant protein and animal proteins: Do they differentially affect cardiovascular disease risk? *Advances in Nutrition, 6*(6), 712–728. doi:10.3945/an.115.009654 PMID:26567196

Righelato, R. C. (1976). Selection of strains of Penicillium chrysogenum with reduced penicillin yields in continuous cultures. *Journal of Chemical Technology and Biotechnology (Oxford, Oxfordshire), 26*(1), 153–159. doi:10.1002/jctb.5020260125

Ritchie, H., & Roser, M. (2017). *Meat and seafood production & consumption.* Our World in Data. Retrieved from https://ourworldindata.org/meat-and-seafood-production-consumption

Ritchie, H., & Roser, M. (2018). *Meat and seafood production & consumption.* Our World in Data. Retrieved September 12, 2018 from https://ourworldindata.org/meat-and-seafood-production-consumption

Ritchie, H., & Roser, M. (2018). *Meat and seafood production & consumption.* Retrieved from https://ourworldindata.org/meat-and-seafood-production-consumption

Ritzer, G. (2017). Can there really be "true" alternatives within the food and drink markets? If so, can they survive as alternative forms? *Journal of Marketing Management, 33*(7-8), 652–661. doi:10.1080/0267257X.2017.1312843

Riviere, J. (1977). Microbial proteins. In M. O. Moss & J. E. Smith (Eds.), *Industrial applications of microbiology* (pp. 105–149). London, UK: Surry University Press.

Roberts, S. (2013). Boys will be boys … Won't they? Change and continuities in contemporary young working-class masculinities. *Sociology, 47*(4), 671–686. doi:10.1177/0038038512453791

Roberts, S. (Ed.). (2015). *Debating modern masculinities: change, continuity, crisis?* Basingstoke, UK: Palgrave Macmillan.

Robinson, J. W. (2010). *William of Ockham's early theory of property rights: Sources, texts, and contexts.* Heritage Branch, Centre for Medieval Studies University of Toronto, Ottawa. Retrieved from http://individual.utoronto.ca/jwrobinson/articles/j Robinson_phd-thesis_ockham.pdf

Rodio, M. (2016). The Vegan diet of American Olympic weightlifter Kendrick Farris. *Men's Journal.* Retrieved from: https://www.mensjournal.com/food-drink/vegan-diet-american-olympic-weightlifter-kendrick-farris/

Rödl, M. B. (2018). *Why France banned meat names for vegetarian alternatives.* The Conversation. Retrieved September 11, 2018 from https://theconversation.com/why-france-banned-meat-names-for-vegetarian-alternatives-95549

Rödl, M. B. (2018). Marketing meat alternatives: Meat myths and their replication in advertising for plant-based meat. In D. Bogueva, D. Marinova, & T. Raphaely (Eds.), *Handbook of research on social marketing and its influence on animal origin food product consumption* (pp. 327–343). Hershey, PA: IGI Global. doi:10.4018/978-1-5225-4757-0.ch022

Rogers, R. (2008). Beasts, burgers and hummers: Meat and the crisis of masculinity in contemporary television advertising. *Environmental Communication, 2*(3), 281–301. doi:10.1080/17524030802390250

Rojas, C., & Mansur, E. (1995). Ecuador: Informaciones generals sobre productos non madereros in Ecuador. In *Memoria, consulta de expertos sobre productor forestales no madereros para America Latina y el Caribe* (pp. 208–223). Santiago, Chile: Regional Office for Latin America and the Caribbean, Food and Agriculture Organization of the United Nations.

Rook, D. (2009). Should great apes have 'human rights'? *Web Journal of Current Legal Issues, 1*, 1-10. Retrieved from http://www.bailii.org/uk/other/journals/WebJCLi/2009/issue1/ rook1.html

Rosa-Magri, M. M., Tauk-Tornisielo, S. M., & Ceccato-Antonini, S. R. (2011). Bioprospection of yeasts as biocontrol agents against phytopathogenic molds. *Brazilian Archives of Biology and Technology, 54*(1), 1–5. doi:10.1590/S1516-89132011000100001

Roser, M., & Ortiz-Ospina, E. (2018). *World population growth.* Our World in Data. Retrieved from https://ourworldindata.org/world-population-growth

Roser, M., & Ritchie, H. (2017). *Yields and land use in agriculture.* Our World in Data. Retrieved from https://ourworldindata.org/yields-and-land-use-in-agriculture

Roser, M., & Ritchie, H. (2018). *Hunger and undernourishment.* Our World in Data. Retrieved from https://ourworldindata.org/hunger-and-undernourishment

Rossi, V., Jovicevic, L., Nistico, V., Orticelli, G., Troiani, M. P., & Marini, S. (1993). In vitro antitumor activity of Lentinus edodes. *Pharmacological Research, 27*, 109–110. doi:10.1006/phrs.1993.1089

Ross, S. M. (2018). The nutrition-brain connection: Nutritional status and cognitive decline. *Holistic Nursing Practice, 32*(3), 169–171. doi:10.1097/HNP.0000000000000270 PMID:29642132

Rothgerber, H. (2013). Real men don't eat (vegetable) quiche: Masculinity and the justification of meat consumption. *Psychology of Men & Masculinity, 14*(4), 363–375. doi:10.1037/a0030379

Roy, F., Boye, J. I., & Simpson, B. K. (2010). Bioactive proteins and peptides in pulse crops: Pea, chickpea and lentil. *Food Research International, 43*(2), 432–442. doi:10.1016/j.foodres.2009.09.002

Rozin, P., Hormes, J. M., Faith, M. S., & Wansink, B. (2012). Is meat male? A quantitative multimethod framework to establish metaphoric relationships. *The Journal of Consumer Research, 39*(3), 629–643. doi:10.1086/664970

Rozin, P., Spranca, M., Krieger, Z., Neuhaus, R., Surillo, D., Swerdlin, A., & Wood, K. (2004). Preference for natural: Instrumental and ideational/moral motivations, and the contrast between foods and medicines. *Appetite, 43*(2), 147–154. doi:10.1016/j.appet.2004.03.005 PMID:15458801

Ruby, M. B., & Heine, S. J. (2011). Meat, morals, and masculinity. *Appetite, 56*(2), 447–450. doi:10.1016/j.appet.2011.01.018 PMID:21256169

Ruby, M. B., Rozin, P., & Chan, C. (2015). Determinants of willingness to eat insects in the USA and India. *Journal of Insects as Food and Feed, 1*(3), 215–225. doi:10.3920/JIFF2015.0029

Rumpold, B. A., & Schlüter, O. K. (2013). Nutritional composition and safety aspects of edible insects. *Molecular Nutrition & Food Research, 57*(5), 802–823. doi:10.1002/mnfr.201200735 PMID:23471778

Rumpold, B. A., & Schluter, O. K. (2013b). Potential and challenges of insects as an innovative source for food and feed production Innovative. *Food Science and Emerging Technologies, 17*, 1–11. doi:10.1016/j.ifset.2012.11.005

Ryu, J. P., Shin, J. T., Kim, J., & Kim, Y. W. (2017). Consumer preference for edible insect-containing cookies determined by conjoint analysis: An exploratory study of Korean consumers. *Entomological Research, 47*(2), 74–83. doi:10.1111/1748-5967.12203

Sabaté, J., & Jehi, T. (2018). The sustainability of vegetarian diets. In W. J. Craig (Ed.), *Vegetarian nutrition and wellness* (pp. 25–44). Boca Raton, FL: CRC Press.

Sabaté, J., & Wien, M. (2010). Vegetarian diets and childhood obesity prevention. *The American Journal of Clinical Nutrition, 91*(5), 1525S–1529S. doi:10.3945/ajcn.2010.28701F PMID:20237136

Sachs, J. D. (2015). *The age of sustainable development*. New York, NY: Columbia University Press. doi:10.7312ach17314

Sadiq, M. B., Ramanoon, S. Z., Mansor, R., Syed-Hussain, S. S., & Mossadeq, W. M. S. (2017). Prevalence of lameness, claw lesions, and associated risk factors in dairy farms in Selangor, Malaysia. *Tropical Animal Health and Production, 49*(8), 1741–1748. doi:10.100711250-017-1387-4 PMID:28856534

Sadler, M. J. (2004). Meat alternatives – market developments and health benefits. *Trends in Food Science & Technology, 15*(5), 250–260. doi:10.1016/j.tifs.2003.09.003

Saima, M. A., Khan, M., Anjum, M., Ahmed, S., Rizwan, M., & Ijaz, M. (2008). Investigation on the availability of amino acids from different animal protein sources in golden cockerels. *Journal of Animal and Plant Sciences, 18*(2-3), 53–55.

Sanda, D. (2018). Vegan trend takes hold in Australia. *The Guardian*. Retrieved September 12, 2018 from http://www.theguardian.com.au/story/5316808/vegan-trend-takes-hold-in-australia/

Santiago–Ramos, D., Figueroa–Cárdenas, J. de D., Véles–Medina, J. J., & Salazar, R. (2018). Physicochemical properties of nixtamalized black bean (Phaseolus vulgaris L.) flours. *Food Chemistry, 240*, 456–462. doi:10.1016/j.foodchem.2017.07.156 PMID:28946298

Santos, A., Navascués, E., Bravo, E., & Marquina, D. (2011). *Ustilago maydis* killer toxin as a new tool for the biocontrol of the wine spoilage yeast *Brettanomyces bruxellensis. International Journal of Food Microbiology, 145*(1), 147–154. doi:10.1016/j.ijfoodmicro.2010.12.005 PMID:21195497

Sara, R., Ralston, R. A., & Walker, K. Z. (2013). Adherence to a Mediterranean-style diet can slow the rate of cognitive decline and decrease the risk of dementia: A systematic review. *Nutrition & Dietetics: the Journal of the Dietitians Association of Australia, 70*, 206–217. doi:10.1111/1747-0080.12016

Sarich, C. (2013). Harvard says reducing red meat consumption can extend life by 20%. *Natural Society*. Retrieved from http://naturalsociety.com/reducing-red-meat-consumption-extend-life-20-percent/#ixzz2zdnAhdzY

Sathasivam, R., Radhakrishnan, R., Hashem, A., & Abd–Allah, E. F. (2017). (in press). Microalgae metabolites: A rich source for food and medicine. *Saudi Journal of Biological Sciences*. doi:10.1016/j.sjbs.2017.11.003

Satheesh, P. V. (2012). Genetically engineered promises & farming realities. In D. Bollier & S. Helfrich (Eds.), *The wealth of the commons: A world beyond market and state* (pp. 141–146). Amherst, MA: The Common Strategies Group.

Satija, A., Bhupathiraju, A. N., Rimm, E. B., Spiegelman, D., Chiuve, S. E., Borgi, L., ... Hu, F. B. (2016). Plant–based dietary patterns and incidence of type 2 diabetes in us men and women: Results from three prospective cohort studies. *PLoS Medicine, 13*(6), e1002039. doi:10.1371/journal.pmed.1002039 PMID:27299701

Satija, A., Bhupathiraju, S. N., Donna Spiegelman, D., Chiuve, S. E., Manson, J. E., Willett, W., ... Hu, F. B. (2017). Healthful and unhealthful plant-based diets and the risk of coronary heart disease in U.S. adults. *Journal of the American College of Cardiology, 70*(4), 411–422. doi:10.1016/j.jacc.2017.05.047 PMID:28728684

Sax, D. (2016). How years of macho food marketing is killing men. *NY Magazine*. Retrieved from http://nymag.com/betamale/2016/06/macho-food-marketing-is-killing-men.htm

Schiller, L. (2015). *Inflection point: The tectonic shift of gender roles*. Kalw Radio, San Francisco. Retrieved from http://kalw.org/post/inflection-point-tectonic-shift-gender-roles#stream/0

Schmidinger, K. (2012). *Worldwide alternatives to animal derived foods – Overview and evaluation models – Solutions to global problems caused by livestock* (Doctoral thesis). University of Natural Resources and Life Sciences, Vienna, Austria.

Schmidinger, K. (2012). *Worldwide alternatives to animal derived foods – overview and evaluation models: Solutions to global problems caused by livestock* (Doctoral dissertation). University of Natural Resources and Life Sciences, Vienna, Austria.

Schmidinger, K., Bogueva, D., & Marinova, D. (2018). New meat without livestock. In D. Bogueva, D. Marinova, & T. Raphaely (Eds.), *Handbook of research on social marketing and its influence on animal origin food product consumption* (pp. 344–361). Hershey, PA: IGI Global. doi:10.4018/978-1-5225-4757-0.ch023

Schonwald, J. (2014). The Frankenburger is coming sooner than you think thanks to Google. *Time Magazine*. Retrieved from http://time.com/3118571/lab-grown-meat-frankenburger-google/

Schösler, H., De Boer, J., & Boersema, J. J. (2012). Can we cut out the meat of the dish? Constructing consumer-oriented pathways towards meat substitution. *Appetite, 58*(1), 39–47. doi:10.1016/j.appet.2011.09.009 PMID:21983048

Schösler, H., de Boer, J., Boersema, J. J., & Aiking, H. (2015). Meat and masculinity among young Chinese, Turkish and Dutch adults in the Netherlands. *Appetite, 89*, 152–159. doi:10.1016/j.appet.2015.02.013 PMID:25681291

Schouteten, J. J., De Steur, H., De Pelsmaeker, S., Lagast, S., Juvinal, J. G., De Bourdeaudhuij, I., ... Gellynck, X. (2016). Emotional and sensory profiling of insect-, plant-and meat-based burgers under blind, expected and informed conditions. *Food Quality and Preference, 52*, 27–31. doi:10.1016/j.foodqual.2016.03.011

Schrock, D., & Schwalbe, M. (2009). Men, masculinity, and manhood acts. *Annual Review of Sociology, 35*(1), 277–295. doi:10.1146/annurev-soc-070308-115933

Schwan, R. F., & Wheals, A. E. (2003). Mixed microbial fermentations of chocolate and coffee. In T. Boekhout & V. Robert (Eds.), *Yeasts in food: Beneficial and detrimental aspects* (pp. 429–449). Hamburg, Germany: Behr's Verlag. doi:10.1533/9781845698485.429

Scotch Porter. (2017). Introducing fresh meat masculine wipes. *The Scotch Porter Journal*. Retrieved from http://journal.scotchporter.com/2017/11/introducing-fresh-meat-masculine-wipes/

Scott, S. E., Inbar, Y., & Rozin, P. (2016). Evidence for absolute moral opposition to genetically modified food in the United States. *Perspectives on Psychological Science, 11*(3), 315–324. doi:10.1177/1745691615621275 PMID:27217243

Scruggs, G. (2018). *'Everything we've heard about global urbanization turns out to be wrong' – researchers*. Place. Retrieved from http://www.thisisplace.org/i/?id=0150beca-e3f5-47e0-bc74-9ccc5ef1db8a&utm_source=Newsletter&utm_medium=Email&utm_campaign=Place

Shaabani, S., Yarmand, M. S., Kiani, H., & Emam–Djomeh, Z. (2018). The effect of chickpea protein isolate in combination with transglutaminase and xanthan on the physical and rheological characteristics of gluten free muffins and batter based on millet flour. *Lebensmittel-Wissenschaft + Technologie*, *90*, 362–372. doi:10.1016/j.lwt.2017.12.023

Shams-White, M. M., Chung, M., Du, M., Fu, Z., Insogna, K. L., Karlsen, M. C., ... Weaver, C. M. (2017). Dietary protein and bone health: A systematic review and meta-analysis from the National Osteoporosis Foundation. *The American Journal of Clinical Nutrition*, *105*(6), 1528–1543. doi:10.3945/ajcn.116.145110 PMID:28404575

Shang, X., Scott, D., Hodge, A. M., English, D. R., Giles, G. G., Ebeling, P. R., & Sanders, K. M. (2016). Dietary protein intake and risk of type 2 diabetes: Results from the Melbourne Collaborative Cohort Study and a meta-analysis of prospective studies. *The American Journal of Clinical Nutrition*, *104*(5), 1352–1365. doi:10.3945/ajcn.116.140954 PMID:27629053

Shankar, B., Agrawal, S., Beadreault, A. R., Avula, L., Martorell, R., Osendarp, S., ... McLean, M. S. (2017). Dietary and nutritional change in India: Implications for strategies, policies, and interventions. *Annals of the New York Academy of Sciences*, *1395*(1), 49–59. doi:10.1111/nyas.13324 PMID:28306140

Shanker, D. (2015). *The US meat industry's wildly successful, 40-year crusade to keep its hold on the American diet.* Quartz Media. Retrieved from https://qz.com/523255/the-us-meat-industrys-wildly-successful-40-year-crusade-to-keep-its-hold-on-the-american-diet/

Shapiro, P. (2018). *Clean meat: How growing meat without animals will revolutionize dinner and the world.* New York, NY: Gallery Books.

Shaw, J. (2012). A diabetes link to meat. *Harvard Magazine*. Retrieved from https://harvardmagazine.com/2012/01/a-diabetes-link-to-meat

Shaw, D., Newholm, T., & Dickinson, R. (2006). Consumption as voting: An exploration of consumer empowerment. *European Journal of Marketing*, *40*(9/10), 1049–1067. doi:10.1108/03090560610681005

Sheldon, I. (2018). Why China's soybean tariffs matter. *The Conversation*. Retrieved September 13, 2018 from http://theconversation.com/why-chinas-soybean-tariffs-matter-94476

Shelomi, M. (2015). Why we still don't eat insects: Assessing entomophagy promotion through a diffusion of innovations framework. *Trends in Food Science & Technology*, *45*(2), 311–318. doi:10.1016/j.tifs.2015.06.008

Shepon, A., Eshel, G., Noor, E., & Milo, R. (2018). The opportunity cost of animal based diets exceeds all food losses. *Proceedings of the National Academy of Sciences of the United States of America*, *15*(15), 3804–3809. doi:10.1073/pnas.1713820115 PMID:29581251

Shestack, J. (1998). The philosophic foundations of human rights. *Human Rights Quarterly*, *20*(2), 201–234. doi:10.1353/hrq.1998.0020

Sheu, F., Chien, P. J., Chien, A. L., Chen, Y. F., & Chin, K. L. (2004). Isolation and characterization of an immunomodulatory protein (APP) from the Jew's Ear mushroom Auricularia polytricha. *Food Chemistry*, *87*(4), 593–600. doi:10.1016/j.foodchem.2004.01.015

Shimamatsu, H. (2004). Mass production of Spirulina, an edible microalga. *Hydrobiologia*, *512*(1–3), 39–44. doi:10.1023/B:HYDR.0000020364.23796.04

Shin, D., & Jeong, D. (2015). Korean traditional fermented soybean products: Jang. *Journal of Ethnic Foods*, *2*(1), 2–7. doi:10.1016/j.jef.2015.02.002

Shoard, C. (2016). Arnold Schwarzenegger and James Cameron urge people to eat less meat. *The Guardian*. Retrieved from https://www.theguardian.com/film/2016/jun/23/arnold-schwarzenegger-james-cameron-eat-less-meat-china

Shoemaker, N. (2017). Price of lab-grown burger falls from $325K to $11.36. *Big Think*. Retrieved from https://bigthink.com/ideafeed/answering-how-a-sausage-gets-made-will-be-more-complicated-in-2020

Shprintzen, A. D. (2012). Looks like meat, smells like meat, tastes like meat: Battle creek, protose and the making of modern American vegetarianism. *Food, Culture, & Society*, *15*(1), 113–128. doi:10.2752/175174412XI3190510222066

Shprintzen, A. D. (2013). *The Vegetarian Crusade: The Rise of an American Reform Movement, 1817–1921*. Chapel Hill, NC: The University of North Carolina Press.

Shurtleff, W., & Aoyagi, A. (2015). *History of Soybeans and Soyfoods in the United Kingdom and Ireland (1613-2015). Includes England, Great Britain, Scotland and Wales*. Lafayette, CA: Soyinfo Center. Retrieved September 10, 2018 from http://www.soyinfocenter.com/pdf/183/UK.pdf

Shurtleff, W., & Aoyagi, A. (2004). *Dr. John Harvey Kellogg and Battle Creek Foods: Work with soy*. Lafayette, CA: Soyinfo Center. Retrieved from http://www.soyinfocenter.com/HSS/john_kellogg_and_battle_creek_foods.php

Shurtleff, W., & Aoyagi, A. (2011). *History of tempeh and tempeh products (1815-2011): Extensively annotated bibliography and sourcebook*. Lafayette, CA: Soyinfo Center.

Shurtleff, W., & Aoyagi, A. (2014). *History of meat alternatives (965 CE to 2014): Extensively annotated bibliography and sourcebook*. Lafayette, CA: Soyinfo Center. Retrieved from http://www.soyinfocenter.com/pdf/179/MAL.pdf

Sialve, B., Bernet, N., & Bernard, O. (2009). Anaerobic digestion of microalgae as a necessary step to make microalgal biodiesel sustainable. *Biotechnology Advances*, *2*(4), 409–416. doi:10.1016/j.biotechadv.2009.03.001 PMID:19289163

Siegrist, M., & Sütterlin, B. (2017). Importance of perceived naturalness for acceptance of food additives and cultured meat. *Appetite*, *113*, 320–326. doi:10.1016/j.appet.2017.03.019 PMID:28315418

Siegrist, M., Sütterlin, B., & Hartmann, C. (2018). Perceived naturalness and evoked disgust influence acceptance of cultured meat. *Meat Science*, *139*, 213–219. doi:10.1016/j.meatsci.2018.02.007 PMID:29459297

Sileshi, G. W., & Kenis, M. (2010). Food security: Farming insects. *Science*, *328*(5978), 568–568. doi:10.1126cience.328.5978.568-a PMID:20430996

Simon, D. R. (2017). Why is government sponsoring meat products that can cause cancer? *The Sydney Morning Herald*. Retrieved from https://www.smh.com.au/ opinion/why-is-government-sponsoring-meat-products-that-can-cause-cancer-20170503-gvy5bd.html

Simon, M. (2018, February 16). Lab-grown meat is coming, whether you like it or not. *Wired*. Retrieved from https://www.wired.com/story/lab-grown-meat/

Simon, D. R. (2013). *Meatonomics: How the rigged economics of meat and dairy make you consume too much–and how to eat better, live longer, and spend smarter*. San Francisco, CA: Conari Press.

Singer, P. (1975). *Animal liberation*. London, UK: Harper Collins.

Singer, P. (1975). *Animal liberation: A new ethics for our treatment of animals*. London, UK: Harper Collins.

Singh, S. (2017). Bill Gates and Richard Branson back startup that grows 'clean meat'. *Bloomberg*. Retrieved from https://www.bloomberg.com/news/articles/2017-08-23/cargill-bill-gates-bet-on-startup-making-meat-without-slaughter

Singh, B. P., Vij, S., & Hati, S. (2014). Functional significance of bioactive peptides derived from soybean. *Peptides*, *54*, 171–179. doi:10.1016/j.peptides.2014.01.022 PMID:24508378

Singh, P. N., & Fraser, G. E. (1998). Dietary risk factors for colon cancer in a low-risk population. *American Journal of Epidemiology*, *148*(8), 761–774. doi:10.1093/oxfordjournals.aje.a009697 PMID:9786231

Siva, N., Thavarajah, D., Johnson, C. R., Duckett, S., Jesch, E. D., & Thavarajah, P. (2017). Can lentil (Lens culinaris Medikus) reduce the risk of obesity? *Journal of Functional Foods*, *38*, 706–715. doi:10.1016/j.jff.2017.02.017

Slade, P. (2018). If you build it, will they eat it? Consumer preferences for plant-based and cultured meat burgers. *Appetite*, *125*, 428–437. doi:10.1016/j.appet.2018.02.030 PMID:29501683

Smil, V. (2002). Worldwide transformation of diets, burdens of meat production and opportunities for novel food proteins. *Enzyme and Microbial Technology*, *30*(3), 305–311. doi:10.1016/S0141-0229(01)00504-X

Smithers, R. (2018, June 10). Sainsbury's to sell vegan 'fake meats' next to real thing in trial. *The Guardian*. Retrieved June 24, 2018, from http://www.theguardian.com/lifeandstyle/2018/jun/11/sainsburys-to-sell-vegan-fake-meats-next-to-real-thing-in-trial

Sobal, J. (2005). Men, meat and marriage: Models of masculinity. *Food & Foodways*, *13*(1-2), 135–158. doi:10.1080/07409710590915409

Sogari, G., & Vantomme, P. (2014). *A tavola con gli insetti*. Fidenza, Italy: Mattioli 1885.

Sogari, G. (2015). Entomophagy and Italian consumers: An exploratory analysis. *Progress in Nutrition*, *17*(4), 311–316.

Sogari, G., Menozzi, D., & Mora, C. (2017). Exploring young foodies' knowledge and attitude regarding entomophagy: A qualitative study in Italy. *International Journal of Gastronomy and Food Science*, *7*, 16–19. doi:10.1016/j.ijgfs.2016.12.002

Sogari, G., Menozzi, D., & Mora, C. (2018). Sensory-liking expectations and perceptions of processed and unprocessed insect products. *International Journal on Food System Dynamics*, *9*(3), 226–239. doi:10.18461/ijfsd.v9i4.942

Solomons, G. L. (1987). Myco-protein: Safety evaluation of a novel food. In C. M. Chambers, P. L. Chambers, & D. S. Davies (Eds.), *Mechanisms and models in toxicology. Archives of Toxicology* (Vol. 11, pp. 191–193). Berlin, Germany: Springer. doi:10.1007/978-3-642-72558-6_30

Soni, R. A., Sudhakar, K., & Rana, R. S. (2017). Spirulina – from growth to nutritional product: A review. *Trends in Food Science & Technology*, *69*, 157–171. doi:10.1016/j.tifs.2017.09.010

Soybeans Association of Heilongjiang Province. (2018). Document No. 1 of the Central Committee in 2017: Several suggestions on deepening the structural reform of the supply side of agriculture and accelerating the cultivation of new momentum of agricultural and rural development. *Economy Daily*. Retrieved September 13, 2018 from http://www.chinasoy.com.cn/Page_Views/AllView.aspx?ID=2716&cateorgy=%e7%89%b9%e5%88%ab%e5%85%b3%e6%b3%a8

Specht, L. (2017). *Mapping emerging industries. Opportunities in clean meat.* The Good Food Institute. Retrived December 13, 2017, from www.gfi.org/images/uploads/2017/06/Mapping-Emerging-Industries.pdf

Spolaore, P., Joannis-Cassan, C., Duran, E., & Isambert, A. (2006). Commercial applications of microalgae. *Journal of Bioscience and Bioengineering*, *101*(2), 87–96. doi:10.1263/jbb.101.87 PMID:16569602

Springmann, M., Godfray, H. C. J., Rayner, M., & Scarborough, P. (2016). Analysis and valuation of the health and climate change cobenefits of dietary change. *Proceedings of the National Academy of Sciences of the United States of America*, *113*(15), 4146–4151. doi:10.1073/pnas.1523119113 PMID:27001851

Springmann, M., Mason-D'Croz, D., Robinson, S., Garnett, T., Godfray, H. C. J., Gollin, D., ... Scarborough, P. (2016). Global and regional health effects of future food production under climate change: A modelling study. *Lancet*, *387*(10031), 1937–1346. doi:10.1016/S0140-6736(15)01156-3 PMID:26947322

Steinbrecher, A., Erber, E., Grandinetti, A., Kolonel, L. N., & Maskarinec, G. (2010). Meat consumption and risk of type 2 diabetes: The Multiethnic Cohort. *Public Health Nutrition*, *14*(4), 568–574. doi:10.1017/S1368980010002004 PMID:20624337

Steinfeld, H., Gerber, P., Wassenaar, T., Castel, V., Rosales, M., & de Haan, C. (2006). *Livestock's long shadow. Environmental issues and options*. Rome, Italy: Food and Agriculture Organization of the United Nations (FAO). Retrieved September 10, 2018 from http://www.fao.org/docrep/010/a0701e/a0701e00.HTM

Steinfeld, H., Gerber, P., Wassenaar, T., Castel, V., Rosales, M., & de Haan, C. (2006). *Livestock's long shadow: Environmental issues and options*. Rome, Italy: Food and Agricultural Organization of the United Nations. Retrieved from www.virtualcentre.org/en/library/key_pub/longshad/A0701E00.pdf

Stryker, J. (2007). Vegansexuality. *New York Times*. Retrieved from https://www.nytimes.com/2007/12/09/magazine/09vegansexuality.html?_r=0

Stubbs, R. J., Scott, S. E., & Duarte, C. (2018). Responding to food, environment and health challenges by changing meat consumption behaviours in consumers. *Nutrition Bulletin*, *43*(2), 125–134. doi:10.1111/nbu.12318

Sumpter, K. C. (2015). Masculinity and meat consumption: An analysis through the theoretical lens of hegemonic masculinity and alternative masculinity theories. *Sociology Compass*, *9*(2), 104–114. doi:10.1111oc4.12241

Sun, X., Guo, Y., Wang, S., & Sun, J. (2007). Social marketing improved the consumption of iron-fortified soy sauce among women in China. *Journal of Nutrition Education and Behavior*, *39*(6), 302–310. doi:10.1016/j.jneb.2007.03.090 PMID:17996625

Sun, X., Li, W., Hu, Y., Zhou, X., Ji, M., Yu, D., ... Luan, G. (2018). Comparison of pregelatinization methods on physicochemical, functional and structural properties of tartary buckwheat flour and noodle quality. *Journal of Cereal Science*, *80*, 63–71. doi:10.1016/j.jcs.2018.01.016

Sustainable Development Knowledge Platform. (n.d.). *Transforming our world: The 2030 Agenda for Sustainable Development*. Retrieved from https://sustainabledevelopment.un.org/post2015/transformingourworld

Sutherland, I. A., & Leathwick, D. M. (2011). Anthelmintic resistance in nematode parasites of cattle: A global issue? *Trends in Parasitology*, *4*(4), 176–181. doi:10.1016/j.pt.2010.11.008 PMID:21168366

Sydney Criminal Lawyers. (2017). *NSW government silences animal rights activists*. Retrieved from https://www.sydneycriminallawyers.com.au/blog/nsw-government-silences-animal-rights-activists/

Sytar, O., Brestic, M., & Rai, M. (2013). Possible ways of fagopyrin biosynthesis and production in buckwheat plants. *Fitoterapia*, *84*(1), 72–79. doi:10.1016/j.fitote.2012.10.008 PMID:23103298

Talaei, M., Wang, Y. L., Yuan, J. M., Pan, A., & Koh, W. P. (2017). Meat, dietary heme iron, and risk of type 2 diabetes mellitus. *American Journal of Epidemiology*, *186*(7), 824–833. doi:10.1093/aje/kwx156 PMID:28535164

Tamang, J. P. (2015). Naturally fermented ethnic soybean foods of India. *Journal of Ethnic Foods*, *2*(1), 8–17. doi:10.1016/j.jef.2015.02.003

Tan, H. S. G., Fischer, A. R. H., Tinchan, P., Stieger, M., Steenbekkers, L. P. A., & Trijp, H. C. M. (2015). Insects as food: Exploring cultural exposure and individual experience as determinants of acceptance. *Food Quality and Preference*, *42*, 78–89. doi:10.1016/j.foodqual.2015.01.013

Tan, H. S. G., van den Berg, E., & Stieger, M. (2016). The influence of product preparation, familiarity and individual traits on the consumer acceptance of insects as food. *Food Quality and Preference*, *52*, 222–231. doi:10.1016/j.foodqual.2016.05.003

Tan, H., Verbaan, Y., & Stieger, M. (2017). How will better products improve the sensory-liking and willingness to buy insect-based foods? *Food Research International*, *92*, 95–105. doi:10.1016/j.foodres.2016.12.021 PMID:28290303

Tappel, A. (2007). Heme of consumed red meat can act as a catalyst of oxidative damage and could initiate colon, breast and prostate cancers, heart disease and other diseases. *Medical Hypotheses*, *68*(3), 562–564. doi:10.1016/j.mehy.2006.08.025 PMID:17045417

Tashjian, L. (2015). *Vegan Strongman Patrik Baboumian breaks his own world record!* Clearly Veg. Retrieved from https://www.clearlyveg.org/blog/2015/09/23/vegan-strongman-patrik-baboumian-breaks-his-own-world-record/

Taylor, E., & Butt, A. (2017). Three charts on Australia's declining taste for beef and growing appetite for chicken. *The Conversation*. Retrieved from https://the conversation.com/three-charts-on-australias-declining-taste-for-beef-and-growing-appetite-for-chicken-78100

Technology, A. I. C. (2017). *How to create a powerful social media strategy for food products*. Retrieved September 13, 2018 from https://aictechnologies.com.au/2017/05/23/selling-food-products-social-media/

Tee, R. D., Gordon, D. J., Welch, J. A., & Taylor, A. N. (1993). Investigation of possible adverse allergic reactions to mycoprotein ('Quorn'). *Clinical and Experimental Allergy*, *23*(4), 257–260. doi:10.1111/j.1365-2222.1993.tb00319.x PMID:8319120

Thavarajah, P., Thavarajah, D., & Vandenberg, A. (2009). Low phytic acid lentils (Lens culinaris L.): A potential solution for increased micronutrient bioavailability. *Journal of Agricultural and Food Chemistry*, *57*(19), 9044–9049. doi:10.1021/jf901636p PMID:19725537

The Dutch Soy Coalition. (2006). *Soy: Big business, big responsibility. Addressing the social- and environmental impact of the soy value chain*. Retrieved September 13, 2018 from http://www.bothends.org/uploaded_files/document/2006_Soy_big_business.pdf

The Economist. (2015). Silicon Valley gets a taste for food. Green Food. *The Economist*. Retrieved from https://www.economist.com/technology-quarterly/2015/03/05/silicon-valley-gets-a-taste-for-food

The European Food Information Council (EUFIC). (2006). *The determinants of food choice*. Retrieved from http://www.eufic.org/en/healthy-living/article/the-determinants-of-food-choice

The Good Food Institute. (n.d.). *The future of meat*. Retrieved from https://cleanmeat.org

The Grocer. (1993a, April 3). Mince Alternative to Beef up Veggie Sales - Haldane Foods. *The Grocer*, p. 33. Retrieved from the Factiva database: https://www.dowjones.com/products/factiva/

The Grocer. (1993b, August 21). McCartney Revamp Targets Bird's Eye's Pole Position. *The Grocer*, p. 35. Retrieved from the Factiva database: https://www.dowjones.com/products/factiva/

The Grocer. (1993c, June 12). Focus on Healthy Food - Tv Backs McCartney Ready Meals - Ross Young's. *The Grocer*, p. 82. Retrieved from the Factiva database: https://www.dowjones.com/products/factiva/

The Grocer. (1994, July 30). Focus on Vegetarian Foods - Lentil and Nut Roast Image Is History. *The Grocer*, p. 41. Retrieved from the Factiva database: https://www.dowjones.com/products/factiva/

The Grocer. (1995, July 29). Marketing - Marlow Foods. *The Grocer*, p. 41. Retrieved from the Factiva database: https://www.dowjones.com/products/factiva/

The Grocer. (1996, October 5). Mainstream Battle for Meat-Free Brands. *The Grocer*, p. 32. Retrieved from the Factiva database: https://www.dowjones.com/products/factiva/

The Grocer. (1999, March 13). The great pretenders. *The Grocer*, p. 47. Retrieved from the Factiva database: https://www.dowjones.com/products/factiva/

The Grocer. (2001, June 2). Steak-loving Americans will soon have a new word to add to t. *The Grocer*, p. 38. Retrieved from the Factiva database: https://www.dowjones.com/products/factiva/

The Grocer. (2003, April 12). Basking in mainstream approval. *The Grocer*, p. 47. Retrieved from the Factiva database: https://www.dowjones.com/products/factiva/

The Grocer. (2006, February 4). Meat-free not just for vegetarians. *The Grocer*, p. 47. Retrieved from the Factiva database: https://www.dowjones.com/products/factiva/

The Guardian. (2005). *Lions save girl from kidnappers*. Retrieved from https://www.theguardian.com/world/2005/jun/22/3

The Guardian. (2013). First lab-grown hamburger gets full marks for 'mouth feel'. *The Guardian Australian Edition*. Retrieved September 10, 2018 from https://www.theguardian.com/science/2013/aug/05/world-first-synthetic-hamburger-mouth-feel

The Pew Charitable Trusts. (2013). *Industrial food animal production in America*. Retrived December 27, 2017 from https://www.jhsph.edu/research/centers-and-institutes/johns-hopkins-center-for-a-livable-future/_pdf/research/clf_reports/CLF-PEW-for%20Web.pdf

The Statistics Portal. (2018). *Import volume of soybeans worldwide from 2013/14 to 2017/18, by country (in million metric tons)*. Statista. Retrieved September 13, 2018 from https://www.statista.com/statistics/612422/soybeans-import-volume-worldwide-by-country/

The Tissue Culture and Art Project. (n.d.). Retrieved Septemver 10, 2018 from http://lab.anhb.uwa.edu.au/tca/

The Vegan Society. (2016, May 17). Find out how many vegans are in Great Britain. *The Vegan Society*. Retrieved March 8, 2018, from https://www.vegansociety.com/whats-new/news/find-out-how-many-vegans-are-great-britain

Thomas, C. S. (2015). *Changing diets: Plant protein sustainability crucial for food security*. Retrieved September 19, 2018 from https://www.foodnavigator.com/Article/2015/01/21/Changing-diets-Plant-protein-sustainability-crucial-for-food-security?utm_source=copyright&utm_medium=OnSite&utm_campaign=copyright

Thompson, C. M. (2012*). The Map of Consciousness – Hawkin's scale*. Stankov's Universal Law Press. Retrieved September 14, 2018 fromhttp://www.stankovuniversallaw.com/2012/01/the-map-of-consciousness-hawkins-scale/

Thrane, U., Andersen, B., Frisvad, J. C., & Smedsgaard, J. (2007). The exo-metabolome in filamentous fungi. In J. Nielsen & J. Hewitt (Eds.), *Metabolomics: A powerful tool in systems biology* (pp. 235–252). Berlin, Germany: Springer. doi:10.1007/4735_2007_0230

Tilman, D., & Clark, M. (2014). Global diets link environmental sustainability and human health. *Nature*, *515*(7528), 518–522. doi:10.1038/nature13959 PMID:25383533

Timilsena, Y. P., Adhikari, R., Barrow, C. J., & Adhikari, B. (2016). Physicochemical and functional properties of protein isolate produced from Australian chia seeds. *Food Chemistry*, *212*, 648–656. doi:10.1016/j.foodchem.2016.06.017 PMID:27374580

Timilsena, Y. P., Vongsvivut, J., Adhikari, R., & Adhikari, B. (2017). Physicochemical and thermal characteristics of Australian chia seed oil. *Food Chemistry, 228*, 394–402. doi:10.1016/j.foodchem.2017.02.021 PMID:28317740

Timke, E., & O'Barr, W. M. (2017). Representation of masculinity and femininity in advertising. *Advertising & Society Review, 17*(3-4). doi:10.1353/asr.2017.0004

Tiplady, C., Walsh, D. B., & Phillips, C. J. C. (2012). Cruelty to Australian cattle in Indonesian abattoirs – how the public responded to media coverage. *Journal of Agricultural & Environmental Ethics, 26*, 869–885. doi:10.100710806-012-9412-0

Tolonen, E. (2018). Meat substitution in the Netherlands: How to further accelerate the Green Protein Transition from an innovation perspective (MSc thesis). Delft University of Technology & Leiden University, The Netherlands.

Tonstad, S., Butler, T., Yan, R., & Fraser, G. E. (2009). Type of vegetarian diet, body weight, and prevalence of type 2 diabetes. *Diabetes Care, 32*(5), 791–796. doi:10.2337/dc08-1886 PMID:19351712

Tosi, E., & Daccordi, M. (1983). Insetti commestibili. Slurp! *Nigrizia, 101*(6), 49–52.

Tovar, D., Zambonino, J., Cahu, C., Gatesoupe, F. J., Vázquez-Juárez, R., & Lésel, R. (2002). Effect of live yeast incorporation in compound diet on digestive enzyme activity in sea bass (*Dicentrarchus labrax*) larvae. *Aquaculture (Amsterdam, Netherlands), 204*(1-2), 113–123. doi:10.1016/S0044-8486(01)00650-0

Trinci, A. P. J. (1992). Myco-protein: A twenty-year overnight success story. *Mycological Research, 96*(1), 1–13. doi:10.1016/S0953-7562(09)80989-1

Tseng, C. K. (2004). The past, present and future of phycology in China. *Hydrobiologia, 512*(1-3), 11–20. doi:10.1023/B:HYDR.0000020363.37807.ec

Tubiello, F. N., Salvatore, M., Cóndor Golec, R. D., Ferrara, A., Rossi, S., Biancalani, R., … Flammin, A. (2014). *Agriculture, forestry and other land use emissions by sources and removals by sinks 1990–2011 analysis*. FAO Statistics Division. Working Paper Series ESS/14-02. Retrieved February 23, 2018 from http://www.fao.org/docrep/019/i3671e/i3671e.pdf

Tucker, C. (2018). Using environmental imperatives to reduce meat consumption: Perspectives from New Zealand. *Kōtuitui: New Zealand Journal of Social Sciences, 13*(1), 99–110. doi:10.1080/1177083X.2018.1452763

Tucker, C. (Ed.). (2018). *Advances in cattle welfare*. Duxford, UK: Woodhead Publishing.

Tucker, C. A. (2014). The significance of sensory appeal for reduced meat consumption. *Appetite, 81*, 168–179. doi:10.1016/j.appet.2014.06.022 PMID:24953197

Tucker, K. L. (2014). Vegetarian diets and bone status. *The American Journal of Clinical Nutrition, 100*(Supplement 1), 329S–335S. doi:10.3945/ajcn.113.071621 PMID:24898237

Tuomisto, H. L., Ellis, M. J., & Haastrup, P. (2014). Environmental impacts of cultured meat: alternative production scenarios. In R. Schenck, & D. Huizenga (Eds.), *Proceedings of the 9th International Conference on Life Cycle Assessment in the Agri-Food Sector (LCA Food 2014), 8-10 October 2014, San Francisco, USA* (pp. 1360–1366). Vashon, WA: ACLCA. Retrieved from https://core.ac.uk/download/pdf/38629617.pdf

Tuomisto, H., Ellis, M., & Haastrup, P. (2014). Environmental impacts of cultured meat: Alternative production scenarios. *Proceedings of the 9th International Conference on Life Cycle Assessment in the Agri-Food Sector*, 1360–1366. Retrieved September 10, 2018 from http://lcafood2014.org/papers/132.pdf

Tuomisto, H., & Teixeira de Mattos, M. J. (2011). Environmental impacts of cultured meat production. *Environmental Science & Technology, 45*(14), 6117–6123. doi:10.1021/es200130u PMID:21682287

Turnbull, W. H., Leeds, A. R., & Edwards, D. G. (1992). Mycoprotein reduces blood lipids in free-living subjects. *The American Journal of Clinical Nutrition*, *55*(2), 415–419. doi:10.1093/ajcn/55.2.415 PMID:1734679

Turnbull, W. H., Walton, J., & Leeds, A. R. (1993). Acute effects of mycoprotein on subsequent energy intake and appetite variables. *The American Journal of Clinical Nutrition*, *58*(4), 507–512. doi:10.1093/ajcn/58.4.507 PMID:8257542

Turner-McGrievy, G., Mandes, T., & Crimarco, A. (2017). A plant-based diet for overweight and obesity prevention and treatment. *Journal of Geriatric Cardiology: JGC*, *14*(5), 369–374. doi:10.11909/j.issn.1671-5411.2017.05.002 PMID:28630616

Turner, N. J., Kuhnlein, H. V., & Egger, K. N. (1987). The cottonwood mushroom (*Tricholoma populinum*): A food resource of the Interior Salish Indian peoples of British Columbia. *Canadian Journal of Botany*, *65*(5), 921–927. doi:10.1139/b87-127

Tuttle, W. (2005). *The world peace diet: Eating for spiritual health and social harmony*. New York: Lantern Books.

U.S. Soybean Export Council. (2006). *U.S. Soy: International Buyers' Guide*. Retrieved September 12, 2018 from https://ussec.org/wp-content/uploads/2015/10/buyers-guide.pdf

Ugalde, U. O., & Castrillo, J. I. (2002). Single cell proteins from fungi and yeasts. In G. G. Khachatourians & D. K. Arora (Eds.), *Applied mycology and biotechnology* (Vol. 2, pp. 123–149). Amsterdam: Elsevier.

Ugwu, C., Ogbonna, J., & Tanaka, H. (2002). Improvement of mass transfer characteristics and productivities of inclined tubular photobioreactors by installation of internal static mixers. *Applied Microbiology and Biotechnology*, *58*(5), 600–607. doi:10.100700253-002-0940-9 PMID:11956741

Uhe, A. M., Collier, G. R., & Odea, K. (1992). A comparison of the effects of beef, chicken and fish-protein on satiety and amino-acid profiles in lean male subjects. *The Journal of Nutrition*, *122*(3), 467–472. doi:10.1093/jn/122.3.467 PMID:1542005

Ullivarri, M. F., Mendoza, L. M., & Raya, R. R. (2014). Killer activity of *Saccharomyces cerevisiae* strains: Partial characterization and strategies to improve the biocontrol efficacy in winemaking. *Antonie van Leeuwenhoek*, *106*(5), 865–878. doi:10.100710482-014-0256-7 PMID:25145824

Ulluwishewa, R. (2015). *Spirituality demystified: Understanding spirituality in rational terms*. Palmerston, New Zealand: Sairon Books.

United Nations. (1948). *The universal declaration of human rights*. General Assembly Resolution 217A (1948, December 10). Retrieved from http://www.un.org/en/universal-declaration-human-rights/

United Nations. (2009). *Food production must double by 2050 to meet demand from world's growing population, innovative strategies needed to combat hunger, experts tell second committee*. Retrieved from https://www.un.org/press/en/2009/gaef3242.doc.htm

United Nations. (2015). *Transforming our world: The 2030 agenda for sustainable development*. General Assembly (2015, September 25). Retrieved from http://www.un.org/en/ development/desa/population/migration/generalassembly/docs/globalcompact/A_RES_70_1_E.pdf

United Nations. (2018). *The Sustainable Development Goals Report 2018*. Statistical Division. Retrieved from https://unstats.un.org/sdgs/report/2018/overview/

United States Department of Agriculture (USDA). (2010). *USDA Climate Change Science Plan*. Retrieved September 19, 2018 from https://www.usda.gov/oce/climate_change/science_plan2010/USDA_CCSPlan_120810.pdf

United States Department of Agriculture (USDA). (2012). *The Pherobase: Database of pheromones and semiochemicals*. Retrieved from http://www.pherobase.com/

United States Department of Agriculture (USDA). (2015). *Dietary guidelines for Americans 2015–2020* (8th ed.). Retrieved April 01, 2018 from https://health.gov/dietaryguidelines/2015/resources/2015-2020_Dietary_Guidelines.pdf

University of California, Riverside. (2016). *Entomophagy (Eating insects)*. San Diego, CA: Ceter for Invasive Species Research. Retrieved from http://cisr.ucr.edu/entomophagy.html

Vagadia, B. H., Vanga, S. K., & Raghavan, V. (2017). Inactivation methods of soybean trypsin inhibitor – A review. *Trends in Food Science & Technology*, *64*, 115–125. doi:10.1016/j.tifs.2017.02.003

Valcheva, R., & Dieleman, L. A. (2016). Prebiotics: Definition and protective mechanisms. *Best Practice & Research. Clinical Gastroenterology*, *30*(1), 27–37. doi:10.1016/j.bpg.2016.02.008 PMID:27048894

Van de Kamp, M. E., Seves, S. M., & Temme, E. H. (2018). Reducing GHG emissions while improving diet quality: Exploring the potential of reduced meat, cheese and alcoholic and soft drinks consumption at specific moments during the day. *BMC Public Health*, *18*(1), 264. doi:10.118612889-018-5132-3 PMID:29458352

Van Den Bulck, H. (2010). Qualitative content analysis. In C. Newbold, O. Boyd-Barrett, & H. Van Den Bulck (Eds.), *The media book* (pp. 83–88). London, UK: Bloomsbury.

van den Ende, H., & Stegwee, D. (1971). Physiology of sex in Mucorales. *Botanical Review*, *37*(1), 22–36. doi:10.1007/BF02860299

Van Der Walt, G. (2008). *Brutal or amazing. You decide…* Retrieved from http://photo-africa.blogspot.com/2008/09/brutal-or-amazing-you-decide.html

Van Eelen, W. F. (2007). *Industrial scale production of meat using cell culture methods*. US Patent Number US7270829B2. Retrieved September 10, 2018 from https://patents.google.com/patent/US7270829B2/en

van Huis, A., Dicke, M., & van Loon, J. J. A. (2015). Insects to feed the world. *1st International Conference Insects to Feed the World - Summary Report*, *1*, 3–5. Retrieved September 12, 2018 from http://www.fao.org/edible-insects/86385/en/

van Huis, A., Itterbeeck, J. V., Klunder, H., Mertens, E., Haloran, A., Muir, G., & Vantomme, P. (2013). *Edible insects: Future prospects for food and feed security*. Rome, Italy: Food and Agriculture Organization of the United Nations (FAO). Retrieved from https://www.researchgate.net/publication/311424459_Edible_insects_Future_prospects_for_food_and_feed_security_Food_and_Agriculture_Organiation_of_the_United_Nations_FAO_Rome_Italy

van Huis, A., Van Itterbeeck, J., Klunder, H., Mertens, E., Halloran, A., Muir, G., & Vantomme, P. (2013). *Edible insects. Future prospects for food and feed security. Food and Agriculture Organization of the United Nations* (Vol. 171). Retrieved September 12, 2018 from http://www.fao.org/docrep/018/i3253e/i3253e.pdf

van Huis, A. (2013). Potential of insects as food and feed in assuring food security. *Annual Review of Entomology*, *58*(1), 563–583. doi:10.1146/annurev-ento-120811-153704 PMID:23020616

van Huis, A. (2016). Edible insects are the future? *The Proceedings of the Nutrition Society*, *75*(03), 294–305. doi:10.1017/S0029665116000069 PMID:26908196

van Huis, A. (2017). Edible insects: Marketing the impossible? *Journal of Insects as Food and Feed*, *3*(2), 67–68. doi:10.3920/JIFF2017.x003

van Huis, A., Dicke, M., & van Loon, J. J. A. (2015). Insects to feed the world. *Journal of Insects as Food and Feed*, *1*(1), 3–5. doi:10.3920/JIFF2015.x002 van Huis, A., Itterbeeck, J. V., Klunder, H., Mertens, E., Halloran, A., Muir, G., & Vantomme, P. (2013). *Edible insects: Future prospects for food and feed security*. Rome, Italy: Food and Agriculture Organisation of the United Nations. Retrieved from http://www.fao.org/docrep/018/i3253e/i3253e.pdf

Van Kleef, E., Otten, K., & van Trijp, H. C. (2012). Healthy snacks at the checkout counter: A lab and field study on the impact of shelf arrangement and assortment structure on consumer choices. *BMC Public Health*, *12*(1), 1072. doi:10.1186/1471-2458-12-1072 PMID:23231863

Van Soest, P. J., & Robertson, J. B. (1977). What is fibre and fibre in food? *Nutrition Reviews*, *35*(3), 12–22. doi:10.1111/j.1753-4887.1977.tb06532.x PMID:840446

Vanhonacker, F., Van Loo, E. J., Gellynck, X., & Verbeke, W. (2013). Flemish consumer attitudes towards more sustainable food choices. *Appetite*, *62*, 7–16. doi:10.1016/j.appet.2012.11.003 PMID:23195711

Vantomme, P. (2015). Way forward to bring insects in the human food chain. *Journal of Insects as Food and Feed*, *1*(2), 121–129. doi:10.3920/JIFF2014.0014

Vartanian, L. R. (2015). Impression management and food intake: Current directions in research. *Appetite*, *86*, 74–80. doi:10.1016/j.appet.2014.08.021 PMID:25149198

Vega-Estrada, J., Montes-Horcasitas, M. C., Domínguez-Bocanegra, A. R., & Canizares-Villanueva, R. O. (2005). Haematococcus pluvialis cultivation in split-cylinder internal-loop airlift photobioreactor under aeration conditions avoiding cell damage. *Applied Microbiology and Biotechnology*, *68*(1), 31–35. doi:10.100700253-004-1863-4 PMID:15657723

Vein, J. (2004). *Method for producing tissue engineered meat for consumption*. US Patent Number US6835390B1. Retrieved September 10, 2018 from https://patents.google.com/patent/US6835390B1/en

Verain, M., Dagevos, H., & Antonides, G. (2015). Flexitarianism: A range of sustainable food styles. In L. A. Reisch & J. Thøgersen (Eds.), *Handbook of research on sustainable consumption* (pp. 209–233). Cheltenham, UK: Edward Elgar Publishing.

Verbeke, W. (2015). Profiling consumers who are ready to adopt insects as a meat substitute in a Western society. *Food Quality and Preference*, *39*, 147–155. doi:10.1016/j.foodqual.2014.07.008

Verbeke, W., Marcu, A., Rutsaert, P., Gaspar, R., Seibt, B., Fletcher, D., & Barnett, J. (2015). 'Would you eat cultured meat?': Consumers' reactions and attitude formation in Belgium, Portugal and the United Kingdom. *Meat Science*, *102*, 49–58. doi:10.1016/j.meatsci.2014.11.013 PMID:25541372

Verbeke, W., Sans, P., & Van Loo, E. J. (2015). Challenges and prospects for consumer acceptance of cultured meat. *Journal of Integrative Agriculture*, *14*(2), 285–294. doi:10.1016/S2095-3119(14)60884-4

Vermeir, I., & Verbeke, W. (2008). Sustainable food consumption among young adults in Belgium: Theory of planned behaviour and the role of confidence and values. *Ecological Economics*, *64*(3), 542–553. doi:10.1016/j.ecolecon.2007.03.007

Verneau, F., La Barbera, F., Kolle, S., Amato, M., Del Giudice, T., & Grunert, K. (2016). The effect of communication and implicit associations on consuming insects: An experiment in Denmark and Italy. *Appetite*, *106*, 30–36. doi:10.1016/j.appet.2016.02.006 PMID:26855371

Véron, O. (2016). From Seitan Bourguignon to Tofu Blanquette: Popularizing veganism in France with food blogs. In J. Catricano & R. R. Simonsen (Eds.), *Critical perspectives on veganism* (pp. 287–305). London, UK: Palgrave Macmillan. doi:10.1007/978-3-319-33419-6_13

Victoria State Government. (2018). Metabolic syndrome. *Better Health Channel*. Retrieved from https://www.betterhealth.vic.gov.au/health/conditionsandtreatments/metabolic-syndrome

Victoria, A. (2015). *How farmers view their animals*. Retrieved from http://agriculture.vic.gov.au/agriculture/livestock/sheep/sheep-notes-newsletters/sheep-notes-autumn-2015/how-farmers-view-their-animals#

Vilcacundo, R., & Hernández–Ledesma, B. (2017). Nutritional and biological value of quinoa (Chenopodium quinoa Willd.). *Current Opinion in Food Science*, *14*, 1–6. doi:10.1016/j.cofs.2016.11.007

Villines, Z. (2018). Top 15 sources of plant-based protein. *Medical News Today*. Retrieved from https://www.medicalnewstoday.com/articles/321474.php

Vincent, W. A. (1969). Algae for food and feed. *Process Biochemistry*, *4*(6), 45–47.

Vinnari, M. (2008). The future of meat consumption—Expert views from Finland. *Technological Forecasting and Social Change*, *75*(6), 893–904. doi:10.1016/j.techfore.2007.02.001

Vivalt, E., & Macdonald, B. (2017). *The impact of new products on ethical beliefs*. Retrieved from https://osf.io/4dczh

Wageningen University and Research. (2017). *List of edible insects of the world (April 1, 2017)*. Retrieved from https://www.wur.nl/en/Expertise-Services/Chair-groups/Plant-Sciences/Laboratory-of-Entomology/Edible-insects/Worldwide-species-list.htm

Wahlen, S., Heiskanen, E., & Aalto, K. (2012). Endorsing sustainable food consumption: Prospects from public catering. *Journal of Consumer Policy*, *35*(1), 7–21. doi:10.100710603-011-9183-4

Wainwright, M. (1992). *An introduction to fungal biotechnology*. Chichester, UK: John Wiley and Sons.

Walker, P., Rhubart-Berg, P., McKenzie, S., & Lawrence, R. S. (2005). Public health implications of meat production and consumption. *Public Health Nutrition*, *8*(4), 348–356. doi:10.1079/PHN2005727 PMID:15975179

Wang, O. (2018). *China orders farmers to grow more soybeans despite deal to buy more produce from US: Officials have ordered farmers to grow more soybeans even though Donald Trump has claimed China will buy 'as much as our farmers can produce'*. US–China Trade Wars. South China Morning Post. Retrieved September 13, 2018 from https://www.scmp.com/news/china/economy/article/2147296/china-orders-farmers-grow-more-soybeans-despite-deal-buy-more

Wang, N., Hatcher, D. W., Toews, R., & Gawalko, E. J. (2009). Influence of cooking and dehulling on nutritional composition of several varieties of lentils (Lens culinaris). *Lebensmittel-Wissenschaft + Technologie*, *42*(4), 842–848. doi:10.1016/j.lwt.2008.10.007

Wang, X., Lin, X., Ouyang, Y. Y., Liu, J., Zhao, G., Pan, A., & Hu, F. B. (2016). Red and processed meat consumption and mortality: Dose–response meta-analysis of prospective cohort studies. *Public Health Nutrition*, *19*(5), 893–905. doi:10.1017/S1368980015002062 PMID:26143683

Wang, X., Ouyang, Y., Liu, J., Zhu, M., Zhao, G., Bao, W., & Hu, F. B. (2014). Fruit and vegetable consumption and mortality from all causes, cardiovascular disease, and cancer: Systematic review and dose-response meta-analysis of prospective cohort studies. *British Medical Journal*, *349*(7969). doi:10.1136/bmj.g4490 PMID:25073782

Wang, Y., & Beydoun, M. A. (2009). Meat consumption is associated with obesity and central obesity among US adults. *International Journal of Obesity*, *33*(6), 621–628. doi:10.1038/ijo.2009.45 PMID:19308071

Wardle, J., Haase, A. M., Steptoe, A., Nillapun, M., Jonwutiwes, K., & Bellisie, F. (2004). Gender differences in food choice: The contribution of health beliefs and dieting. *Annals of Behavioral Medicine*, *27*(2), 107–116. doi:10.120715324796abm2702_5 PMID:15053018

Warensjö, E., Byberg, L., Melhus, H., Gedeborg, R., Mallmin, H., Wolk, A., & Michaëlsson, K. (2011). Dietary calcium intake and risk of fracture and osteoporosis: Prospective longitudinal cohort study. *BMJ (Clinical Research Ed.)*, *342*(1), d1473. doi:10.1136/bmj.d1473 PMID:21610048

Warren, M. A. (1973). On the moral and legal status of abortion. *The Monist*, *57*(4), 1–9. PMID:11661013

Watrous, M. (2014). *The curious case of Hungry Man*. Baking Business. Retrieved from https://www.bakingbusiness.com/articles/26125-the-curious-case-of-hungry-man

Weary, D. M., & von Keyserlingk, M. A. G. (2017). Public concerns about dairy-cow welfare: How should the industry respond? *Animal Production Science*, *57*(7), 1201–1209. doi:10.1071/AN16680

Weaver, C. M. (2013). Potassium and health. *Advances in Nutrition*, *4*(3), 368S–377S. doi:10.3945/an.112.003533 PMID:23674806

Webber, J. (2018). *Woody Harrelson credits vegan diet for his high energy levels*. Live kindly. Retrieved from https://www.livekindly.co/woody-harrelson-credits-vegan-diet-for-his-high-energy-levels/

Weedon, C. (1987). *Feminist practice and poststructuralist theory*. Oxford, UK: Blackwell.

Wei, H., Gao, Z., Liang, R., Li, Z., Hao, H., & Liu, X. (2016). Whole-grain consumption and the risk of all-cause, CVD and cancer mortality: A meta-analysis of prospective cohort studies. *British Journal of Nutrition*, *116*(03), 514–525. doi:10.1017/S0007114516001975 PMID:27215285

Weinrich, R. (2018). Cross-cultural comparison between German, French, and Dutch consumer preferences for meat substitutes. *Sustainability*, *10*(6), 1819. doi:10.3390u10061819

Wellesley, L., Happer, C., & Froggatt, A. (2015). *Changing climate, changing diets pathways to lower meat consumption*. Chatham House Report. Retrieved March 13, 2018 from: https://www.chathamhouse.org/sites/files/chathamhouse/publications/research/CHHJ3820%20Diet%20and%20climate%20change%2018.11.15_WEB_NEW.pdf

Went, F. A. F. C. (1895). Monascus purpureus, le champignon de l'âng-quac. Une nouvelle thélébolée. *Annales des Sciences Naturelles. Botanique et Biologie*, *8*, 1–18.

White, A., & Peretz, T. (2009). Emotions and redefining black masculinity: Movement narratives of two profeminist organizers. *Men and Masculinities*, *12*(4), 403–424. doi:10.1177/1097184X08326007

White, B. A., Horwath, C. C., & Conner, T. S. (2013). Many apples a day keep the blues away--daily experiences of negative and positive affect and food consumption in young adults. *British Journal of Health Psychology*, *18*(4), 782–798. doi:10.1111/bjhp.12021 PMID:23347122

Whyte, L. (2018). *Lyn Whyte (Animals Australia)*. Retrieved September 14, 2018 from https://www.facebook.com/Lyn-White-Animals-Australia--162121637186671/

Wikfors, G. H., Ferris, G. E., & Smith, B. C. (1992). The relationship between gross biochemical composition of cultured algal foods and growth of the hard clam, Mercenaria mercenaria (L.). *Aquaculture (Amsterdam, Netherlands)*, *108*(1–2), 135–154. doi:10.1016/0044-8486(92)90324-E

Wild, F., Czerny, M., Janssen, A. M., Kole, A. P., Zunabovic, M., & Domig, K. J. (2014). The evolution of a plant-based alternative to meat. From niche markets to widely accepted meat alternatives. *Agro Food Industry Hi-Tech*, *25*(1), 45–49.

Wildlife African Foundation. (2002). *Kenyan lioness adopts another oryx*. Retrieved from https://www.awf.org/news/kenyan-lioness-adopts-another-oryx

Wildlife Films. (2014). *Incredible leopard and baby baboon interaction.* Retrieved from https://www.youtube.com/watch?v=ugi4x8kZJzk

Wilks, M., & Phillips, C. J. (2017). Attitudes to in vitro meat: A survey of potential consumers in the United States. *PLoS One*, *12*(2), e0171904. doi:10.1371/journal.pone.0171904 PMID:28207878

Willis, O. (2018). Omega-3: What are the health benefits and how should you eat it? *ABC News Health*. Retrieved from http://www.abc.net.au/news/health/2018-02-15/what-are-the-health-benefits-of-omega-3/9446528

Wilson, A. (2012). Meat and cancer – country comparisons. *Cancer News, Articles, & Information*. Retrieved from http://healthhubs.net/cancer/meat-and-cancer-country-comparisons/

Wilson, A. E., Chislock, M. F., Yang, Z., Barros, M. U. G., & Roberts, J. F. (2018). Pond bank access as an approach for managing toxic cyanobacteria in beef cattle pasture drinking water ponds. *Environmental Monitoring and Assessment*, *190*(4), 247. doi:10.100710661-018-6595-5 PMID:29574498

Wise, S. (2010). Legal personhood and the Nonhuman Rights Project. *Animal Law*, *17*(1), 1–12.

Wise, S., Pennebaker, D. A., & Hegedus, C. (2016). *Unlocking the cage. Documentary Film.* New York, NY: HBO Documentary Films.

Woldometers. (2018). *Current world population.* Retrieved from http://www.worldometers.info/world-population/

Wolka, K., Sterk, G., Biazin, B., & Negash, M. (2012). Benefits, limitations and sustainability of soil and water conservation structures in Omo-Gibe basin, Southwest Ethiopia. *Land Use Policy*, *73*, 1–10. doi:10.1016/j.landusepol.2018.01.025

Woll, S., & Böhm, I. (2018). In-vitro meat: A solution for problems of meat production and meat consumption? *Ernaehrungs Umschau International, 65*(1), 12–21. Retrieved September 10, 2018 from https://www.ernaehrungs-umschau.de/fileadmin/Ernaehrungs-Umschau/pdfs/pdf_2018/01_18/EU01_2018_Special_invitro_englisch.pdf

Womeni, H. M., Linder, M., Tiencheu, B., Mbiapo, F. T., Villeneuve, P., Fanni, J., & Parmentier, M. (2009). Oils of insects and larvae consumed in Africa: Potential sources of polyunsaturated fatty acids. *OCL – Oleagineux, Corps Gras, Lipides, 16*(4), 230-235. Retrieved from https://www.ocl-journal.org/articles/ocl/abs/2009/04/ocl2009164p230/ocl2009164p230.html

Wong, J. (2017). *Man made: A gendered food culture.* The Isthmus. Retrieved from http://www.theisthmus.com.au/2017/03/man-made-a-gendered-food-culture/

Wong, A. Y., & Chan, A. W. (2016). Genetically modified foods in China and the United States: A primer of regulation and intellectual property protection. *Food Science and Human Wellness*, *5*(3), 124–140. doi:10.1016/j.fshw.2016.03.002

Wood, J. (2015). *Gendered lives: Communication, gender and culture.* Stanford, CT: Cengage Learning.

World Cancer Research Fund and American Institute for Cancer Research (WCRF/AICR). (2007). *Food, nutrition, physical activity and the prevention of cancer: A global perspective*. Washington, DC: AICR. Retrieved from http://www.aicr.org/assets/docs/pdf/reports/Second_Expert_Report.pdf

World Cancer Research Fund. (2012). *Worldwide data.* Retrieved from https://www.wcrf.org/int/cancer-facts-figures/worldwide-data

World Cancer Research Fund/American Institute for Cancer Research (WCRF/AICR). (2018). *Diet, nutrition, physical activity and cancer: A global perspective*. Continuous Update Project Expert Report 2018. Retrieved from https://www.dietandcancerreport.org

World Health Organisation (WHO). (2015). *Q&A on the carcinogenicity of the consumption of red meat and processed meat.* Retrieved from http://www.who.int/features/qa/cancer-red-meat/en/

World Health Organization (WHO). (2003). *Diet, nutrition and the prevention of chronic diseases: Report of a joint WHO/FAO Expert Consultation.* WHO Technical Report Series 916. Retrieved from http://www.who.int/dietphysicalactivity/publications/trs916/en/

World Health Organization (WHO). (2014). *Increasing fruit and vegetable consumption to reduce the risk of noncommunicable diseases: Biological, behavioural and contextual rationale.* Retrieved from http://www.who.int/elena/titles/bbc/fruit_vegetables_ncds/en/

World Health Organization (WHO). (2015). *Q&A on the carcinogenicity of the consumption of red meat and processed meat.* Retrieved from http://www.who.int/features/qa/cancer-red-meat/en/

World Health Organization (WHO). (2015). *Q&A on the carcinogenicity of the consumption of red meat and processed meat.* Retrieved from http://www.who.int/features/qa/cancer–red–meat/en/

World Health Organization (WHO). (2016). *Increasing potassium intake to reduce blood pressure and risk of cardiovascular diseases in adults.* Retrieved from http://www.who.int/elena/titles/potassium_cvd_adults/en/

World Health Organization (WHO). (2017a). *Cardiovascular diseases (CDVs): Key facts.* Retrieved from http://www.who.int/en/news-room/fact-sheets/detail/cardiovascular-diseases-(cvds)

World Health Organization (WHO). (2017b). *Reducing sodium intake to reduce blood pressure and risk of cardiovascular diseases in adults.* Retrieved from http://www.who.int/elena/titles/sodium_cvd_adults/en/

World Health Organization (WHO). (2018). *Noncommunicable diseases: Fact sheet.* Retrieved from http://www.who.int/news-room/fact-sheets/detail/noncommunicable-diseases

World Health Organization (WHO). (2018a). *Cancer.* Retrieved from: http://www.who.int/en/news-room/fact-sheets/detail/cancer

World Health Organization (WHO). (2018b). *World Health Statistics 2018. Monitoring Health for the SDGs, sustainable development goals.* Retrieved from http://apps.who.int/iris/bitstream/handle/10665/272596/9789241565585-eng.pdf?ua=1

World Society for the Protection of Animals. (2007). *Universal declaration on animal welfare: An agreement among people and nations to recognize the animals are sentient and can suffer, to respect their welfare needs and to end animal cruelty – for good.* Retrieved from https://www.worldanimalprotection.ca/sites/default/files/ca_-_en_files/case_for_a_udaw_tcm22-8305.pdf

World Wide Fund for Nature (WWF). (2012). *Climate change on your plate.* WWF Germany. Retrieved September 11, 2018 from https://www.wwf.de/fileadmin/fm-wwf/Publikationen-PDF/Climate_change_on_your_plate.pdf

Wu, J., Zeng, R., Huang, J., Li, X., Zhang, J., Ho, J. C.-M., & Zheng, Y. (2016). Dietary protein sources and incidence of breast cancer: A dose–response meta–analysis of prospective studies. *Nutrients, 8*(11), 730. doi:10.3390/nu8110730 PMID:27869663

Xinhua. (2018). China soybean production forecast to recover. *China Daily.* Retrieved September 13, 2018 fromhttp://www.chinadaily.com.cn/a/201804/23/WS5add59d9a3105cdcf6519e50.html

Yamaguchi, K. (1997). Recent advances in microalgal bioscience in Japan, with special reference to utilization of biomass and metabolites: A review. *Journal of Applied Phycology, 8*(6), 487–502. doi:10.1007/BF02186327

Yen, A. L. (2010). Edible insects and other invertebrates in Australia: future prospects. In P. B. Durst, D. V. Johnson, R. N. Leslie & K. Shono (Eds), *Forest insects as food: Humans bite back* (pp. 65–84). Bangkok, Thailand: Food and Agriculture Organization of the United Nations. Retrieved from http://www.fao.org/docrep/012/i1380e/i1380e00.pdf

Yen, A. L. (2009). Edible insects: Traditional knowledge or western phobia? *Entomological Research, 39*(5), 289–298. doi:10.1111/j.1748-5967.2009.00239.x

Yen, A. L. (2015). Insects as food and feed in the Asia Pacific region: Current perspectives and future directions. *Journal of Insects as Food and Feed, 1*(1), 33–55. doi:10.3920/JIFF2014.0017

Yhoung–Aree, J., Puwastien, P., & Attig, A. G. (2010). Edible insects in Thailand: An unconventional protein source? *Ecology of Food and Nutrition, 36*(2-4), 133–149. doi:10.1080/03670244.1997.9991511

Yi, Y., & La, S. (2004). What influences the relationship between customer satisfaction and repurchase intention? Investigating the effects of adjusted expectations and customer loyalty. *Psychology and Marketing, 21*(5), 351–373. doi:10.1002/mar.20009

Yoon, E. (2018). *How China is working hard to be less reliant on US farmers*. Retrieved September 13, 2018 from https://www.cnbc.com/2018/05/11/china-looking-to-rely-less-on-us-for-soybeans.html

YouTube VegSource. (2006). *Carl Lewis: Olympic medals through the vegan diet*. Retrieved from https://www.youtube.com/watch?v=bOTETXwfIaY

You, W., & Henneberg, M. (2016). Meat consumption providing a surplus energy in modern diet contributes to obesity prevalence: An ecological analysis. *BMC. Nutrition (Burbank, Los Angeles County, Calif.), 2*, 22. doi:10.118640795-016-0063-9

Zagrobelny, M., Dreon, A. L., Gomiero, T., Marcazzan, G. L., Glaring, M. A., MøLler, B. L., & Paoletti, M. G. (2009). Toxic moths: Source of a truly safe delicacy. *Journal of Ethnobiology, 29*(1), 64–76. doi:10.2993/0278-0771-29.1.64

Zahran, H. H. (1999). *Rhizobium*-legume symbiosis and nitrogen fixation under severe conditions and in an arid climate. *Microbiology and Molecular Biology Reviews, 63*(4), 968–989. PMID:10585971

Zaraska, M. (2016). *This is why you crave beef: Inside secrets of Big Meat's billion-dollar ad and lobbying campaigns*. Salon.com. Retrieved from https://www.salon.com/2016/04/03/this_is_why_you_crave_beef_inside_secrets_of_big_meats_billion_dollar_ad_and_lobbying_campaigns/

Zebeli, Q., Ghareeb, K., Humer, E., Metzler-Zebeli, B. U., & Besenfelder, U. (2015). Nutrition, rumen health and inflammation in the transition period and their role on overall health and fertility in dairy cows. *Research in Veterinary Science, 103*, 126–136. doi:10.1016/j.rvsc.2015.09.020 PMID:26679807

Zechner, E. (2017). Inflammatory disease caused by intestinal pathobionts. *Current Opinion in Microbiology, 35*, 64–69. doi:10.1016/j.mib.2017.01.011 PMID:28189956

Zhang, J., Tang, Q., Zimmerman-Kordmann, M., Reutter, W., & Fan, H. (2002). Activation of B lymphocytes by GLIS, a bioactive proteoglycan from Ganoderma lucidum. *Life Sciences, 71*(6), 623–638. doi:10.1016/S0024-3205(02)01690-9 PMID:12072151

Zhang, L., Li, X., Ma, B., Gao, Q., Du, H., Han, Y., ... Qiao, Z. (2017). The tartary buckwheat genome provides insights into rutin biosynthesis and abiotic stress tolerance. *Molecular Plant, 10*(9), 1224–1237. doi:10.1016/j.molp.2017.08.013 PMID:28866080

Zhang, N., Ju, Z., & Zuo, T. (2018). Time for food: The impact of diet on gut microbiota and human health. *Nutrition (Burbank, Los Angeles County, Calif.), 51-52*, 80–85. doi:10.1016/j.nut.2017.12.005 PMID:29621737

Zhang, Z. L., Zhou, M. L., Tang, Y., Li, F. L., Tang, Y. X., Shao, J. R., ... Wu, Y. M. (2012). Bioactive compounds in functional buckwheat food. *Food Research International*, *49*(1), 389–395. doi:10.1016/j.foodres.2012.07.035

Zhu, F. (2016a). Buckwheat starch: Structures, properties, and applications. *Trends in Food Science & Technology*, *49*, 121–135. doi:10.1016/j.tifs.2015.12.002

Zhu, F. (2016b). Chemical composition and health effects of Tartary buckwheat. *Food Chemistry*, *203*, 231–245. doi:10.1016/j.foodchem.2016.02.050 PMID:26948610

Zhu, F. (2018). Chemical composition and food uses of teff (Eragrostis tef). *Food Chemistry*, *239*, 402–415. doi:10.1016/j.foodchem.2017.06.101 PMID:28873585

Zhu, F., Du, B., Li, R., & Li, J. (2014). Effect of micronization technology on physicochemical and antioxidant properties of dietary fibre from buckwheat hulls. *Biocatalysis and Agricultural Biotechnology*, *3*(3), 30–34. doi:10.1016/j.bcab.2013.12.009

Zhu, L. L., Brescoll, V. L., Newman, G. E., & Uhlmann, E. L. (2015). Macho nachos. The implicit effects of gendered food packaging on preferences for healthy and unhealthy foods. *Social Psychology*, *46*(4), 182–196. doi:10.1027/1864-9335/a000226

Zielińska, E., Baraniak, B., Karaś, M., Rybczyńska, K., & Jakubczyk, A. (2015). Selected species of edible insects as a source of nutrient composition. *Food Research International*, *77*, 460–466. doi:10.1016/j.foodres.2015.09.008

Ziemba, J. V. (1966). Let soy proteins work wonders for you. *Food Engineering*, *38*(5), 82–84, 87–90, 93.

Zorpette, G. (2013). Muscling out meat. *IEEE Spectrum*, *50*(6), 64–70. doi:10.1109/MSPEC.2013.6521036

Zubi, W. (2005). *Production of single cell protein from base hydrolyzed of date extract by product by the fungus Fusarium graminearum* (Master thesis). Garyounis University, Benghazi, Libya.

Zukav, G. (2004). *The seat of the soul*. Gdynia, Poland: Centrum.

About the Contributors

Diana Bogueva has more than 20 years of international communication and media experience as a professional radio journalist working with several worldwide recognised news media, including the Bulgarian National Radio, Bulgarian Section of BBC, London and Special Broadcasting Services (SBS), Sydney, Australia. She is currently working for SBS radio as a program producer and is Director of Market Research with Sydney Focus Groups. Diana completed her PhD studies at Curtin University, Australia, focusing on reduction in meat consumption, social marketing and sustainability. She is a member of the International Federations of Journalists and a Wood Badge member of the World Scout Organization.

Dora Marinova is Professor of Sustainability at the Curtin University Sustainability Policy Institute. Dora has over 400 referred publications and has successfully supervised 60 PhD. She served as a member of the Australian National Health and Medical Research Council's Panels on Centres of Research Excellence in Population Health and in Health Services Research. Her research interests cover innovation models, including the evolving global green system of innovation and the emerging area of sustainometrics. Together with Talia Raphaely, Dora co-edited the book *Impact of Meat Consumption on Health and Environmental Sustainability* which was awarded World Best Book at the prestigious 22nd Gourmand Awards.

Talia Raphaely, originally from Cape Town, South Africa, has 30 years of experience in behavioural and attitudinal change, communications and diverse media, sustainability awareness and conscientiousness, collaboration and partnership building for sustainability. She has worked with multicultural and heterogeneous groups in diverse organizational settings, including academia, media, research-based organisations, government bodies, non-government organisations, community-based organisations and industry. Talia undertakes sustainability consultancy research and is the Editor of the Customs Homes Magazine. The book she co-edited with Dora Marinova *Impact of Meat Consumption on Health and Environmental Sustainability* won World Best Book at the prestigious 22nd Gourmand Awards.

Kurt Schmidinger, graduated geophysicist from Graz University of Technology, Austria and doctor in food science from Vienna University, Austria, works for various animal protection and vegetation and environmental associations, software development companies and is a project leader of www.futurefood.org in Austria, Europe. As a geophysicist, food scientist and animal protectionist, Kurt Schmidinger is interested in the global consequences of human nutritional habits, climatic stability of food, global livestock farming and global meat consumption including environment and climate, world food, human health and animal welfare.

* * *

Harun Aksu graduated from the Veterinary Faculty of Istanbul University in 1987. In 1994, he completed his PhD at the Food Hygiene and Technology Department, Health Sciences Institute, Istanbul University on the topic "Presence and importance of Bacillus cereus in various ready-to-eat meals in our country". He became a Professor in 2005, has worked in various organisations in the food area, including national and international projects and competed many studies on food. As a professor at the Food Hygiene and Technology Department at Istanbul University – Cerrahpaşa, Turkey, his expertise is in food biotechnology, food processing and food microbiology.

Mehmet Alpaslan graduated from Trakya University (University of Thrace), Turkey and completed his PhD in Food Engineering. He has conducted research in the fields of food chemistry and technology and was Head of the Department of Food Engineering and Dean of the Faculty of Engineering at Inönü University, Turkey. Prof. Alpaslan is now Head of the Department of Nutrition and Dietetics of Namık Kemal University, Turkey.

Chrysostomos Apostolidis is a Senior Lecturer of Marketing at Newcastle Business School, Northumbria University, which he joined in 2012. As an academic, he teaches and specializes in social and ethical marketing and sustainable consumer behaviour and has published several articles, book chapters, refereed conference papers and reports in this area. His main research interest is in the impact of new technologies on sustainable and ethical consumer behaviour and he has received funding and awards for his research. Chrysostomos has several food marketing and policy publications mainly relating to encouraging meat substitution and more sustainable meat consumption patterns and he is particularly interested in the role that digital marketing and new technologies can play in supporting more sustainable food consumption.

Ramazan Mert Atan is working as a researcher in the Division of Dietetics, Faculty of Health Sciences at Bandırma Onyedi Eylül University, Turkey. He completed his degrees in Nutrition and Dietetics from Kirklareli University and Tekirdağ Namik Kemal University in Turkey.

Hans Dagevos (PhD) is senior scientist at Wageningen University & Research in the Netherlands. He approaches and analyses the world of food from the perspective of the sociology of consumption. This has resulted in numerous publications ranging from scholarly papers and research reports, books and book chapters, to book reviews and weblogs that he (co)authored and (co)edited. His latest research interests, next to reduction of meat consumption and food transition towards sustainable diets, are in the fields of social innovation, circular consumption, and sustainable food policy.

Levent Gülüm is a researcher in the Department of Chemistry at Abant İzzet Baysal University, Turkey. His research expertise is in statistical methods and statistical analysis.

Xiumei Guo is a senior research fellow at Curtin University Sustainability Policy (CUSP) Institute, Curtin University and is an accomplished supervisor of PhD students. Her research interests are in the area of global sustainability, urban and regional development, demography, economic development, and energy efficiency with a focus on Asia and the Pacific, particularly on China. Dr Guo has been conducting research on low carbon cities, city smog issues, urban and transport planning and Indigenous participation in a low-carbon economy in countries, including Australia, China, Bangladesh and India. She has published widely in international peer-reviewed journals.

Amzad Hossain is a guest teacher at the Institute of Environmental Science, Rajshahi University, Bangladesh and has also been a university associate at Curtin University. His research interests are in the area of sustainability, including the implementation of the UN Sustainable Development Goals in Bangladesh, values education, indigenous culture, self-reliance, appropriate technology, climate change and spirituality with relevance to individual, social, technological and environmental sustainability. He has published over 100 research papers, including 17 commentaries on http://newsfrombangladesh.net and 20 popular articles on the mystic Baul tradition of Bangladesh. Amzad has a MCom from Rajshahi University, MA (by research) in religious studies from Deakin University, Australia and two PhDs (in Anthropology and in Sustainability) from Murdoch University, Australia.

Nursena Kağan is a researcher at Tekirdağ Namık Kemal University in Turkey.

Weronika Kalamus graduated with a Bachelor degree from Florence University in Italy majoring in Modern Languages and Cultures. Following her interest in creating justice in the world, she undertook an International Master in Peace, Conflicts and Development. Her passion for more equal world led her to become an animal activist and plant-based diet promotor. She has presented various public talks about the impact the meat and livestock industry has on the planet, human health and society, including at Caledonian University in Glasgow in 2015 and 2016. In April 2017, she organised and presented at the Intercultural Seminar "The real cost of meat" at the UNESCO Jaume I University in Spain.

Jie Li is a research associate at the Dyson School of Applied Economics and Management at Cornell University. She received both her Master and PhD degrees in applied economics and management from Cornell University. Her research interests cover agricultural and food marketing as well as using econometrics and experimental methods to explore questions in the consumer behaviors of various industries. She was awarded the inaugural Bob Kalik scholarship from the Wine Market Council in recognition of her research on wine marketing. In addition to her research interests, she also works closely with farmers in upstate New York to encourage their adoption of sustainable farming systems. Dr Li also serves as the director of program development at Cornell Institute for China economic research (CICER). In her position at CICER, she oversees the training programs between CICER and other research institutes, government agencies and universities in China.

Aijun Liu received his Master degree in 2002 and PhD degree in 2007 from Nanjing Agricultural University, China. He also studied in Lareinstein University of Professional Education, the Netherlands for one year and was a visiting scholar for one year at Curtin University, Australia. Aijun's areas of research are agribusiness and resource management. He is a lecturer in Agricultural Market Research & Analysis and E-commerce & Logistic Management. He is also a supervisor of master students in Agricultural Economics & Business Administration. The aim of his research is to help the agricultural sector sell their value-added products to the market by segmenting the market. One of his projects analyses fresh produce consumers' confidence in food safety. He has also participated in projects related to resource management, including household waste sorting and reduction.

Patricia Marshall is an Accredited Practising Dietitian with the Dietitians Association of Australia and a Credentialed Diabetes Educator with the Australian Diabetes Educators Association. She has 40 years of experience as a public health nutritionist, dietitian and diabetes educator, working in community centres, worksites, child health centres, schools, Aboriginal communities, hospitals and universities in remote, rural and metropolitan settings in New South Wales, the Northern Territory and Western Australia and also in the Solomon Islands. Tricia has training and extensive practical experience in chronic disease prevention and self-management, health promotion, research and evaluation. She has a degree in Agricultural Science, Graduate Certificate in Nutrition and Dietetics, and Graduate Diploma in Diabetes Education. She is lecturing in Public Health Nutrition and coordinating the Graduate Certificate in Diabetes course at Curtin University.

Alexis J. Nagy graduated with a Bachelor degree from The University of Sydney majoring in cultural anthropology with minors in psychology and archaeology. Her interest in animal welfare and developmental psychology drove her passion for animal cognisance and human rights. As a result, she has taken on a Master's degree with Curtin University in sustainability and climate policy to further her enthusiasm for seeking policy initiatives which represent a wider global community that is not just human-centric, but that can also incorporate nonhuman animals.

Navneet completed his PhD from Kurukshetra University in India under the supervision of Professor R. A. Mehrotra and at present is working as a professor in the Department of Botany and Microbiology at Gurukula Kangri University, India. He has completed many research projects funded by the University Grants Commission of India and the Uttarakhand State Council for Science and Technology (UCOST). Navneet is a Fellow of the Indian Phytopathological Society and a Fellow of the Indian Botanical Society.

Clive Phillips is Foundation Professor of Animal Welfare at the University of Queensland, where he directs the Centre for Animal Welfare and Ethics in the School of Veterinary Science. He is well known for raising concerns about the rapidly intensifying global animal trade, the subject of his recent book, The Animal Trade (CABI, 2015). Other books include The Welfare of Animals – the Silent Majority, Principles of Cattle Production, Cattle Behaviour and Welfare, and Animal Welfare in Australia. He has also authored over 300 scientific journal articles. In 2012, Clive was awarded the Australian Museum prize for scientific research for the protection of animals.

Jaco Quist (PhD) is assistant professor in Sustainable Innovation and Transitions at the Faculty of Technology, Policy, Management of Delft University of Technology, the Netherlands. His work evolves around sustainable futures, visions, backcasting, transition management and innovative niches for sustainable consumption and a circular economy, which includes a special interest in meat alternatives. He is the chair of the European Roundtable for Sustainable Consumption and Production (ERSCP).

Malte B. Rödl received his Master's degree in Industrial Ecology from Chalmers University of Technology, Gothenburg, Sweden, in 2015 as part of the Erasmus Mundus MIND programme. Afterwards, he started his PhD at, and generously funded by, the Sustainable Consumption Institute at The University of Manchester. His research revolves around the relations of meat and meat alternatives in a cultural context.

Akash Saklani completed his Bachelor and Master degrees in Botany from Hemvati Nandan Bahuguna Garhwal University, Uttarakhand, India and is currently pursuing his PhD studies in Botany at Gurukul Kangri University under the supervision of Professor Navneet. His research work is on the estimation of phytodiversity in the Rajaji Tiger reserve in India. He has published many research papers on the conservation of phytodiversity in the Rajaji Tiger reserve and the adjacent to it Shivalik Hill.

Xiaoling Shao is working at Nanjing Audit University, China. Her research interests include resource and environmental economics, pro-environmental behaviour and water pollution treatment projects, and has published many papers in these fields. She has hosted or participated in scientific research projects, such as the China Natural Science Foundation Project and the Humanities and Social Science Fund Project of the Ministry of Education of China. Her teaching interests include environmental health, resource and environmental economics.

Giovanni Sogari has a PhD in Agricultural Economics (www.giovannisogari.com). He is a postdoctoral researcher at the Department of Food and Drugs at the University of Parma (Italy) and a visiting fellow at the Dyson School of Applied Economics and Management of Cornell University (USA). Since 2017, he is also a Marie Skłodowska-Curie Fellow with the European project "CONSUMEHealth. Using consumer science to improve healthy eating habits". His main research interests involve consumers behavior in the food system, especially on healthy diet, fruit and vegetable consumption, sustainability, novel food and edible insects. In particular, in 2014 Dr Sogari wrote one of the first books for the general public about the prospective of edible insects as food and feed, with Mr. Paul Vantomme (Food and Agriculture Organization of the United Nations) as a co-author. Since then he has been presenting the outcomes of his research in many conferences worldwide and published manuscripts in several peer-review journals.

Anusha Thakur has completed her MBA degree in International Business from University of Petroleum & Energy Studies, Dehradun, India. She is also pursuing a Post Graduation course in Market Research & Data Analytics from MICA, Ahmedabad. She has been working as a Senior Research Analyst, wherein her duties and responsibilities catered to the secondary research segment of Market Research, particularly of the energy and food & beverage domain.

Ella Tolonen (MSc) holds a Master Degree of Industrial Ecology from Delft University of Technology and Leiden University in the Netherlands. For her thesis research project, she studied the Dutch meat substitute innovation system. Currently, she is in Finland as an intern at the Good Food Institute, a US non-profit organisation which promotes in vitro meat and plant-based substitutes for animal food products. Ella is assisting the organization's expansion into Europe. She aspires to continue working on the topic of meat alternatives consumption in the future.

Shagufta M. Trishna is pursuing PhD studies at the Curtin University Sustainability Policy Institute in the area of sustainable development with a strong interest in Bangladesh. She graduated in architecture from the University of Asia Pacific, Dhaka (Bangladesh). After completing her internship, she served in a few companies as a full-time architect and designer for more than 12 years. Her research focuses on the deteriorating living conditions in Dhaka and exploring opportunities for sustainable solutions. She is passionate about writing, reading and gathering of new knowledge.

Matti Wilks completed her PhD in psychology at the University of Queensland, studying moral development and attitudes to clean meat. She was awarded a postdoctoral research fellowship to continue her study of clean meat at Princeton University in 2018, and in 2019 will commence a postdoctoral research position at Yale University, USA. She is the author of several scientific journal articles and book chapters exploring attitudes to clean meat and has engaged extensively with the public on this topic, including interviews for ABC Radio National, Nature Biotechnology and the Conversation.

Seydi Yıkmış graduated from the Department of Food Engineering, Faculty of Engineering, Inönü University, Turkey in 2005. In 2013, he graduated from the Department of Food Engineering, Institute of Science, Süleyman Demirel University and completed his PhD at the Department of Food Hygiene and Technology, Institute of Health Sciences, Istanbul University, Turkey. He is currently working as a lecturer in the Nutrition and Dietetics Department of Namık Kemal University in Turkey. His skills and expertise are in food science and technology, food quality, food and nutrition, food microbiology, food safety and food analysis.

Anna K. Żołnierczyk is an Assistant Professor at the Faculty of Biotechnology and Food Science of the Wrocław University of Environmental and Life Sciences, where she completed her PhD. Her research interests lie in the area of the following topics: use of living organisms for biotransformation; Ezymatic biocatalysis in the synthesis of organic compounds; Hop flavonoids (xanthohumol, isoxantohumol, 8-prenylnaringenin), modification of their structure to produce compounds with interesting biological activity (anticancer, antioxidant, antimicrobial, antidiabetic); Insects, their nutritional properties; *Caenorhabditis elegans* as a model organism in *in vivo* studies of the biological activity of compounds and plant extracts. She is an author or co-author of over a dozen scientific publications and numerous popular science lectures. She is a tutor at the Student Scientific Circle of Molecular Cuisine whose activity covers running workshops and educational demonstrations related to molecular cuisine and participation in numerous educational projects.

Index

T

U

V

W

Y

Ensure Quality Research is Introduced to the Academic Community

Become an IGI Global Reviewer for Authored Book Projects

The overall success of an authored book project is dependent on quality and timely reviews.

In this competitive age of scholarly publishing, constructive and timely feedback significantly expedites the turnaround time of manuscripts from submission to acceptance, allowing the publication and discovery of forward-thinking research at a much more expeditious rate. Several IGI Global authored book projects are currently seeking highly qualified experts in the field to fill vacancies on their respective editorial review boards:

Applications may be sent to:
development@igi-global.com

Applicants must have a doctorate (or an equivalent degree) as well as publishing and reviewing experience. Reviewers are asked to write reviews in a timely, collegial, and constructive manner. All reviewers will begin their role on an ad-hoc basis for a period of one year, and upon successful completion of this term can be considered for full editorial review board status, with the potential for a subsequent promotion to Associate Editor.

If you have a colleague that may be interested in this opportunity, we encourage you to share this information with them.

www.ingramcontent.com/pod-product-compliance
Lightning Source LLC
LaVergne TN
LVHW062115300926
849161LV00010B/2464

* 9 7 8 1 5 2 2 5 7 3 5 0 0 *